IET CONTROL, ROBOTICS AND SENSORS SERIES 107

Solved Problems in Dynamical Systems and Control

Other volumes in this series:

Volume 8	**A History of Control Engineering, 1800–1930** S. Bennett
Volume 18	**Applied Control Theory, 2nd Edition** J.R. Leigh
Volume 20	**Design of Modern Control Systems** D.J. Bell, P.A. Cook and N. Munro (Editors)
Volume 28	**Robots and Automated Manufacture** J. Billingsley (Editor)
Volume 33	**Temperature Measurement and Control** J.R. Leigh
Volume 34	**Singular Perturbation Methodology in Control Systems** D.S. Naidu
Volume 35	**Implementation of Self-tuning Controllers** K. Warwick (Editor)
Volume 37	**Industrial Digital Control Systems, 2nd Edition** K. Warwick and D. Rees (Editors)
Volume 39	**Continuous Time Controller Design** R. Balasubramanian
Volume 40	**Deterministic Control of Uncertain Systems** A.S.I. Zinober (Editor)
Volume 41	**Computer Control of Real-time Processes** S. Bennett and G.S. Virk (Editors)
Volume 42	**Digital Signal Processing: Principles, devices and applications** N.B. Jones and J.D. McK. Watson (Editors)
Volume 44	**Knowledge-based Systems for Industrial Control** J. McGhee, M.J. Grimble and A. Mowforth (Editors)
Volume 47	**A History of Control Engineering, 1930–1956** S. Bennett
Volume 49	**Polynomial Methods in Optimal Control and Filtering** K.J. Hunt (Editor)
Volume 50	**Programming Industrial Control Systems Using IEC 1131-3** R.W. Lewis
Volume 51	**Advanced Robotics and Intelligent Machines** J.O. Gray and D.G. Caldwell (Editors)
Volume 52	**Adaptive Prediction and Predictive Control** P.P. Kanjilal
Volume 53	**Neural Network Applications in Control** G.W. Irwin, K. Warwick and K.J. Hunt (Editors)
Volume 54	**Control Engineering Solutions: A practical approach** P. Albertos, R. Strietzel and N. Mort (Editors)
Volume 55	**Genetic Algorithms in Engineering Systems** A.M.S. Zalzala and P.J. Fleming (Editors)
Volume 56	**Symbolic Methods in Control System Analysis and Design** N. Munro (Editor)
Volume 57	**Flight Control Systems** R.W. Pratt (Editor)
Volume 58	**Power-plant Control and Instrumentation: The control of boilers and HRSG systems** D. Lindsley
Volume 59	**Modelling Control Systems Using IEC 61499** R. Lewis
Volume 60	**People in Control: Human factors in control room design** J. Noyes and M. Bransby (Editors)
Volume 61	**Nonlinear Predictive Control: Theory and practice** B. Kouvaritakis and M. Cannon (Editors)
Volume 62	**Active Sound and Vibration Control** M.O. Tokhi and S.M. Veres
Volume 63	**Stepping Motors, 4th Edition** P.P. Acarnley
Volume 64	**Control Theory, 2nd Edition** J.R. Leigh
Volume 65	**Modelling and Parameter Estimation of Dynamic Systems** J.R. Raol, G. Girija and J. Singh
Volume 66	**Variable Structure Systems: From principles to implementation** A. Sabanovic, L. Fridman and S. Spurgeon (Editors)
Volume 67	**Motion Vision: Design of compact motion sensing solution for autonomous systems** J. Kolodko and L. Vlacic
Volume 68	**Flexible Robot Manipulators: Modelling, simulation and control** M.O. Tokhi and A.K.M. Azad (Editors)
Volume 69	**Advances in Unmanned Marine Vehicles** G. Roberts and R. Sutton (Editors)
Volume 70	**Intelligent Control Systems Using Computational Intelligence Techniques** A. Ruano (Editor)
Volume 71	**Advances in Cognitive Systems** S. Nefti and J. Gray (Editors)
Volume 72	**Control Theory: A guided tour, 3rd Edition** J.R. Leigh
Volume 73	**Adaptive Sampling with Mobile WSN** K. Sreenath, M.F. Mysorewala, D.O. Popa and F.L. Lewis
Volume 74	**Eigenstructure Control Algorithms: Applications to aircraft/rotorcraft handling qualities design** S. Srinathkumar
Volume 75	**Advanced Control for Constrained Processes and Systems** F. Garelli, R.J. Mantz and H. De Battista
Volume 76	**Developments in Control Theory towards Glocal Control** L. Qiu, J. Chen, T. Iwasaki and H. Fujioka (Editors)
Volume 77	**Further Advances in Unmanned Marine Vehicles** G.N. Roberts and R. Sutton (Editors)
Volume 78	**Frequency-Domain Control Design for High-Performance Systems** J. O'Brien
Volume 80	**Control-oriented Modelling and Identification: Theory and practice** M. Lovera (Editor)
Volume 81	**Optimal Adaptive Control and Differential Games by Reinforcement Learning Principles** D. Vrabie, K. Vamvoudakis and F. Lewis
Volume 83	**Robust and Adaptive Model Predictive Control of Nonlinear Systems** M. Guay, V. Adetola and D. DeHaan
Volume 84	**Nonlinear and Adaptive Control Systems** Z. Ding
Volume 88	**Distributed Control and Filtering for Industrial Systems** M. Mahmoud
Volume 89	**Control-based Operating System Design** A. Leva *et al.*
Volume 90	**Application of Dimensional Analysis in Systems Modelling and Control Design** P. Balaguer
Volume 91	**An Introduction to Fractional Control** D. Valério and J. Costa
Volume 92	**Handbook of Vehicle Suspension Control Systems** H. Liu, H. Gao and P. Li
Volume 93	**Design and Development of Multi-Lane Smart Electromechanical Actuators** F.Y. Annaz
Volume 94	**Analysis and Design of Reset Control Systems** Y. Guo, L. Xie and Y. Wang
Volume 95	**Modelling Control Systems Using IEC 61499, 2nd Edition** R. Lewis and A. Zoitl
Volume 96	**Cyber-Physical System Design with Sensor Networking Technologies** S. Zeadally and N. Jabeur (Editors)
Volume 99	**Practical Robotics and Mechatronics: Marine, space and medical applications** I. Yamamoto
Volume 102	**Recent Trends in Sliding Mode Control** L. Fridman, J.-P. Barbot and F. Plestan (Editors)
Volume 105	**Mechatronic Hands: Prosthetic and robotic design** P.H. Chappell

Solved Problems in Dynamical Systems and Control

J. Tenreiro Machado, António M. Lopes, Duarte Valério and Alexandra M. Galhano

The Institution of Engineering and Technology

Published by The Institution of Engineering and Technology, London, United Kingdom

The Institution of Engineering and Technology is registered as a Charity in England & Wales (no. 211014) and Scotland (no. SC038698).

First published 2016

The Institution of Engineering and Technology
Michael Faraday House
Six Hills Way, Stevenage
Herts, SG1 2AY, United Kingdom

www.theiet.org

British Library Cataloguing in Publication Data
A catalogue record for this product is available from the British Library

ISBN 978-1-78561-174-2 (hardback)
ISBN 978-1-78561-175-9 (PDF)

Typeset in India by MPS Limited
Printed in the UK by CPI Group (UK) Ltd, Croydon

Contents

1 Block diagram algebra and system transfer functions **1**
1.1 Fundamentals 1
1.1.1 List of symbols 1
1.1.2 Laplace transform and Laplace domain 1
1.1.3 Transfer function 2
1.1.4 Block diagram 2
1.1.5 Block diagram algebra 3
1.2 Worked examples 3
1.3 Proposed exercises 9
1.4 Block diagram analysis using computer packages 20
1.4.1 MATLAB 20
1.4.2 SCILAB 24
1.4.3 OCTAVE 26

2 Mathematical models **29**
2.1 Fundamentals 29
2.1.1 List of symbols 29
2.1.2 Modeling of electrical systems 31
2.1.3 Modeling of mechanical systems 32
2.1.4 Modeling of liquid-level systems 36
2.1.5 Modeling of thermal systems 38
2.2 Worked examples 39
2.2.1 Electrical systems 39
2.2.2 Mechanical systems 40
2.2.3 Liquid-level systems 42
2.2.4 Thermal systems 45
2.3 Proposed exercises 47
2.3.1 Electrical systems 47
2.3.2 Mechanical systems 53
2.3.3 Liquid-level systems 60
2.3.4 Thermal systems 68

3 Analysis of continuous systems in the time domain **73**
3.1 Fundamentals 73
3.1.1 List of symbols 73
3.1.2 Time response of a continuous LTI system 74
3.1.3 Time response of first-order systems 74

3.1.4 Time response of second-order systems 77
3.1.5 Routh's stability criterion 84
3.1.6 Steady-state errors 85
3.2 Worked examples 86
3.2.1 Routh–Hurwitz criterion 86
3.2.2 Transient response 87
3.2.3 Steady-state errors 89
3.3 Proposed exercises 89
3.3.1 Routh–Hurwitz criterion 89
3.3.2 Transient response 94
3.3.3 Steady-state errors 102
3.4 Time response analysis using computer packages 107
3.4.1 MATLAB 108
3.4.2 SCILAB 110
3.4.3 OCTAVE 112

4 Root-locus analysis 115
4.1 Fundamentals 115
4.1.1 List of symbols 115
4.1.2 Root-locus preliminaries 115
4.1.3 Root-locus practical sketching rules ($K \geq 0$) 117
4.1.4 Root-locus practical sketching rules ($K \leq 0$) 118
4.2 Solved problems 119
4.3 Proposed problems 124
4.4 Root-locus analysis using computer packages 131
4.4.1 MATLAB 131
4.4.2 SCILAB 132
4.4.3 OCTAVE 133

5 Frequency domain analysis 135
5.1 Fundamentals 135
5.1.1 List of symbols 135
5.1.2 Frequency response preliminaries 136
5.1.3 Bode diagram 136
5.1.4 Nyquist diagram 137
5.1.5 Nichols diagram 139
5.1.6 Nyquist stability 139
5.1.7 Relative stability 140
5.2 Solved problems 142
5.2.1 Bode diagram and phase margins 142
5.2.2 Nyquist and Nichols diagrams 145
5.3 Proposed problems 148
5.3.1 Bode diagram and phase margins 148
5.3.2 Nyquist and Nichols diagrams 160

5.3.3 Root-locus and frequency domain analysis 164
5.4 Frequency domain analysis using computer packages 173
5.4.1 MATLAB 173
5.4.2 SCILAB 177
5.4.3 OCTAVE 181

6 PID controller synthesis **185**
6.1 Fundamentals 185
6.1.1 List of symbols 185
6.1.2 The PID controller 185
6.1.3 PID tuning 187
6.2 Solved problems 190
6.3 Proposed problems 191

7 State space analysis of continuous systems **195**
7.1 Fundamentals 195
7.1.1 List of symbols 195
7.1.2 State space representation 196
7.1.3 The Cayley–Hamilton theorem 203
7.1.4 Matrix exponential 203
7.1.5 Computation of the matrix exponential 204
7.1.6 Solution of the state-space equation 206
7.1.7 Controllability 207
7.1.8 Observability 207
7.2 Solved problems 207
7.3 Proposed problems 215
7.4 State space analysis of continuous systems using computer packages 237
7.4.1 MATLAB 237
7.4.2 SCILAB 240
7.4.3 OCTAVE 242

8 Controller synthesis by pole placement **245**
8.1 Fundamentals 245
8.1.1 List of symbols 245
8.1.2 Pole placement using an input–output representation 246
8.1.3 Preliminaries of pole placement in state space 248
8.1.4 Calculation of the feedback gain 250
8.1.5 Estimating the system state 250
8.1.6 Calculation of the state estimator gain 252
8.1.7 Simultaneous pole placement and state estimation 253
8.2 Solved problems 254
8.2.1 Pole placement using an input–output representation 254
8.2.2 Pole placement in state space 256

8.3 Proposed problems 257
8.3.1 Pole placement using an input–output representation 257
8.3.2 Pole placement in state space 262

9 Discrete-time systems and $\mathscr{Z}$-transform 265
9.1 Fundamentals 265
9.1.1 List of symbols 265
9.1.2 Discrete-time systems preliminaries 266
9.1.3 The $\mathscr{Z}$-transform 267
9.1.4 Discrete-time models 268
9.1.5 Controllability and observability 271
9.1.6 Stability and the Routh–Hurwitz criterion 272
9.2 Solved problems 272
9.3 Proposed problems 276
9.4 Discrete-time systems and $\mathscr{Z}$-transform analysis using computer packages 283
9.4.1 MATLAB 283
9.4.2 SCILAB 285
9.4.3 OCTAVE 286

10 Analysis of nonlinear systems with the describing function method 287
10.1 Fundamentals 287
10.1.1 List of symbols 287
10.1.2 The describing function 287
10.1.3 Describing functions of common nonlinearities 288
10.1.4 Nonlinear systems analysis 288
10.2 Solved problems 290
10.3 Proposed problems 292
10.4 Describing function method using computer packages 310
10.4.1 MATLAB 311
10.4.2 SCILAB 313
10.4.3 OCTAVE 315

11 Analysis of nonlinear systems with the phase plane method 317
11.1 Fundamentals 317
11.1.1 List of symbols 317
11.1.2 Phase plane method preliminaries 317
11.1.3 Singular points 318
11.1.4 Limit cycles 319
11.2 Solved problems 320
11.3 Proposed problems 324
11.4 Phase plane analysis using computer packages 336
11.4.1 MATLAB 336
11.4.2 SCILAB 337
11.4.3 OCTAVE 339

12 Fractional order systems and controllers **341**
12.1 Fundamentals 341
12.1.1 List of symbols 341
12.1.2 Grünwald–Letnikov definition 341
12.1.3 Riemann–Liouville definition 342
12.1.4 Equivalence of definitions and Laplace transforms 343
12.1.5 Caputo definition 343
12.1.6 Fractional transfer functions 344
12.1.7 Fractional controllers 346
12.1.8 Integer approximations 346
12.2 Solved problems 347
12.3 Proposed problems 352
12.4 Fractional control using computer packages 358
12.4.1 MATLAB 358
12.4.2 SCILAB 361
12.4.3 OCTAVE 363

Appendix A **365**
Solutions **379**
References **433**
Index **435**

Chapter 1
Block diagram algebra and system transfer functions

1.1 Fundamentals

We introduce the Laplace transform as a method of converting differential equations in time into algebraic equations in a complex variable. Afterward, we present the concepts of transfer function and block diagram as a means to represent linear time-invariant (LTI) dynamical systems.

1.1.1 List of symbols

$b(t)$	feedback signal
$e(t)$	actuating error signal
$G(s)$, $H(s)$	transfer function
$\mathscr{L}$	Laplace operator
$r(t)$	time-domain input
s	Laplace variable
t	time
$y(t)$	time-domain output

1.1.2 Laplace transform and Laplace domain

If $f(t)$ is a piece-wise continuous function in time-domain, then its Laplace transform is given by [1]:

$$F(s) = \mathscr{L}[f(t)] = \int_0^\infty e^{-st} f(t)dt \tag{1.1}$$

where $F(s)$ denotes a complex-valued function of the complex variable s and t represents time.

The Laplace transform converts linear time-domain differential equations into Laplace-domain (or complex s-domain) algebraic equations.

The inverse Laplace transform is:

$$\mathscr{L}^{-1}[F(s)] = f(t) = \frac{1}{2\pi j} \lim_{T\to\infty} \int_{\sigma-jT}^{\sigma+jT} e^{st} F(s)ds \tag{1.2}$$

where $j = \sqrt{-1}$. The integration is done along the vertical line $Re(s) = \sigma$ in the complex plane, such that σ is greater than the real part of all singularities of $F(s)$. If $F(s)$ is a smooth function on $-\infty < Re(s) < \infty$, then σ can be set to zero.

Usually, given a rational function, $F(s)$, we adopt the partial fraction decomposition, or partial fraction expansion, method for expressing $F(s)$ as a sum of a polynomial and one, or several, fractions with simpler denominators. We then use the tables of Laplace transforms [2,3] to obtain $f(t)$ (see Appendix A).

1.1.3 Transfer function

A transfer function, $G(s)$, represents the dynamics of a LTI system [4]. Mathematically, it is the ratio between system output, $Y(s)$, and input, $R(s)$, in the Laplace domain, considering that all initial conditions and point equilibrium are zero: $G(s) = \frac{Y(s)}{R(s)}$.

The transfer function is an intrinsic property of a LTI system that represents the differential equation relating the system input to the output. Based on the transfer function, the system dynamic response can be determined for different inputs [2,5,6].

1.1.4 Block diagram

A block diagram is a graphic representation of a system dynamic model that includes no information about the physical construction of the system. Consequently, a block diagram representing a given system is not unique. Moreover, the main source of energy, as well as the energy flow in the system is not explicitly shown. Each individual block establishes a unilateral relationship between an input and an output signal, being assumed that there is no interaction between blocks.

Figure 1.1 depicts an example of a system block diagram. The ratio $G_{FF} = \frac{Y(s)}{E(s)} = G(s)$ between the output $Y(s)$ and the actuating error $E(s)$ is the feed-forward transfer function. The ratio $G_{OL} = \frac{B(s)}{E(s)} = G(s)H(s)$ between the feedback $B(s)$ and actuating error $E(s)$ signals represents the open-loop transfer function. Finally, the ratio between output $Y(s)$ and input $R(s)$ denotes the system closed-loop transfer function:

$$G_{CL} = \frac{Y(s)}{R(s)} = \frac{G(s)}{1 + G(s)H(s)} \tag{1.3}$$

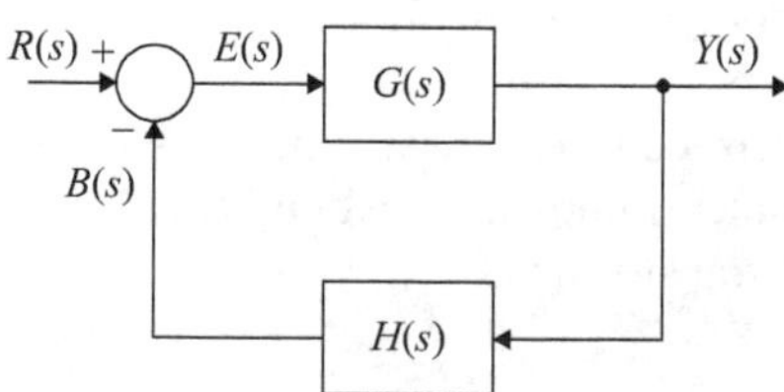

Figure 1.1 Example of a system block diagram

1.1.5 Block diagram algebra

Block diagrams can be systematically simplified by means of a general procedure that involves the identification of sub-diagrams, followed by block reduction according to given rules [3]. In Appendix A, we show several block diagrams that frequently occur in control systems, as well as the corresponding simplified blocks.

1.1.5.1 Mason rule

In complex block diagrams the transfer function, G, can be calculated by means of the Mason's rule [4]:

$$G = \frac{\sum_{k=1}^{N} G_k \Delta_k}{\Delta} \tag{1.4}$$

where

- $\Delta = 1 - \sum L_i + \sum L_i L_j - \sum L_i L_j L_k + \cdots + (-1)^m \sum \cdots + \cdots$.
- N: total number of forward paths between input and output.
- G_k: path gain of the kth forward path between input and output.
- L_i: loop gain of each closed loop in the system.
- $L_i L_j$: product of the loop gains of any two non-touching loops (no common nodes).
- $L_i L_j L_k$: product of the loop gains of any three pairwise non-touching loops.
- Δ_k: cofactor value of Δ for the kth forward path, with the loops touching the kth forward path removed.

1.2 Worked examples

Problem 1.1 Consider the block diagram of a control system in Figure 1.2. $R(s) = \mathscr{L}[r(t)]$ is the Laplace transform of the input and $Y(s) = \mathscr{L}[y(t)]$ is the Laplace transform of the output. The transfer function of the system $\frac{Y(s)}{R(s)}$ is:

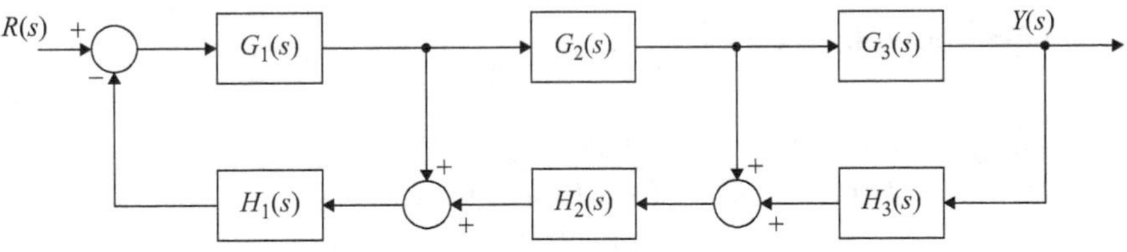

Figure 1.2 Block diagram of Problem 1.1

A) $\frac{Y(s)}{R(s)} = \frac{G_1(s)G_2(s)G_3(s)}{1 + G_1(s)G_2(s)G_3(s)H_1(s)H_2(s)H_3(s)}$

B) $\frac{Y(s)}{R(s)} = \frac{G_1(s)}{1 + G_1(s)H_1(s)} \cdot \frac{G_2(s)}{1 + G_2(s)H_2(s)} \cdot \frac{G_3(s)}{1 + G_3(s)H_3(s)}$

C) $\dfrac{Y(s)}{R(s)} = \dfrac{G_1(s)G_2(s)G_3(s)}{1 + G_1(s)H_1(s)\{1 + G_2(s)H_2(s)[1 + G_3(s)H_3(s)]\}}$

D) None of the above.

Resolution Simplifying and combining blocks as in Figure 1.3 we get the system transfer function.

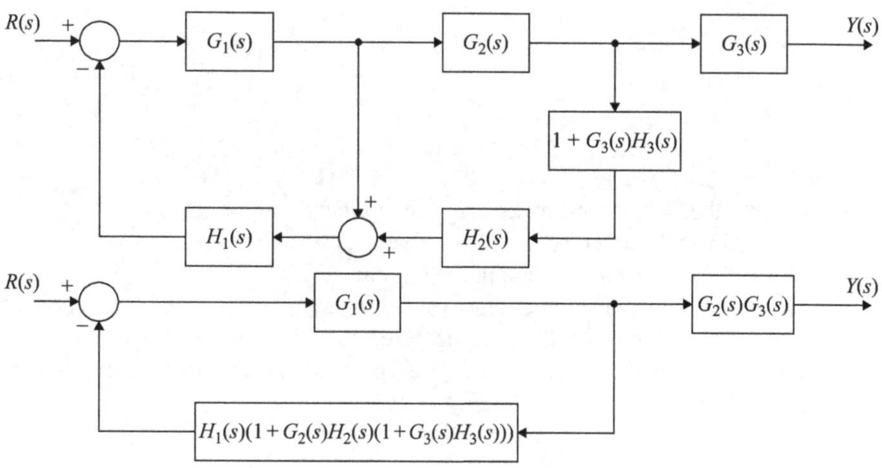

Figure 1.3 Resolution of Problem 1.1

Thus, the correct answer is option **C)**.

Problem 1.2 Consider the block diagram of a control system in Figure 1.4. The transfer function of the system $\dfrac{Y(s)}{R(s)}$ is:

A) $\dfrac{Y(s)}{R(s)} = \dfrac{G_1(s)G_2(s)G_3(s) + H_2(s)}{1 + G_1(s)G_2(s)G_3(s)H_1(s)H_3(s)}$

B) $\dfrac{Y(s)}{R(s)} = \dfrac{G_1(s)[G_2(s)G_3(s) + H_2(s)]}{1 + G_1(s)G_2(s)G_3(s)H_1(s)H_3(s)}$

C) $\dfrac{Y(s)}{R(s)} = \dfrac{G_1(s)G_2(s)G_3(s) + H_2(s)}{1 + G_1(s)H_1(s)[1 + G_2(s)G_3(s)H_3(s)]}$

D) None of the above.

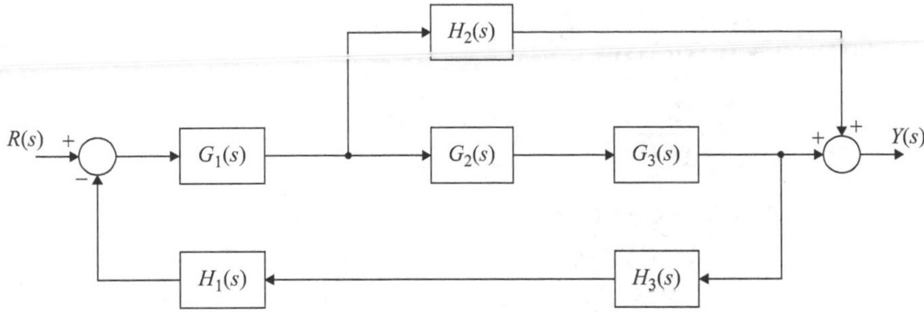

Figure 1.4 Block diagram of Problem 1.2

Resolution Simplifying and combining blocks as in Figure 1.5 we get the system transfer function.

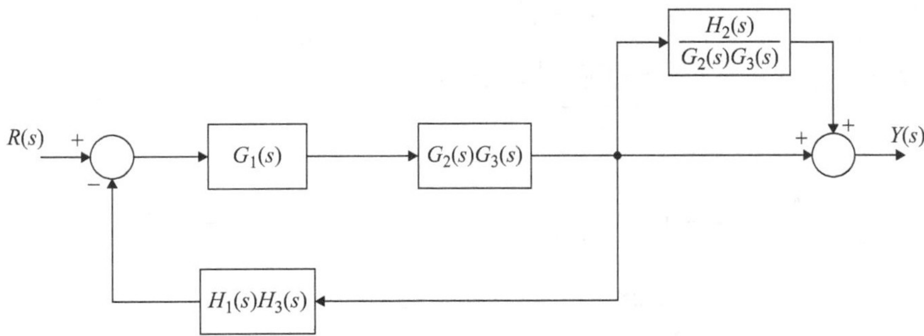

Figure 1.5 Resolution of Problem 1.2

Thus, the correct answer is option **B)**.

Problem 1.3 Consider the block diagram of a control system in Figure 1.6. The transfer function of the system $\dfrac{Y(s)}{R(s)}$ is:

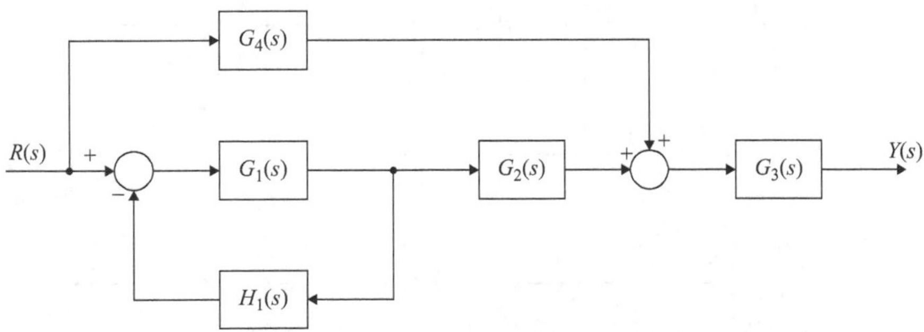

Figure 1.6 Block diagram of Problem 1.3

A) $\dfrac{Y(s)}{R(s)} = \dfrac{G_1(s)G_2(s)}{1 + G_1(s)H_1(s)} + G_3(s)G_4(s)$

B) $\dfrac{Y(s)}{R(s)} = \left[\dfrac{G_1(s)G_2(s)}{1 + G_1(s)H_1(s)} + G_4(s)\right] G_3(s)$

C) $\dfrac{Y(s)}{R(s)} = \dfrac{G_1(s)G_2(s)G_3(s)G_4(s)}{1 + G_1(s)H_1(s)}$

D) $\dfrac{Y(s)}{R(s)} = \dfrac{G_1(s)G_2(s) + G_3(s)G_4(s)}{1 + G_1(s)H_1(s)}.$

Resolution Simplifying and combining blocks as in Figure 1.7 we get the system transfer function.

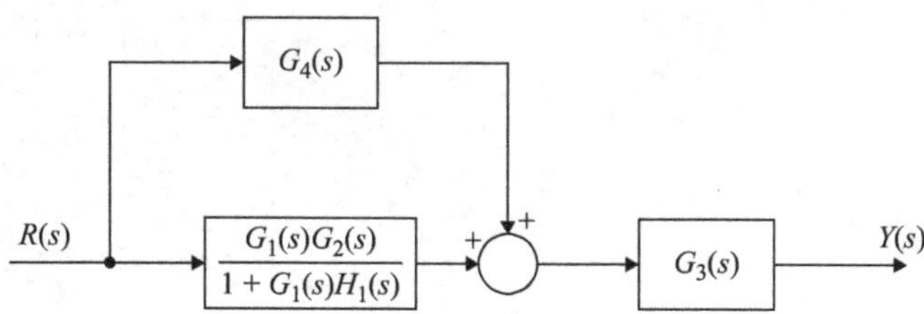

Figure 1.7 Resolution of Problem 1.3

Thus, the correct answer is option **B)**.

Problem 1.4 Consider the block diagram of a control system in Figure 1.8. The transfer function of the system $\dfrac{Y(s)}{R(s)}$ is:

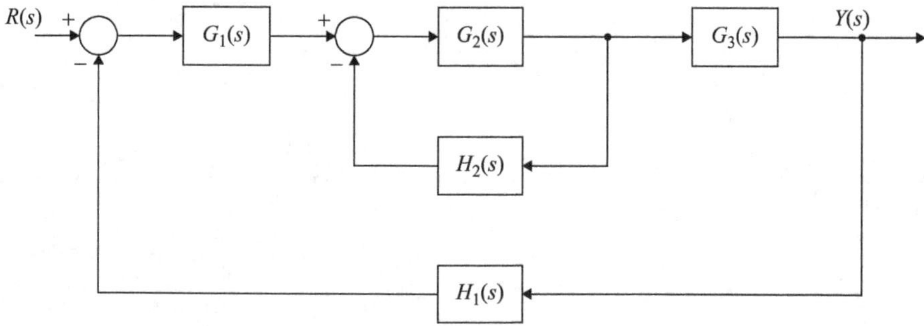

Figure 1.8 Block diagram of Problem 1.4

A) $$\frac{Y(s)}{R(s)} = \frac{G_1(s)G_2(s)G_3(s)}{1 + G_2(s)\,[H_2(s) + G_1(s)G_3(s)H_1(s)]}$$

B) $$\frac{Y(s)}{R(s)} = \frac{G_1(s)G_2(s)G_3(s)}{1 + G_2(s)\,[H_2(s) + G_1(s)H_1(s)]}$$

C) $$\frac{Y(s)}{R(s)} = \frac{G_1(s)G_2(s)G_3(s)}{1 + G_1(s)H_1(s) + G_2(s)H_2(s)}$$

D) $$\frac{Y(s)}{R(s)} = \frac{G_1(s)G_2(s)G_3(s)}{1 + G_2(s)H_2(s) + G_1(s)H_1(s)G_3(s)H_2(s)}.$$

Resolution Simplifying and combining blocks as in Figure 1.9 we get the system transfer function.

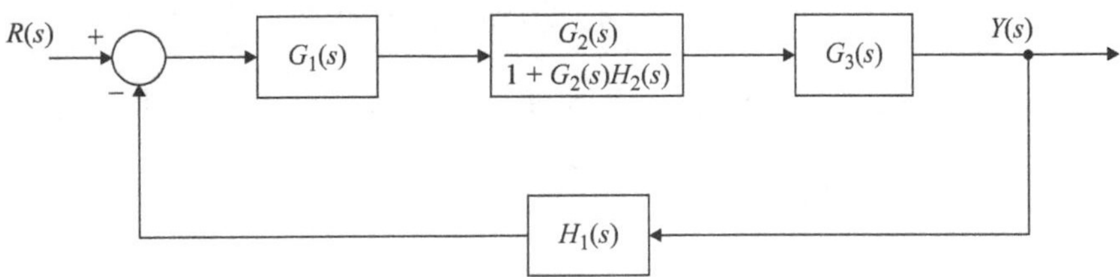

Figure 1.9 Resolution of Problem 1.4

Thus, the correct answer is option **A)**.

Problem 1.5 Consider the block diagram of a control system in Figure 1.10. Find the transfer function $\frac{Y(s)}{R(s)}$.

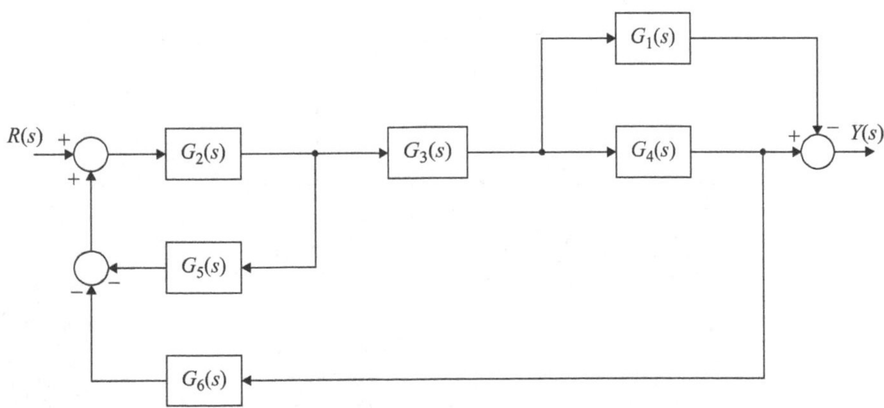

Figure 1.10 Block diagram of Problem 1.5

Resolution Simplifying and combining blocks as in Figure 1.11 we get the system transfer function.

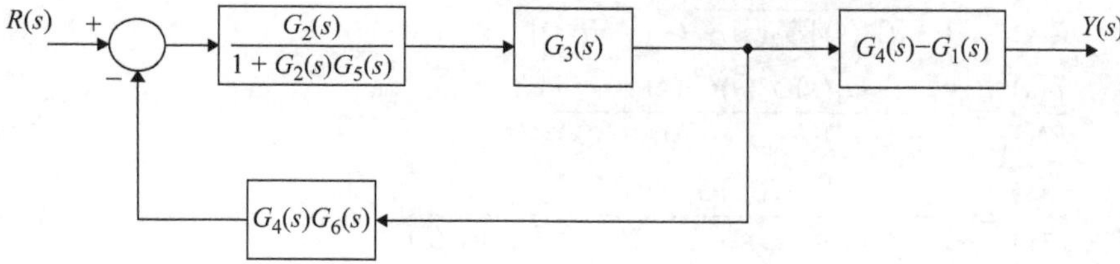

Figure 1.11 Resolution of Problem 1.5

Thus, $\dfrac{Y(s)}{R(s)} = \dfrac{G_2(s)G_3(s)\,[G_4(s) - G_1(s)]}{1 + G_2(s)G_5(s) + G_2(s)G_3(s)G_4(s)G_6(s)}$

Problem 1.6 Consider the block diagram of a system in Figure 1.12. The transfer function of the system $\dfrac{Y(s)}{R(s)}$ is:

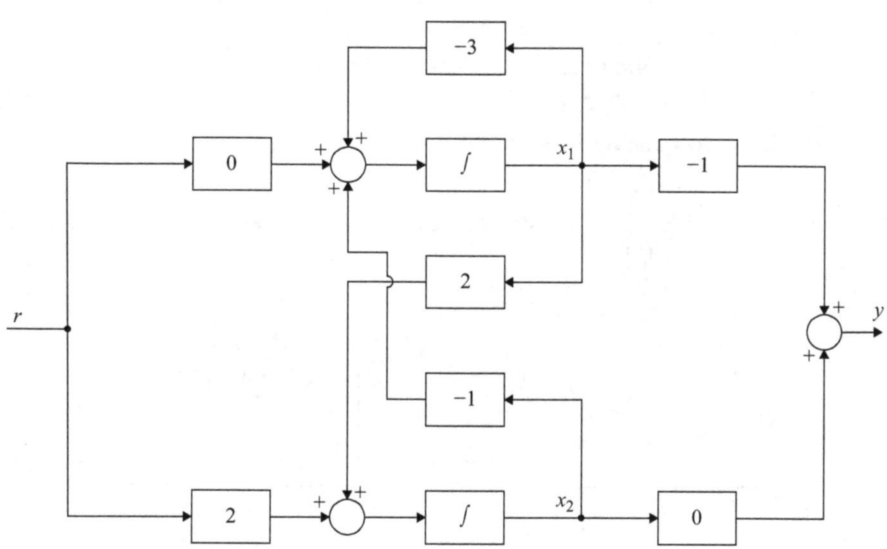

Figure 1.12 Block diagram of Problem 1.6

A) $\dfrac{Y(s)}{R(s)} = \dfrac{1}{(s+1)^2}$

B) $\dfrac{Y(s)}{R(s)} = \dfrac{2}{(s+1)(s+2)}$

C) $\dfrac{Y(s)}{R(s)} = \dfrac{3}{(s+1)(s+3)}$

D) None of the above.

Resolution Simplifying and combining blocks as in Figure 1.13 we get the system transfer function.

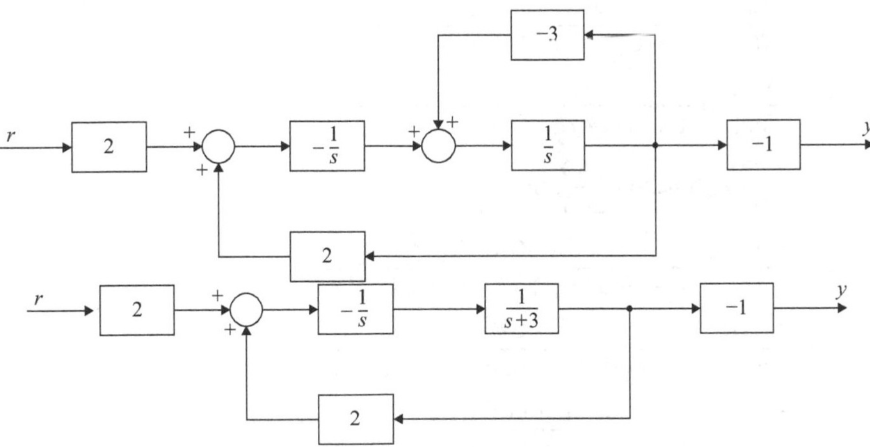

Figure 1.13 Resolution of Problem 1.6

Thus, the correct answer is option **B)**.

1.3 Proposed exercises

Exercise 1.1 Consider the block diagram in Figure 1.14.

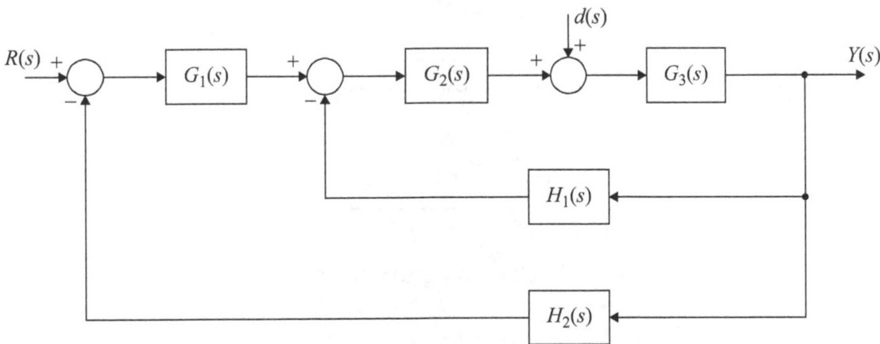

Figure 1.14 Block diagram of Exercise 1.1

1. The transfer function $\dfrac{Y(s)}{R(s)}$ is:

A) $$\frac{Y(s)}{R(s)} = \frac{G_1(s)G_2(s)G_3(s)}{1 + G_2(s)G_3(s)\left[H_1(s) + G_1(s)H_2(s)\right]}$$

B) $$\frac{Y(s)}{R(s)} = \frac{G_3(s)}{1 + G_2(s)G_3(s)\left[H_1(s) + G_1(s)H_2(s)\right]}$$

C) $$\frac{Y(s)}{R(s)} = \frac{G_1(s)G_2(s)G_3(s)}{1 + G_2(s)G_3(s)\left[H_1(s) + H_2(s)\right]}$$

D) $$\frac{Y(s)}{R(s)} = \frac{G_3(s)}{1 + G_2(s)G_3(s)\left[H_1(s) + H_2(s)\right]}.$$

2. The transfer function $\dfrac{Y(s)}{d(s)}$ is:

E) $$\frac{Y(s)}{d(s)} = \frac{G_1(s)G_2(s)G_3(s)}{1 + G_2(s)G_3(s)\left[H_1(s) + G_1(s)H_2(s)\right]}$$

F) $$\frac{Y(s)}{d(s)} = \frac{G_3(s)}{1 + G_2(s)G_3(s)\left[H_1(s) + G_1(s)H_2(s)\right]}$$

G) $$\frac{Y(s)}{d(s)} = \frac{G_1(s)G_2(s)G_3(s)}{1 + G_2(s)G_3(s)\left[H_1(s) + H_2(s)\right]}$$

H) $$\frac{Y(s)}{d(s)} = \frac{G_3(s)}{1 + G_2(s)G_3(s)\left[H_1(s) + H_2(s)\right]}.$$

Exercise 1.2 Consider the block diagram in Figure 1.15. The transfer function $\dfrac{Y(s)}{R(s)}$ is:

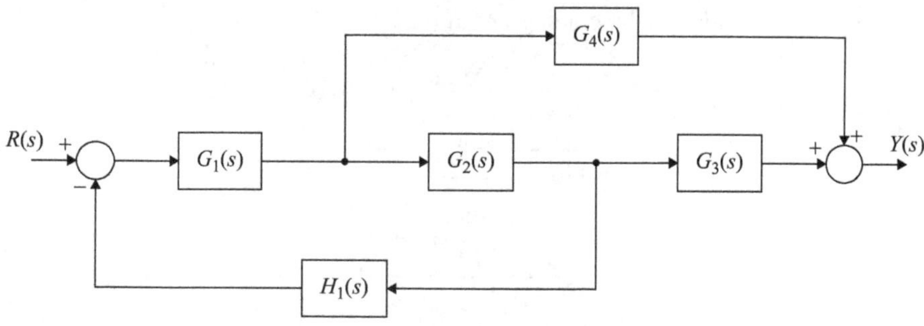

Figure 1.15 Block diagram of Exercise 1.2

A) $$\frac{Y(s)}{R(s)} = \frac{G_1(s)\left[G_2(s)G_3(s) + G_4(s)\right]}{1 + G_1(s)G_2(s)H_1(s)}$$

B) $\dfrac{Y(s)}{R(s)} = \dfrac{G_1(s)G_2(s)\,[G_3(s) + G_4(s)]}{1 + G_1(s)G_2(s)H_1(s)}$

C) $\dfrac{Y(s)}{R(s)} = \dfrac{G_1(s)G_2(s)G_3(s) + G_4(s)}{1 + G_1(s)G_2(s)H_1(s)}$

D) $\dfrac{Y(s)}{R(s)} = \dfrac{G_1(s)G_4(s)\,[G_2(s) + G_3(s)]}{1 + G_1(s)G_2(s)H_1(s)}.$

Exercise 1.3 Consider the block diagram in Figure 1.16. The transfer function $\dfrac{Y(s)}{R(s)}$ is:

A) $\dfrac{Y(s)}{R(s)} = \dfrac{G_1(s)G_2(s) + G_4(s)}{1 + G_2(s)G_3(s)H(s)}$

B) $\dfrac{Y(s)}{R(s)} = \dfrac{G_3(s)\,[G_1(s)G_2(s) + G_4(s)]}{1 + G_2(s)G_3(s)H(s)}$

C) $\dfrac{Y(s)}{R(s)} = \dfrac{G_1(s)G_2(s)G_3(s) + G_4(s)}{1 + G_2(s)G_3(s)H(s)}$

D) None of the above.

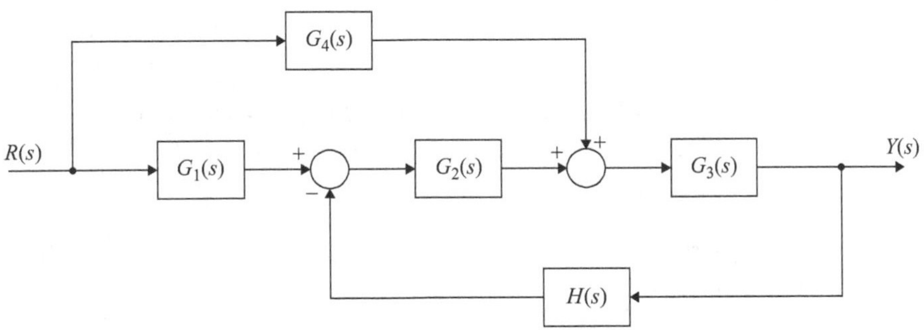

Figure 1.16 Block diagram of Exercise 1.3

Exercise 1.4 Consider the block diagram in Figure 1.17. The transfer function $\dfrac{Y(s)}{R(s)}$ is:

A) $\dfrac{Y(s)}{R(s)} = \dfrac{G_1(s)G_2(s)G_3(s)}{1 + G_1(s)G_2(s)G_3(s)H(s)}$

B) $\dfrac{Y(s)}{R(s)} = \dfrac{G_2(s)\,[G_1(s) + G_3(s)]}{1 + G_1(s)G_2(s)G_3(s)H(s)}$

C) $\dfrac{Y(s)}{R(s)} = \dfrac{G_1(s)G_2(s)G_3(s)}{1 + G_2(s)H(s)}$

D) None of the above.

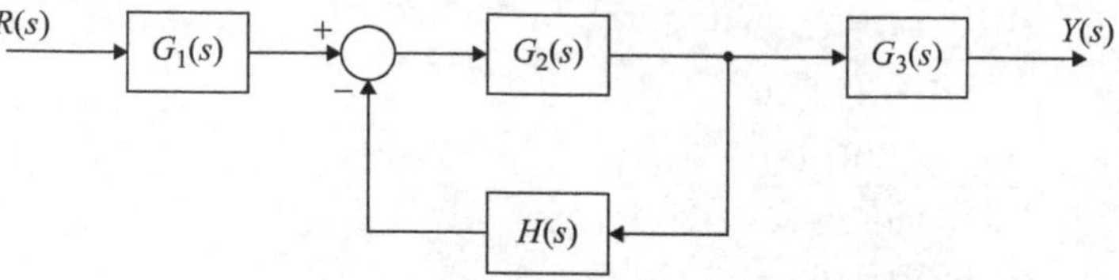

Figure 1.17 Block diagram of Exercise 1.4

Exercise 1.5 Consider the block diagram in Figure 1.18. The transfer function $\frac{Y(s)}{R(s)}$ is:

A) $\frac{Y(s)}{R(s)} = \frac{G_1(s)G_2(s)}{1 + G_1(s)G_2(s)H_1(s)H_2(s)}$

B) $\frac{Y(s)}{R(s)} = \frac{G_1(s) + G_2(s)}{1 + [G_1(s) + G_2(s)]\,[H_1(s) + H_2(s)]}$

C) $\frac{Y(s)}{R(s)} = \frac{G_1(s)G_2(s)}{1 + G_1(s)G_2(s)\,[H_1(s) + H_2(s)]}$

D) None of the above.

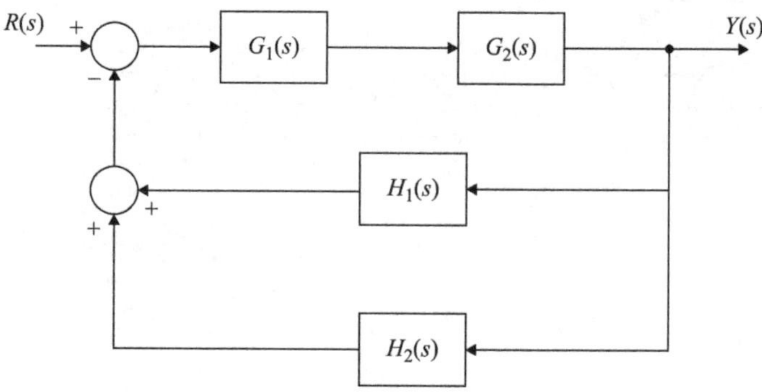

Figure 1.18 Block diagram of Exercise 1.5

Exercise 1.6 Consider the two block diagrams in Figure 1.19. The two systems are equivalent if and only if:

A) $W(s) = \frac{1}{G(s)}$

B) $W(s) = \dfrac{G(s)}{1 + G(s)}$

C) $W(s) = G(s)$

D) None of the above.

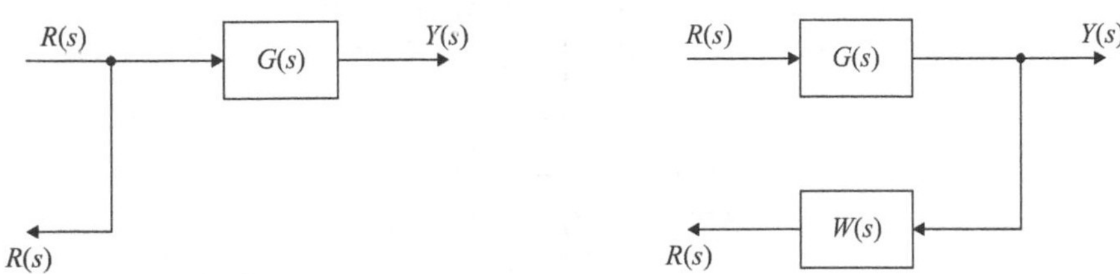

Figure 1.19 Block diagrams of Exercise 1.6

Exercise 1.7 Consider the two block diagrams shown in Figure 1.20. The transfer function $\dfrac{Y(s)}{R(s)}$ is identical for the two cases, if and only if:

A) $W(s) = \dfrac{1}{G(s)}$

B) $W(s) = \dfrac{G(s)}{1 + G(s)H(s)}$

C) $W(s) = \dfrac{G(s)}{1 + G(s)}$

D) $W(s) = \dfrac{1}{H(s)}$.

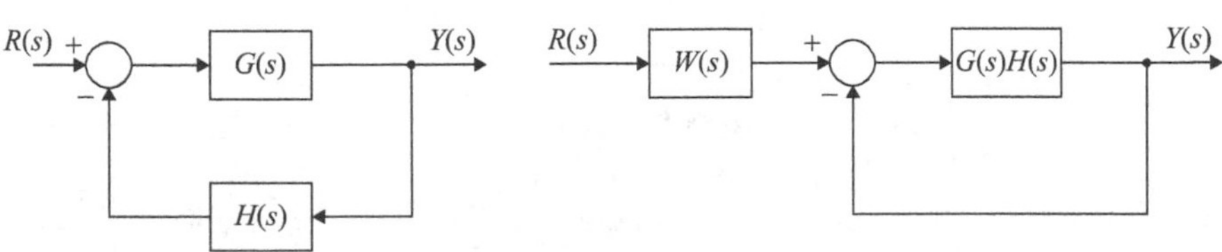

Figure 1.20 Block diagrams of Exercise 1.7

Exercise 1.8 Simplifying the block diagram in Figure 1.21, obtain the transfer function $\frac{Y(s)}{R(s)}$.

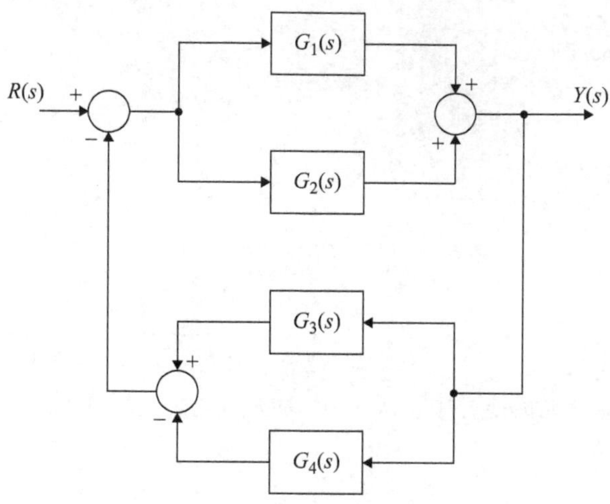

Figure 1.21 Block diagram of Exercise 1.8

Exercise 1.9 Simplify the block diagram of Figure 1.22 to obtain the transfer function $\frac{Y(s)}{R(s)}$.

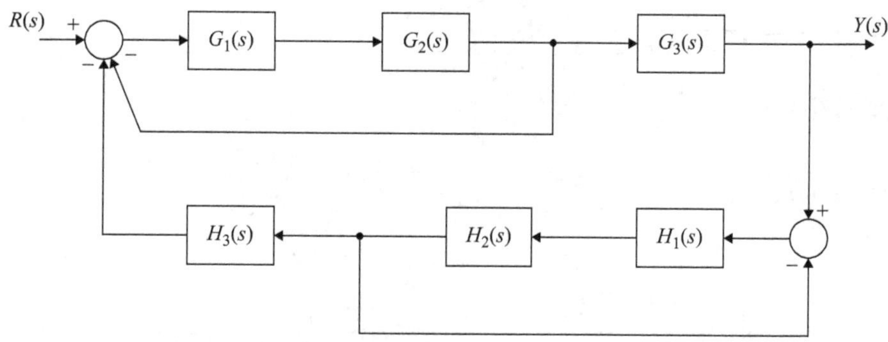

Figure 1.22 Block diagram of Exercise 1.9

Exercise 1.10 For the block diagram in Figure 1.23, the transfer function $\frac{Y(s)}{R(s)}$ is:

A) $\frac{Y(s)}{R(s)} = \frac{G(s)}{1+G(s)H(s)}$

B) $\dfrac{Y(s)}{R(s)} = \dfrac{G(s)H(s)}{1 + G(s)H(s)}$

C) $\dfrac{Y(s)}{R(s)} = \dfrac{1}{1 + G(s)H(s)}$

D) None of the above.

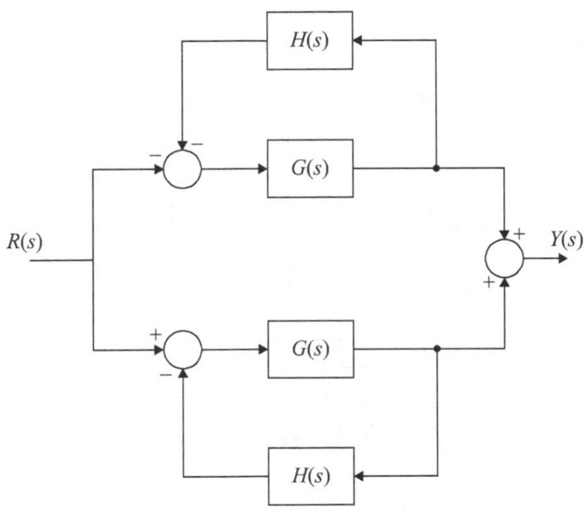

Figure 1.23 Block diagram of Exercise 1.10

Exercise 1.11 For the block diagram in Figure 1.24, the transfer function $\dfrac{Y(s)}{R(s)}$ is:

A) $\dfrac{Y(s)}{R(s)} = \dfrac{b^2}{(s + a)^2 + b^2}$

B) $\dfrac{Y(s)}{R(s)} = \dfrac{b}{(s + a)^2 + b^2}$

C) $\dfrac{Y(s)}{R(s)} = \dfrac{1}{(s + a)^2 + b^2}$

D) $\dfrac{Y(s)}{R(s)} = \dfrac{a^2 + b^2}{(s + a)^2 + b^2}.$

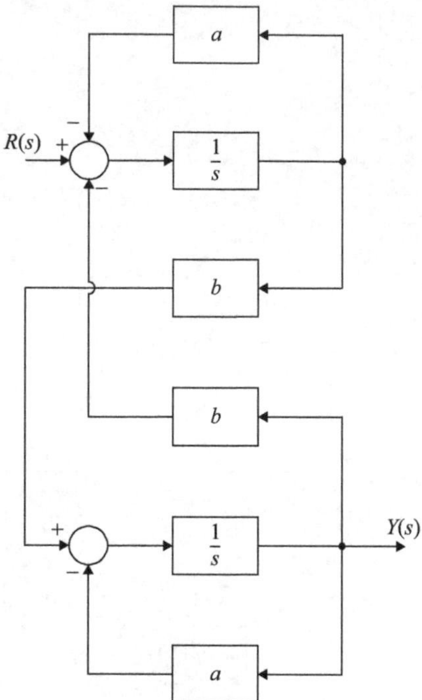

Figure 1.24 Block diagram of Exercise 1.11

Exercise 1.12 Consider the block diagram in Figure 1.25.

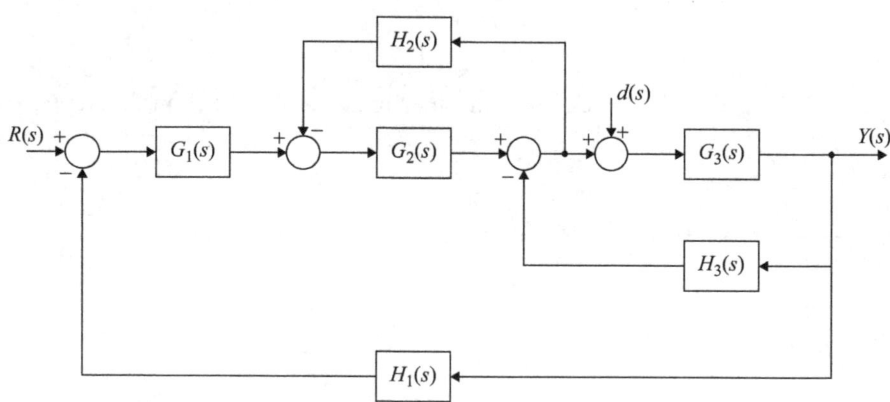

Figure 1.25 Block diagram of Exercise 1.12

The output $Y(s)$ with inputs $R(s)$ and $d(s)$ acting simultaneously is:

A) $$Y(s) = \frac{G_1(s)G_2(s)G_3(s)R(s) + G_3(s)\left[1 + G_2(s)H_2(s)\right]d(s)}{1 + G_2(s)H_2(s) + G_3(s)H_3(s) + G_1(s)G_2(s)G_3(s)H_1(s)}$$

B) $Y(s) = \dfrac{G_1(s)G_2(s)G_3(s)R(s) + G_2(s)\,[1 + G_3(s)H_2(s)]\,d(s)}{1 + G_2(s)H_2(s) + G_3(s)H_3(s) + G_1(s)G_2(s)G_3(s)H_1(s)}$

C) $Y(s) = \dfrac{G_1(s)G_2(s)G_3(s)R(s) + G_3(s)d(s)}{1 + G_2(s)H_2(s) + G_3(s)H_3(s) + G_1(s)G_2(s)G_3(s)H_1(s)}$

D) None of the above.

Exercise 1.13 Consider the block diagram in Figure 1.26. The transfer function $\dfrac{Y(s)}{R(s)}$ is:

A) $\dfrac{Y(s)}{R(s)} = \dfrac{G_1(s)\,[G_2(s) + G_3(s)]\,G_4(s)}{1 + G_1(s)\,[G_2(s) + G_3(s)]\,G_4(s)H_1(s)}$

B) $\dfrac{Y(s)}{R(s)} = \dfrac{G_1(s)\,[G_2(s) + G_3(s)]\,G_4(s)}{1 + G_4(s)H_2(s) + G_1(s)\,[G_2(s) + G_3(s)]\,G_4(s)H_1(s)}$

C) $\dfrac{Y(s)}{R(s)} = \dfrac{G_1(s)G_4(s)}{1 + G_1(s)G_4(s)H_1(s)}$

D) $\dfrac{Y(s)}{R(s)} = \dfrac{G_1(s)\,[G_2(s) + G_3(s)]\,G_4(s)}{1 - G_4(s)H_2(s) + G_1(s)\,[G_2(s) + G_3(s)]\,G_4(s)H_1(s)}$.

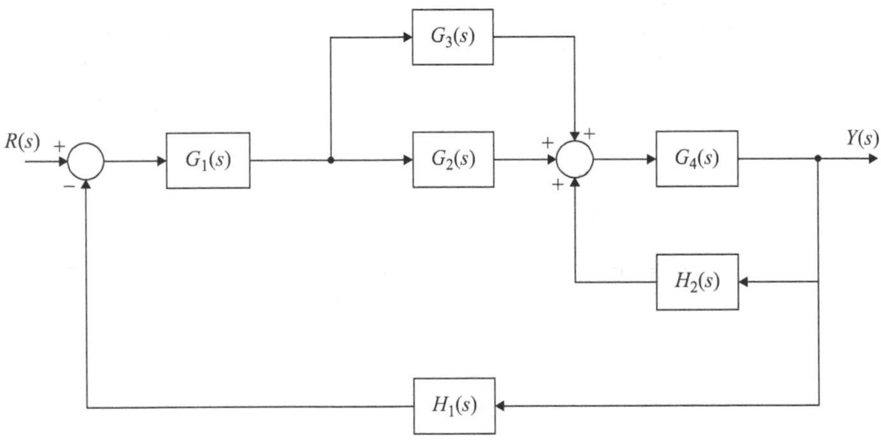

Figure 1.26 Block diagram of Exercise 1.13

Exercise 1.14 Consider the block diagram in Figure 1.27.

1. The transfer function $\dfrac{Y(s)}{R(s)}$ is:

   A) $\dfrac{Y(s)}{R(s)} = \dfrac{G_1(s) + G_2(s)}{1 + G_3(s)G_4(s)H_1(s)}$

B) $\dfrac{Y(s)}{R(s)} = \dfrac{G_3(s)G_4(s)}{1 + G_3(s)G_4(s)H_1(s)}$

C) $\dfrac{Y(s)}{R(s)} = \dfrac{[G_1(s) + G_2(s)]\, G_3(s)G_4(s)}{1 + G_3(s)G_4(s)H_1(s)}$

D) None of the above.

2. The transfer function $\dfrac{Y(s)}{d(s)}$ is:

E) $\dfrac{Y(s)}{d(s)} = \dfrac{G_1(s) + G_2(s)}{1 + G_3(s)G_4(s)H_1(s)}$

F) $\dfrac{Y(s)}{d(s)} = \dfrac{G_3(s)G_4(s)}{1 + G_3(s)G_4(s)H_1(s)}$

G) $\dfrac{Y(s)}{d(s)} = \dfrac{[G_1(s) + G_2(s)]\, G_3(s)G_4(s)}{1 + G_3(s)G_4(s)H_1(s)}$

H) None of the above.

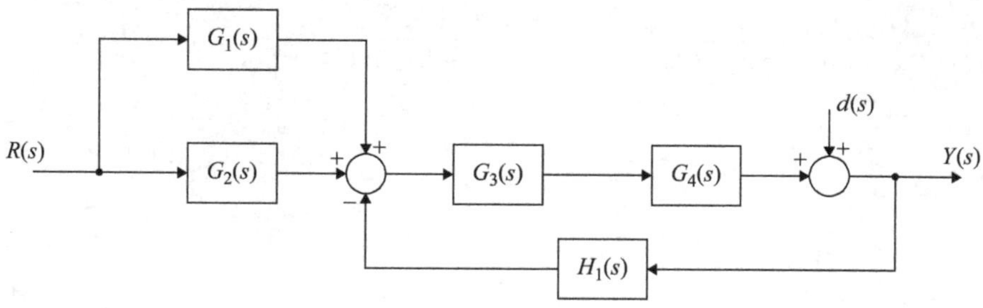

Figure 1.27 Block diagram of Exercise 1.14

Exercise 1.15 Consider the block diagram in Figure 1.28. Determine the transfer function $\dfrac{Y(s)}{R(s)}$.

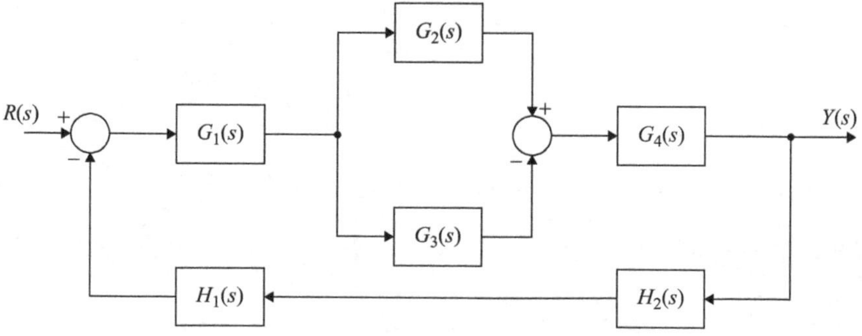

Figure 1.28 Block diagram of Exercise 1.15

Exercise 1.16 Consider the block diagram in Figure 1.29. Determine the transfer function $\dfrac{Y(s)}{R(s)}$.

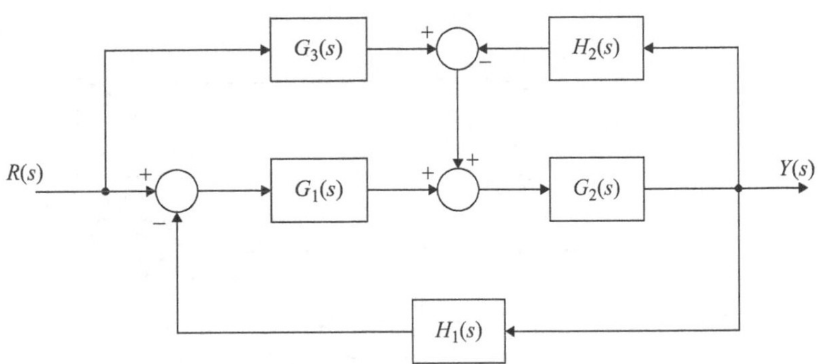

Figure 1.29 Block diagram of Exercise 1.16

Exercise 1.17 Consider the block diagram in Figure 1.30. Determine the transfer function $\dfrac{Y(s)}{R(s)}$.

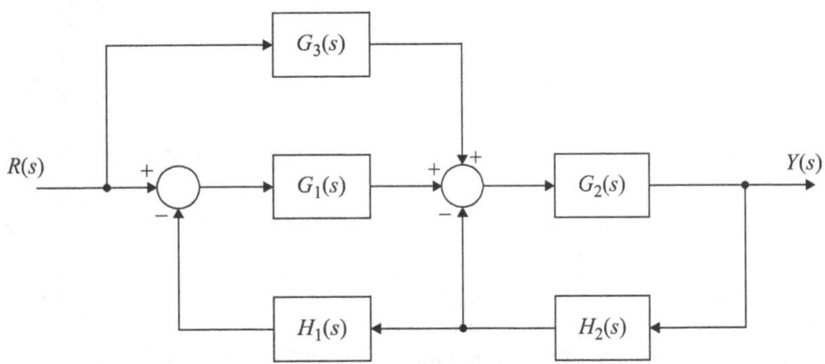

Figure 1.30 Block diagram of Exercise 1.17

Exercise 1.18 Consider the block diagram in Figure 1.31. Determine the transfer function $\dfrac{Y(s)}{R(s)}$.

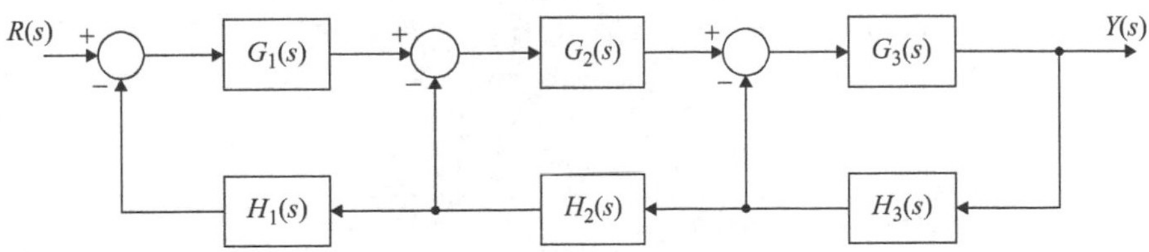

Figure 1.31 Block diagram of Exercise 1.18

Exercise 1.19 Consider the block diagram in Figure 1.32. Determine the transfer function $\frac{Y(s)}{R(s)}$.

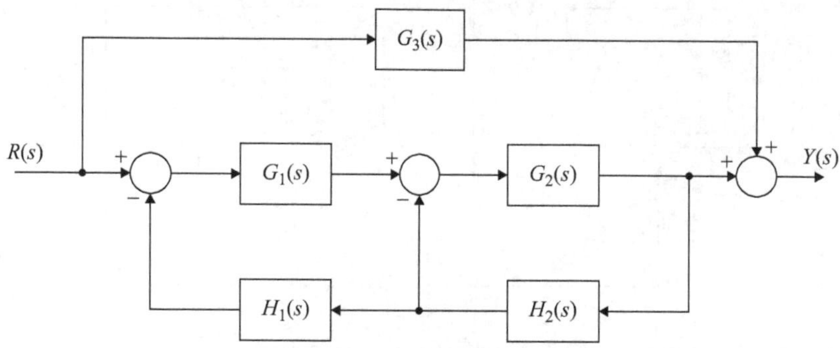

Figure 1.32 Block diagram of Exercise 1.19

1.4 Block diagram analysis using computer packages

This section presents several commands for handling block diagrams using the computer packages MATLAB®, SCILAB™ and OCTAVE©.

1.4.1 MATLAB

This subsection describes some basic commands that can be adopted with the package MATLAB.

We consider the series, parallel and feedback connection of two models $G(s)$ and $H(s)$.

The transfer functions $G(s) = \frac{s+1}{s(s+2)}$ and $H(s) = \frac{s+3}{s+4}$ are represented by the arrays `numG` and `denG` and `numH` and `denH`, respectively. The command `tf` converts to transfer function mode. The commands `G*H` and `G+H` establish the series and parallel connection of two models, respectively. The command `feedback` converts the open-loop transfer function into a closed-loop one via H.

The complete code is as follows.

```
%%%% Feedback connection of two models %%%%
% Numerator and denominator of transfer function G(s)
numG = [0 1 1];
denG = [1 2 0];
```

SCILAB™ is a trademark of INRIA.
OCTAVE© is copyright John W. Eaton and others.
MATLAB® is registered trademark of The MathWorks, Inc.

```
% Numerator and denominator of transfer function H(s)
numH = [1 3];
denH = [1 4];

% Create transfer function model, convert to
% transfer function model
G = tf(numG,denG)
H = tf(numH,denH)

% Transfer function of series connection of G and H
T1 = G*H

% Transfer function of parallel connection of G and H
T2 = G+H

% Closed-loop transfer function with feedback via H
T3 = feedback(G,H)
```

MATLAB creates the following Command Window.

```
G =

    s + 1
  ---------
  s^2 + 2 s

Continuous-time transfer function.


H =

  s + 3
  -----
  s + 4

Continuous-time transfer function.


T1 =
```

```
     s^2 + 4 s + 3
  -----------------
  s^3 + 6 s^2 + 8 s

Continuous-time transfer function.


T2 =

  s^3 + 6 s^2 + 11 s + 4
  ----------------------
    s^3 + 6 s^2 + 8 s

Continuous-time transfer function.


T3 =

        s^2 + 5 s + 4
  ----------------------
  s^3 + 7 s^2 + 12 s + 3

Continuous-time transfer function.
```

The command connect leads to the connection of several block diagram elements based on signal names. The previous example can be programmed with the following code.

```
%%%% Feedback connection of two models %%%%
% Numerator and denominator of transfer function G(s)
numG = [0 1 1];
denG = [1 2 0];

% Numerator and denominator of transfer function H(s)
numH = [1 3];
denH = [1 4];

% Create transfer function model, convert to
% transfer function model
G = tf(numG,denG)
H = tf(numH,denH)

% Name the inputs and outputs
```

```
G.u = 'e';
G.y = 'c';
H.u = 'c';
H.y = 'b';

% Create the summing junction
Sum = sumblk('e = r - b');

% Combine G, H, and the summing junction to create
% the aggregate model from r to c
T = connect(G,H,Sum,'r','c')
```

MATLAB creates the following Command Window.

```
G =

     s + 1
  ---------
  s^2 + 2 s

Continuous-time transfer function.


H =

  s + 3
  -----
  s + 4

Continuous-time transfer function.


T =

  From input "r" to output "c":
      s^2 + 5 s + 4
  ----------------------
  s^3 + 7 s^2 + 12 s + 3

Continuous-time transfer function.
```

1.4.2 SCILAB

This subsection describes some basic commands that can be adopted with the package SCILAB.

We consider the series, parallel and feedback connection of two models $G(s)$ and $H(s)$.

The transfer functions $G(s) = \frac{s+1}{s(s+2)}$ and $H(s) = \frac{s+3}{s+4}$ are represented by the arrays `numG` and `denG` and `numH` and `denH`, respectively. The command `tf` converts to transfer function mode. The commands `G*H` and `G+H` establish the series and parallel connection of two models, respectively. The command `feedback` converts the open-loop transfer function into a closed-loop one via H.

The complete code is as follows.

```
//// Feedback connection of two models ////

// Numerator and denominator of transfer function G(s)
numG = poly([1 1 0],'s','c');
denG = poly([0 2 1],'s','c');

// Numerator and denominator of transfer function H(s)
numH = poly([3 1],'s','c');
denH = poly([4 1],'s','c');

// Define continuous LTI systems
G = syslin('c',numG,denG)
H = syslin('c',numH,denH)

// Transfer function of series connection of G and H
T1 = G*H;

// Transfer function of parallel connection of G and H
T2 = G+H;

// Closed-loop transfer function with feedback via H
T3 = G/.H; // feedback is via H

disp (T1,'T1=');
disp (T2,'T2=');
disp (T3,'T3=');
```

SCILAB creates the following Console.

```
G =

     s + 1
  ---------
  s^2 + 2 s

Continuous-time transfer function.


H =

  s + 3
  -----
  s + 4

Continuous-time transfer function.


T1 =

     s^2 + 4 s + 3
  -----------------
  s^3 + 6 s^2 + 8 s

Continuous-time transfer function.


T2 =

  s^3 + 6 s^2 + 11 s + 4
  ----------------------
    s^3 + 6 s^2 + 8 s

Continuous-time transfer function.


T3 =

        s^2 + 5 s + 4
  ----------------------
  s^3 + 7 s^2 + 12 s + 3

Continuous-time transfer function.
```

1.4.3 OCTAVE

This subsection describes some basic commands that can be adopted with the package OCTAVE. It is required to load package `control`.

We consider the series, parallel and feedback connection of two models $G(s)$ and $H(s)$.

The transfer functions $G(s) = \frac{s+1}{s(s+2)}$ and $H(s) = \frac{s+3}{s+4}$ are represented by the arrays `numG` and `denG` and `numH` and `denH`, respectively. The command `tf` converts to transfer function mode. The commands `G*H` and `G+H` establish the series and parallel connection of two models, respectively. The command `feedback` converts the open-loop transfer function into a closed-loop one via H.

The complete code is as follows.

```
%%%% Feedback connection of two models %%%%
% Numerator and denominator of transfer function G(s)
numG = [0 1 1];
denG = [1 2 0];

% Numerator and denominator of transfer function H(s)
numH = [1 3];
denH = [1 4];

% Create transfer function model, convert to
% transfer function model
G = tf(numG,denG)
H = tf(numH,denH)

% Transfer function of series connection of G and H
T1 = G*H

% Transfer function of parallel connection of G and H
T2 = G+H

% Closed-loop transfer function with feedback via H
T3 = feedback(G,H)
```

OCTAVE creates the following Command Window.

```
Transfer function 'G' from input 'u1' to output ...

         s + 1
 y1:  ---------
      s^2 + 2 s
```

```
Continuous-time model.

Transfer function 'H' from input 'u1' to output ...

      s + 3
 y1:  -----
      s + 4

Continuous-time model.

Transfer function 'T1' from input 'u1' to output ...

        s^2 + 4 s + 3
 y1:  -----------------
      s^3 + 6 s^2 + 8 s

Continuous-time model.

Transfer function 'T2' from input 'u1' to output ...

      s^3 + 6 s^2 + 11 s + 4
 y1:  ----------------------
        s^3 + 6 s^2 + 8 s

Continuous-time model.

Transfer function 'T3' from input 'u1' to output ...

          s^2 + 5 s + 4
 y1:  ----------------------
      s^3 + 7 s^2 + 12 s + 3

Continuous-time model.
```

Chapter 2
Mathematical models

2.1 Fundamentals

A mathematical model is a description of a system by means of mathematical concepts and language. The mathematical model can be used to predict the system behavior, to explain the effect of individual components, and to decide about the changes needed to achieve the system specifications.

2.1.1 List of symbols

$\mathbf{a}_c$ absolute linear acceleration vector of point c
$\mathbf{a}_{cj}$ absolute acceleration vector of the center of mass of the jth rigid body in a multi-body mechanical system
A area normal to the heat flow
$A_{V(OL)}$ open-loop gain of an operational amplifier
B_d viscous friction coefficient
c specific heat of a material; system center of mass
C capacitance
e_b back electromotive force induced by the rotation of the armature windings in a magnetic field
E energy
f force
$\mathbf{F}_i$ ith force vector
g acceleration due to gravity
h equivalent head
$\dot{\mathbf{H}}_c$ time derivative of the angular moment vector with respect to the center of mass c
$\dot{\mathbf{H}}_p$ time derivative of the angular moment vector in point p
H convection coefficient
i_{in} input current
i_k current flowing in the kth branch of an electrical circuit
I_j inertia of the jth rigid body with respect to its center of mass c_j
$\mathbf{I}_p$ inertia matrix of a rigid body about point p
J_k inertia of the kth wheel in a mechanical transmission
J_T total inertia referred to the motor axis of a multi-level mechanical transmission

k	thermal conductivity
K_s	spring stiffness constant
K_T	motor torque constant
$\mathscr{L}$	Laplace operator
L	inductance
L_h	inertance defined in terms of head
L_p	inertance defined in terms of pressure
m_j	mass of the jth rigid body
M	mass
$\mathbf{M}_{i/p}$	ith actuating moment vector with reference to p
n_1, n_2	radii of the input and output wheels in a mechanical transmission
n_k	radius of the kth wheel of a mechanical transmission
N	transmission ratio
p	arbitrary point
P	pitch of a lead-screw mechanical transmission
q	heat flow rate
q_v	volume flow rate
$\mathbf{r}_{c/p}$	position vector of the center of mass with respect to p
R	resistance
R_h	hydraulic resistance in terms of equivalent head h
R_{in}	input resistance
R_{out}	output resistance
R_p	hydraulic resistance in terms of pressure drop Δp
s	Laplace variable
t	time
T	torque, temperature
T_1, T_2	torque at the input and output wheels of a mechanical transmission
v_k	voltage at the kth component of an electrical circuit
v_{in}	input voltage
v_{out}	output voltage
V	volume
x	linear displacement
$\boldsymbol{\alpha}$	angular acceleration vector
α_1, α_2	angular acceleration of the input and output wheels in a mechanical transmission
δ_j	distance between points c_j and p
ΔX	thickness of a thermal conductor
ΔT	temperature difference
η	efficiency of a transmission composed of two wheels
η_k	efficiency of the kth level transmission in a multi-level chain
θ	angular displacement
ν	efficiency of a lead-screw mechanical transmission
ρ	density of a fluid
$\boldsymbol{\omega}$	angular velocity vector
ω_1, ω_2	angular velocities of the input and output wheels in a mechanical transmission

2.1.2 Modeling of electrical systems

2.1.2.1 RLC electrical circuits

Electrical circuits are governed by Kirchhoff's laws [7]. The current law states that the algebraic sum of all currents flowing towards or away from a node is zero:

$$\sum_{k=1}^{n} i_k(t) = 0 \tag{2.1}$$

where n represents the total number of branches connected to the node, and i_k is the current flowing in the kth branch.

Kirchhoff's voltage law states that the algebraic sum of the voltages around a loop in an electrical circuit is zero:

$$\sum_{k=1}^{n} v_k(t) = 0 \tag{2.2}$$

where n represents the total number of voltages (or potential differences), v_k, around the loop.

The equations relating voltage and current across the circuit basic components, namely resistance, R, capacitance, C, and inductance, L, are:

$$v(t) = Ri(t) \tag{2.3}$$

$$v(t) = L\frac{di(t)}{dt} \tag{2.4}$$

$$i(t) = C\frac{dv(t)}{dt} \tag{2.5}$$

By means of the Kirchhoff's laws and the expressions given above, we can obtain the differential equations of any electrical circuit in the time domain.

Alternatively, if we are interested in the system transfer function, we can analyze the circuit in the Laplace domain, by considering:

$$V(s) = RI(s) \tag{2.6}$$

$$V(s) = sLI(s) \tag{2.7}$$

$$I(s) = sCV(s) \tag{2.8}$$

2.1.2.2 Electrical circuits with operational amplifiers

Operational amplifiers are electronic circuits that implement mathematical operations on voltage signals, namely amplification, algebraic sum, differentiation, integration, and filtering [7]. An "ideal" operational amplifier (Figure 2.1) has infinite input resistance, R_{in}, (i.e., no current flows into the input terminals, $i_{in} = 0$), zero output resistance, R_{out}, (i.e., the output voltage, $v_{out} = (v^+ - v^-) \cdot A_{V(OL)}$, is not affected by the load connected to the output terminal), and infinite open-loop gain, $A_{V(OL)}$.

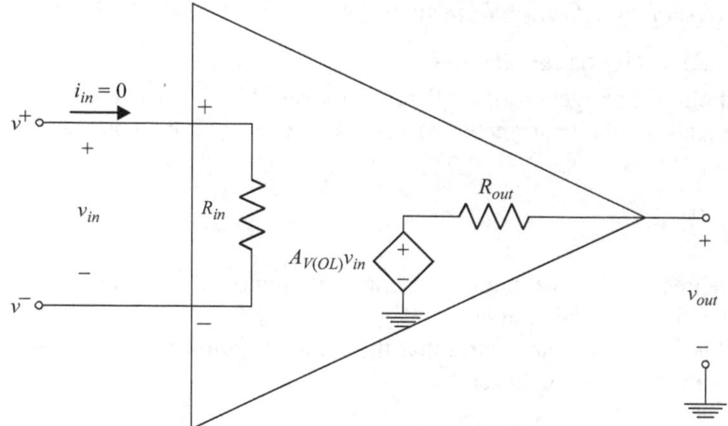

Figure 2.1 Operational amplifier model

When a negative feedback loop is included in the circuit, we can assume that the voltages at the input terminals are equal ($v^+ = v^-$). With these assumptions, and using the basic tools for analyzing electric circuits that were presented above, we can obtain the mathematical model of any system based on operational amplifiers.

2.1.3 Modeling of mechanical systems

Mechanical systems are governed by Newton's second law and by the general moment equation [3,4].

Given a n rigid bodies system, submitted to N external forces, Newton's second law states that:

$$\sum_{i=1}^{N} \mathbf{F}_i = \sum_{j=1}^{n} m_j \mathbf{a}_{cj} = m\mathbf{a}_c \tag{2.9}$$

where $\mathbf{F}_i$ ($i = 1, \ldots, N$) is the ith external actuating force vector, m_j represents the mass of the jth ($j = 1, \ldots, n$) rigid body, $m = \sum_{j=1}^{n} m_j$ is the total mass, $\mathbf{a}_c$ represents the absolute linear acceleration vector of the system center of mass, and $\mathbf{a}_{cj}$ is the absolute linear acceleration vector of the center of mass of the jth rigid body.

For the same system, submitted to N external moments, the general moment equation states that:

$$\sum_{i=1}^{N} \mathbf{M}_{i/p} = \dot{\mathbf{H}}_p + m\mathbf{r}_{c/p} \times \mathbf{a}_p \tag{2.10}$$

where p is an arbitrary point, c denotes the system center of mass, $\mathbf{M}_{i/p}$ is the ith actuating moment vector with reference to p, $\dot{\mathbf{H}}_p$ represents the time derivative of the angular moment vector about point p, $\mathbf{r}_{c/p}$ is the position vector of the center of mass with respect to p, and $\mathbf{a}_p$ is the absolute linear acceleration vector of point p.

Alternatively, we can write:

$$\sum_{i=1}^{N} \mathbf{M}_{i/p} = \dot{\mathbf{H}}_c + m\mathbf{r}_{c/p} \times \mathbf{a}_c \tag{2.11}$$

with $\dot{\mathbf{H}}_c$ representing the time derivative of the angular moment vector with respect to the center of mass c, and $\mathbf{a}_c$ denoting the absolute linear acceleration vector of c.

For a rigid body moving in the three-dimensional space we have:

$$\mathbf{H}_p = \mathbf{I}_p\boldsymbol{\omega} = \begin{bmatrix} H_x \\ H_y \\ H_z \end{bmatrix}_p = \begin{bmatrix} I_{xx} & I_{xy} & I_{xz} \\ I_{yx} & I_{yy} & I_{yz} \\ I_{zx} & I_{zy} & I_{zz} \end{bmatrix}_p \begin{bmatrix} \omega_x \\ \omega_y \\ \omega_z \end{bmatrix} \tag{2.12}$$

where $\mathbf{I}_p$ denotes the inertia matrix of the rigid body about point p, and $\boldsymbol{\omega}$ represents the angular velocity vector.

For rigid body motion in the two-dimensional space, the general moment equation yields:

$$\sum_{i=1}^{N} \mathbf{M}_{i/p} = I_p\boldsymbol{\alpha} + m\mathbf{r}_{c/p} \times \mathbf{a}_p \tag{2.13}$$

where I_p is the inertia of the rigid body about point p, and $\boldsymbol{\alpha}$ represents the rigid body angular acceleration vector.

According to Steiner's theorem, if the rigid body is composed of n separated masses, then its inertia is given by:

$$I_p = \sum_{j=1}^{n} I_j + m_j\delta_j^2 \tag{2.14}$$

where I_j is the inertia of the jth rigid body with respect to its center of mass, c_j, and δ_j represents the distance between c_j and p.

2.1.3.1 Mechanical transmissions

Mechanical systems often include mechanical transmissions between the motor and the load axes [3,4].

The screw transmission is used for angular/linear motion conversion. The pitch parameter, P, characterizes the screw transmission, defining the relationship between translation and rotation:

$$x = \theta \frac{p}{2\pi} \tag{2.15}$$

where x and θ represent linear and angular displacement, respectively.

The relationship between torque and force is:

$$F = T\frac{2\pi\eta}{p} \tag{2.16}$$

where η represents the efficiency of the transmission.

In other cases, mechanical transmissions are used for increasing/decreasing torque (or decreasing/increasing angular velocity).

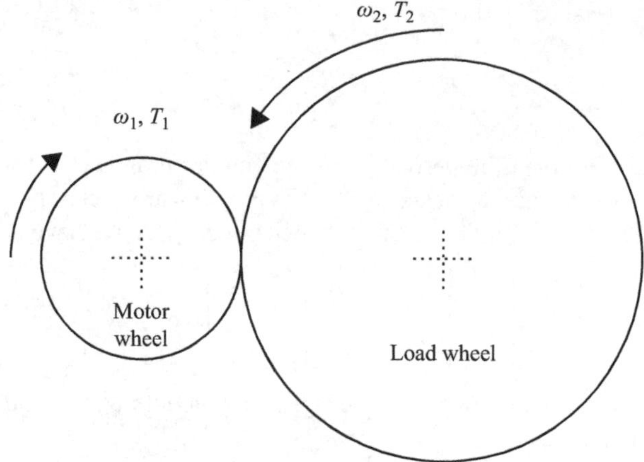

Figure 2.2 Simple mechanical transmission composed of two wheels

For a two-wheel gear (Figure 2.2), the transmission ratio, N, and the efficiency, η, are given by:

$$N = \frac{n_1}{n_2} = \frac{\omega_1}{\omega_2} = \frac{\alpha_1}{\alpha_2} \tag{2.17}$$

$$\eta = \frac{T_2\omega_2}{T_1\omega_1} \tag{2.18}$$

where n_1 and n_2 represent the radii, T_1 and T_2 are the torques, and ω_1 and ω_2 (α_1 and α_2) correspond to the angular velocities (accelerations) of the motor and load wheels, respectively.

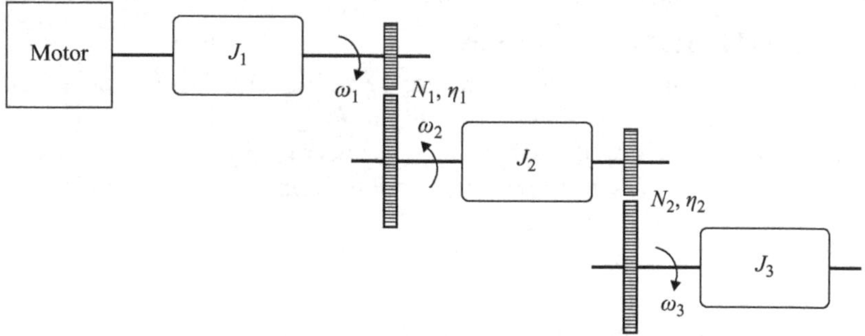

Figure 2.3 Multi-level mechanical transmission chain

For a multi-level transmission chain (Figure 2.3) the total inertia, J_T, referred to the motor axis is:

$$J_T = \left[J_1 + \frac{J_2}{\eta_1}\left(\frac{1}{N_1}\right)^2 + \frac{J_3}{\eta_1\eta_2}\left(\frac{1}{N_1N_2}\right)^2 + \cdots + \frac{J_{n+1}}{\eta_1\ldots\eta_n}\left(\frac{1}{N_1\ldots N_n}\right)^2\right] \tag{2.19}$$

where J_k ($k = 1, \ldots, n$) is the inertia of the kth wheel, and N_k and η_k denote the transmission ratio and efficiency of the kth transmission, respectively.

2.1.3.2 Mechanical spring and damper

Mechanical systems usually comprise linear springs and dampers [3,4]. The relationship between linear displacement/velocity and force on these elements is given by:

$$\frac{dx}{dt} = \frac{1}{K_s}\frac{dF}{dt} \tag{2.20}$$

$$\frac{dx}{dt} = \frac{1}{B_d}F \tag{2.21}$$

For angular displacement and torque, we have:

$$\frac{d\theta}{dt} = \frac{1}{K_s}\frac{dT}{dt} \tag{2.22}$$

$$\frac{d\theta}{dt} = \frac{1}{B_d}T \tag{2.23}$$

where x and θ represent the relative linear and angular displacement of the element terminals, and the constants K_s and B_d represent the spring stiffness and the viscous friction coefficient.

It should be noted that the spring and damper are elements analogous to the electrical resistance and inductance, respectively.

2.1.3.3 Armature current controlled DC motor

In an armature current controlled DC motor (Figure 2.4) the field current i_f is held constant, and the armature current, i_a, is controlled through the armature voltage e_a [3,4].

The motor torque, $T(t)$, increases linearly with the armature current:

$$T(t) = K_T i_a(t) \tag{2.24}$$

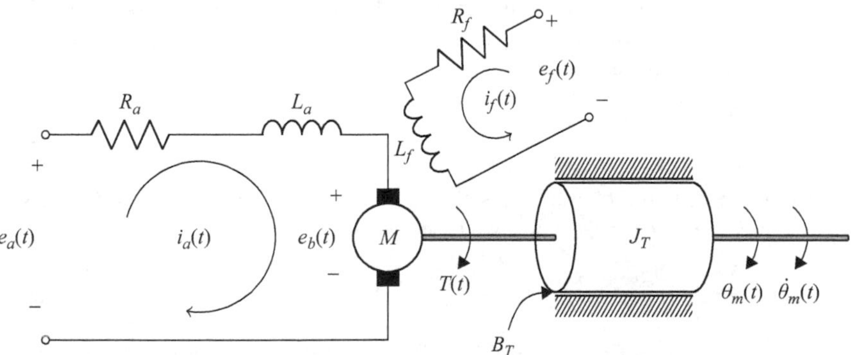

Figure 2.4 Armature current controlled permanent magnet DC motor

where K_T is the motor torque constant.

Applying Kirchhoff's voltage law to the armature loop we have:

$$e_a(t) - R_a i_a(t) - L_a \frac{di_a(t)}{dt} - e_b(t) = 0 \tag{2.25}$$

with e_b denoting the "back electromotive force" induced by the rotation of the armature windings in a magnetic field, and being proportional to the rotation speed $\dot{\theta}_m(t)$:

$$e_b(t) = K_e \frac{d\theta_m(t)}{dt} \tag{2.26}$$

where K_e is a constant.

Finally, using the general moment equation, we obtain:

$$T(t) = J_T \frac{d^2\theta_m(t)}{dt^2} + B_T \frac{d\theta_m(t)}{dt} \tag{2.27}$$

2.1.4 Modeling of liquid-level systems

Many industrial processes involve systems consisting of liquid-filled tanks, connected by pipes with valves, orifices, and other restrictions to fluid flow [3,4].

Fluid flow regime depends on the Reynolds number [2]. Turbulent flow occurs for Reynolds numbers greater than about 3000, while laminar flow is observed when the Reynolds number is lesser than about 2000. Systems involving laminar flow are represented by linear differential equations. Systems involving turbulent flow are modeled by nonlinear differential equations. Usually, these equations can be linearized if the system region of operation is limited.

2.1.4.1 Liquid-level system resistance, capacitance, and inertance

In a liquid-level system, a valve, a change in a pipe diameter, or a short pipe connecting two tanks, act as energy dissipative elements, or hydraulic resistances. If the flow is laminar, the resistance is given by:

$$R_p = \frac{\Delta p}{q_v} = \frac{p_1 - p_2}{q_v} \tag{2.28}$$

where R_p represents the hydraulic resistance defined in terms of pressure drop, $\Delta p = p_1 - p_2$, and q_v is the volume flow rate.

Alternatively, we can write:

$$R_h = \frac{h}{q_v} \tag{2.29}$$

where R_h represents the hydraulic resistance in terms of equivalent head (or height), h, that is, difference of level. Thus, we have:

$$R_p = \frac{\Delta p}{q_v} = \frac{\rho g h}{q_v} = \rho g R_h \tag{2.30}$$

where ρ and g denote the density of the fluid, and the acceleration due to gravity, respectively.

For turbulent flow, the relationship between pressure drop and volume flow rate is:

$$\Delta p = kq_v^2 \tag{2.31}$$

where k is a constant determined experimentally.

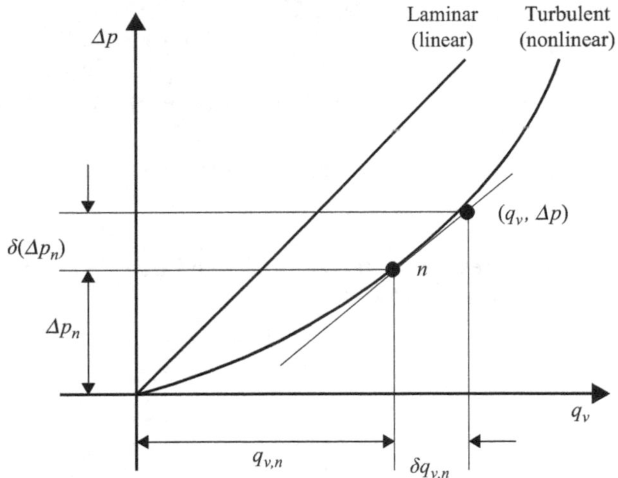

Figure 2.5 Laminar and turbulent regime relationships between pressure drop and volume flow rate

This relationship can be linearized by means of a Taylor series expansion about a given operational point, n (Figure 2.5). Thus, keeping the first two linear terms results in:

$$\Delta p = \Delta p_n + \left.\frac{\partial(\Delta p)}{\partial q_v}\right|_{\dot{q}_v=\dot{q}_{v,n}} (q_v - q_{v,n}) \tag{2.32}$$

where the expressions

$$\Delta p - \Delta p_n = \delta(\Delta p) \tag{2.33}$$

$$q_v - q_{v,n} = \delta q_{v,n} \tag{2.34}$$

denote small variations in pressure and in volume flow rate, respectively.

Thus, we have:

$$\delta(\Delta p_n) = \left.\frac{\partial(\Delta p)}{\partial q_v}\right|_{q_v=q_{v,n}} \delta q_{v,n} \tag{2.35}$$

and the linearized resistance is given by:

$$R_p = \left.\frac{\partial(\Delta p)}{\partial q_v}\right|_{q_v=q_{v,n}} = \frac{\delta(\Delta p)}{\delta q_{v,n}} = 2kq_{v,n} = 2\left(\frac{\Delta p}{q_v}\right)_{q_v=q_{v,n}} \tag{2.36}$$

This means that the linearized resistance for turbulent flow is twice the resistance of laminar flow.

The hydraulic capacitance reflects the storage capacity of a reservoir-type device, and is given by:

$$C_p = \frac{q_v}{\dot{p}} = \frac{dV}{dp} \tag{2.37}$$

$$C_h = \frac{q_v}{\dot{h}} = \frac{dV}{dh} \tag{2.38}$$

when expressed in terms of pressure and head, respectively. Variable V represents volume.

The relationship between head and static pressure yields:

$$C_h = \rho g C_p \tag{2.39}$$

The inertance measures the inertia effect of liquid flow, being important when dealing with long pipes. For laminar flow, we have:

$$L_p = \frac{\Delta p}{\dot{q}_v} \tag{2.40}$$

$$L_h = \frac{\Delta h}{\dot{q}_v} \tag{2.41}$$

$$L_p = \rho g L_h \tag{2.42}$$

where L_p and L_h represent the inertance in terms of pressure and head, respectively.

The equations shown above are usually used with the law of conservation of mass to obtain the differential equations that govern liquid-level systems.

The law of conservation of mass states that:

$$\rho \sum_{k=1}^{n} q_k(t) = \frac{dm}{dt} \tag{2.43}$$

meaning that the time rate of change of the fluid mass in a tank equals the mass flow rate towards or away from the tank.

2.1.5 *Modeling of thermal systems*

Thermal systems involve transfer of heat between materials. Heat can flow from one material to another by conduction, convection, and radiation [2,8].

Considering that a thermal system can be represented by a lumped-parameter model, for conduction or convection heat transfer, we have:

$$q = K\Delta T \tag{2.44}$$

where q represents the heat flow rate, ΔT is the temperature difference, and the coefficient K is given by:

$$K = \frac{kA}{\Delta X} \tag{2.45}$$

$$K = HA \tag{2.46}$$

for conduction and convection, respectively. The parameter k represents the thermal conductivity, A is the area normal to the heat flow, ΔX represents the thickness of the conductor, and H is the convection coefficient.

2.1.5.1 Thermal resistance and capacitance

We define the thermal resistance R for heat transfer between two materials as the ratio:

$$R = \frac{d\Delta T}{dq} = \frac{1}{K} \tag{2.47}$$

The thermal capacitance C is defined by:

$$C = mc \tag{2.48}$$

where m represents the mass, and c denotes the specific heat of the material.

For thermal systems, the inertance is usually neglected.

The equations listed above are usually used with the law of conservation of energy to obtain the differential equations that govern thermal systems.

The law of conservation of energy is given by:

$$\sum_{k=1}^{n} q_k(t) = \frac{dE}{dt} \tag{2.49}$$

meaning that the time rate of change of energy, E, in a thermal system equals the heat flow rate into the system minus the heat flow rate out of the system.

2.2 Worked examples

2.2.1 Electrical systems

Problem 2.1 Consider the electrical circuit shown in Figure 2.6. Variables V_{in} and V_{out} denote the input and output voltages. Determine the transfer function $G(s) = \dfrac{V_{out}(s)}{V_{in}(s)}$. Analyze the result when $R = 0$.

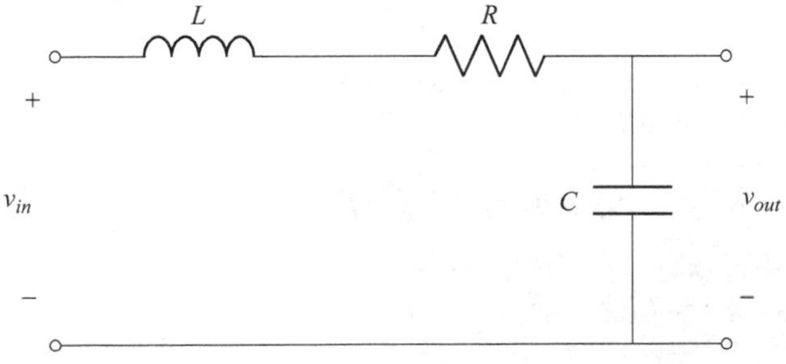

Figure 2.6 Electrical circuit of Problem 2.1

Resolution All elements are in series, and thus the same current i flows throughout the circuit. From the definition of impedance,

$$\begin{cases} Ls + R + \frac{1}{Cs} = \frac{V_{in}}{i} \\ \frac{1}{Cs} = \frac{V_{out}}{i} \end{cases} \Rightarrow \begin{cases} i = \frac{V_{in}}{\frac{CLs^2+RCs+1}{Cs}} \\ i = \frac{V_{out}}{\frac{1}{Cs}} \end{cases}$$

Equaling i,

$$\frac{V_{out}}{V_{in}} = \frac{1}{CLs^2 + RCs + 1}$$

If $R = 0$, the roots of the denominator (the poles of the transfer function) are pure imaginary numbers: the output of the plant will present sustained oscillations. (In practice, noise always makes these oscillations increase with time.)

2.2.2 Mechanical systems

Problem 2.2 Consider the mechanical system depicted in Figure 2.7. Displacements are denoted by x_1, x_2, x_3, x_4 and applied forces by f_1, f_2. The mathematical model of this system is:

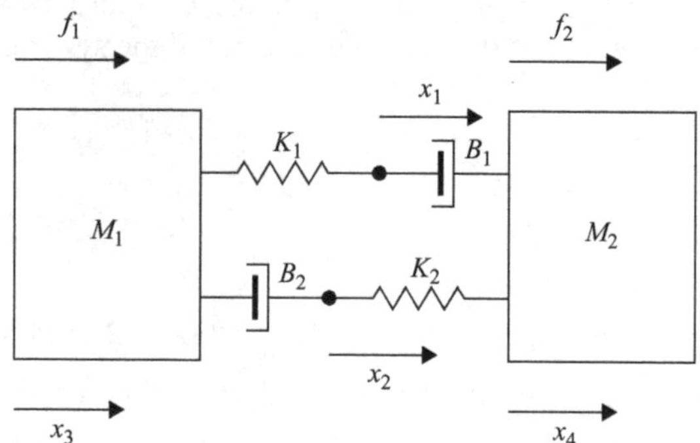

Figure 2.7 Mechanical system of Problem 2.2

A)
$$\begin{cases} M_1\ddot{x}_1 + K_1(x_3 - x_1) + B_2(\dot{x}_3 - \dot{x}_2) = f_1 \\ K_1(x_3 - x_1) = B_1(\dot{x}_1 - \dot{x}_3) \\ K_2(x_2 - x_4) = B_2(\dot{x}_3 - \dot{x}_2) \\ M_2\ddot{x}_2 + K_2(x_4 - x_2) + B_1(\dot{x}_4 - \dot{x}_2) = f_2 \end{cases}$$

B)
$$\begin{cases} M_1\ddot{x}_3 + K_1(x_3 - x_2) + B_2(\dot{x}_3 - \dot{x}_1) = f_1 \\ K_1(x_2 - x_1) = B_1(\dot{x}_4 - \dot{x}_1) \\ K_2(x_4 - x_2) = B_2(\dot{x}_3 - \dot{x}_2) \\ M_2\ddot{x}_4 + K_2(x_4 - x_1) + B_1(\dot{x}_4 - \dot{x}_1) = f_2 \end{cases}$$

C)
$$\begin{cases} M_1\ddot{x}_3 + K_1(x_3 - x_1) + B_2(\dot{x}_3 - \dot{x}_2) = f_1 \\ K_1(x_3 - x_1) = B_1(\dot{x}_1 - \dot{x}_4) \\ K_2(x_2 - x_4) = B_2(\dot{x}_3 - \dot{x}_2) \\ M_2\ddot{x}_4 + K_2(x_4 - x_2) + B_1(\dot{x}_4 - \dot{x}_1) = f_2 \end{cases}$$

D) None of the above.

Resolution Because there are no masses in-between the springs and the dampers, the forces these elements exert have to cancel each other:

$$\begin{cases} K_1(x_1 - x_3) = B_1(\dot{x}_4 - \dot{x}_1) \\ B_2(\dot{x}_2 - \dot{x}_3) = K_2(x_4 - x_2) \end{cases}$$

As an alternative, we can think that there *are* masses (let them be M_{x_1} and M_{x_2}), and then let them be 0. We will obtain the same result:

$$\begin{cases} K_1(x_3 - x_1) + B_1(\dot{x}_4 - \dot{x}_1) = M_{x_1}\ddot{x}_1 = 0 \\ B_2(\dot{x}_3 - \dot{x}_2) + K_2(x_4 - x_2) = M_{x_2}\ddot{x}_2 = 0 \end{cases}$$

Newton's law is applied to masses M_1 and M_2:

$$\begin{cases} f_1 + K_1(x_1 - x_3) + B_2(\dot{x}_2 - \dot{x}_3) = M_1\ddot{x}_3 \\ f_2 - K_2(x_4 - x_2) - B_1(\dot{x}_4 - \dot{x}_1) = M_2\ddot{x}_4 \end{cases}$$

Thus, the correct answer is option **C)**.

Problem 2.3 Consider the mechanical system shown in Figure 2.8, where x_1, x_2 and x_3 are the horizontal displacements of masses M_1, M_2 and M_3, respectively. Assume

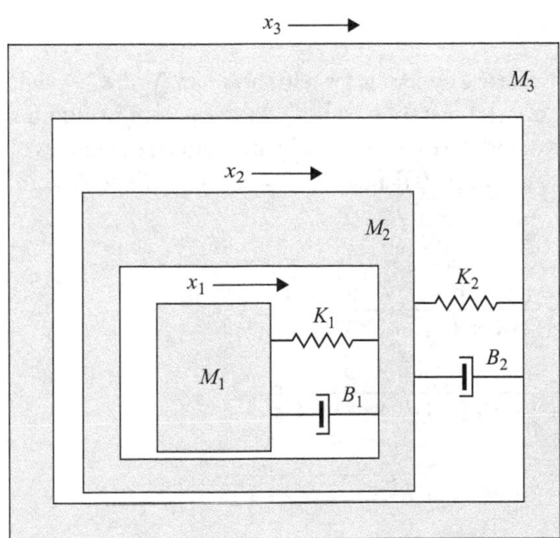

Figure 2.8 Mechanical system of Problem 2.3

that the masses slid on each other with neglectable friction. The mathematical model of this system is:

A) $$\begin{cases} M_1\ddot{x}_1 + K_1(x_1 - x_2) + B_1(\dot{x}_1 - \dot{x}_2) = 0 \\ M_2\ddot{x}_2 + K_1(x_2 - x_1) + B_1(\dot{x}_2 - \dot{x}_1) + K_2(x_2 - x_3) + B_2(\dot{x}_2 - \dot{x}_3) = 0 \\ M_3\ddot{x}_3 + K_2(x_3 - x_2) + B_2(\dot{x}_3 - \dot{x}_2) = 0 \end{cases}$$

B) $$\begin{cases} M_1\ddot{x}_1 + K_1x_1 + B_1\dot{x}_1 = 0 \\ M_2\ddot{x}_2 + K_1x_2 + B_1\dot{x}_2 + K_2x_2 + B_2\dot{x}_2 = 0 \\ M_3\ddot{x}_3 + K_2x_3 + B_2\dot{x}_3 = 0 \end{cases}$$

C) $$\begin{cases} M_1\ddot{x}_1 = K_1(x_1 - x_2) + B_1(\dot{x}_1 - \dot{x}_2) \\ M_2\ddot{x}_2 = K_1(x_2 - x_1) + B_1(\dot{x}_2 - \dot{x}_1) - K_2(x_2 - x_3) - B_2(\dot{x}_2 - \dot{x}_3) \\ M_3\ddot{x}_3 = K_2(x_3 - x_2) + B_2(\dot{x}_3 - \dot{x}_2) \end{cases}$$

D) None of the above.

Resolution The forces exerted by springs and dampers will be

$$\begin{cases} f_{K_1} = K_1(x_2 - x_1) \\ f_{K_2} = K_2(x_3 - x_2) \\ f_{B_1} = B_1(\dot{x}_2 - \dot{x}_1) \\ f_{B_2} = B_2(\dot{x}_3 - \dot{x}_2) \end{cases}$$

Newton's law is applied to masses M_1, M_2 and M_3:

$$\begin{cases} K_1(x_2 - x_1) + B_1(\dot{x}_2 - \dot{x}_1) = M_1\dot{x}_1 \\ K_2(x_3 - x_2) + B_2(\dot{x}_3 - \dot{x}_2) - K_1(x_2 - x_1) - B_1(\dot{x}_2 - \dot{x}_1) = M_2\dot{x}_2 \\ -K_2(x_3 - x_2) - B_2(\dot{x}_3 - \dot{x}_2) = M_3\dot{x}_3 \end{cases}$$

Thus, the correct answer is option **A)**.

2.2.3 *Liquid-level systems*

Problem 2.4 Consider the hydraulic system shown in Figure 2.9, where $q_1(t)$ and $q_2(t)$ represent flows. Let $h_1(t)$, $h_2(t)$, A_1 and A_2 be the height of liquid and cross section areas of reservoirs 1 and 2, respectively. The hydraulic resistances are represented by R_1 and R_2. Let $Q_1(s) = \mathscr{L}[q_1(t)]$ and $H_2(s) = \mathscr{L}[h_2(t)]$. The system transfer function $\dfrac{H_2(s)}{Q_1(s)}$ is:

A) $$\frac{H_2(s)}{Q_1(s)} = \frac{1}{s(A_1A_2R_2s + A_1 + A_2)}$$

B) $$\frac{H_2(s)}{Q_1(s)} = \frac{1}{1 + (A_1R_1 + A_2R_2)s + A_1A_2s^2}$$

C) $$\frac{H_2(s)}{Q_1(s)} = \frac{1}{1 + (A_1R_1 + A_2R_2)s + A_1R_1A_2R_2s^2}$$

D) $$\frac{H_2(s)}{Q_1(s)} = \frac{1}{s[1 + (A_1R_1 + A_2R_2)s]}.$$

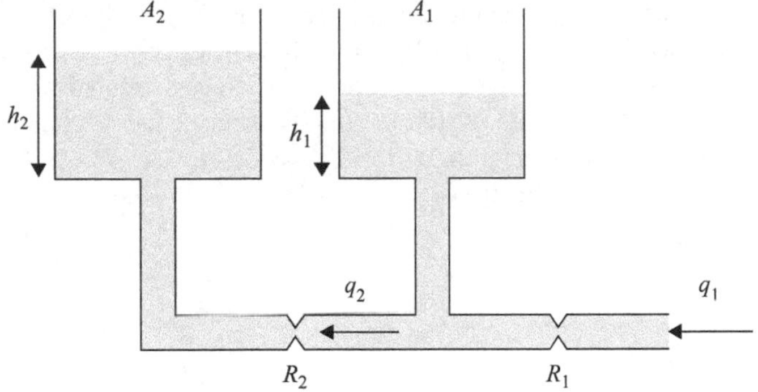

Figure 2.9 Hydraulic system of Problem 2.4

Resolution We will assume that the fluid is incompressible. Flow q_1 is split into two components:

$$q_1 = A_1\dot{h}_1 + q_2$$

Flow q_2 is integrally converted into an increase of h_2:

$$q_2 = A_2\dot{h}_2$$

Applying the Laplace transformation, we get

$$\begin{cases} Q_1 = sA_1H_1 + Q_2 \\ Q_2 = sA_2H_2 \end{cases}$$

We also know that hydraulic resistance R_2 relates flow q_2 with $h_1 - h_2$. (Notice that $h_1 - h_2$ is a difference of two potentials; considering the electrical analogy should help.)

$$q_2 = \frac{h_1 - h_2}{R_2}$$

Replacing the Laplace transform of this in the two Laplace transforms above, we get:

$$\begin{cases} Q_1R_2 = sA_1R_2H_1 + H_1 - H_2 \\ H_1 - H_2 = sA_2R_2H_2 \end{cases} \Rightarrow \begin{cases} Q_1R_2 = (sA_1R_2 + 1)H_1 - H_2 \\ H_1 = (sA_2R_2 + 1)H_2 \end{cases}$$

Replacing the second equation in the first, to eliminate H_1,

$$\begin{aligned} Q_1R_2 &= (sA_1R_2 + 1)(sA_2R_2 + 1)H_2 - H_2 \\ &= [(sA_1R_2 + 1)(sA_2R_2 + 1) - 1]H_2 \\ &= (s^2A_1A_2R_2^2 + sA_1R_2 + sA_2R_2)H_2 \\ &= s(sA_1A_2R_2 + A_1 + A_2)H_2R_2 \end{aligned}$$

Thus, the correct answer is option **A)**.

Problem 2.5 Consider the hydraulic system shown in Figure 2.10. Let $Q_i(s) = \mathscr{L}[q_i(t)]$, $Q_d(s) = \mathscr{L}[q_d(t)]$, $Q_1(s) = \mathscr{L}[q_1(t)]$, $Q_o(s) = \mathscr{L}[q_o(t)]$, $H_1(s) = \mathscr{L}[h_1(t)]$, $H_2(s) = \mathscr{L}[h_2(t)]$. A_1 and A_2 denote the cross-sectional areas of reservoirs 1 and 2; R_1 and R_2 are hydraulic resistances. Determine the mathematical model and the corresponding block diagram with $H_2(s) = G_1(s)Q_i(s) + G_2(s)Q_d(s)$, where $G_1(s)$ and $G_2(s)$ are transfer functions.

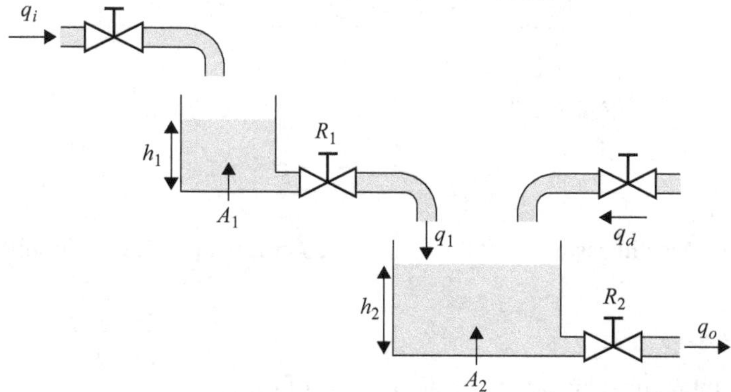

Figure 2.10 Hydraulic system of Problem 2.5

Resolution Variations of height in each tank are due to the flows that fill and empty them:

$$\begin{cases} A_1\dot{h}_1 = q_i - q_1 \\ A_2\dot{h}_2 = q_1 + q_d - q_o \end{cases}$$

Flows through the resistances depend on water heights (that determine potential energy available to overcome the resistance):

$$\begin{cases} q_1 = \frac{h_1}{R_1} \\ q_o = \frac{h_2}{R_2} \end{cases}$$

All that is left to do is applying the Laplace transform and rearranging terms:

$$\begin{cases} Q_i = sA_1H_1 + Q_1 \\ Q_1 = \frac{H_1}{R_1} \\ Q_o = \frac{H_2}{R_2} \\ Q_1 + Q_d = sA_2H_2 + Q_o \end{cases} \Rightarrow \begin{cases} Q_i = sA_1R_1Q_1 + Q_1 = (sA_1R_1 + 1)Q_1 \\ H_1 = Q_1R_1 \\ H_2 = Q_oR_2 \\ Q_1 + Q_d = sA_2R_2Q_o + Q_o = (sA_2R_2 + 1)Q_o \end{cases}$$

We can now draw the block diagram of Figure 2.11:

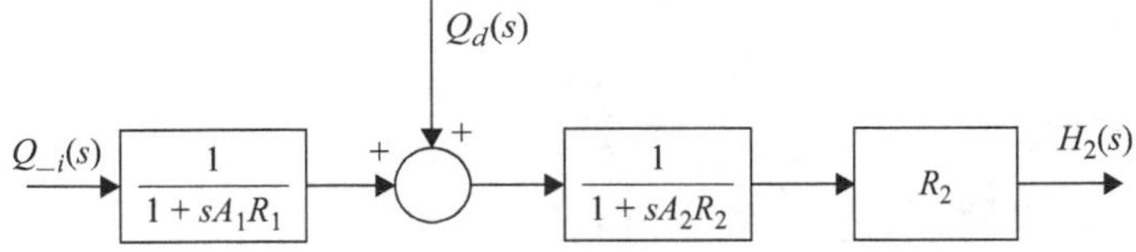

Figure 2.11 Block diagram of Problem 2.6

2.2.4 Thermal systems

Problem 2.6 Consider the electric oven shown in Figure 2.12. Variable q_i denotes the heat flux produced by the heating element and T_1, T_2 and T_0 are the temperatures of the heating element, the air inside the oven and the external environment, respectively. Symbols C_1 and C_2 represent the thermal capacities of the heating element and the air in the furnace; R_1 and q_1 are respectively the thermal resistance and the heat flux between the heating element and the air inside the oven; R_2 and q_2 are the thermal resistance and the heat flux between the air inside the oven and the outside environment. The thermal capacity of the outside environment is very high, therefore its temperature T_0 does not suffer significant changes.

1. Determine the mathematical model of the system and sketch the corresponding block diagram (hint: sketch the corresponding electric circuit).
2. Find transfer functions $G_1(s)$, $G_2(s)$, $G_3(s)$ and $G_4(s)$ such that $T_1(s) = G_1(s)Q_i(s) + G_2(s)T_0(s)$ and $T_2(s) = G_3(s)Q_i(s) + G_4(s)T_0(s)$.

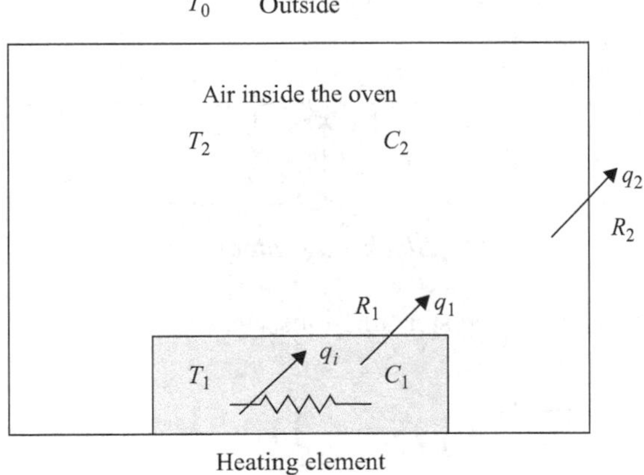

Figure 2.12 Electric oven of Problem 2.6

Resolution The resolution is very similar to that of problem 2.5 above. Variations of temperature in the heating element and in the tank are due to the heat flows affecting each of them:

$$\begin{cases} C_1\dot{T}_1 = q_i - q_1 \\ C_2\dot{T}_2 = q_1 - q_2 \end{cases}$$

Heat flows through thermal resistances depend on temperature differences (that determine potential energy available to overcome the resistance):

$$\begin{cases} q_1 = \frac{T_1 - T_2}{R_1} \\ q_2 = \frac{T_2 - T_0}{R_2} \end{cases}$$

Applying the Laplace transform, we get

$$\begin{cases} Q_i = sC_1T_1 + Q_1 \\ Q_1 = \frac{T_1 - T_2}{R_1} \\ Q_1 = sC_2T_2 + Q_2 \\ Q_2 = \frac{T_2 - T_0}{R_2} \end{cases} \tag{2.50}$$

The corresponding electric circuit is shown in Figure 2.13.

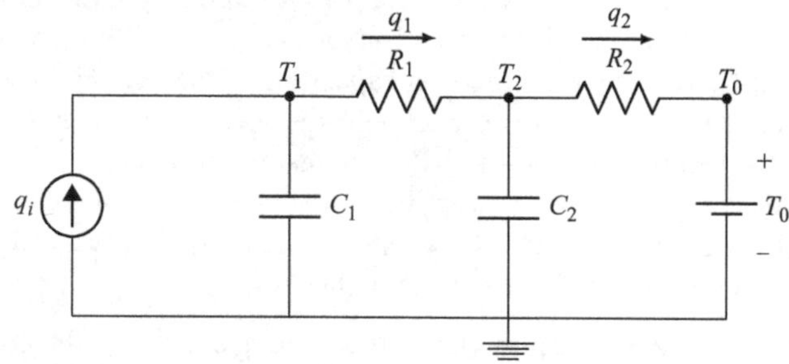

Figure 2.13 Electric circuit of Exercise 2.6

We can now draw the block diagram of Figure 2.14:

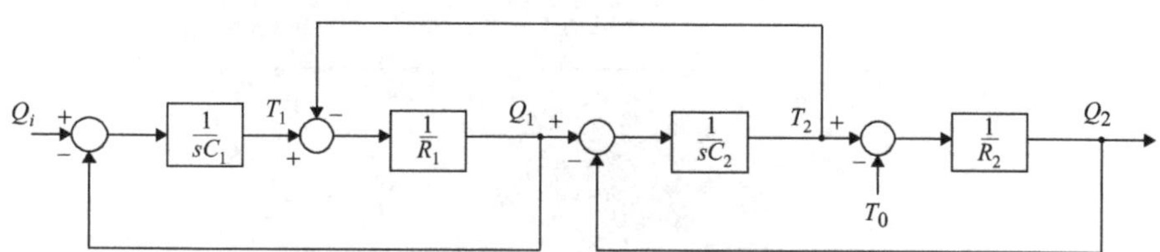

Figure 2.14 Block diagram of Exercise 2.6

To obtain the requested transfer functions, we first replace the second equation of (2.50) in the first:

$$Q_i = sC_1T_1 + \frac{T_1 - T_2}{R_1} = \left(sC_1 + \frac{1}{R_1}\right)T_1 - \frac{1}{R_2}T_2$$

We then replace the fourth equation in the third, and the result in the first:

$$Q_i = sC_1T_1 + sC_2T_2 + \frac{T_2 - T_0}{R_2} = sC_1T_1 + \left(sC_2 + \frac{1}{R_2}\right)T_2 - \frac{1}{R_2}T_0$$

We can put this into matrix form:

$$\begin{bmatrix} Q_i \\ Q_i + \frac{1}{R_2}T_0 \end{bmatrix} = \begin{bmatrix} sC_1 + \frac{1}{R_1} & -\frac{1}{R_2} \\ sC_1 & sC_2 + \frac{1}{R_2} \end{bmatrix} \begin{bmatrix} T_1 \\ T_2 \end{bmatrix} \tag{2.51}$$

Using Cramer's rule we can find the desired solution:

$$T_1 = \frac{\begin{vmatrix} Q_i & -\frac{1}{R_2} \\ Q_i + \frac{1}{R_2}T_0 & sC_2 + \frac{1}{R_2} \end{vmatrix}}{\begin{vmatrix} sC_1 + \frac{1}{R_1} & -\frac{1}{R_2} \\ sC_1 & sC_2 + \frac{1}{R_2} \end{vmatrix}}$$

$$T_2 = \frac{\begin{vmatrix} sC_1 + \frac{1}{R_1} & Q_i \\ sC_1 & Q_i + \frac{1}{R_2}T_0 \end{vmatrix}}{\begin{vmatrix} sC_1 + \frac{1}{R_1} & -\frac{1}{R_2} \\ sC_1 & sC_2 + \frac{1}{R_2} \end{vmatrix}}$$

Tedious but straightforward calculations will lead to

$$\begin{cases} T_1 = \overbrace{\frac{C_2 s + \frac{1}{R_1} + \frac{1}{R_2}}{C_1C_2s^2 + \left(\frac{C_1}{R_2} + \frac{C_2}{R_1} + \frac{C_1}{R_1}\right)s + \frac{1}{R_1R_2}}}^{G_1(s)} Q_i + \overbrace{\frac{\frac{1}{R_1R_2}}{C_1C_2s^2 + \left(\frac{C_1}{R_2} + \frac{C_2}{R_1} + \frac{C_1}{R_1}\right)s + \frac{1}{R_1R_2}}}^{G_2(s)} T_0 \\ T_2 = \underbrace{\frac{\frac{1}{R_1}}{C_1C_2s^2 + \left(\frac{C_1}{R_2} + \frac{C_2}{R_1} + \frac{C_1}{R_1}\right)s + \frac{1}{R_1R_2}}}_{G_3(s)} Q_i + \underbrace{\frac{\frac{C_1}{R_2}s + \frac{1}{R_1R_2}}{C_1C_2s^2 + \left(\frac{C_1}{R_2} + \frac{C_2}{R_1} + \frac{C_1}{R_1}\right)s + \frac{1}{R_1R_2}}}_{G_4(s)} T_0 \end{cases}$$

2.3 Proposed exercises

2.3.1 Electrical systems

Exercise 2.1 Consider the circuit in Figure 2.15.

1. Present the equations that model the circuit.
2. Obtain the transfer function $\frac{I_3(s)}{V_i(s)}$.

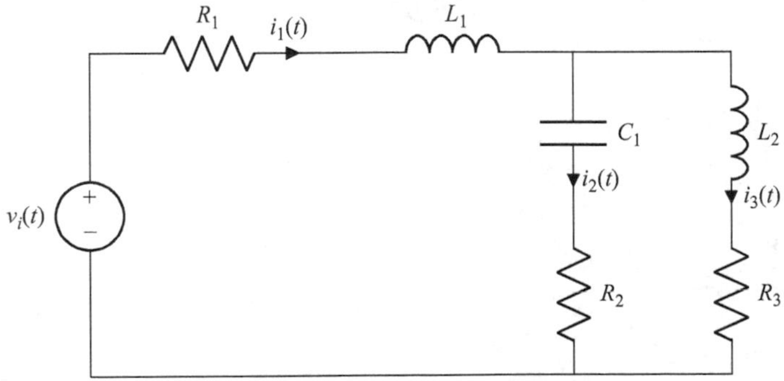

Figure 2.15 Circuit of Exercise 2.1

Exercise 2.2 Consider the circuit shown in Figure 2.16. The transfer function $V_o(s)/V_i(s)$ is:

A) $V_o(s)/V_i(s) = \dfrac{sCR(2sCR+1)}{(sCR)^2 + 3sCR + 1}$

B) $V_o(s)/V_i(s) = \dfrac{sCR}{(sCR)^2 + 2sCR + 1}$

C) $V_o(s)/V_i(s) = \dfrac{sCR(sCR+1)}{(sCR)^2 + 2sCR + 1}$

D) $V_o(s)/V_i(s) = \dfrac{sCR(sCR+1)}{(sCR)^2 + 2sCR + 1}$.

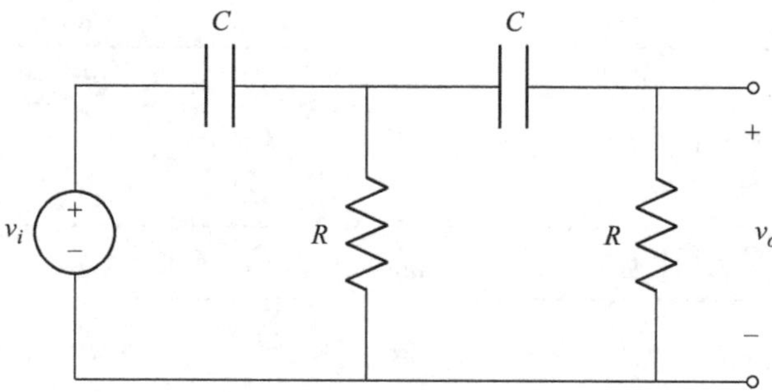

Figure 2.16 Circuit of Exercise 2.2

Exercise 2.3 Consider the electrical system in Figure 2.17. Let $\omega_0 = R/L$. It follows that:

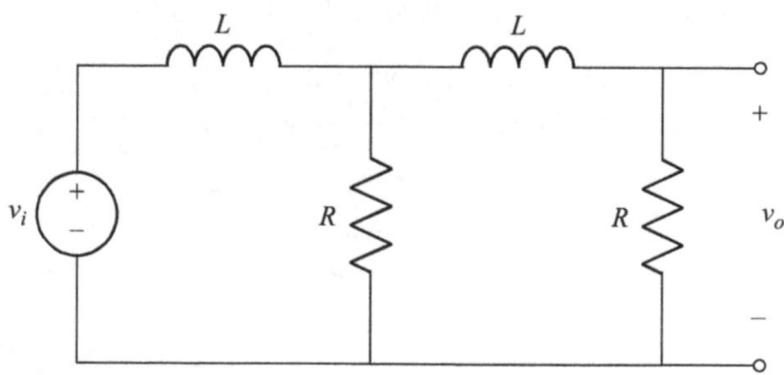

Figure 2.17 Circuit of Exercise 2.3

A) $\dfrac{V_o(s)}{V_i(s)} = \dfrac{\omega_0^2}{s^2 + 2s\omega_0 + \omega_0^2}$

B) $\dfrac{V_o(s)}{V_i(s)} = \dfrac{\omega_0^2}{s^2 + s\omega_0 + \omega_0^2}$

C) $\dfrac{V_o(s)}{V_i(s)} = \dfrac{\omega_0^2}{s^2 + 3s\omega_0 + \omega_0^2}$

D) None of the above.

Exercise 2.4 Consider the circuit shown in Figure 2.18. The transfer function $V_o(s)/V_i(s)$ is:

A) $\dfrac{V_o(s)}{V_i(s)} = \dfrac{sCR}{(sCR)^2 + 3sCR + 1}$

B) $\dfrac{V_o(s)}{V_i(s)} = \dfrac{sCR}{(sCR)^2 + 2sCR + 1}$

C) $\dfrac{V_o(s)}{V_i(s)} = \dfrac{sCR}{(sCR)^2 + sCR + 1}$

D) None of the above.

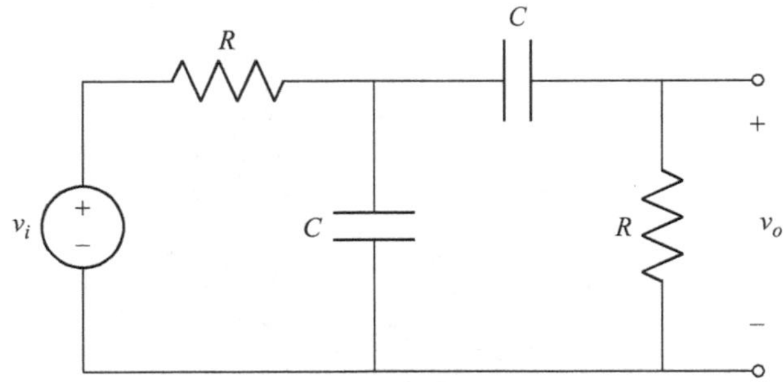

Figure 2.18 Circuit of Exercise 2.4

Exercise 2.5 Consider the electrical circuit shown in Figure 2.19. Find the model relating the current i_2 in capacitor C_2 with voltage v.

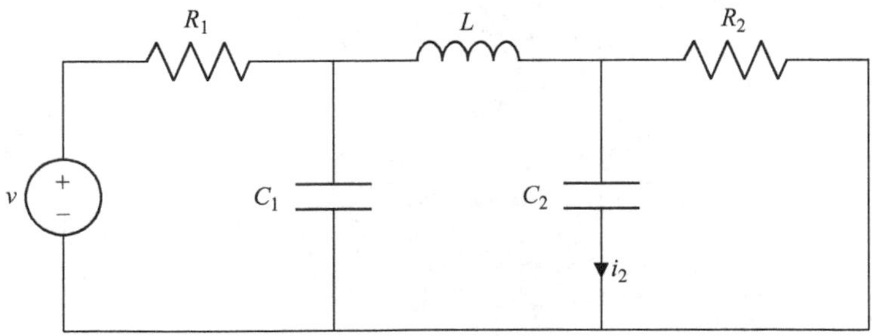

Figure 2.19 Circuit of Exercise 2.5

Exercise 2.6 Obtain transfer function $V_0(s)/V_i(s)$ for the system shown in Figure 2.20.

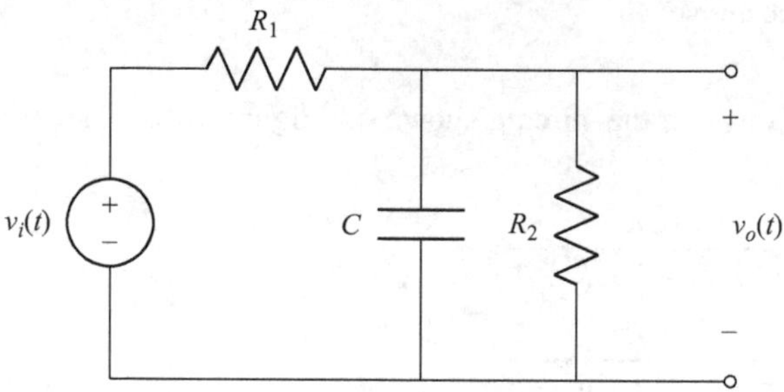

Figure 2.20 Circuit of Exercise 2.6

Exercise 2.7 Consider the circuit shown in Figure 2.21. The transfer function $V_o(s)/V_i(s)$ is:

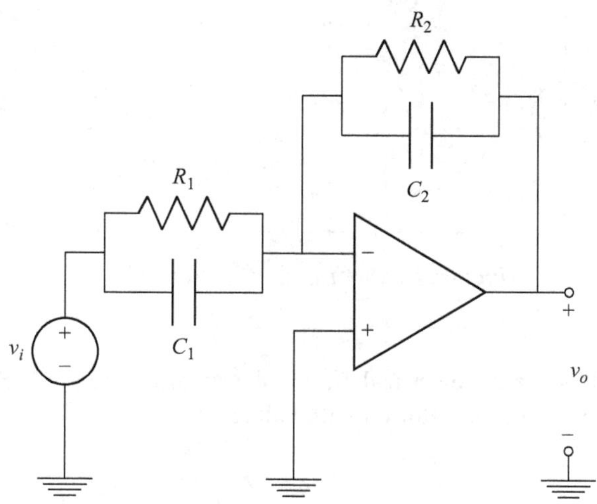

Figure 2.21 Circuit of Exercise 2.7

A) $$\frac{V_o(s)}{V_i(s)} = -\frac{R_2}{R_1}\frac{sC_1R_1+1}{sC_2R_2+1}$$

B) $$\frac{V_o(s)}{V_i(s)} = -\frac{R_1}{R_2}\frac{sC_1R_1+1}{sC_2R_2+1}$$

C) $$\frac{V_o(s)}{V_i(s)} = -\frac{R_2}{R_1}\frac{sC_2R_2+1}{sC_1R_1+1}$$

D) $$\frac{V_o(s)}{V_i(s)} = -\frac{R_1}{R_2}\frac{sC_2R_2+1}{sC_1R_1+1}.$$

Exercise 2.8 Consider the circuit shown in Figure 2.22. The transfer function $V_o(s)/V_i(s)$ is:

A) $$\frac{V_o(s)}{V_i(s)} = \frac{R_4(C_1R_1+C_2R_2)}{R_1R_3C_2}\left[1+\frac{1}{s}+s\right]$$

B) $$\frac{V_o(s)}{V_i(s)} = \frac{R_4}{R_1R_3C_2}\left[1+\frac{1}{C_1R_1+C_2R_2}\cdot\frac{1}{s}+\frac{1}{C_1R_1+C_2R_2}\cdot s\right]$$

C) $$\frac{V_o(s)}{V_i(s)} = \frac{R_4(C_1R_1+C_2R_2)}{R_1R_3C_2}\left[1+\frac{1}{C_1R_1+C_2R_2}\cdot\frac{1}{s}+\frac{C_1R_1C_2R_2}{C_1R_1+C_2R_2}\cdot s\right]$$

D) None of the above.

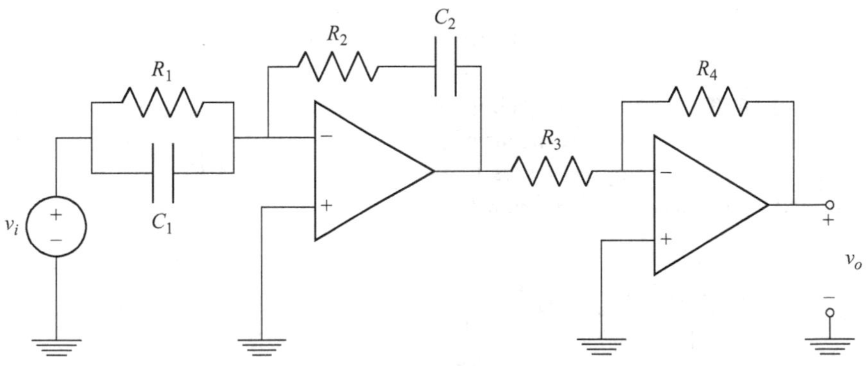

Figure 2.22 Circuit of Exercise 2.8

Exercise 2.9 Consider the circuit represented in Figure 2.23. Assume the operational amplifier is ideal (i.e., infinite input resistance, infinite gain, and zero output resistance). The transfer function $V_o(s)/V_i(s)$ is:

A) $$\frac{V_o(s)}{V_i(s)} = 1+\frac{R_1}{R_2(1+sCR_2)}$$

B) $$\frac{V_o(s)}{V_i(s)} = 1+\frac{R_2}{R_1(1+sCR_2)}$$

C) $\dfrac{V_o(s)}{V_i(s)} = R_2 + \dfrac{R_1}{R_2(1+sCR_2)}$

D) $\dfrac{V_o(s)}{V_i(s)} = R_1 + \dfrac{R_2}{R_1(1+sCR_2)}$.

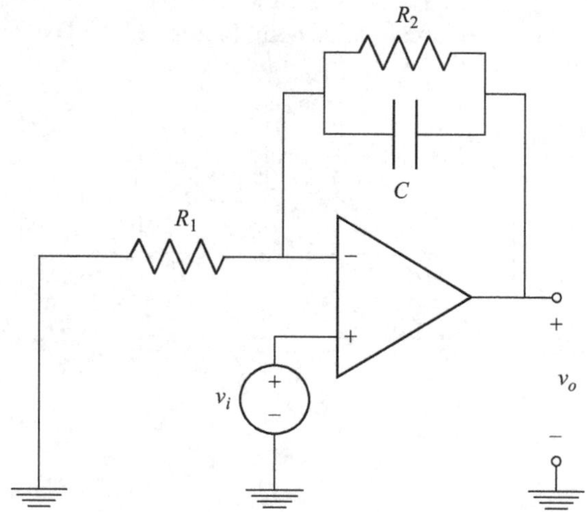

Figure 2.23 Circuit of Exercise 2.9

Exercise 2.10 In the system shown in Figure 2.24, the DC motor is coupled to a disc (with inertia J and radius r) by means of a shaft drive with negligible mass. Find the dynamical models relating velocities v_c and v_M with the input V, knowing that the motor can be modeled as in Section 2.1.3.3 above.

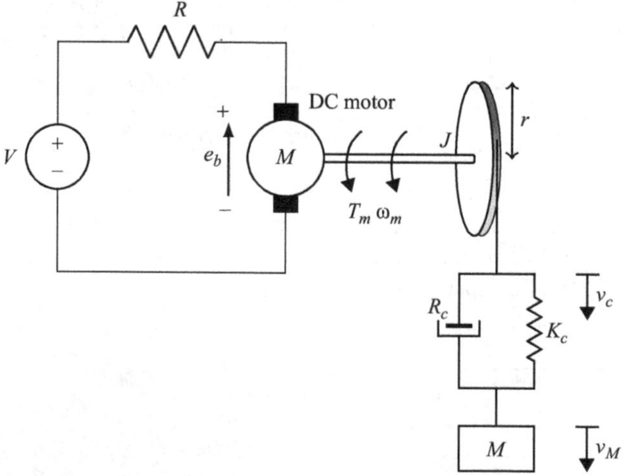

Figure 2.24 Mechatronic system of Exercise 2.10

2.3.2 Mechanical systems

Exercise 2.11 Derive the transfer function $X_2(s)/F(s)$ for the mechanical system shown in Figure 2.25.

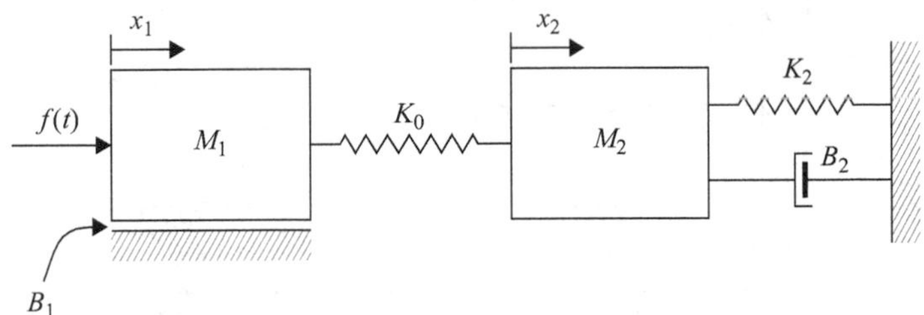

Figure 2.25 Mechanical system of Exercise 2.11

Exercise 2.12 Obtain the transfer function $G(s) = X_2(s)/F(s)$ of the system in Figure 2.26.

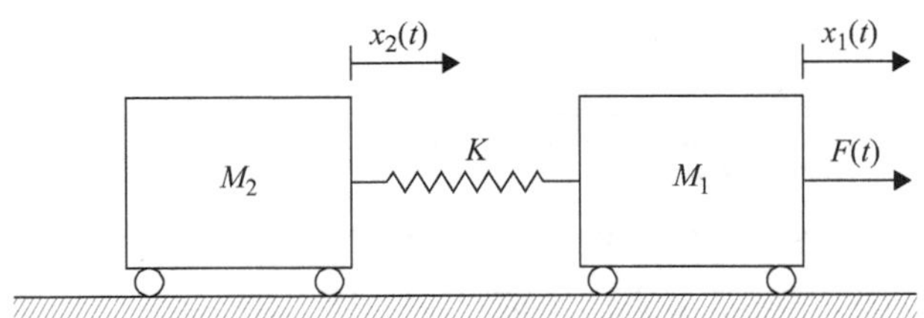

Figure 2.26 Mechanical system of Exercise 2.12

Exercise 2.13 Consider the mechanical system shown in Figure 2.27. The transfer function $Y(s)/U(s)$ is:

A) $\dfrac{Y(s)}{U(s)} = \dfrac{1}{Ms^2 + Bs + K}$

B) $\dfrac{Y(s)}{U(s)} = \dfrac{Bs + K}{Ms^2 + Bs + K}$

C) $\dfrac{Y(s)}{U(s)} = \dfrac{K}{Ms^2 + Bs + K}$

D) None of the above.

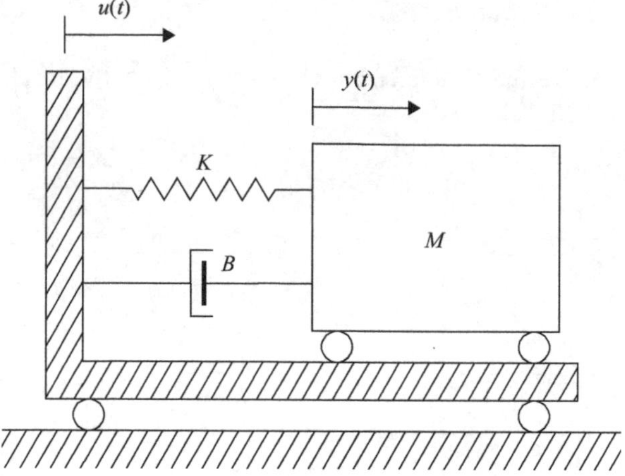

Figure 2.27 Mechanical system of Exercise 2.13

Exercise 2.14 Consider the mechanical system shown in Figure 2.28. The transfer function $X_1(s)/F(s)$ is:

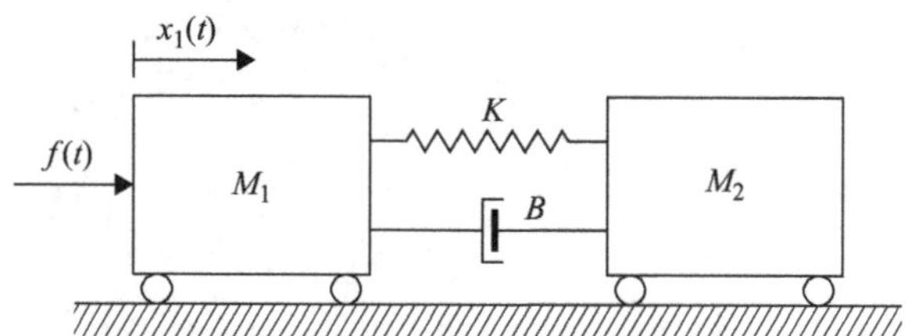

Figure 2.28 Mechanical system of Exercise 2.14

A) $$\frac{X_1(s)}{F(s)} = \frac{M_2s^2 + Bs + K}{M_1M_2s^4 + (M_1 + M_2)s^2 + Bs + K}$$

B) $$\frac{X_1(s)}{F(s)} = \frac{M_1s^2 + Bs + K}{(M_2s^2 + Bs + K)s^2}$$

C) $$\frac{X_1(s)}{F(s)} = \frac{M_1s^2 + Bs + K}{\left[(M_1 + M_2)s^2 + Bs + K\right]s^2}$$

D) $$\frac{X_1(s)}{F(s)} = \frac{M_2s^2 + Bs + K}{\left[M_1M_2s^2 + (M_1 + M_2)\,(Bs + K)\right]s^2}.$$

Exercise 2.15 Consider the translation mechanical system shown in Figure 2.29, where B_2 is a viscous friction coefficient.

1. Write the dynamic equations that describe the system's behavior.
2. Determine the system's transfer function $G(s) = X_2(s)/F(s)$.

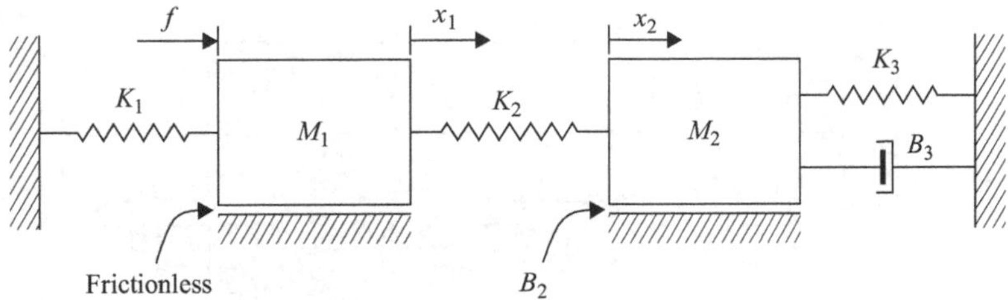

Figure 2.29 Mechanical system of Exercise 2.15

Exercise 2.16 Consider the translation mechanical system shown in Figure 2.30, where B_1 and B_2 are viscous friction coefficients.

1. Write the dynamic equations that describe the system's behavior.
2. Obtain the system's transfer function $G(s) = X_2(s)/F(s)$.

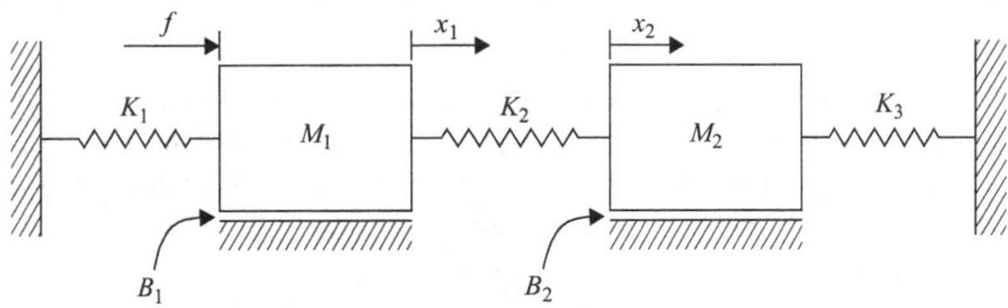

Figure 2.30 Mechanical system of Exercise 2.16

Exercise 2.17 Obtain the transfer function $G(s) = X(s)/F(s)$ of the system in Figure 2.31.

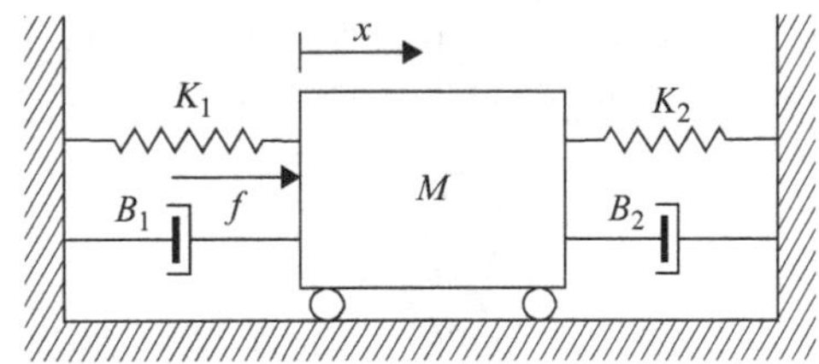

Figure 2.31 Mechanical system of Exercise 2.17

Exercise 2.18 Consider the mechanical system shown in Figure 2.32, where x_1, x_2 and x_3 are the displacements of the masses M_1, M_2 and M_3, respectively, and f is the applied force. Determine the differential equations that model this system.

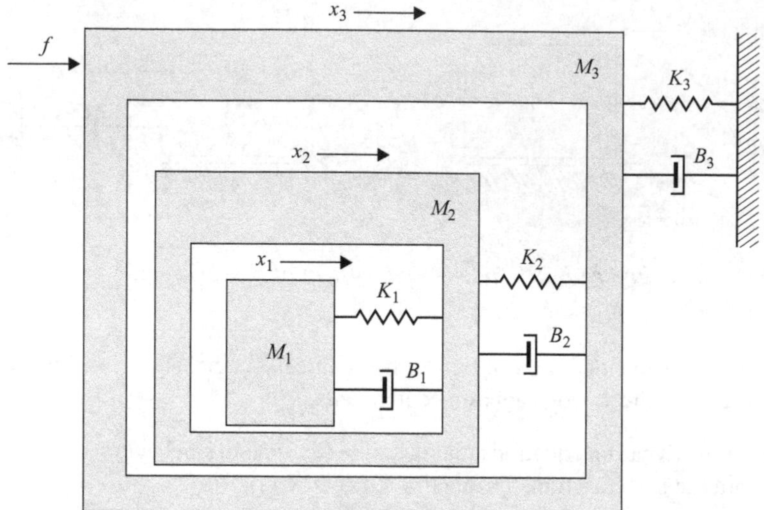

Figure 2.32 Mechanical system of Exercise 2.18

Exercise 2.19 Consider the belt-pulley drive system depicted in Figure 2.33 where a mass M is moved by means of a conveyor of negligible mass driven by a motor torque T. The pulley has a radius R and the friction between the belt and pulley is negligible. The variable $x(t)$ represents the linear displacement of the mass M.

1. Determine the mathematical model of the system.
2. Determine the transfer function $\dfrac{X(s)}{T(s)}$.
3. A second mass m is suspended over mass M, connected to its ends by means of two linear springs with stiffness $k/2$ each (see the figure). Variable $y(t)$ represents the linear displacement of mass m. Determine the transfer function $\dfrac{Y(s)}{T(s)}$.

Exercise 2.20 Consider the gear system represented in Figure 2.34, where T_1 represents the applied torque, J the inertia and ω_2 the angular velocity of the load. Then, it follows that:

A) $T_1 = J\left(\dfrac{N_1}{N_2}\right)\dot{\omega}_2$

B) $T_1 = J\left(\dfrac{N_2}{N_1}\right)\dot{\omega}_2$

C) $T_1 = J\left(\dfrac{N_1}{N_2}\right)^2\dot{\omega}_2$

D) $T_1 = J\left(\dfrac{N_2}{N_1}\right)^2\dot{\omega}_2.$

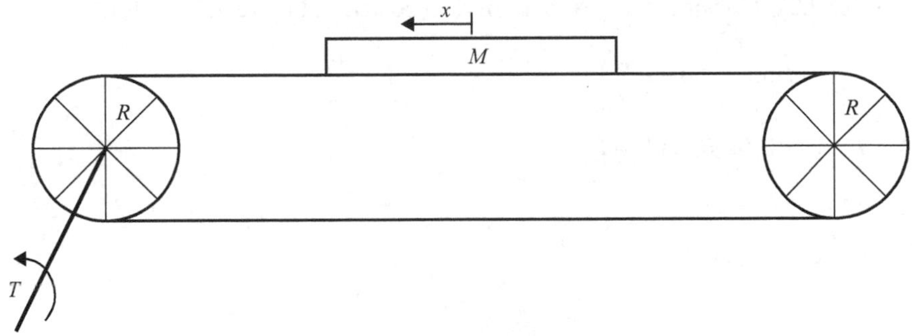

(a) Belt-pulley drive system with mass M

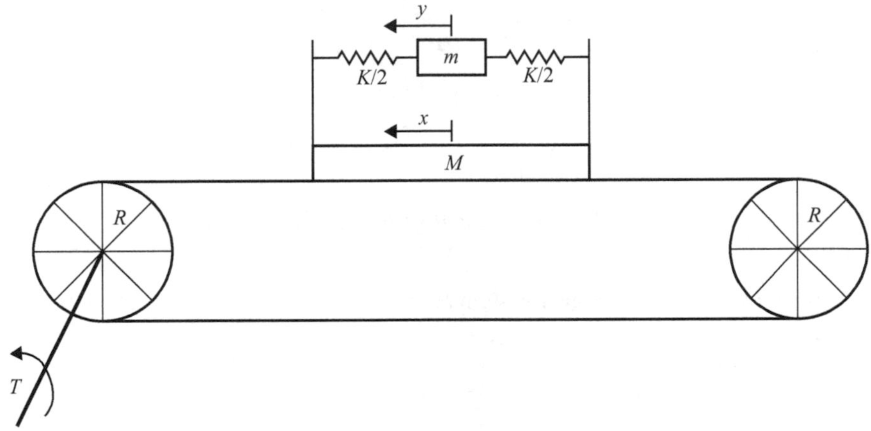

(b) Belt-pulley drive system with mass m connected to the ends of M by means of two springs

Figure 2.33 Belt pulley drive system of Exercise 2.19

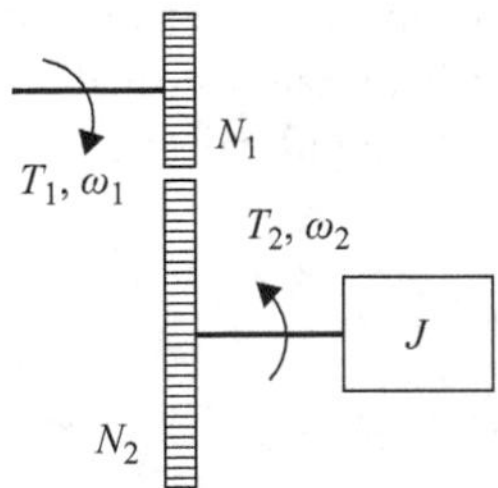

Figure 2.34 Mechanical system of Exercise 2.20

Exercise 2.21 Consider the gear system represented in Figure 2.35. It follows that:

A) $T_1 = (J\dot{\omega}_1 + B\omega_1)\left(\frac{N_2}{N_1}\right)^2$

B) $T_1 = (J\dot{\omega}_1 + B\omega_1)\left(\frac{N_1}{N_2}\right)^2$

C) $T_1 = J\dot{\omega}_1 + B\omega_1$

D) None of the above.

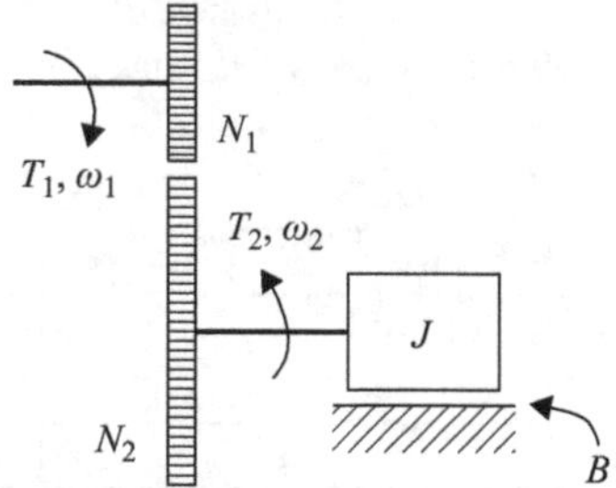

Figure 2.35 Mechanical system of Exercise 2.21

Exercise 2.22 Consider the gear system represented in Figure 2.36. It follows that:

A) $\dfrac{T_1}{T_4} = \dfrac{N_1N_2}{N_3N_4}$

B) $\dfrac{T_1}{T_4} = \dfrac{N_1N_3}{N_2N_4}$

C) $\dfrac{T_1}{T_4} = \dfrac{N_3N_4}{N_1N_2}$

D) $\dfrac{T_1}{T_4} = \dfrac{N_2N_4}{N_1N_3}.$

Exercise 2.23 Consider the mechanical system shown in Figure 2.37, with inertia J, stiffness K and viscous friction B. The mathematical model describing the relationship between the input torque T and the angular displacements θ_1 and θ_2 is:

A) $\begin{cases} T = K(\theta_1 - \theta_2) \\ K(\theta_1 - \theta_2) = J\ddot{\theta}_2 + B\dot{\theta}_2 \end{cases}$

B) $T = K(\theta_1 - \theta_2) + J\ddot{\theta}_2 + B\dot{\theta}_2$

C) $\begin{cases} T = K(\theta_1 - \theta_2) \\ T = J\ddot{\theta}_2 + B\dot{\theta}_2 \end{cases}$

D) None of the above.

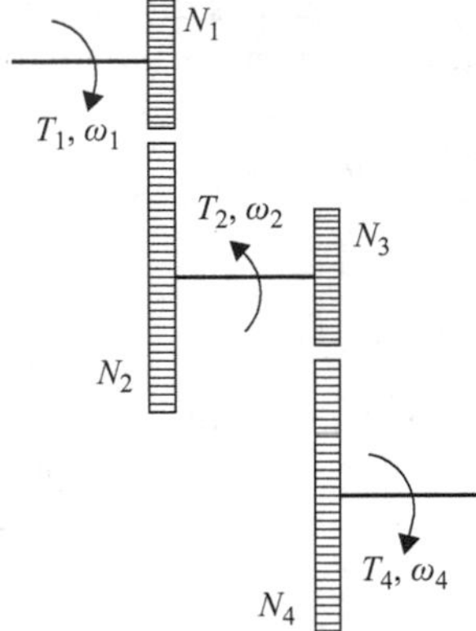

Figure 2.36 Gear train of Exercise 2.22

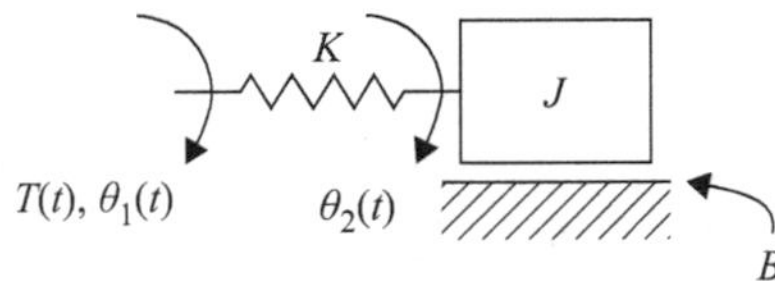

Figure 2.37 Mechanical system of Exercise 2.23

Exercise 2.24 Consider the system in Figure 2.38 comprising a mass M, a damper with viscous friction coefficient B and a spring with stiffness K. These elements are connected to a fixed pulley of radius r with inertia J. The system is driven by a torque T.

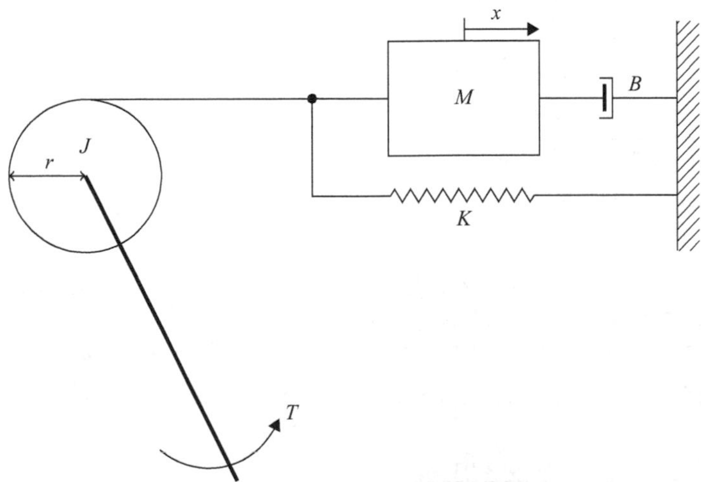

Figure 2.38 Mechanical system of Exercise 2.24

1. Model the system, writing the equations that describe its dynamics.
2. Obtain the transfer function considering the torque T as input and the mass position x as output.
3. Sketch the system block diagram.

2.3.3 Liquid-level systems

Exercise 2.25 Consider the system shown in Figure 2.39, consisting of two tanks, with sections A_1 and A_2. Knowing that the linear approximation to the relationship between the flow rate and the height is given by $q(t) = h(t)/R$ (where R is the hydraulic resistance), then transfer function $\frac{Q_0(s)}{Q_i(s)}$ is:

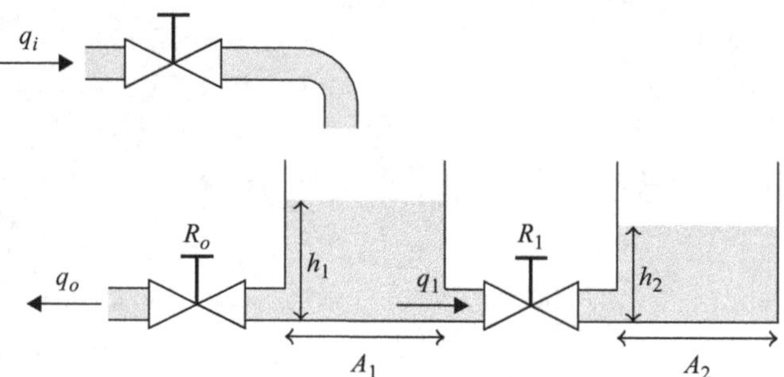

Figure 2.39 Hydraulic system of Exercise 2.25

A) $\dfrac{Q_o(s)}{Q_i(s)} = \dfrac{A_2R_1s + 1}{A_1A_2R_1R_0s^2 + (A_1R_0 + A_2R_1 + A_2R_0)s + 1}$

B) $\dfrac{Q_o(s)}{Q_i(s)} = \dfrac{1}{A_1A_2R_1R_0s^2 + A_2R_1s + 1}$

C) $\dfrac{Q_o(s)}{Q_i(s)} = \dfrac{A_1A_2R_oR_1s + 1}{A_1A_2R_1R_0s^2 + A_1R_0s + 1}$

D) None of the above.

Exercise 2.26 Consider the hydraulic system shown in Figure 2.40, where $q_0(t)$, $q_1(t)$ and $q_2(t)$ represent flows. Let $h_1(t)$ and $h_2(t)$ be the height of liquid in reservoirs 1 and 2, respectively, and their areas be A_1 and A_2. The hydraulic resistances are represented by R_0, R_1 and R_2, respectively. Transfer function $\frac{Q_2(s)}{Q_0(s)}$ is:

A) $\dfrac{Q_2(s)}{Q_0(s)} = \dfrac{R_1}{1 + (A_1R_1 + A_2R_2)s + A_1R_1A_2R_2s^2}$

B) $$\frac{Q_2(s)}{Q_0(s)} = \frac{1}{1+(A_1R_1+A_2R_2+A_1R_2)s+A_1R_1A_2R_2s^2}$$

C) $$\frac{Q_2(s)}{Q_0(s)} = \frac{1}{1+(A_1R_1+A_2R_2)s+A_1R_1A_2R_2s^2}$$

D) $$\frac{Q_2(s)}{Q_0(s)} = \frac{R_2}{R_1\left[1+(A_1R_1+2A_2R_2)s+A_1R_1A_2R_2s^2\right]}.$$

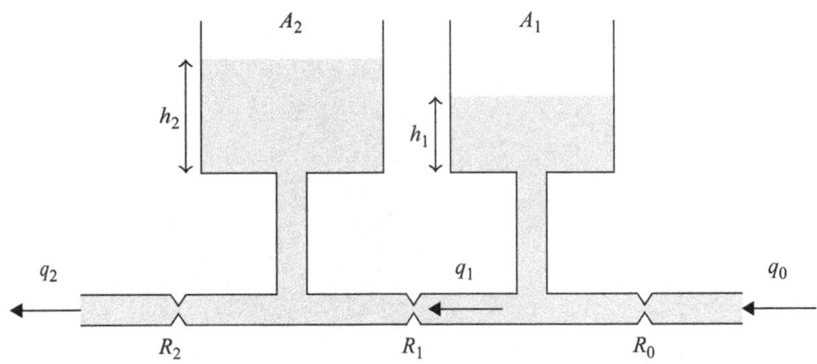

Figure 2.40 Hydraulic system of Exercise 2.26

Exercise 2.27 Consider the hydraulic system in Figure 2.41, where $q_i(t)$ and $q_o(t)$ are the input and output flows. Flow $q_i(t)$ passes through a pipe with neglectable resistance and length L, with velocity V. Let $h_1(t)$, $h_2(t)$ and $h_3(t)$ be liquid levels in reservoirs

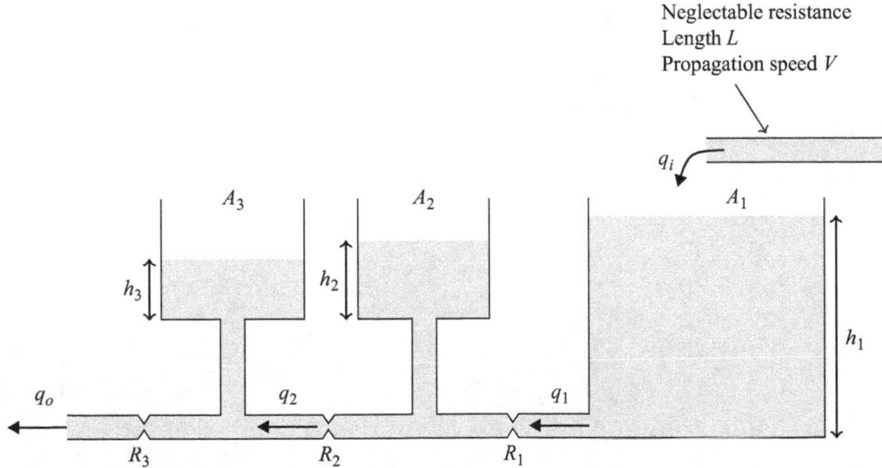

Figure 2.41 System of Exercise 2.27

1, 2 and 3, with cross-sections A_1, A_2 and A_3. Flow resistances are represented as R_1, R_2 and R_3. The plant's model is:

1. in the time domain

A)
$$\begin{cases} q_i = q_1 \\ \frac{h_1-h_2}{R_1} + A_1\frac{dh_1}{dt} = q_1 \\ q_i = q_1 + A_2\frac{dh_2}{dt} \\ \frac{h_2-h_3}{R_2} + A_2\frac{dh_2}{dt} = q_2 \\ q_1 = q_2 + A_3\frac{dh_3}{dt} \\ \frac{h_3}{R_3} = q_0 \end{cases}$$

B)
$$\begin{cases} q_i = A_1\frac{dh_1}{dt} \\ \frac{h_1-h_2}{R_1} = q_1 \\ q_1 = A_2\frac{dh_2}{dt} \\ \frac{h_2-h_3}{R_2} = q_2 \\ q_2 = A_3\frac{dh_3}{dt} \\ \frac{h_3}{R_3} = q_0 \end{cases}$$

C)
$$\begin{cases} q_i = A_1\frac{dh_1}{dt} + q_1 \\ \frac{h_1-h_2}{R_1} = q_1 \\ q_1 = q_2 + A_2\frac{dh_2}{dt} \\ \frac{h_2-h_3}{R_2} = q_2 \\ q_2 = q_0 + A_3\frac{dh_3}{dt} \\ \frac{h_3}{R_3} = q_0 \end{cases}$$

D) None of the above.

2. in the frequency domain

E)
$$\begin{cases} Q_i e^{-s\frac{L}{V}} = Q_1 + sA_1H_1 \\ H_1 - H_2 = R_1Q_1 \\ Q_1 = Q_2 + sA_2H_2 \\ H_2 - H_3 = R_2Q_2 \\ Q_2 = Q_o + sA_3H_3 \\ H_3 = R_3Q_o \end{cases}$$

F)
$$\begin{cases} Q_i = Q_1 + sA_1H_1e^{-s\frac{L}{V}} \\ H_1 - H_2 = R_1Q_1 \\ Q_1 = Q_2 + sA_2H_2 \\ H_2 - H_3 = R_2Q_2 \\ Q_2 = Q_o + sA_3H_3 \\ H_3 = R_3Q_o \end{cases}$$

G)
$$\begin{cases} Q_i = Q_1e^{-s\frac{L}{V}} + sA_1H_1 \\ H_1 - H_2 = R_1Q_1 \\ Q_1 = Q_2 + sA_2H_2 \\ H_2 - H_3 = R_2Q_2 \\ Q_2 = Q_o + sA_3H_3 \\ H_3 = R_3Q_o \end{cases}$$

H) None of the above.

Exercise 2.28 Consider the hydraulic system in Figure 2.42. Let $h_1(t)$, $h_2(t)$ and $h_3(t)$ be liquid levels in reservoirs 1, 2 and 3, with cross-sections A_1, A_2 and A_3. Flow resistances are represented as R_1, R_2 and R_3. The plant's model is:

A)
$$\begin{cases} q_i = A_1\dot{h}_1 + q_1 \\ h_1 = R_1 q_1 \\ q_1 = A_2\dot{h}_2 - q_1 + q_2 \\ h_2 - h_1 = R_2 q_2 \\ q_2 = A_3\dot{h}_3 + q_0 \\ h_3 = R_3 q_0 \end{cases}$$

B)
$$\begin{cases} q_i = A_1\dot{h}_1 + q_1 \\ h_1 - h_2 = R_1 q_1 \\ q_1 = A_2\dot{h}_2 + q_2 \\ h_2 = R_2 q_2 \\ q_2 = A_3\dot{h}_3 + q_0 \\ h_3 = R_3 q_0 \end{cases}$$

C)
$$\begin{cases} q_i = A_1\dot{h}_1 - q_1 \\ h_1 = R_1 q_1 \\ q_1 = A_2\dot{h}_2 - q_2 \\ h_1 - h_2 = R_2 q_2 \\ q_2 = A_3\dot{h}_3 + q_0 \\ h_2 - h_3 = R_3 q_0 \end{cases}$$

D) None of the above.

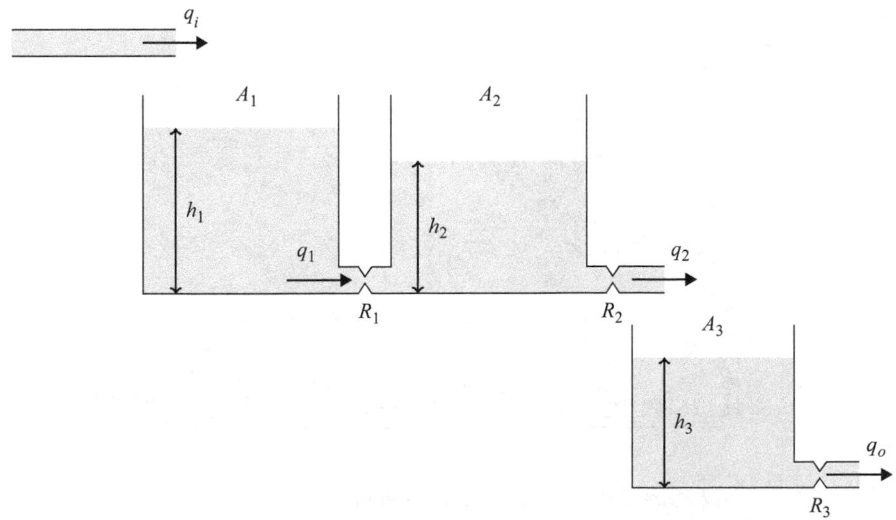

Figure 2.42 System of Exercise 2.28

Exercise 2.29 Consider the hydraulic system in Figure 2.43. Let $h_1(t)$ and $h_2(t)$ be liquid levels in reservoirs 1 and 2, with cross-sections A_1 and A_2. Flow resistances are represented as R_1 and R_2. The plant's model is:

A) $$\begin{cases} \frac{dh_1}{dt} = \frac{1}{A_1}\left(q_{i1} + \frac{h_1-h_2}{R}\right) \\ \frac{dh_2}{dt} = \frac{1}{A_2}\left(q_{i2} - \frac{h_1-h_2}{R}\right) \end{cases}$$

B) $$\begin{cases} \frac{dh_1}{dt} = \frac{1}{A_1}\left(q_{i1} + \frac{h_1-h_2}{R}\right) \\ \frac{dh_2}{dt} = \frac{1}{A_2}\left(q_{i2} + \frac{h_1-h_2}{R}\right) \end{cases}$$

C) $$\begin{cases} \frac{dh_1}{dt} = \frac{1}{A_1}\left(q_{i1} - \frac{h_1-h_2}{R}\right) \\ \frac{dh_2}{dt} = \frac{1}{A_2}\left(q_{i2} + \frac{h_1-h_2}{R}\right) \end{cases}$$

D) None of the above.

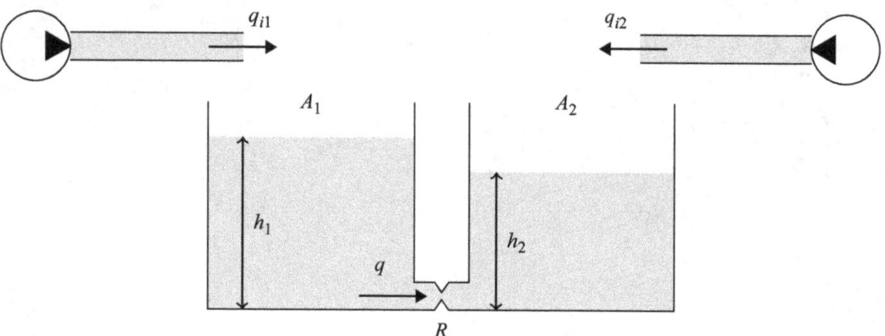

Figure 2.43 System of Exercise 2.29

Exercise 2.30 Consider the hydraulic system in Figure 2.44.

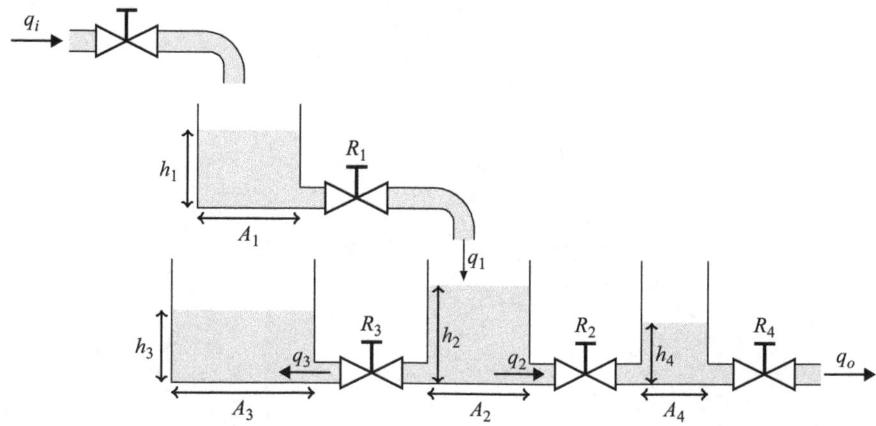

Figure 2.44 Hydraulic system of Exercise 2.30

1. Write the equations of the system dynamics in terms of the parameters shown in the figure.
2. Build a block diagram for this system, taking as input the flow Q_i and as output the flow Q_o.
3. Simplify the block diagram above, to obtain the transfer function of the system $G(s) = Q_o(s)/Q_i(s)$.

Exercise 2.31 Consider the hydraulic system in Figure 2.45, where $q_i(t)$ and $q_o(t)$ are the input and output flows. Let $h_1(t)$, $h_2(t)$ and $h_3(t)$ be the liquid levels in reservoirs 1, 2 and 3, with cross-sections A_1, A_2 and A_3. Flow resistances are represented as R_1 to R_5. The plant's model is:

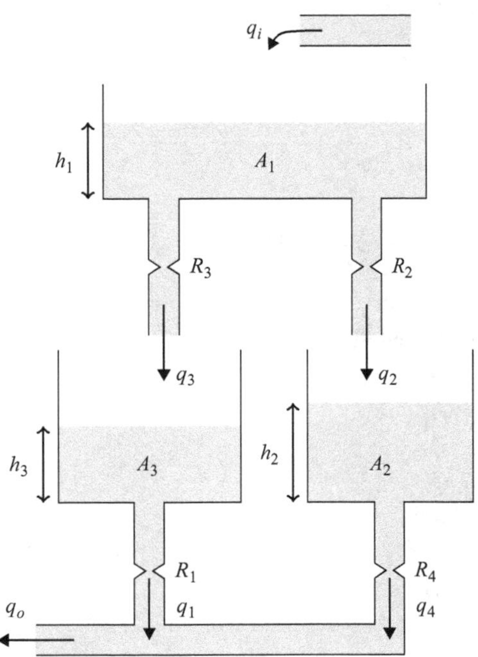

Figure 2.45 System of Exercise 2.31

A)
$$\begin{cases} Q_i = Q_2 + Q_3 + sA_1H_1 \\ H_1 = R_3Q_3 \\ H_1 = R_2Q_2 \\ Q_3 = Q_1 + sA_3H_3 \\ Q_2 = Q_4 + sA_2H_2 \\ H_3 = R_1Q_1 \\ H_2 = R_4Q_4 \\ Q_1 + Q_4 = Q_o \end{cases}$$

B)
$$\begin{cases} Q_i = sA_1H_1 \\ H_1 = R_3Q_3 \\ H_1 = R_2Q_2 \\ Q_3 = Q_1 + sA_3H_3 \\ Q_2 = Q_4 + sA_2H_2 \\ H_3 = R_1Q_1 \\ H_2 = R_4Q_4 \\ Q_1 - Q_4 = Q_o \end{cases}$$

C)
$$\begin{cases} Q_i + Q_2 + Q_3 = sA_1H_1 \\ H_1 = R_3Q_3 \\ H_1 = R_2Q_2 \\ Q_3 + Q_1 = sA_3H_3 \\ Q_2+Q_4 = sA_2H_2 \\ H_3 = R_1Q_1 \\ H_2 = R_4Q_4 \\ Q_1 + Q_4 + Q_o = 0 \end{cases}$$

D) None of the above.

Exercise 2.32 Consider the hydraulic system in Figure 2.46, where $q_i(t)$ and $q_o(t)$ are the input and output flows. Let $h_1(t)$, $h_2(t)$ and $h_3(t)$ be liquid levels in reservoirs 1, 2 and 3, with cross-sections A_1, A_2 and A_3. Flow resistances are represented as R_1, R_2 and R_3. The plant's model is:

A)
$$\begin{cases} Q_i = Q_1 + sA_1H_1 \\ H_1 - H_2 = R_1Q_1 \\ Q_1 = Q_2 + sA_2H_2 \\ H_2 - H_3 = R_2Q_2 \\ Q_2 = Q_o + sA_3H_3 \\ H_3 = R_3Q_o \end{cases}$$

B)
$$\begin{cases} Q_i = Q_1 + sA_1H_1 \\ H_1 - H_2 = R_1Q_1 \\ Q_1 = Q_2 + Q_o + sA_2H_2 \\ H_2 - H_3 = R_2Q_2 \\ Q_2 = sA_3H_3 \\ H_2 = R_3Q_o \end{cases}$$

C)
$$\begin{cases} Q_i = Q_1 + sA_1H_1 \\ H_1 - H_2 = R_1Q_1 \\ Q_1 = Q_2 + sA_2H_2 \\ H_2 - H_3 = R_2Q_2 \\ Q_2 = Q_o + sA_3H_3 \\ H_3 = R_3Q_o \end{cases}$$

D) None of the above.

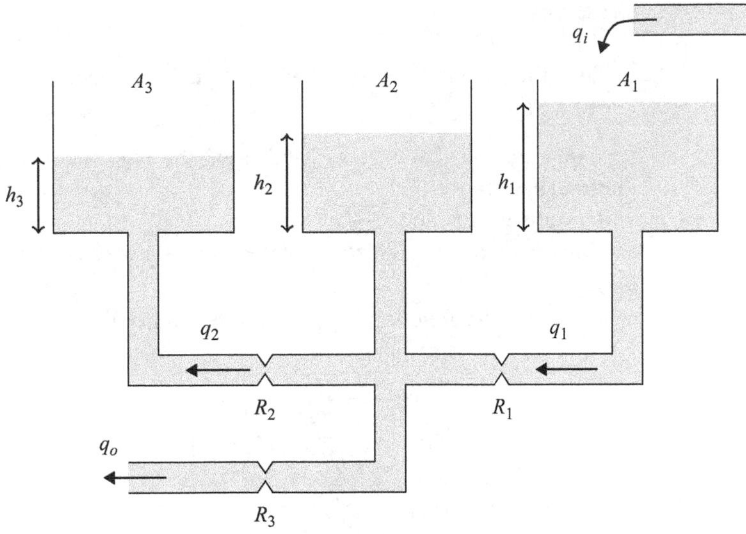

Figure 2.46 System of Exercise 2.32

Exercise 2.33 Consider the hydraulic system in Figure 2.47, where $q_i(t)$ and $q_o(t)$ are the input and output flows. Let q_1, q_2 and q_3 be liquid flows, $h_1(t)$, $h_2(t)$ and $h_3(t)$ be liquid levels in reservoirs 1, 2 and 3, with cross-sections A_1, A_2 and A_3, and R_1 and R_3 be flow resistances. The output flow of the hydraulic pump into a pipe of length L is $q_2 = kh_2$ and has a constant speed of V. Find the plant's model in the Laplace domain.

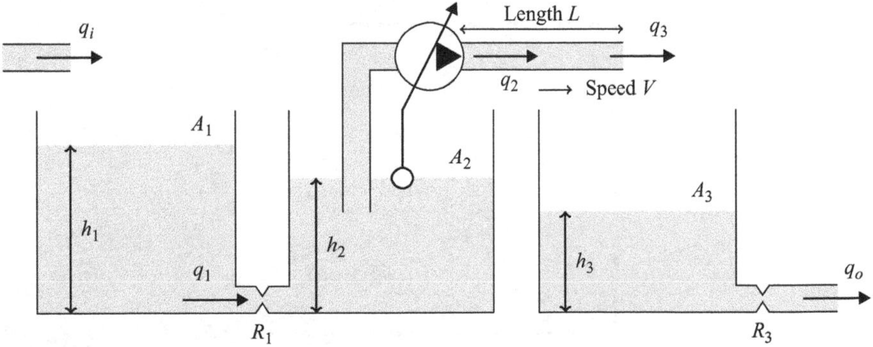

Figure 2.47 System of Exercise 2.33

2.3.4 Thermal systems

Exercise 2.34 Consider a thermometer reading temperature T_m immersed in fluid at temperature T_0 as seen on Figure 2.48.

1. Neglect the thermal capacity of the glass, and draw the equivalent electrical circuit of this thermal system.
2. Find the first-order transfer function $\frac{T_m(s)}{T_0(s)}$.
3. Modify the electrical circuit to take into account the thermal capacity of the glass, at temperature T_g.
4. Find the second-order transfer function $\frac{T_m(s)}{T_0(s)}$ for these conditions.

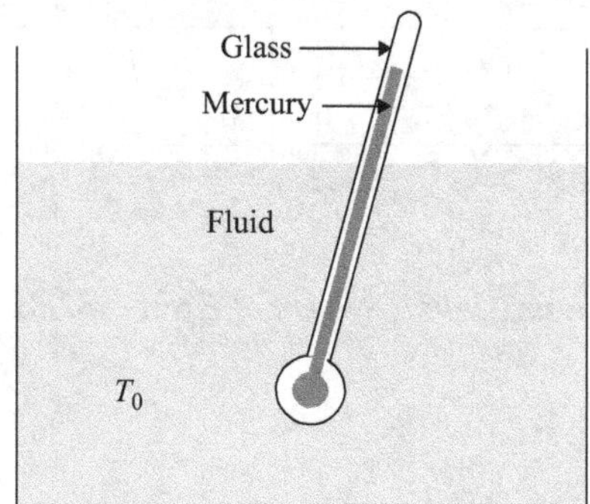

Figure 2.48 Thermometer of Exercise 2.34

Exercise 2.35 Consider an electrical circuit heating a box to a temperature T_i with thermal capacitance C_T, cooled by a constant external temperature T_o, as seen in Figure 2.49. Resistance R dissipates energy with power P, and q is the heat flux cooling the box, which takes place due to a thermal resistance R_T.

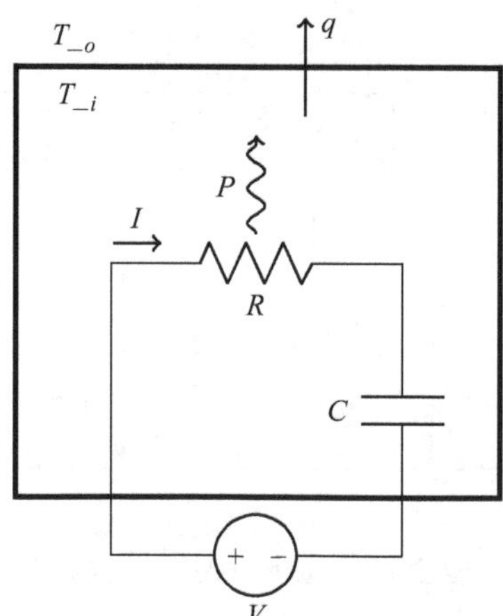

Figure 2.49 Diagram of Exercise 2.35

1. Find the transfer function $\frac{T(s)}{P(s)}$, where $T = T_i - T_o$.
2. Find the transfer function $\frac{I(s)}{V(s)}$.
3. Can you find a transfer function $\frac{T(s)}{V(s)}$?
4. Could you find transfer function $\frac{T(s)}{V(s)}$ if the electrical circuit had no capacitor?

Exercise 2.36 Consider a room and a wall separating it from outside street, as represented in the Figure 2.50. The temperature inside the room is denoted by T_i, while the temperature at the street is T_o. The wall is made of bricks and has a window made of glass. The temperatures at the middle of their widths are T_b and T_g, respectively. Furthermore, their thermal resistances are R_b and R_g, for the left and for the right sides (making, therefore, a total of $2R_b$ and $2R_g$) and their thermal capacitances are C_b and C_g, respectively. Derive the mathematical model of the thermal system.

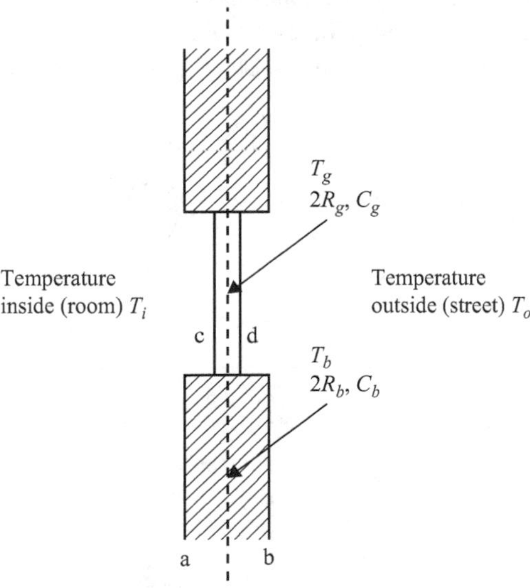

Figure 2.50 Wall and window of Exercise 2.36

Exercise 2.37 Consider a room and a wall separating it from outside street, as represented in the Figure 2.51. The temperature inside the room is denoted by T_i, while the temperature at the street is T_o. The wall is made of 3 distinct materials. The temperatures at the middle of their widths are T_1, T_2 and T_3, respectively. Furthermore, their thermal resistances are R_1, R_2 and R_3, for the left and for the right sides (making, therefore, a total of $2R_1$, $2R_2$ and $2R_3$) and their thermal capacitances are C_1, C_2 and C_3, respectively. Derive the mathematical model of the thermal system.

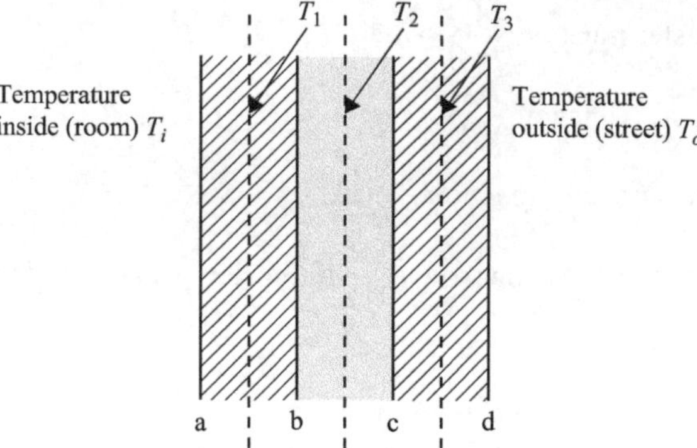

Figure 2.51 Wall of Exercise 2.37

Exercise 2.38 Consider a submerged Wave Energy Converter (WEC), fixed to the bottom of the sea, producing electricity from the sea waves, as seen in Figure 2.52. It is filled with air, and its upper part oscillates under the varying pressure of the waves. Electricity is produced in the Power Take-Off (PTO) mechanism from the movement of the upper part relative to the sea bottom. Let T_{sea} be the temperature of the sea, T_{air}

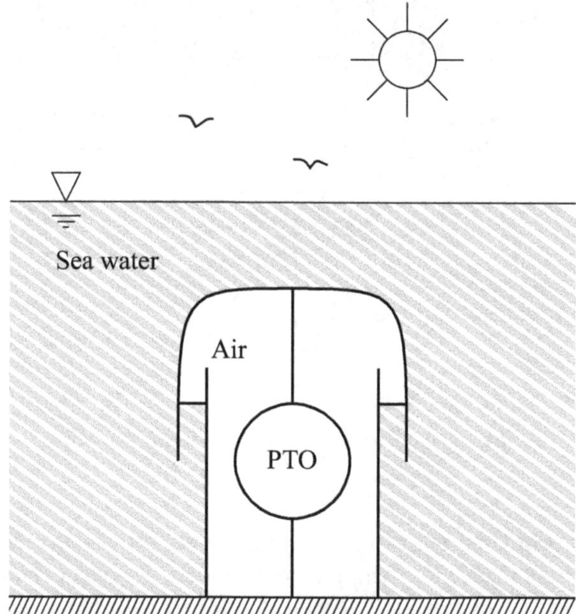

Figure 2.52 Wave Energy Converter of Exercise 2.38

the temperature of the air, and T_{PTO} the temperature of the PTO; T_{sea} can be assumed constant with time, while the two others vary with time but are spatially homogeneous. Let C_{PTO} and A_{PTO} be the thermal capacity and the external area of the PTO, and let C_{air} and A_{air} be those of the air inside the WEC. Neglect the thermal capacity of the WEC itself, separating air from sea water, and assume that convection always takes place with the same coefficient h.

1. Suppose that the PTO delivers an electrical power P, and has an efficiency η. Find the differential equations that govern heat transfer in this system.
2. Let $T_1 = T_{PTO} - T_{air}$ and $T_2 = T_{air} - T_{sea}$. Find transfer function $\frac{T_1}{P}$.

Chapter 3
Analysis of continuous systems in the time domain

3.1 Fundamentals

The time response of a control system to typical test input signals, namely unit-impulse, unit step, and unit ramp functions, is an important design criterion. In fact, given the system response to these test inputs, we can infer about the system behavior in response to more general real signals [2–5,9].

3.1.1 List of symbols

A amplitude of the input
e_{ss} steady-state error
$g(t)$ impulse time response
$G(s)$ transfer function
$\mathscr{L}$ Laplace operator
M_p maximum overshoot
$r(t)$ time-domain input
s Laplace variable
t time
t_p peak time
t_r rise time (0%–100% criterion)
t_{r1} rise time (10%–90% criterion)
t_s settling time
T time constant
$y(t)$ time-domain output function
$y(t_p)$ maximum output
y_{ss} steady-state output
$u(t)$ unit-step function
$\delta(t)$ unit-impulse function
ζ damping coefficient
τ system type
ω_n undamped natural frequency
ω_d damped natural frequency

3.1.2 Time response of a continuous LTI system

A continuous LTI system can be represented by linear constant-coefficient differential equations in continuous time.

The system time response includes two components:

1. The transient response, which determines the system behavior during a short time period after a change in system equilibrium, and vanishes as time increases (supposing the system is asymptotically stable),
2. The steady-state response, which imposes the system behavior after the transient time period.

If we know the system differential equations and the initial conditions, then we can determine the system time behavior.

For a LTI system with impulse time response $g(t)$, the output $y(t)$ to an input $r(t)$ is given by the time convolution:

$$y(t) = g(t) \star r(t) \tag{3.1}$$

In the Laplace domain, we have:

$$Y(s) = G(s) \cdot R(s) \tag{3.2}$$

where $G(s)$ is the Laplace transform of $g(t)$, that is, the transfer function of the system. The roots of the denominator (numerator) of $G(s)$ are the closed-loop poles (zeros). The system order corresponds to the highest power of s in the denominator of $G(s)$.

The stability of a LTI system can be determined from the location of the closed-loop poles in the s plane. The system is stable if all poles lie in the left-half s plane. Thus, stability is a characteristic of the system, and does not depend on the input signal.

3.1.3 Time response of first-order systems

The transfer function of a first-order system is given by:

$$G(s) = \frac{1}{1 + Ts} \tag{3.3}$$

where T represents the time constant.

3.1.3.1 Time response to the unit impulse

The input is $r(t) = \delta(t)$, which corresponds to $R(s) = 1$, in the Laplace domain. The system output, in the Laplace domain, results in $Y(s) = G(s)$ and, in the time domain, we have:

$$y(t) = \mathscr{L}^{-1}[Y(s)] = \frac{1}{T}e^{-\frac{t}{T}},\ t \geq 0 \tag{3.4}$$

Figure 3.1 illustrates the system time response, where $y(t)$ decays exponentially from its maximum value $y(0) = \frac{1}{T}$ towards zero. The slope of the tangent to $y(t)$ at

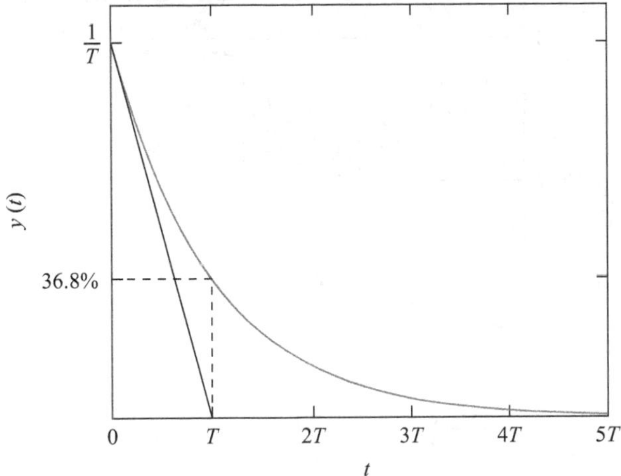

Figure 3.1 Time response of a first-order system to the unit impulse

$t=0$ is $-\frac{1}{T^2}$. At time $t=T$ the response has decayed to approximately 36.8% of its initial value.

3.1.3.2 Time response to the unit step

The input is $r(t)=u(t)$ and, in the Laplace domain, $R(s)=\frac{1}{s}$. The system output, in the Laplace domain, results in $Y(s)=\frac{G(s)}{s}$ and, in the time domain, we have:

$$y(t) = \mathscr{L}^{-1}[Y(s)] = \mathscr{L}^{-1}\left[\frac{1}{s(1+Ts)}\right] = 1 - e^{-\frac{t}{T}},\ t \geq 0 \tag{3.5}$$

Figure 3.2 illustrates the system time response. We see that $y(t)$ grows exponentially, starting from zero and asymptotically converging to one as time tends to infinity, $\lim_{t\to\infty} y(t)=1$. The slope of the tangent to $y(t)$ at $t=0$ is $\frac{1}{T}$. At time $t=T$ and $t=4T$ (i.e., the time instants corresponding to one and four time constants) the system response has reached approximately 63.2% and 98.2% of its final value, respectively. For $t\geq 4T$, $y(t)$ always remains within 2% of its final value. The settling time is defined as $t_s=4T$ and the steady-state error is:

$$e_{ss} = \lim_{t\to\infty}[y(t) - r(t)] = 0 \tag{3.6}$$

3.1.3.3 Time response to the unit ramp

In this case the input is $r(t)=tu(t)$ and in the Laplace domain, $R(s)=\frac{1}{s^2}$. The system output, in the Laplace domain, is given by $Y(s)=\frac{G(s)}{s^2}$. In the time domain, we have:

$$y(t) = \mathscr{L}^{-1}[Y(s)] = \mathscr{L}^{-1}\left[\frac{1}{s^2(1+Ts)}\right] = t - T + Te^{-\frac{t}{T}},\ t \geq 0 \tag{3.7}$$

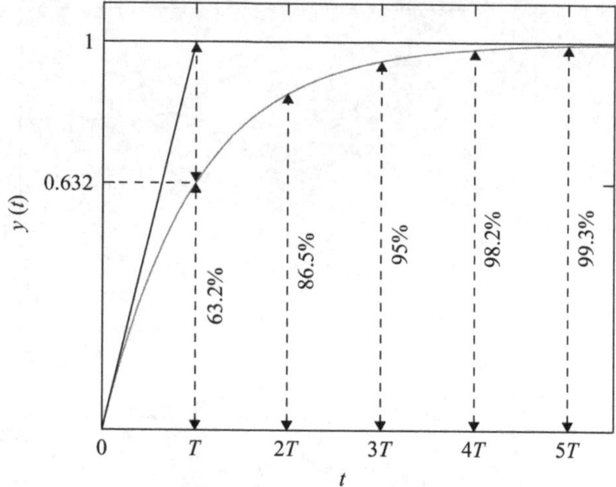

Figure 3.2 Time response of a first-order system to the unit step

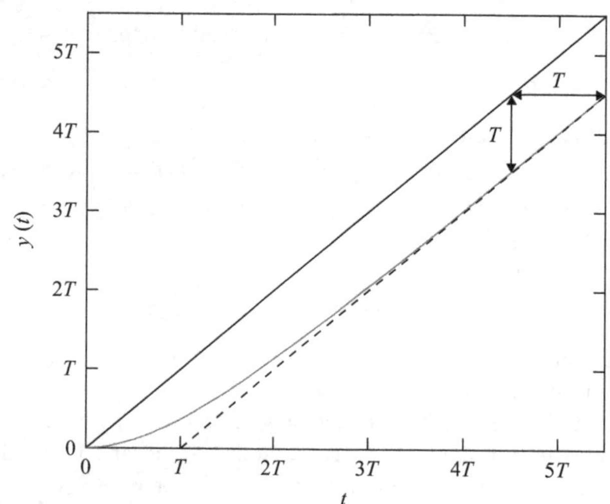

Figure 3.3 Time response of a first-order system to the unit ramp

When $t \to \infty$, $y(t) \approx t - T$. Figure 3.3 illustrates the system time response. In this case, it is interesting to note that the steady-state error is:

$$e_{ss} = \lim_{t \to \infty} [y(t) - r(t)] = T \tag{3.8}$$

By differentiating the response of the system to a given input, we obtain the response to the derivative of that input. Similarly, by integrating the response of the system to a given signal, we obtain the response to the integral of that signal.

The corresponding integration constant can be calculated from the zero-output initial condition.

3.1.4 Time response of second-order systems

The system transfer function of a second-order system is given by:

$$G(s) = \frac{\omega_n^2}{s^2 + 2\zeta\omega_n s + \omega_n^2} \tag{3.9}$$

where ω_n is the undamped natural frequency and ζ represents the damping coefficient.

3.1.4.1 Time response to the unit impulse

The input is $r(t) = \delta(t)$, which corresponds to $R(s) = 1$, in the Laplace domain. The system output, in the Laplace domain, results in $Y(s) = G(s)$.

Undamped system ($\zeta = 0$)
In this case, we have:

$$y(t) = \mathscr{L}^{-1}[Y(s)] = \mathscr{L}^{-1}\left[\frac{\omega_n^2}{s^2 + \omega_n^2}\right] = \omega_n \sin(\omega_n t),\ t \geq 0 \tag{3.10}$$

Figure 3.4 illustrates the system normalized responsc, $y(t)/\omega_n$, where the abscissa is the dimensionless variable $\omega_n t$. For $\zeta = 0$ the system output oscillates with constant amplitude and frequency ω_n.

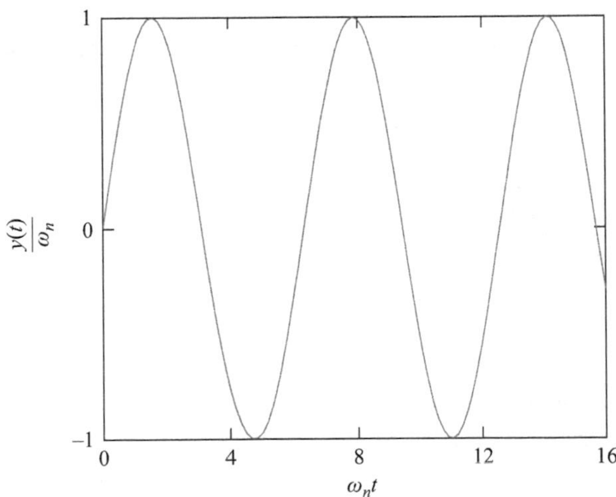

Figure 3.4 Time response of a second-order system to the unit impulse for $\zeta = 0$

Under-damped system $(0 < \zeta < 1)$
For the under-damped system the output is:

$$y(t) = \mathscr{L}^{-1}[Y(s)] = \mathscr{L}^{-1}\left[\frac{\omega_n^2}{(s+\zeta\omega_n)^2 + \omega_d^2}\right] = \frac{\omega_n}{\sqrt{1-\zeta^2}} e^{-\zeta\omega_n t}\sin(\omega_d t),\ t \geq 0 \tag{3.11}$$

where $\omega_d = \omega_n\sqrt{1-\zeta^2}$ represents the damped natural frequency.

Figure 3.5 depicts the system response, $y(t)$ versus $\omega_n t$, for several values of ζ $(0 < \zeta < 1)$. We can see that the system output has damped oscillations with frequency ω_d.

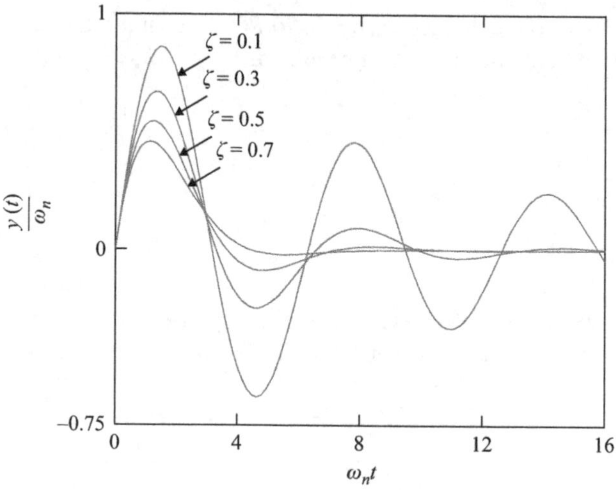

Figure 3.5 Time response of a second-order system to the unit impulse for $0 \leq \zeta < 1$

The peak time, t_p, and maximum output value, y_{max}, are:

$$t_p = \frac{\tan^{-1}\left(\sqrt{1-\zeta^2}/\zeta\right)}{\omega_d} \tag{3.12}$$

$$y(t_p) = \omega_n e^{-\frac{\zeta}{\sqrt{1-\zeta^2}}\tan^{-1}\left(\frac{\sqrt{1-\zeta^2}}{\zeta}\right)} \tag{3.13}$$

Critically damped system $(\zeta = 1)$
For the critically damped case the response is:

$$y(t) = \mathscr{L}^{-1}[Y(s)] = \mathscr{L}^{-1}\left[\frac{\omega_n^2}{(s+\omega_n)^2}\right] = k\omega_n^2 t e^{-\omega_n t},\ t \geq 0 \tag{3.14}$$

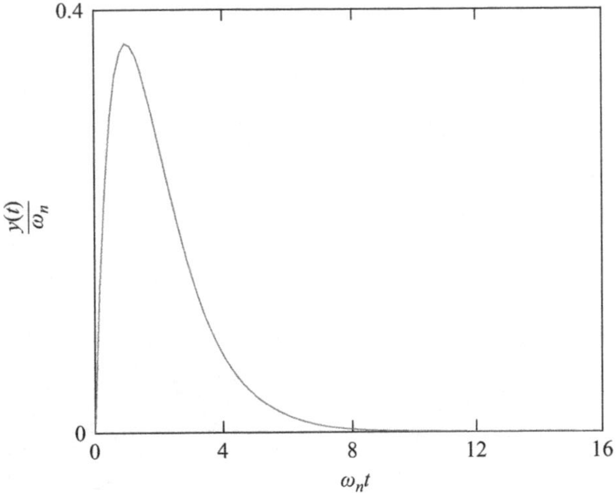

Figure 3.6 Time response of a second-order system to the unit impulse for $\zeta = 1$

The peak time is $t_p = 1/\omega_n$ and the maximum output value corresponds to $y(t_p) = \omega_n^2 e^{-1}$ (Figure 3.6).

Over-damped system ($\zeta > 1$)
In the over-damped case, we have:

$$y(t) = \mathscr{L}^{-1}[Y(s)] = \frac{\omega_n}{2\sqrt{\zeta^2 - 1}}\left(e^{-\left(\zeta - \sqrt{\zeta^2-1}\right)\omega_n t} - e^{-\left(\zeta + \sqrt{\zeta^2-1}\right)\omega_n t}\right), \ t \geq 0 \tag{3.15}$$

Figure 3.7 shows the system output for several values of $\zeta > 1$. For $\zeta \geq 1$, the system response to the unit-impulse is always positive or zero.

In this case the peak time, t_p, is:

$$t_p = \frac{1}{2\omega_n\sqrt{\zeta^2 - 1}} \ln\left(\frac{\zeta + \sqrt{\zeta^2 - 1}}{\zeta - \sqrt{\zeta^2 - 1}}\right) \tag{3.16}$$

3.1.4.2 Time response to the unit step

The input is $r(t) = u(t)$, which corresponds to $R(s) = 1/s$, in the Laplace domain. The system output, in the Laplace domain, results in $Y(s) = G(s)/s$.

Undamped system ($\zeta = 0$)
In this case, we have:

$$y(t) = \mathscr{L}^{-1}[Y(s)] = \mathscr{L}^{-1}\left[\frac{k\omega_n^2}{s(s^2 + \omega_n^2)}\right] = 1 - \cos(\omega_n t), \ t \geq 0 \tag{3.17}$$

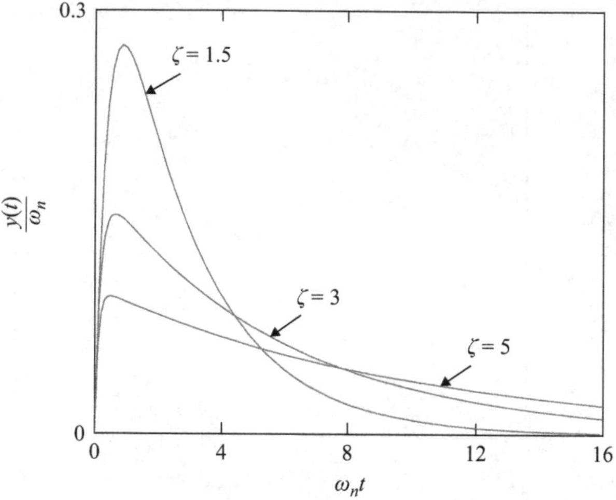

Figure 3.7 Time response of a second-order system to the unit-impulse for $\zeta > 1$

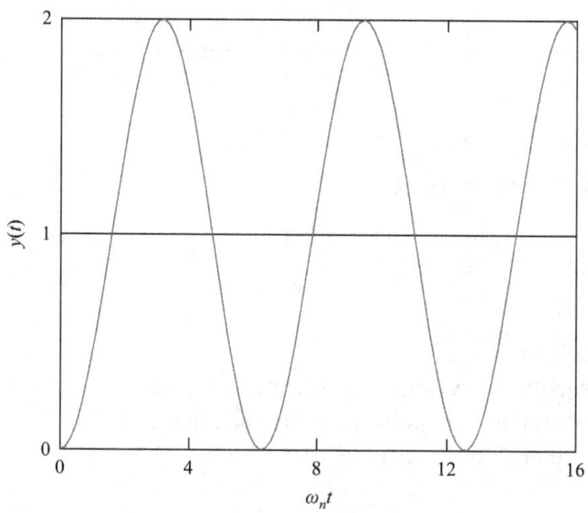

Figure 3.8 Time response of a second-order system to the unit step for $\zeta = 0$

The system time response, $y(t)$, oscillates with constant amplitude and natural frequency ω_n (Figure 3.8).

Under-damped system ($0 < \zeta < 1$)
For this case the closed-loop poles are $s_1 = -\zeta\omega_n - j\omega_n\sqrt{1-\zeta^2}$ and $s_2 = -\zeta\omega_n + j\omega_n\sqrt{1-\zeta^2}$. The system response is:

$$y(t) = \mathscr{L}^{-1}[Y(s)] = \mathscr{L}^{-1}\left[\frac{k\omega_n^2}{s(s+\zeta\omega_n)^2+\omega_d^2)}\right] \tag{3.18}$$

$$y(t) = 1 - e^{-\zeta\omega_n t}\left(\cos(\omega_d t) + \frac{\zeta}{\sqrt{1-\zeta^2}}\sin(\omega_d t)\right),\ t \geq 0 \tag{3.19}$$

or,

$$y(t) = 1 - \frac{e^{-\zeta\omega_n t}}{\sqrt{1-\zeta^2}}\sin(\omega_d t + \theta),\ t \geq 0 \tag{3.20}$$

where

$$\theta = \tan^{-1}\frac{\sqrt{1-\zeta^2}}{\zeta} \tag{3.21}$$

and

$$\zeta = \cos(\theta) \tag{3.22}$$

Figure 3.9 depicts $y(t)$ for various values of ζ versus the dimensionless variable $\omega_n t$. For all cases the slope of the tangent to $y(t)$ at $t=0$ is zero. The system response oscillates with frequency ω_d.

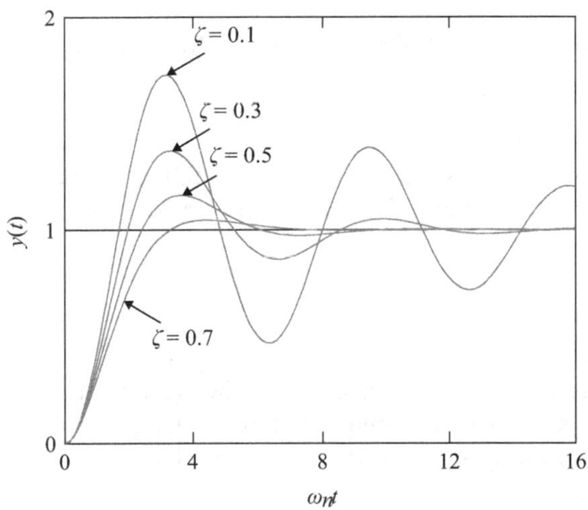

Figure 3.9 Time response of a second-order system to the unit step for $0 < \zeta < 1$

The main transient response specifications are given by the following parameters (Figure 3.10).

- Peak time—is the time required for the system response to reach its maximum value:

$$t_p = \frac{\pi}{\omega_d} \tag{3.23}$$

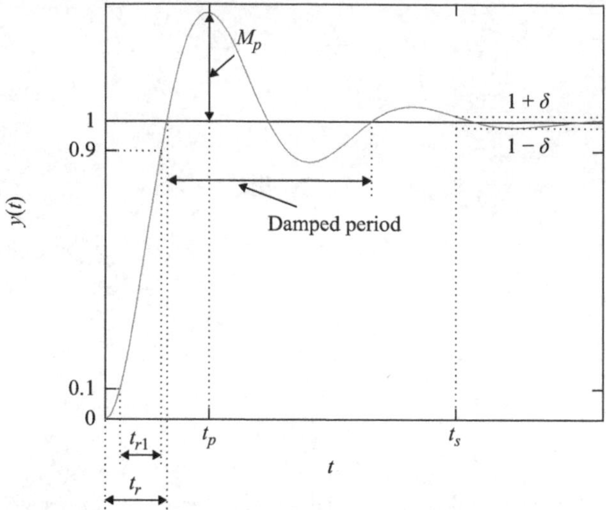

Figure 3.10 Transient response parameters of the second-order system response to the unit step ($0 < \zeta < 1$)

- Maximum output value—is the peak value of the system output:

$$y(t_p) = 1 + e^{-\frac{\zeta\pi}{\sqrt{1-\zeta^2}}} \tag{3.24}$$

- Maximum overshoot—is the difference between the maximum and the steady-state output values. Often it is represented as a percentage of the steady-state system response:

$$M_p = \frac{y(t_p) - y(\infty)}{y(\infty)} = e^{-\frac{\zeta\pi}{\sqrt{1-\zeta^2}}} \tag{3.25}$$

- Rise time—is the time, t_r (t_{r1}), required for the system output to rise from 0% to 100% (10%–90%) of its final value. Usually t_r is used for under-damped systems and t_{r1} is for critically and over-damped systems:

$$t_r = \frac{\pi - \theta}{\omega_d} \tag{3.26a}$$

$$t_{r1} = \frac{\exp\left(\frac{\theta}{\tan\theta}\right)}{\omega_n} \tag{3.26b}$$

- Settling time—is the time required for the system response to settle within a certain percentage δ of its final value.
  $\delta = 5\%$ criterion:

$$t_s \approx \frac{3}{\zeta\omega_n} = 3T \tag{3.27}$$

$\delta = 2\%$ criterion:

$$t_s \approx \frac{4}{\zeta \omega_n} = 4T \tag{3.28}$$

Critically damped system ($\zeta = 1$)
For $\zeta = 1$ the system response is:

$$y(t) = \mathscr{L}^{-1}[Y(s)] = \mathscr{L}^{-1}\left[\frac{\omega_n^2}{s(s+\omega_n)^2}\right] \tag{3.29}$$

$$y(t) = 1 - e^{-\omega_n t}(1 + \omega_n t), \ t \geq 0 \tag{3.30}$$

A critically damped system exhibits the fastest response without oscillation (Figure 3.11).

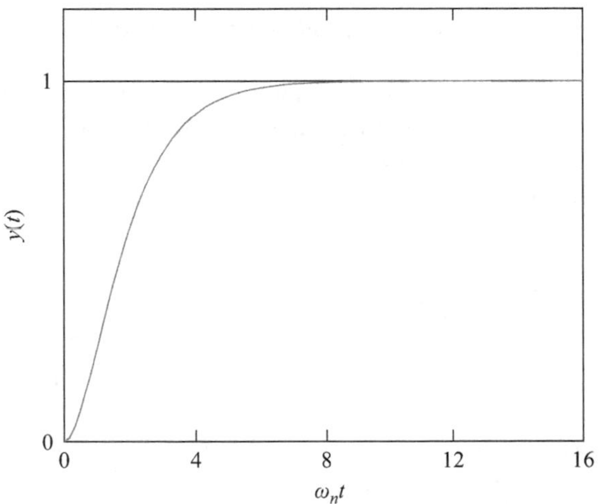

Figure 3.11 Time response of a second-order system to the unit step for $\zeta = 1$

Over-damped system ($\zeta > 1$)
For this case the poles are $s_1 = -\zeta\omega_n - \omega_n\sqrt{\zeta^2 - 1}$ and $s_2 = -\zeta\omega_n + \omega_n\sqrt{\zeta^2 - 1}$. The system response is:

$$y(t) = \mathscr{L}^{-1}[Y(s)] = \mathscr{L}^{-1}\left[\frac{\omega_n^2}{s(s-s_1)(s-s_2)}\right] \tag{3.31}$$

$$y(t) = 1 - \frac{\omega_n}{2\sqrt{\zeta^2 - 1}}\left(\frac{e^{s_1 t}}{s_1} - \frac{e^{s_2 t}}{s_2}\right), \ t \geq 0 \tag{3.32}$$

Figure 3.12 depicts the system response for several values of $\zeta > 1$.

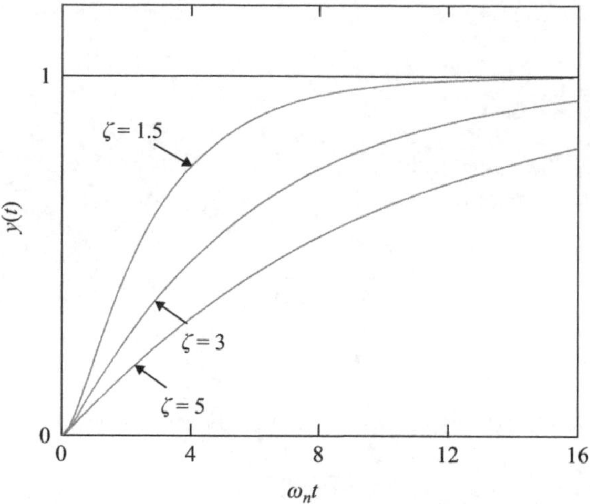

Figure 3.12 Time response of a second-order system to the unit step for $\zeta > 1$

As mentioned before, by differentiating the response of the system to a given input, we obtain the response to the derivative of that input. Similarly, by integrating the response of the system to a given signal, we obtain the response to the integral of that signal. The corresponding integration constant can be calculated from the zero-output initial condition.

3.1.5 Routh's stability criterion

The Routh's stability criterion leads to the determination of the number of non-negative closed-loop poles of the characteristic polynomial without actually solving the equation.

For an nth-degree polynomial $D(s) = a_n s^n + a_{n-1}s^{n-1} + \cdots + a_1 s + a_0$ we construct a table with $n + 1$ rows and the following structure:

$$
\begin{array}{c|cccc}
n & a_n & a_{n-2} & a_{n-4} & \cdots \\
n-1 & a_{n-1} & a_{n-3} & a_{n-5} & \cdots \\
n-2 & b_1 & b_2 & b_3 & \cdots \\
n-3 & c_1 & c_2 & c_3 & \cdots \\
\cdots & \cdots & \cdots & \cdots & \\
0 & h_1 & & &
\end{array}
$$

where the elements b_i and c_i can be computed as follows:

$$b_i = \frac{a_{n-1}a_{n-2i} - a_n a_{n-2i-1}}{a_{n-1}} \tag{3.33a}$$

$$c_i = \frac{b_1 a_{n-2i-1} - b_{i+1}a_{n-1}}{b_1} \tag{3.33b}$$

When completed, the number of sign changes in the first column will be the number of non-negative poles.

There are two special cases:

- Zero in first column element: if the first term in a row is zero, but the remaining terms are not zero, the zero element is replaced by a small positive constant ϵ and the calculation continues as shown above.
- Zero row: if all the coefficients in a row are zero, then there are roots of equal magnitude and opposite sign. The zero row is replaced by taking the coefficients of the auxiliary polynomial $dP(s)/ds$, where $P(s)$ is obtained from the values in the row above the zero row.

3.1.6 Steady-state errors

The steady-state error, e_{ss}, is given by:

$$e_{ss} = \lim_{t\to\infty} e(t) = \lim_{s\to 0} sE(s) = \lim_{s\to 0} \frac{R(s)}{1 + G(s)H(s)} \tag{3.34}$$

The steady-state error to a given input $r(t)$ depends on the system type, that is, the number, τ, of open-loop poles located at the origin of the s plane. Table 3.1 summarizes the steady-state errors for various inputs and values of τ, where A is the amplitude of the input function and K_p, K_v, K_a and K_j are the position, velocity, acceleration and jerk error constants, respectively:

$$K_p = \lim_{s\to 0} G(s)H(s) \tag{3.35}$$

$$K_v = \lim_{s\to 0} sG(s)H(s) \tag{3.36}$$

$$K_a = \lim_{s\to 0} s^2G(s)H(s) \tag{3.37}$$

$$K_j = \lim_{s\to 0} s^3G(s)H(s) \tag{3.38}$$

Table 3.1 Steady-state errors

System type	**Step**	**Input ramp**	**Parabolic**	**Cubic**
0	$\frac{A}{1+K_p}$	∞	∞	∞
1	0	$\frac{A}{K_v}$	∞	∞
2	0	0	$\frac{A}{K_a}$	∞
3	0	0	0	$\frac{A}{K_j}$

3.2 Worked examples

3.2.1 Routh–Hurwitz criterion

Problem 3.1 Consider a closed-loop system with characteristic equation $s^4 - 2s^3 - 13s^2 + 14s + 24 = 0$. The closed-loop system presents:

A) Three poles in the right half-plane
B) Two poles in the right half-plane
C) One pole in the right half-plane
D) Zero poles in the right half-plane.

Resolution There are negative coefficients, so the system cannot be stable; option **D)** is ruled out at the outset. To know the exact number of unstable (right half-plane) poles, we use the Routh–Hurwitz criterion:

$$\begin{array}{r|ccc} s^4 & 1 & -13 & 24 \\ s^3 & -2 & 14 & \\ \hline s^2 & b_1 & b_2 & \\ s & c_1 & & \\ 1 & d_1 & & \end{array}$$

where

$$b_1 = \frac{-2 \times (-13) - 14 \times 1}{-2} = -6 \tag{3.39}$$

$$b_2 = \frac{14 \times 24 - 0 \times 13}{14} = 24 \tag{3.40}$$

$$c_1 = \frac{-6 \times 14 - 24 \times (-2)}{-6} = 6 \tag{3.41}$$

$$d_1 = \frac{6 \times 24 - 0 \times (-6)}{6} = 24 \tag{3.42}$$

All other elements are zeros.

We now consider the leftmost column, $[1 \;\; -2 \;\; -6 \;\; 6 \;\; 24]^T$. As there are two changes of signal (from 1 to −2 and from −6 to 6), there are two unstable poles. Thus, the correct answer is option **B)**.

Problem 3.2 Consider a system with characteristic equation $Q(s) = s^4 + 4s^3 + 6s^2 + 4s + 5$. Using the Routh–Hurwitz criterion we conclude that the number of roots n in the right half s plane is:

A) Zero
B) One
C) Two
D) Three.

Resolution The correct answer is option **A)**: zero, as can be seen from the Routh–Hurwitz table,

s^4	1	6	5
s^3	4	4	
s^2	5	4	
s	ϵ		
1	5		

which, depending on the sign of $\epsilon \approx 0$, may or may not have changes of sign in the first column, which means that there is a pair of imaginary roots, but none with positive real part.

3.2.2 *Transient response*

Problem 3.3 Let $U(s) = \mathscr{L}[u(t)]$, $Y(s) = \mathscr{L}[y(t)]$. Consider the time response $y(t)$ of a second-order system $\dfrac{Y(s)}{U(s)} = \dfrac{10}{s^2 + s + 10}$, when the input is a unit step $u(t) = 1$, $t \geq 0$. The peak time t_p and corresponding peak value of the time response $y(t_p)$ are:

A) $t_p = 1.006$ s and $y(t_p) = 1.605$
B) $t_p = 1.600$ s and $y(t_p) = 1.806$
C) $t_p = 1.905$ s and $y(t_p) = 1.725$
D) None of the above.

Resolution We have $\omega_n = \sqrt{10} = 3.16$ rad/s and $2\zeta\omega_n = 1 \Rightarrow \zeta = \frac{1}{2\sqrt{10}} = 0.158$; so we get $t_p = \frac{\pi}{\omega_n\sqrt{1-\zeta^2}} = 1.006$ s and $y(t_p) = 1 + e^{-\frac{\zeta\pi}{\sqrt{1-\zeta^2}}} = 1.605$. Thus, the correct answer is option **A)**.

Problem 3.4 Consider a second-order system described by the transfer function $\dfrac{Y(s)}{U(s)} = \dfrac{\omega_n^2}{s^2 + 2\zeta\omega_n s + \omega_n^2}$ where ζ and ω_n are the damping coefficient and the undamped natural frequency, respectively. The system time response $y(t)$ for a unit step $u(t) = 1$, $t \geq 0$, is represented in Figure 3.13. Therefore, we can conclude that:

A) $\omega_n = 7$ rad/s
B) $\omega_n = 9$ rad/s
C) $\omega_n = 8$ rad/s
D) None of the above.

Resolution From $y(t_p) = 1 + e^{-\frac{\zeta\pi}{\sqrt{1-\zeta^2}}}$ we get $\zeta = \frac{1}{2}$; from here and from $t_p = \frac{\pi}{\omega_n\sqrt{1-\zeta^2}}$ s we get $\omega_n = 9$ rad/s. Thus, the correct answer is option **B)**.

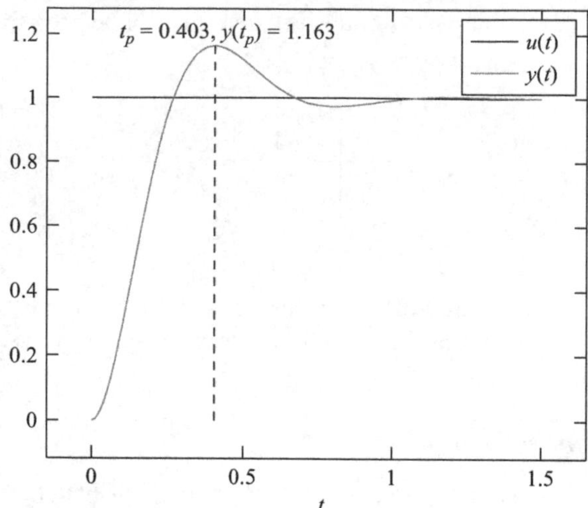

Figure 3.13 Unit step response of Problem 3.4

Problem 3.5 Consider a second-order system and the output time response $y(t)$ represented in Figure 3.14 for a unit step $u(t) = 1,\ t \geq 0$. The system is described by the transfer function $G(s) = \frac{Y(s)}{U(s)}$:

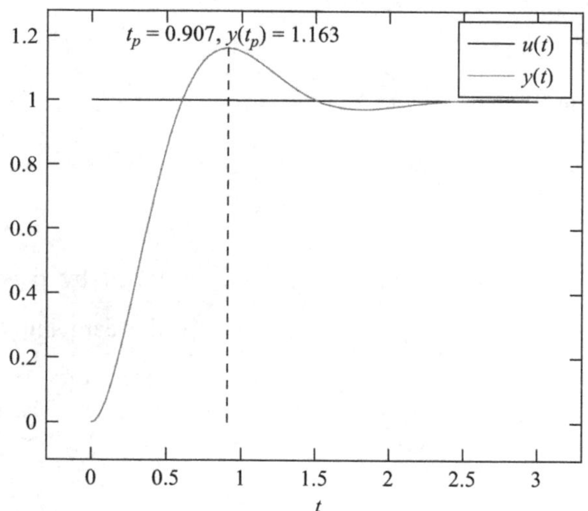

Figure 3.14 Unit step response of Problem 3.5

A) $G(s) = \dfrac{16}{s^2 + 4s + 16}$

B) $G(s) = \dfrac{16}{s^2 + 4s + 4}$

C) $G(s) = \dfrac{16}{s^2 + 8s + 16}$

D) $G(s) = \dfrac{16}{s^2 + 4s + 8}$.

Resolution The correct answer is option **A)**.

3.2.3 Steady-state errors

Problem 3.6 Let $R(s) = \mathscr{L}[r(t)]$ and $Y(s) = \mathscr{L}[y(t)]$, where $r(t)$ and $y(t)$ are the input and output of a closed-loop system with transfer function $\dfrac{Y(s)}{R(s)} = \dfrac{G(s)}{1 + G(s)H(s)}$, where $G(s) = \dfrac{1}{s\left(s^2 + s + 10\right)}$ and $H(s) = 1$. The error in the time response is given by $e(t) = r(t) - y(t)$. For a unit ramp signal input $r(t) = t$, $t \geq 0$, the steady-state error e_{ss} becomes:

A) $e_{ss} = 0$
B) $e_{ss} = 10$
C) $e_{ss} = \frac{1}{10}$
D) None of the above.

Resolution Since $H(s) = 1$, the error is given by $E(s) = R(s)\frac{1}{1+G(s)}$ and thus, by the final value theorem, the steady-state error is $e_{ss} = \lim_{t \to +\infty} e(t) = \lim_{s \to 0} sE(s) = \lim_{s \to 0} s\dfrac{1}{s^2}\dfrac{1}{1 + \frac{1}{s(s^2+s+10)}} = \lim_{s \to 0} \dfrac{1}{s}\dfrac{s(s^2 + s + 10)}{s(s^2 + s + 10) + 1} = 10$. Thus, the correct answer is option **B)**.

3.3 Proposed exercises

3.3.1 Routh–Hurwitz criterion

Exercise 3.1 Consider a closed-loop system with characteristic equation $s^4 + 2s^3 + 3s^2 + 4s + K = 0$, $K \in \mathbb{R}$. The system is stable for a gain K in the range:

A) $0 < K < 1$
B) $0 < K < 3$
C) $0 < K < 2$
D) None of the above.

Exercise 3.2 Consider a closed-loop system with characteristic equation $s^3 + Ks^2 + 5s + 10 = 0$, $K \in \mathbb{R}$. The system is stable for a gain K in the range:

A) $0 < K < 1$
B) $1 < K < 2$
C) $0 < K < 2$
D) None of the above.

Exercise 3.3 Consider a closed-loop system with characteristic equation $s^3 + 34.5s^2 + 7\,500s + 7\,500K = 0$, $K \in \mathbb{R}$. The system is stable for a gain K in the range:

A) $0 < K < 34.5$
B) $-7500 < K < -34.5$
C) $34.5 < K < 7500$
D) None of the above.

Exercise 3.4 Consider a closed-loop system with characteristic equation $s^3 + 2s^2 + Ks + 1 = 0$, $K \in \mathbb{R}$. The closed-loop system presents all the roots in the left half-plane if:

A) $0 < K < 1$
B) $1 < K < 2$
C) $0 < K < 2$
D) None of the above.

Exercise 3.5 Consider a system with characteristic equation $2s^3 - 3s^2 + s + 4 = 0$. The closed-loop system presents:

A) Three poles in the right half-plane
B) Two poles in the right half-plane
C) One pole in the right half-plane
D) Zero poles in the right half-plane.

Exercise 3.6 Consider a system with characteristic equation $s^4 + 14s^3 + 12s^2 - 4s - 13 = 0$. The closed-loop system presents:

A) Two poles in the right half-plane
B) Zero poles in the right half-plane
C) One pole in the right half-plane
D) None of the above.

Exercise 3.7 Consider a system with characteristic equation $s^4 + 4s^3 + 6s^2 + 4s + 1 = 0$. The closed-loop system presents:

A) Two poles in the right half-plane
B) Zero poles in the right half-plane

C) One pole in the right half-plane
D) None of the above.

Exercise 3.8 Consider the feedback control system represented in Figure 3.15. Is there any value of gain K leading to the time response $y(t)$ for a unit step input signal $u(t) = 1, t \geq 0$?

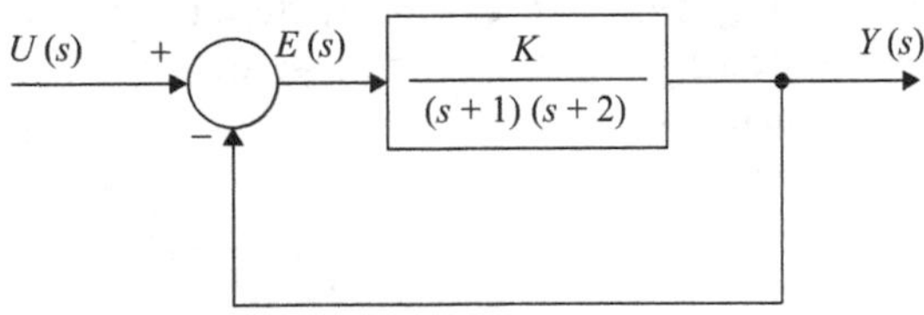

(a) System block diagram

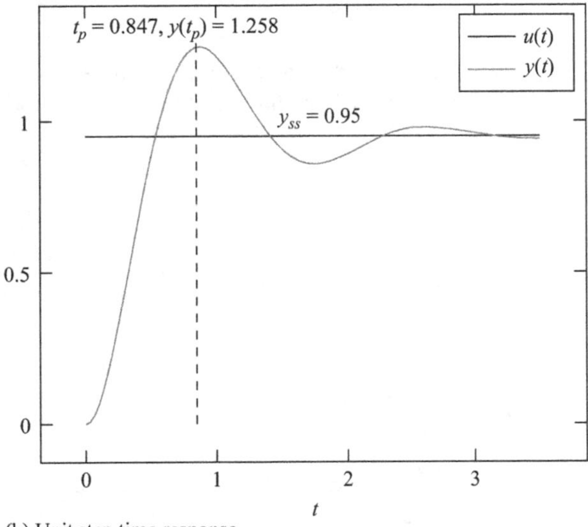

(b) Unit step time response

Figure 3.15 Control system of Exercise 3.8

Exercise 3.9 Consider a system with characteristic equation $Q(s) = 2s^5 + 4s^4 + 4s^3 + 8s^2 + 2s + 2$. Using the Routh–Hurwitz criterion we conclude that the number of roots n in the right half s plane is:

A) Zero
B) One
C) Two
D) Three.

Exercise 3.10 Consider a system with characteristic equation $Q(s) = s^3 + 3s^2 + 10s + K = 0$, $K \in \mathbb{R}$. The system is stable for gain K such that:

A) $0 < K < 30$
B) $0 < K < 3$
C) $0 < K < 10$
D) None of the above.

Exercise 3.11 Consider a system with characteristic equation $Q(s) = s^4 + 2s^3 + 4s^2 + 2s + 6$. Using the Routh–Hurwitz criterion we conclude that the number of roots n in the right half s plane is:

A) Zero
B) One
C) Two
D) Three.

Exercise 3.12 Consider a system with characteristic equation $Q(s) = s^3 - s^2 + 3s + 5$. Using the Routh–Hurwitz criterion we conclude that the number of roots n in the right half s plane is:

A) Zero
B) One
C) Two
D) Three.

Exercise 3.13 Consider the closed-loop system shown in the Figure 3.16. Find the range of values for gain K such that the system is stable.

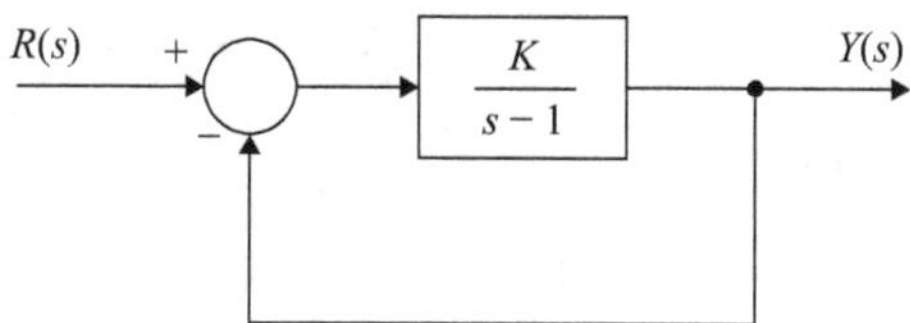

Figure 3.16 Closed loop of Exercise 3.13

Exercise 3.14 Consider a system with characteristic equation $Q(s) = s^3 + Ks^2 + 10s + 5 = 0$, $K \in \mathbb{R}$. Therefore, the system is stable for values of the gain K:

A) $0 < K < 1$
B) $0 < K < 5$
C) $0 < K < 2$
D) None of the above.

Exercise 3.15 Consider a system with characteristic polynomial $s^4 + s^3 + 6s^2 + 26s + 20$. Using the Routh–Hurwitz criterion we conclude that the number of roots n in the right half s plane is:

A) Zero
B) One
C) Two
D) Three.

Exercise 3.16 Consider a system with characteristic polynomial $Q(s) = s^5 + 3s^4 + s^2 + 1$. Using the Routh–Hurwitz criterion we conclude that the number of roots n in the right half s plane is:

A) Zero
B) One
C) Two
D) Three.

Exercise 3.17 Consider a system with characteristic polynomial $s^4 + s^3 + 2s^2 + s + K$, $K \in \mathbb{R}$. Using the Routh–Hurwitz criterion we conclude that the system is stable for:

A) $0 < K < 1$
B) $K > 1$
C) $1 < K < 2$
D) $K > 0$.

Exercise 3.18 A system has characteristic equation $Q(s) = s^3 + 2s^2 + (K_1 + 1)\, s + K_2$, $K_1, K_2 \in \mathbb{R}$. Using the Routh–Hurwitz criterion we conclude that the system is stable for:

A) $K_2 > 1 \wedge K_2 < 2K_1$
B) $K_2 > 0 \wedge K_2 < 2K_1 + 2$
C) $K_1 > 2 \wedge K_2 < K_1$
D) $K_1 > K_2 \wedge K_2 < 2$.

Exercise 3.19 Consider a closed-loop system with characteristic equation $Q(s) = s^4 + 6s^3 + Ks^2 + 5s + 3 = 0$, $K \in \mathbb{R}$. Using the Routh–Hurwitz criterion we conclude that the system is stable for:

A) $K > 0.83$
B) $K > 4.43$
C) $0.83 < K < 4.43$
D) None of the above.

Exercise 3.20 Consider a closed-loop system with characteristic equation $Q(s) = 2s^5 + 4s^4 + 4s^3 + 8s^2 + 2s + 2$. Using the Routh–Hurwitz criterion we conclude that the number n of roots in the right half-plane is:

A) $n = 0$
B) $n = 1$
C) $n = 2$
D) The Routh–Hurwitz criterion cannot be used.

Exercise 3.21 Find the range of values of K for which the closed-loop system in Figure 3.17 is stable, when:

1. $H(s) = 1$,
2. $H(s) = s(s + 1)$.

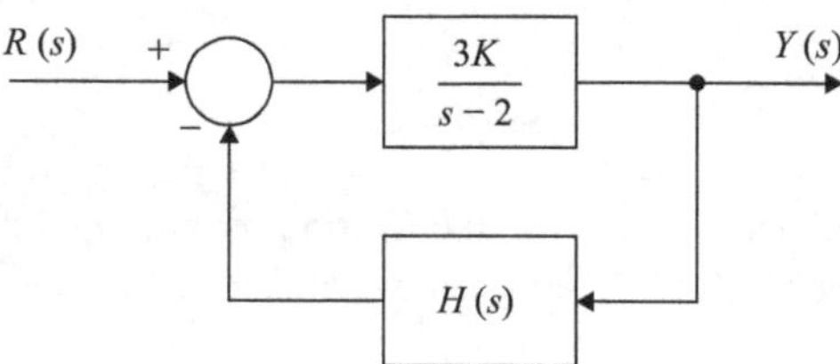

Figure 3.17 Closed loop of Exercise 3.21

3.3.2 Transient response

Exercise 3.22 Consider a second-order system with damping coefficient $\zeta = 0.55$ and undamped natural frequency $\omega_n = 10$ rad/s. The unit step response has a rise time t_r, a settling time t_s and an overshoot M_p given by:

A) $t_r = 0.727$ s
B) $t_r = 0.126$ s
C) $t_r = 0.192$ s
D) None of the above.

E) $t_s = 0.727$ s
F) $t_s = 0.126$ s
G) $t_s = 0.192$ s
H) None of the above.

I) $M_p = 0.727$
J) $M_p = 0.126$
K) $M_p = 0.192$
L) None of the above.

Exercise 3.23 The peak time, t_p, and the overshoot, $y(t_p)$, in the response to a unit step input of a second-order system with transfer function $\frac{Y(s)}{U(s)} = \frac{18}{s^2 + 2s + 9}$ are:

A) $t_p = 1.238$ s, $y(t_p) = 1.906$
B) $t_p = 1.305$ s, $y(t_p) = 2.125$
C) $t_p = 1.111$ s, $y(t_p) = 2.659$
D) None of the above.

Exercise 3.24 Consider the output time response $y(t)$ of a second-order system with transfer function $\frac{Y(s)}{U(s)} = \frac{2}{s^2 + 2s + 16}$ for a unit step input signal $u(t) = 1,\ t \geq 0$. Find the peak time t_p, the overshoot of the output $y(t_p)$ and the output steady-state value $y_{ss} = y(t \to \infty)$.

Exercise 3.25 Consider the output time response $y(t)$ of a second-order system $\frac{Y(s)}{U(s)} = \frac{10}{s^2 + s + 4}$ for a unit step input signal $u(t) = 1, t \geq 0$.

1. The damping coefficient ζ and the undamped natural frequency ω_n are:
   A) $\zeta = 1.0,\ \omega_n = 4.0$ rad/s
   B) $\zeta = 0.1,\ \omega_n = 0.63$ rad/s
   C) $\zeta = 0.25,\ \omega_n = 2.0$ rad/s
   D) $\zeta = 0.5,\ \omega_n = 0.4$ rad/s.
2. The peak time t_p and the output overshoot $y(t_p)$ are:
   E) $t_p = 0.972$ s, $y(t_p) = 1.023$
   F) $t_p = 0.556$ s, $y(t_p) = 1.982$
   G) $t_p = 1.622$ s, $y(t_p) = 3.611$
   H) $t_p = 2.016$ s, $y(t_p) = 0.806$.

Exercise 3.26 Consider the output time response $y(t)$ of a second-order system $\frac{Y(s)}{U(s)} = \frac{K\omega_n^2}{s^2 + 2\zeta\omega_n s + \omega_n^2}$ for a unit step input signal $u(t) = 1,\ t \geq 0$. If the peak time is $t_p = 0.5$ s, output overshoot of the time response is $y(t_p) = 1.76$ and the steady-state output is $y_{ss} = \lim_{t\to\infty} y(t) = 1.1$, then the system's damping coefficient ζ and undamped natural frequency ω_n are:

A) $K = 1.0,\ \omega_n = 6.666$ rad/s, $\zeta = 0.260$
B) $K = 1.1,\ \omega_n = 6.366$ rad/s, $\zeta = 0.160$
C) $K = 1.2,\ \omega_n = 5.566$ rad/s, $\zeta = 0.166$
D) $K = 1.3,\ \omega_n = 3.666$ rad/s, $\zeta = 0.666$.

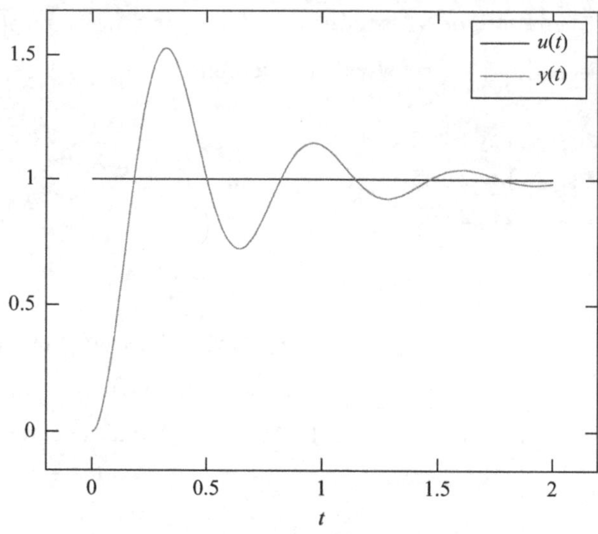

Figure 3.18 Time response of Exercise 3.27

Exercise 3.27 Consider a second-order system with damping coefficient ζ and the undamped natural frequency ω_n. Figure 3.18 depicts the time response for a unit step input signal $u(t) = 1,\ t \geq 0$, where we observe a rise time $t_r = 0.132$ s and a 2% settling time $t_s = 2.0$ s. Therefore, we conclude that:

A) $\zeta = 0.2,\ \omega_n = 5$ rad/s
B) $\zeta = 0.2,\ \omega_n = 10$ rad/s
C) $\zeta = 0.1,\ \omega_n = 5$ rad/s
D) $\zeta = 0.1,\ \omega_n = 10$ rad/s.

Exercise 3.28 Consider a unit feedback control system with transfer function in the direct loop $\dfrac{Y(s)}{U(s)} = \dfrac{\omega_n^2}{s^2 + 2\zeta\omega_n s + \omega_n^2}$ where ζ and ω_n denote the damping coefficient and the undamped natural frequency, respectively. Figure 3.19 depicts the time response $y(t)$ for a unit step input signal $u(t) = 1,\ t \geq 0$. If $t_p = 0.4117$ s and $y(t_p) = 1.3723$ then we conclude that:

A) $\zeta = 0.20$
B) $\zeta = 0.25$
C) $\zeta = 0.30$
D) $\zeta = 0.35$.

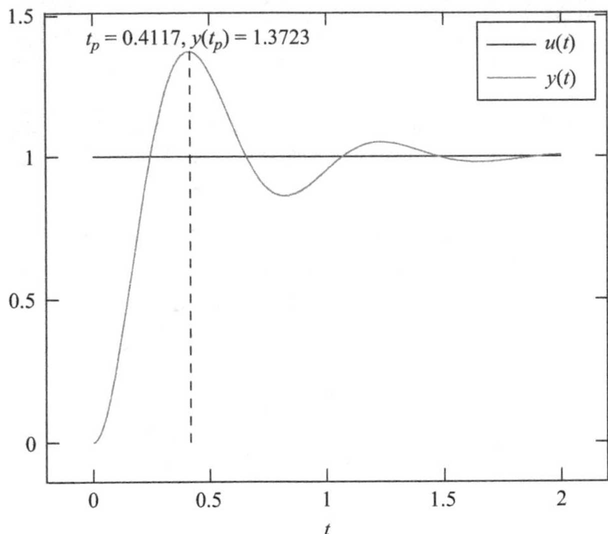

Figure 3.19 Time response of Exercise 3.28

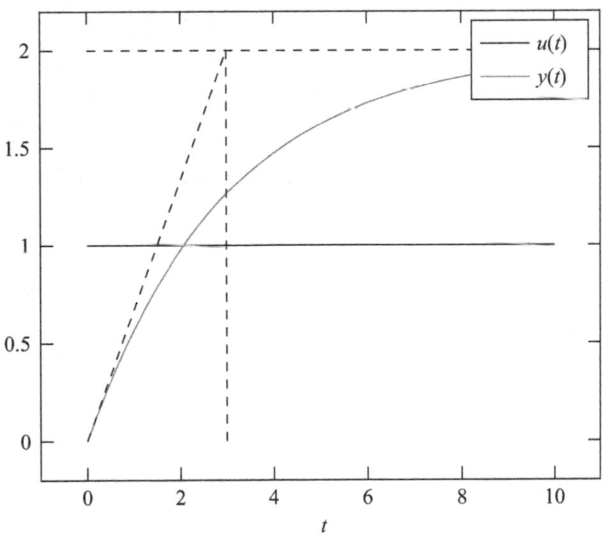

Figure 3.20 Time response of Exercise 3.29

Exercise 3.29 Consider a first-order system with transfer function $\dfrac{Y(s)}{U(s)} = \dfrac{K}{Ts+1}$, $K > 0$ and $T > 0$. The time response $y(t)$ for a step input signal $u(t) = 1$, $t \geq 0$, is shown in Figure 3.20. Therefore, we conclude that:

A) $K = 2, T = 3$ s
B) $K = 3, T = 3$ s

C) $K = 3, T = 2$ s
D) None of the above.

Exercise 3.30 Consider the mechanical system depicted in Figure 3.21. For the input force $f(t) = 8.9$ N, $t \geq 0$, the output time response $x(t)$ (i.e., the displacement) has the evolution in Figure 3.22.

1. Find the values of mass M, viscous friction coefficient B and spring stiffness K.
2. Suppose we want the same steady-state regime and the same settling time, but a maximum overshoot of 0.15%. What should the values of M, B and K be then?

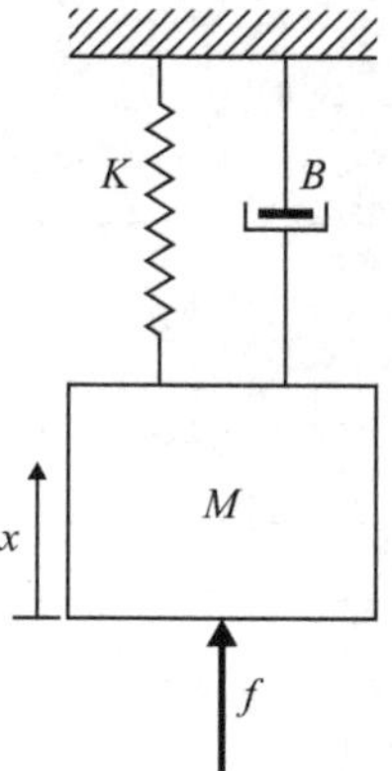

Figure 3.21 Mechanical system of Exercise 3.30

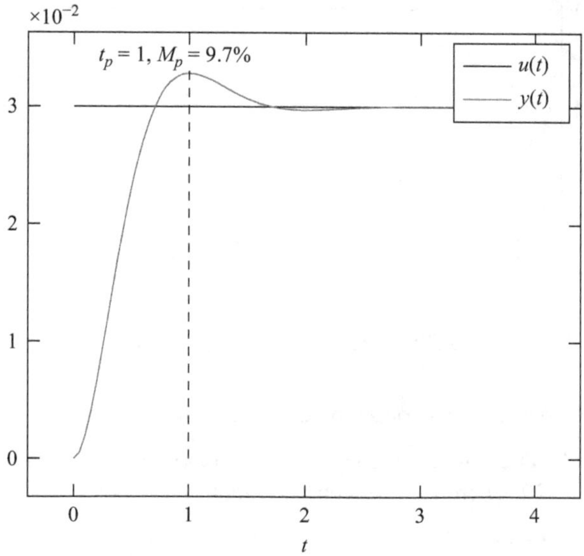

Figure 3.22 Time response of the mechanical system of Exercise 3.30

Exercise 3.31 Consider a system with transfer function $\frac{Y(s)}{U(s)} = \frac{b^2}{(s+a)^2+b^2}$. The damping coefficient ζ is:

A) $\zeta = \frac{2a}{a^2+b^2}$
B) $\zeta = a^2 + b^2$
C) $\zeta = 2a$
D) None of the above.

Exercise 3.32 Consider a second-order system with transfer function $\frac{Y(s)}{U(s)} = \frac{25}{s^2+4s+25}$. Its time response $y(t)$ for a unit step input $u(t) = 1$, $t \geq 0$ has a rise time t_r given by:

A) $t_r = 0.332$ s
B) $t_r = 0.123$ s
C) $t_r = 0.233$ s
D) None of the above.

Exercise 3.33 Consider a second-order system with transfer function $\frac{Y(s)}{U(s)} = \frac{\omega_n^2}{s^2+2\zeta\omega_n s+\omega_n^2}$. Figure 3.23 shows its time response $y(t)$ for a unit step input $u(t) = 1$, $t \geq 0$. Therefore:

A) $\zeta = 1.0$ and $\omega_n = 10$ rad/s
B) $\zeta = 0.5$ and $\omega_n = 15$ rad/s
C) $\zeta = 0.707$ and $\omega_n = 5$ rad/s
D) None of the above.

Exercise 3.34 Figure 3.24 shows the time response $y(t)$ of a second-order system for a unit step input $u(t) = 1$, $t \geq 0$. Find its transfer function.

Exercise 3.35 Consider a second-order system with transfer function $\frac{Y(s)}{U(s)} = \frac{\omega_n^2}{s^2+2\zeta\omega_n s+\omega_n^2}$ where ζ and ω_n denote the damping coefficient and the undamped natural frequency, respectively. The time response $y(t)$ of the system for a unit step input signal $u(t) = 1$, $t \geq 0$, is depicted in Figure 3.25. We can conclude that:

A) $\omega_n = 4$ rad/s
B) $\omega_n = 2$ rad/s
C) $\omega_n = 3$ rad/s
D) None of the above.

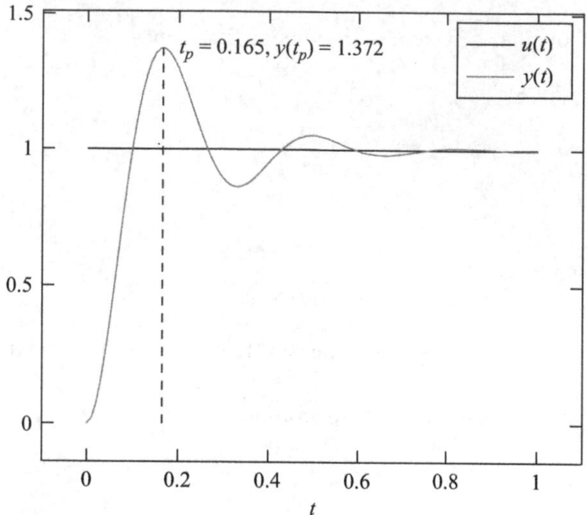

Figure 3.23 Time response of Exercise 3.33

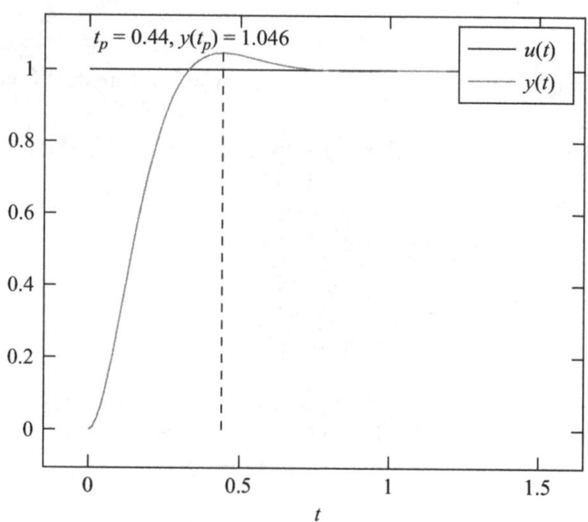

Figure 3.24 Time response of Exercise 3.34

Exercise 3.36 Consider a second-order system with transfer function $\frac{Y(s)}{U(s)} = \frac{\omega_n^2}{s^2 + 2\zeta\omega_n s + \omega_n^2}$. Figure 3.26 shows its time response $y(t)$ for a unit step input signal $u(t) = 1,\ t \geq 0$, with peak time $t_p = 1.622$ s and output peak value $y(t_p) = 1.444$. Find the values of the undamped natural frequency ω_n and the damping coefficient ζ.

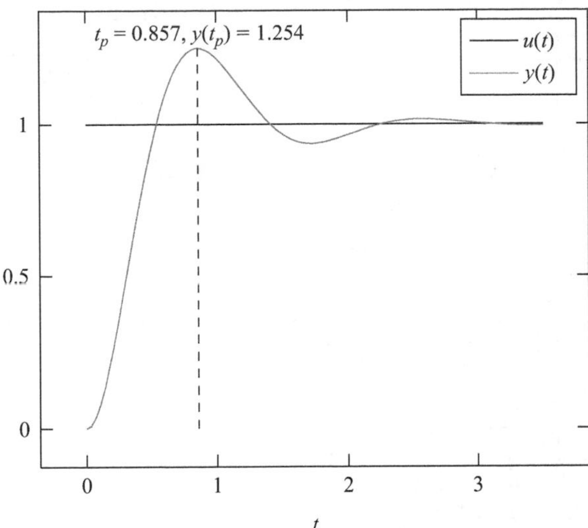

Figure 3.25 Time response of Exercise 3.35

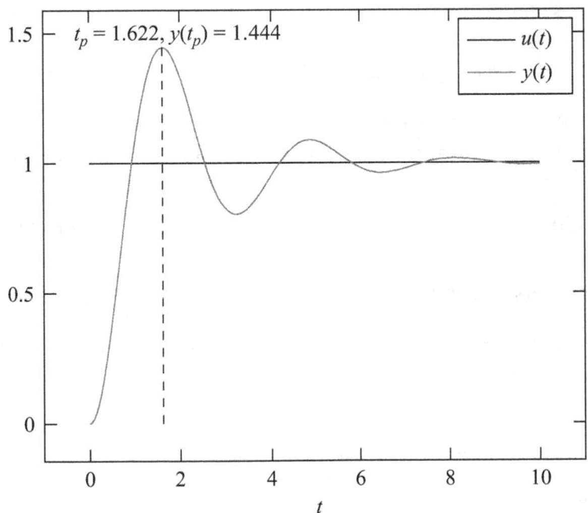

Figure 3.26 Time response of Exercise 3.36

Exercise 3.37 Consider a second-order system with transfer function $\dfrac{Y(s)}{U(s)} = \dfrac{\omega_n^2}{s^2 + 2\zeta\omega_n s + \omega_n^2}$. Figure 3.27 shows its time response $y(t)$ for a unit step input signal $u(t) = 1$, $t \geq 0$. Find the values of the undamped natural frequency ω_n and the damping coefficient ζ.

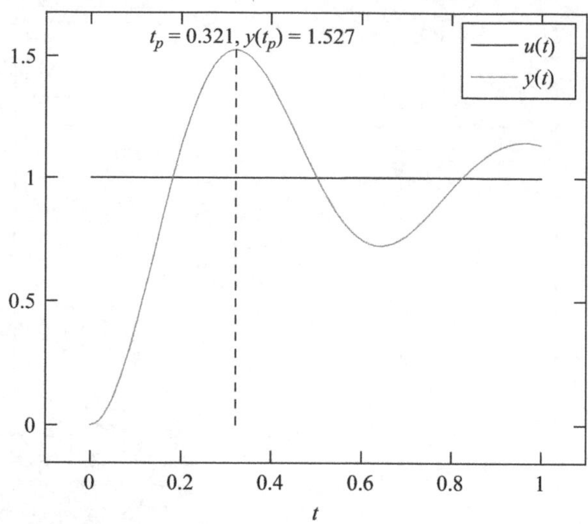

Figure 3.27 Time response of Exercise 3.37

Exercise 3.38 Consider a second-order system with transfer function $\frac{Y(s)}{U(s)} = \frac{\omega_n^2}{s^2 + 2\zeta\omega_n s + \omega_n^2}$. Its time response $y(t)$ for a unit step input signal $u(t) = 1$, $t \geq 0$ has a peak time of $t_p = 1.20$ s and an output peak value of $y(t_p) = 1.30$. Find the values of the undamped natural frequency ω_n and the damping coefficient ζ.

Exercise 3.39 Consider a second-order system with transfer function $\frac{Y(s)}{U(s)} = \frac{\omega_n^2}{s^2 + 2\zeta\omega_n s + \omega_n^2}$. Its frequency response $G(j\omega)$ has a resonance frequency of $\omega_r = 3.50$ rad/s and a peak gain of $M_r = 1.25$. Find the values of the undamped natural frequency ω_n and the damping coefficient ζ.

3.3.3 Steady-state errors

Exercise 3.40 Consider a closed-loop system with unit feedback and transfer function in the direct loop $G(s) = \frac{Y(s)}{E(s)} = \frac{8}{(s+2)(s+4)}$.

1. For the closed loop's unit step response, find:
   (a) the peak time t_p
   (b) the maximum overshoot $y(t_p)$
   (c) the rise time t_r
   (d) the settling time t_s
2. For a unit ramp input signal $u(t) = t$, $t \geq 0$, what is this plant's steady-state error $e_{ss} = \lim_{t \to +\infty} u(t) - y(t)$?

Exercise 3.41 The output time response $y(t)$ of a second-order system for a unit step input signal $u(t) = 1,\ t \geq 0$, is shown in Figure 3.28. Notice that the steady-state response is $y_{ss} = \lim_{t\to\infty} y(t) = 4.0$. Find the system's transfer function and its zeros and poles.

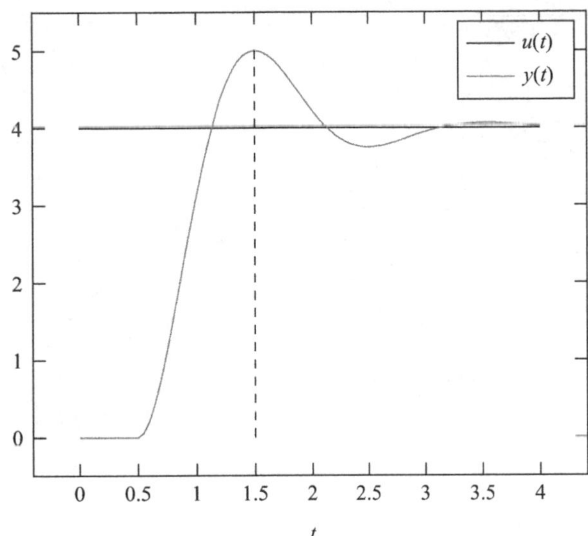

Figure 3.28 Time response of Exercise 3.41

Exercise 3.42 Consider a unit feedback control system with transfer function in the direct loop $G(s) = \dfrac{4}{s(s+2)}$.

1. For the closed loop's unit step response, find:
   (a) the peak time t_p
   (b) the maximum overshoot $y(t_p)$
   (c) the rise time t_r
   (d) the settling time t_s
2. For a unit ramp input signal $u(t) = t, t \geq 0$, what is this plant's steady-state error $e_{ss} = \lim_{t\to+\infty} u(t) - y(t)$?

Exercise 3.43 Consider a unit feedback control system with transfer function in the direct loop $G(s) = 10\dfrac{(s+2)}{(s+3)(s+5)}$. For a unit step input signal $u(t) = 1, t \geq 0$, the time response reveals a steady-state error e_{ss} given by:

A) $e_{ss} = \frac{3}{7}$
B) $e_{ss} = \frac{4}{3}$
C) $e_{ss} = 0$
D) None of the above.

Exercise 3.44 Consider a system with transfer function $Y(s)/R(s) = G(s)$. The time response $y(t)$ for a ramp input signal $r(t) = 10t$, $t \geq 0$ has a steady-state error $e_{ss} = 2$ as shown in Figure 3.29. Therefore, for $m \leq n$, we conclude that:

A) $G(s) = \dfrac{1}{2} \cdot \dfrac{1 + b_1 s + \cdots + b_m s^m}{1 + a_1 s + \cdots + a_n s^n}$

B) $G(s) = 5\dfrac{1 + b_1 s + \cdots + b_m s^m}{1 + a_1 s + \cdots + a_n s^n}$

C) $G(s) = \dfrac{1}{2} \cdot \dfrac{1 + b_1 s + \cdots + b_m s^m}{s(1 + a_1 s + \cdots + a_n s^n)}$

D) $G(s) = 5\dfrac{1 + b_1 s + \cdots + b_m s^m}{s(1 + a_1 s + \cdots + a_n s^n)}$.

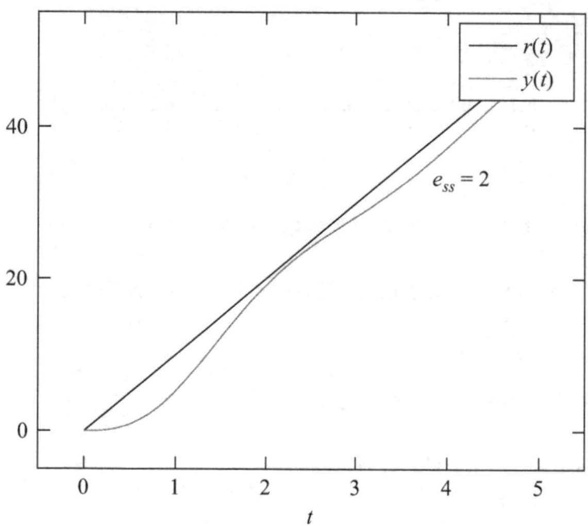

Figure 3.29 Time response of Exercise 3.44

Exercise 3.45 Consider a system with transfer function $Y(s)/U(s) = G(s)$. The output time response $y(t)$ for a step input $u(t) = 10$, $t \geq 0$, exhibits the steady-state error $e_{ss} = 1$ shown in Figure 3.30. Therefore, for $m \leq n$ we conclude that:

A) $G(s) = 10\dfrac{1 + b_1 s + \cdots + b_m s^m}{1 + a_1 s + \cdots + a_n s^n}$

B) $G(s) = 9\dfrac{1 + b_1 s + \cdots + b_m s^m}{1 + a_1 s + \cdots + a_n s^n}$

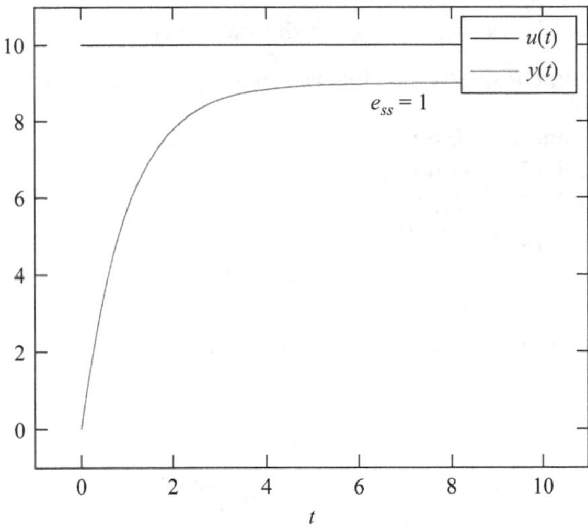

Figure 3.30 Time response of Exercise 3.45

C) $G(s) = 10\dfrac{1 + b_1 s + \cdots + b_m s^m}{s(1 + a_1 s + \cdots + a_n s^n)}$

D) $G(s) = 9\dfrac{1 + b_1 s + \cdots + b_m s^m}{s(1 + a_1 s + \cdots + a_n s^n)}$.

Exercise 3.46 Consider a system with open-loop transfer function $GH(s) = \dfrac{K(s+5)}{s(s+1)(s+2)}$, $K \in \mathbb{R}^+$.

1. The open-loop system is stable when:
   A) $0 < K < 3$
   B) $3 < K < 9$
   C) $0 < K < 5$
   D) None of the above.
2. When $K = 20$, the steady-state error e_{ss} of the closed-loop system for a unit step input signal is:
   E) $e_{ss} = \frac{2}{7}$
   F) $e_{ss} = \infty$
   G) $e_{ss} = \frac{1}{20}$
   H) $e_{ss} = 0$.

Exercise 3.47 Consider a system with transfer function $G(s) = \dfrac{Y(s)}{U(s)} = \dfrac{10}{(s+1)(s+5)}$. The steady-state time response $y(t)$ for the input signal $u(t) = 2\sin(2t)$ is:

A) $y(t) = 0.33\sin(2t - 10.00°)$
B) $y(t) = 1.50\sin(2t - 35.00°)$
C) $y(t) = 1.66\sin(2t - 85.24°)$
D) $y(t) = 3.33\sin(2t - 23.30°)$.

Exercise 3.48 Consider a second-order system with transfer function $\dfrac{Y(s)}{U(s)} = \dfrac{2}{s^2 + 2s + 4}$. Find:

1. the damping coefficient,
2. the undamped natural frequency,
3. the step-response response overshoot,
4. the step-response peak time,
5. the unit-step response steady-state output.

Exercise 3.49 The block diagram in Figure 3.31 models the pen in a plotter and the corresponding control system. Figure 3.32 shows the unit-step response.

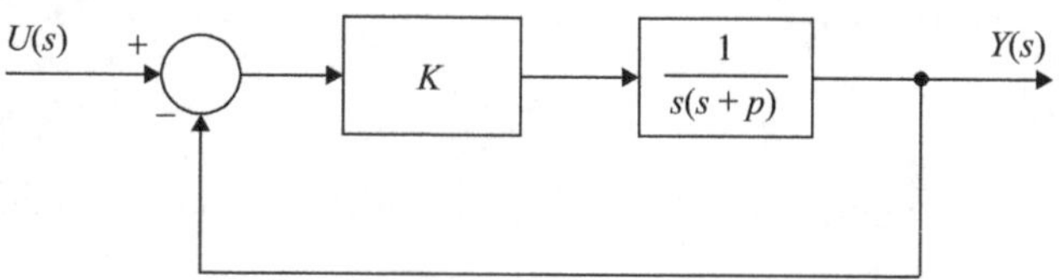

Figure 3.31 Block diagram of Exercise 3.49

1. Find from the step-response the system's damping coefficient ζ and natural frequency ω_n.
2. We want to improve the time response $y(t)$, so as to have a percent overshoot $M_p \leq 5\%$ and a 2% settling time $t_s < 0.1$ s. Find, if possible, a value of gain K for these specifications.

Exercise 3.50 Consider the closed-loop system in Figure 3.33, where $G(s) = \dfrac{s+2}{s(s+1)(s+4)}$, $H(s) = 1$, $K > 0$.

1. Find the range of values of K for which the closed loop is stable.
2. Find K so that the steady-state error for a unit ramp input is 10%.

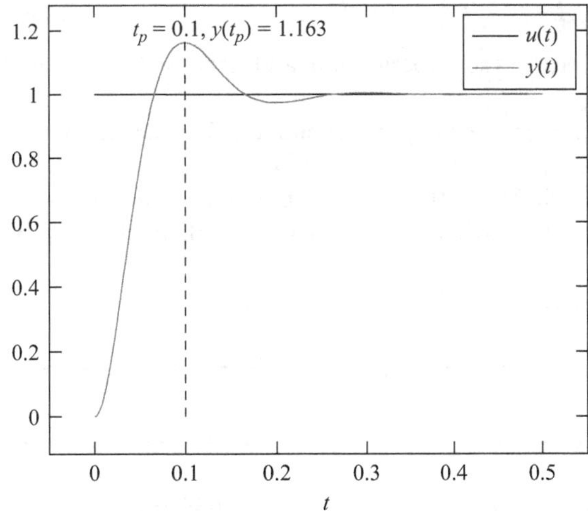

Figure 3.32 Time response of Exercise 3.49

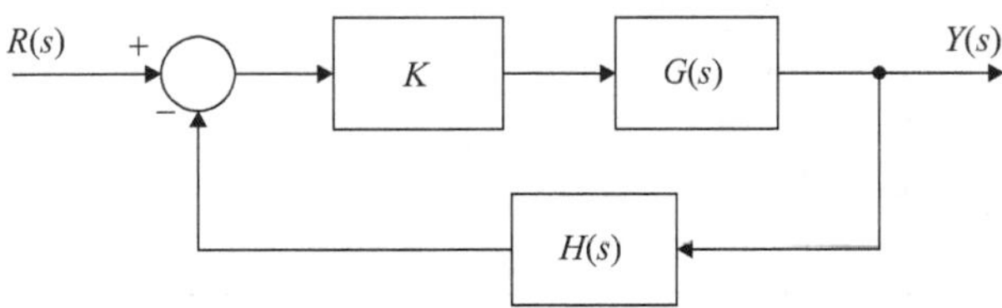

Figure 3.33 Block diagram of Exercise 3.50

Exercise 3.51 Let $y(t)$ be the output of a closed-loop system, with open-loop transfer function $G(s) = \dfrac{1}{s(s^2 + s + 10)}$ and unit feedback, for a unit step input $u(t) = 1$, $t \geq 0$. Let the error of the closed-loop be $e(t) = u(t) - y(t)$. The steady-state error $e_{ss} = \lim_{t \to \infty} e(t)$ is given by:

A) $e_{ss} = 0$
B) $e_{ss} = 10$
C) $e_{ss} = \frac{1}{10}$
D) None of the above.

3.4 Time response analysis using computer packages

This section presents several commands for handling the time response using the computer packages MATLAB®, SCILAB™ and OCTAVE©.

3.4.1 MATLAB

This subsection describes some basic commands that can be adopted with the package MATLAB.

For the Routh–Hurwitz criterion consider the Problem 3.2 with the characteristic equation $Q(s) = s^4 + 4s^3 + 6s^2 + 4s + 5 = 0$.

We start by defining the vector containing the polynomial coefficients `[1 4 6 4 5]`. The roots of the equation are then obtained by means of the command `root`.

The complete code is as follows.

```
roots([1 4 6 4 5])
```

MATLAB creates the following Command Window.

```
ans =

  -2.0000 + 1.0000i
  -2.0000 - 1.0000i
   0.0000 + 1.0000i
   0.0000 - 1.0000i
```

We are calculating the roots, but not using the Routh–Hurwitz criterion. For that we can use the script `Routh_Hurwitz_Stability.m` by Ramin Shamshiri.

Save the file and type `Routh_Hurwitz_Stability` in the MATLAB Command Window. We get the message:

`Input coefficients of characteristic equation, i.e: [an an-1 an-2 ... a0]` = After inserting the vector `[1 4 6 4 5]` we obtain:

```
----------------------------------------
Roots of characteristic equation is:

ans =

  -2.0000 + 1.0000i
  -2.0000 - 1.0000i
   0.0000 + 1.0000i
```

```
     0.0000 - 1.0000i

--------The Routh--Hurwitz array is:--------

m =

      1.0000     6.0000     5.0000
      4.0000     4.0000          0
      5.0000     5.0000          0
      0.0010          0          0
      5.0000          0          0

                ----> System is Stable <----
```

We consider now the time response and Problem 3.3. The transfer function $G(s) = \frac{10}{s^2+s+10}$ is represented by two arrays num and den. The unit step response is accomplished by means of the command `step`.

The complete code is as follows.

```
%%%% Time response %%%%
% Numerator and denominator of transfer function
num = [0 0 10];
den = [1 1 10];

% Unit step input
step(num,den)

% Formatting the chart
grid
title('Unit step time response of G(s) = 10/(s^2+s+10)')

% Rise time, settling time, and other step response
% characteristics
sys = tf(num,den);
S = stepinfo(sys)
```

The command `stepinfo` gives information about rise time, settling time, and other step response characteristics.

MATLAB creates the figure window represented in Figure 3.34 and the following Command Window.

```
S =

          RiseTime: 0.3738
      SettlingTime: 7.3148
       SettlingMin: 0.6347
       SettlingMax: 1.6045
         Overshoot: 60.4530
        Undershoot: 0
              Peak: 1.6045
          PeakTime: 1.0131
```

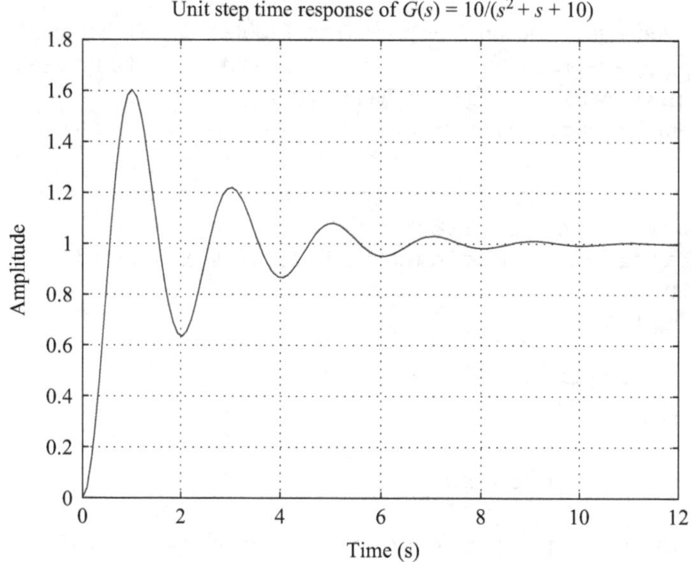

Figure 3.34 Unit step time response of $G(s) = \frac{10}{s^2+s+10}$ *using MATLAB*

To obtain the impulse response we just substitute command `step` by `impulse`.

To simulate the time response of a dynamic system with arbitrary inputs we can use command `lsim`.

3.4.2 SCILAB

This subsection describes some basic commands that can be adopted with the package SCILAB.

For the Routh–Hurwitz criterion consider Problem 3.2 with the characteristic equation $Q(s) = s^4 + 4s^3 + 6s^2 + 4s + 5 = 0$.

We start by defining the vector containing the polynomial coefficients `[1 4 6 4 5]`. The roots of the equation are then obtained by means of the command `root`.

The complete code is as follows.

```
roots([1 4 6 4 5])
```

SCILAB creates the following Console.

```
 ans  =

  - 2. + i
  - 2. - i
    1.874D-16 + i
    1.874D-16 - i
```

We consider now the time response and Problem 3.2. The transfer function $G(s) = \frac{10}{s^2+s+10}$ is represented by two arrays num and den. The unit step response is accomplished by means of the command `step`.

The complete code is as follows.

```
//// Time response ////
// Numerator and denominator of transfer function
num = poly([10 0 0],'s','coeff');
den = poly([10 1 1],'s','coeff');

//create a scilab continuous system LTI object
TF = syslin('c',num,den)

//time: linearly spaced vector
t = linspace(0,12,120);

// Unit step input
step_res = csim('step',t,TF)
plot(t,step_res)

// Formatting the chart
xgrid()
xtitle('Unit step time response of G(s) = 10/(s^2+s+10)')
xlabel('Time (seconds)')
ylabel('Amplitude')
```

SCILAB creates the figure window represented in Figure 3.35.

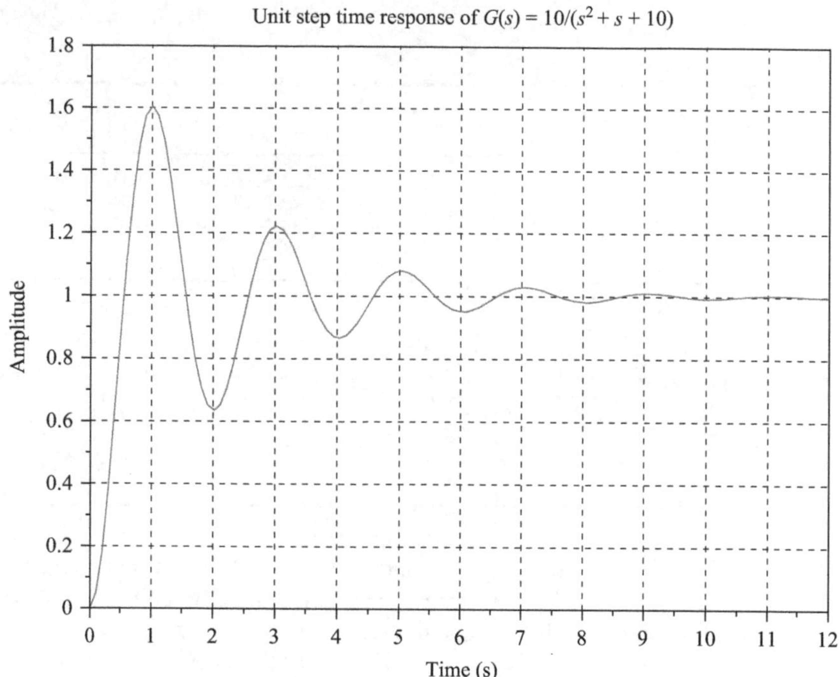

Figure 3.35 Unit step time response of $G(s) = \frac{10}{s^2+s+10}$ *using SCILAB*

To obtain the impulse response we just substitute '`step`' by '`imp`'.

3.4.3 OCTAVE

This subsection describes some basic commands that can be adopted with the package OCTAVE. It is required to load package `control`.

For the Routh–Hurwitz criterion consider the Problem 3.2 with the characteristic equation $Q(s) = s^4 + 4s^3 + 6s^2 + 4s + 5 = 0$.

We start by defining the vector containing the polynomial coefficients `[1 4 6 4 5]`. The roots of the equation are then obtained by means of the command `root`.

The complete code is as follows.

```
roots([1 4 6 4 5])
```

OCTAVE creates the following Command Window.

```
ans =

  -2.00000 + 1.00000i
  -2.00000 - 1.00000i
   0.00000 + 1.00000i
   0.00000 - 1.00000i
```

We are calculating the roots, but not using the Routh–Hurwitz criterion. For that we can use the script `Routh_Hurwitz_Stability.m` by Ramin Shamshiri.

Save the file and type `Routh_Hurwitz_Stability` in the OCTAVE Command Window. We get the message:
`Input coefficients of characteristic equation, i.e: [an an-1 an-2 ... a0]` = After inserting the vector `[1 4 6 4 5]` we obtain:

```
----------------------------------------
Roots of characteristic equation is:

ans =

  -2.0000 + 1.0000i
  -2.0000 - 1.0000i
   0.0000 + 1.0000i
   0.0000 - 1.0000i

--------The Routh--Hurwitz array is:--------

m =

    1.0000    6.0000    5.0000
    4.0000    4.0000         0
    5.0000    5.0000         0
    0.0010         0         0
    5.0000         0         0

          ----> System is Stable <----
```

We consider now the time response and Problem 3.2. The transfer function $G(s) = \frac{10}{s^2+s+10}$ is represented by two arrays `num` and `den` followed by the command `tf`. The unit step response (Figure 3.36) is accomplished by means of the command `step`.

The complete code is as follows.

```
%%%% Time response %%%%
% Numerator and denominator of transfer function
num = [0 0 10];
den = [1 1 10];

% Transfer function
G = tf(num,den)

% Unit step input
step(G)

% Formatting the chart
grid
title('Unit step time response of G(s) = 10/(s^2+s+10)')
```

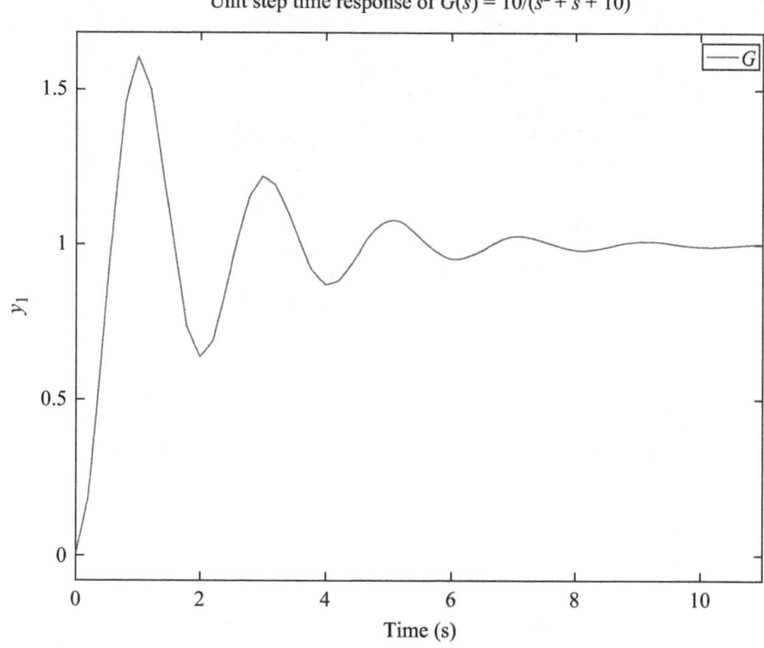

Figure 3.36 Unit step time response of $G(s) = \frac{10}{s^2+s+10}$ *using OCTAVE*

To obtain the impulse response we just substitute command `step` by `impulse`.

To simulate the time response of a dynamic system with arbitrary inputs we can use command `lsim`.

Chapter 4
Root-locus analysis

4.1 Fundamentals

We introduce the root-locus method as an important tool for the analysis of closed-loop feedback systems. In fact, the relative stability and the transient performance are directly related to the location of the closed-loop roots of the characteristic equation in the s plane [2–5,9]. We present the main properties of the root-locus, which are also used as practical sketching rules for quickly obtaining the root-locus chart by hand.

4.1.1 List of symbols

d	number of poles
$D(s)$	denominator of the open-loop transfer function
$G_{CL}(s)$	closed-loop transfer function
$G_{OL}(s)$	open-loop transfer function
K	gain
n	number of zeros
$N(s)$	numerator of the open-loop transfer function
p_i	pole of a transfer function
s	Laplace variable
s_0	test point in the s plane
z_i	zero of a transfer function
θ	angles of the asymptotes with respect to the real axis
σ	intersection of the asymptotes in the real axis
σ_B	breakaway or break-in point of a root-locus

4.1.2 Root-locus preliminaries

Root-locus analysis is a graphical method that shows how the poles of a closed-loop transfer function change with relation to a given system parameter [2,10–11]. Usually, the chosen parameter is a proportional gain, $K \geq 0$, included in a feedback closed-loop controlled system (Figure 4.1).

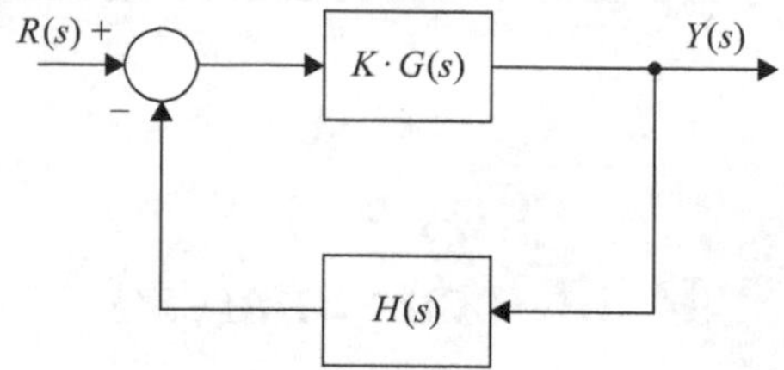

Figure 4.1 Feedback closed-loop controlled system

The open- and closed-loop transfer functions are given by $G_{OL}(s) = K \cdot G(s)H(s)$ and $G_{CL}(s) = K \cdot G(s)/[1 + K \cdot G(s)H(s)]$, respectively. The denominator of $G_{CL}(s)$ is the characteristic equation, and its roots are the system closed-loop poles.

Every point of the root-locus simultaneously satisfies the magnitude and the argument (angle) conditions, given by:

$$|K \cdot G(s)H(s)| = 1 \tag{4.1a}$$

$$\underline{/K \cdot G(s)H(s)} = (2l + 1) \cdot 180^\circ, \quad l = 0, \pm 1, \pm 2, \ldots \tag{4.1b}$$

4.1.2.1 Geometric interpretation

Given $G_{OL}(s)$ factored into the pole-zero form:

$$G_{OL}(s) = K \cdot G(s)H(s) = K \cdot \frac{N(s)}{D(s)} = K \cdot \frac{\prod_{i=1}^{n}(s - z_i)}{\prod_{i=1}^{d}(s - p_i)}, \quad d \geq n \tag{4.2}$$

where p_i and z_i denote the poles and zeros of the open-loop transfer function, respectively, then the magnitude and argument of $G_{OL}(s_0)$ for a given test point s_0 can be calculated by drawing vectors from the poles and zeros to the point s_0:

$$|G_{OL}(s_0)| = K \cdot \frac{\prod_{i=1}^{n}|(s_0 - z_i)|}{\prod_{i=1}^{d}|(s_0 - p_i)|} \tag{4.3a}$$

$$\underline{/G_{OL}(s_0)} = \sum_{i=1}^{n}\underline{/(s_0 - z_i)} - \sum_{i=1}^{d}\underline{/(s_0 - p_i)} \tag{4.3b}$$

The root-locus is a classical and powerful tool for the dynamical analysis and design of integer-order LTI systems [12–15]. Nowadays, there are efficient numerical algorithms, implemented in several software packages (e.g., MATLAB®, OCTAVE©, SCILAB™) [16–18] that take advantage of the powerful digital processors of modern computers to perform root-locus analysis. However, the ability to quickly sketch root-locus by hand is important in making fundamental decisions early in the design process.

4.1.3 Root-locus practical sketching rules ($K \geq 0$)

The root-locus is always symmetrical with respect to the real axis. For $K=0$ the closed- and open-loop poles are the same. As K increases from zero to infinity, the loci of the closed-loop poles originate from the open-loop poles and terminate at the open-loop zeros (finite zeros or zeros at infinity). If the number of open-loop poles, d, is equal, or greater than the number of open-loop zeros, n, then the number or loci, or number of branches of the root-locus, is equal to d.

Before applying the sketching rules, we should express the characteristic equation as:

$$1 + K\frac{(s - z_1)(s - z_2)\cdots(s - z_n)}{(s - p_1)(s - p_2)\cdots(s - p_d)} = 0 \tag{4.4}$$

The main steps for sketching the root-locus are as follows [2,9].

1. **Locate the open-loop poles and zeros in the complex plane**. Given the characteristic equation, the locations of the open-loop poles are indicated by crosses. The locations of the open-loop zeros are indicated by circles.
2. **Determine the root loci on the real axis**. Choose a test point in the real axis. If the total number of real poles and real zeros to the right of the test point is *odd*, then the test point lies on a root-locus.
3. **Determine the asymptotes of root loci**. The root loci for values of s located far from the origin of the complex plane approach a set of straight-line asymptotes. A root-locus branch may lie on one side, or may cross the corresponding asymptote. The angles of the asymptotes with respect to the real axis are given by:

$$\theta = 180^\circ\frac{(2l + 1)}{d - n}, \quad l = 0, \pm 1, \pm 2, \ldots \tag{4.5}$$

All asymptotes intersect at one point on the real axis given by:

$$\sigma = \frac{\sum_{i=1}^{d} p_i - \sum_{i=1}^{n} z_i}{d - n} \tag{4.6}$$

4. **Calculate the breakaway and break-in points**. Breakaway and break-in points are points on the real axis where two or more branches of the root-locus depart from, or arrive at, the real axis. If a root-locus lies between two adjacent open-loop poles on the real axis, then there exists at least one breakaway point between the two poles. Similarly, if the root-locus lies between two adjacent zeros, then there exists at least one break-in point between the two zeros.

   The breakaway or break-in points correspond to maximums or minimums of the function $K(s) = -\frac{1}{G(s)H(s)}$, for $s = \sigma_B$, meaning that they can be calculated by solving:

$$\frac{dK(\sigma_B)}{d\sigma_B} = 0 \tag{4.7}$$

Alternatively, σ_B can be determined by solving the equation:

$$\sum_{i=1}^{d} \frac{1}{\sigma_B + p_i} = \sum_{i=1}^{n} \frac{1}{\sigma_B + z_i} \tag{4.8}$$

5. **Determine departure and arrival angles.** The angle of departure from a complex pole and the angle of arrival at a complex zero correspond to the directions of the root loci near the complex poles and zeros, respectively. These angles are determined by using the argument condition. The angle of the root-locus departure from a pole (arrival to a zero) is the deference between the net angle due to all other poles and zeros and the criterion $(2l+1)\cdot 180°$, $l = 0, \pm 1, \pm 2, \ldots$.

   For a pole (zero), p_k (z_k), we consider a test point, s_0, close to p_k (z_k) and calculate:

$$\angle(s_0 - p_k) = (2l+1)\cdot 180° + \sum_{i} \angle(s_0 - z_i) - \sum_{i \neq k} \angle(s_0 - p_i) \tag{4.9a}$$

$$\angle(s_0 - z_k) = (2l+1)\cdot 180° + \sum_{i \neq k} \angle(s_0 - z_i) - \sum_{i} \angle(s_0 - p_i) \tag{4.9b}$$

6. **Find the points where the root loci intersect the imaginary axis.** The points where the root loci cross the imaginary axis are determined by making $s = j\omega$ in the characteristic equation, and solving for ω and K. These values give the frequencies at which root loci cross the imaginary axis, and the corresponding gain, respectively.

   Alternatively, we can use the Routh–Hurwitz criterion as follows. First, we find the gain at the $j\omega$-axis crossing points by forcing a row of zeros in the Routh–Hurwitz table. Secondly, we go back one row to the even polynomial equation and solve for the roots. This yields the frequency at the imaginary axis crossing.

4.1.4 Root-locus practical sketching rules ($K \leq 0$)

If the characteristic equation cannot be expressed as (4.4), but rather as

$$1 - K\frac{(s - z_1)(s - z_2)\cdots(s - z_n)}{(s - p_1)(s - p_2)\cdots(s - p_d)} = 0 \tag{4.10}$$

then the rules 2, 3 and 5 above are modified as follows:

2. **Determine the root loci on the real axis.** Choose a test point in the real axis. If the total number of real poles and real zeros to the right of the test point is *even*, then the test point lies on a root-locus.
3. **Determine the asymptotes of root loci.** The angles of the asymptotes with respect to the real axis are given by:

$$\theta = 360° \frac{(2l+1)}{d-n}, \quad l = 0, \pm 1, \pm 2, \ldots \tag{4.11}$$

5. **Determine departure and arrival angles**. The angle of departure from a complex pole and the angle of arrival at a complex zero are given by:

$$\angle(s_0 - p_k) = (2l + 1) \cdot 360° + \sum_i \angle(s_0 - z_i) - \sum_{i \neq k} \angle(s_0 - p_i) \tag{4.12a}$$

$$\angle(s_0 - z_k) = (2l + 1) \cdot 360° + \sum_{i \neq k} \angle(s_0 - z_i) - \sum_i \angle(s_0 - p_i) \tag{4.12b}$$

4.2 Solved problems

Problem 4.1 Consider the plants with the following open-loop transfer functions. Admit both cases $K > 0$ and $K < 0$. Plot their root-locus, determining all the relevant points, the asymptotes, the departure and arrival angles, and the values of K ensuring stability.

1. $G_1(s) = \dfrac{K(s+1)}{s^2(s+2)}$
2. $G_2(s) = \dfrac{K(s-2)}{s^2(s+1)(s+3)}$
3. $G_3(s) = \dfrac{K}{s(s^2+0.2s+1)}$
4. $G_4(s) = \dfrac{K(s+1)}{s(s^2+3s+9)}$
5. $G_5(s) = \dfrac{K}{s(s+1)^3}$
6. $G_6(s) = \dfrac{K(s^2+2,8s+4)}{s(s+3)(s^2+2s+4)}$
7. $G_7(s) = \dfrac{K(s+1)}{s^2(s+3)(s+4)}$
8. $G_8(s) = \dfrac{K(s+1)}{s^2(s+10)}$
9. $G_9(s) = \dfrac{K(s+2)}{s(s^2+2s+2)}$
10. $G_6(s) = \dfrac{-10s+40}{s(s^2+40s+1025)}$

Resolution

1. There are $3 - 1 = 2$ asymptotes, vertical when $K > 0$, horizontal when $K < 0$, intersecting at $\sigma = \frac{0+0-2+1}{2} = -0.5$. All poles are stable for $K > 0$. See Figure 4.2.

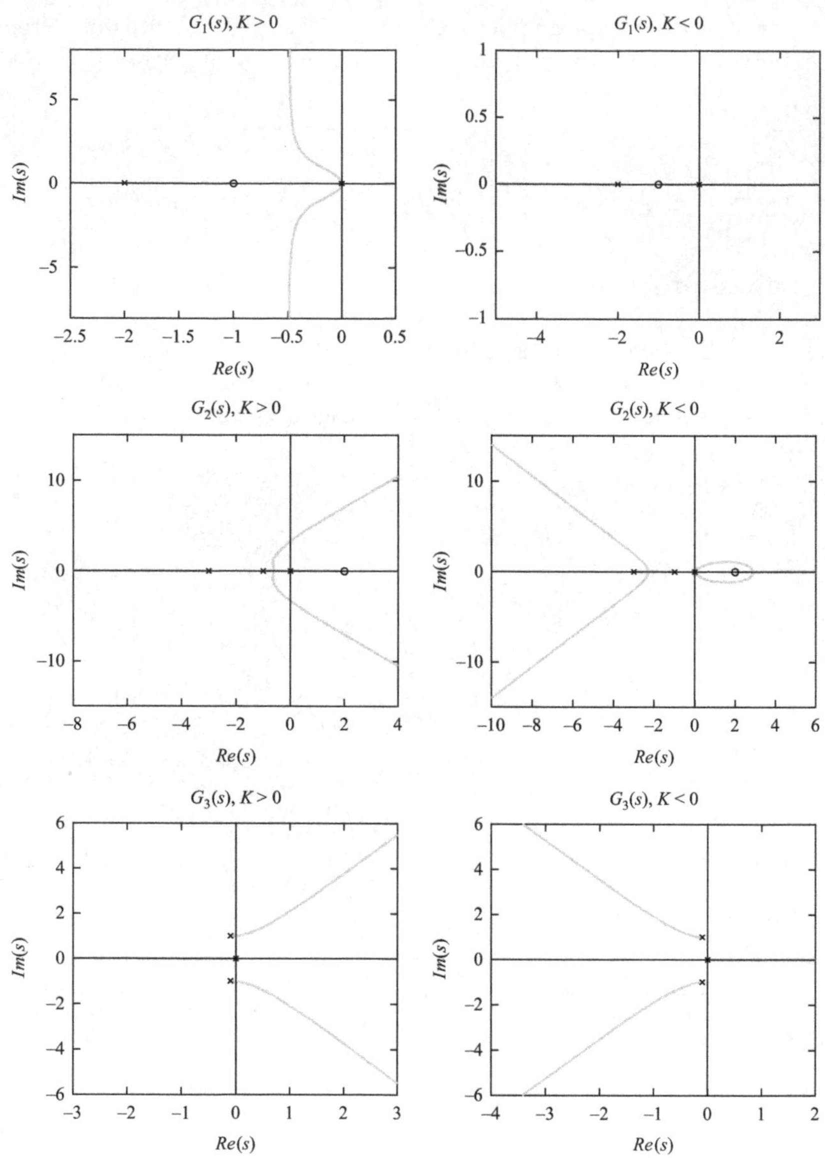

Figure 4.2 Root-locus plots of Problem 4.1

2. There are $4-1=3$ asymptotes, with $\theta=\pm 60°$ and $\theta=180°$ when $K>0$, and $\theta=\pm 120°$ and $\theta=0°$ when $K<0$, intersecting at $\sigma=\frac{0+0-1-3-2}{3}=-2$. Convergence/divergence points are at $\frac{\mathrm{d}}{\mathrm{d}s}\frac{s^4+4s^3+3s^2}{s-2}=0 \Leftrightarrow s=0 \vee s=2.9 \vee s=-2.3 \vee s=-0.6$. It is clear that there are always unstable poles. See Figure 4.2.
3. There are 3 asymptotes, with $\theta=\pm 60°$ and $\theta=180°$ when $K>0$, and $\theta=\pm 120°$ and $\theta=0°$ when $K<0$, intersecting at $\sigma=\frac{0-0.1-0.1}{3}=-0.067$.

The departure angle of the upmost pole for $K > 0$ is $\phi = 180° - 90° - (90° + \arctan \frac{0.1}{0.995}) = -5.7°$. Using the Routh–Hurwitz criterion we find that all poles are stable for $0 < K < 0.2$. See Figure 4.2.

4. There are $3 - 1 = 2$ asymptotes, vertical when $K > 0$, horizontal when $K < 0$, intersecting at $\sigma = \frac{0-1.5-1.5+1}{2} = -1$. The departure angle of the upmost pole for $K > 0$ is $\phi = 180° - 90° - (90° + \arctan \frac{1.5}{2.6}) + (90° + \arctan \frac{0.5}{2.6}) = 70.9°$. When $K < 0$, the root-locus converges at $\frac{\mathrm{d}}{\mathrm{d}s} \frac{s^3+3s^2+9s}{s+1} = 0 \Leftrightarrow s = -2.5$ (only real solution). All poles are stable for $K > 0$. See Figure 4.3.

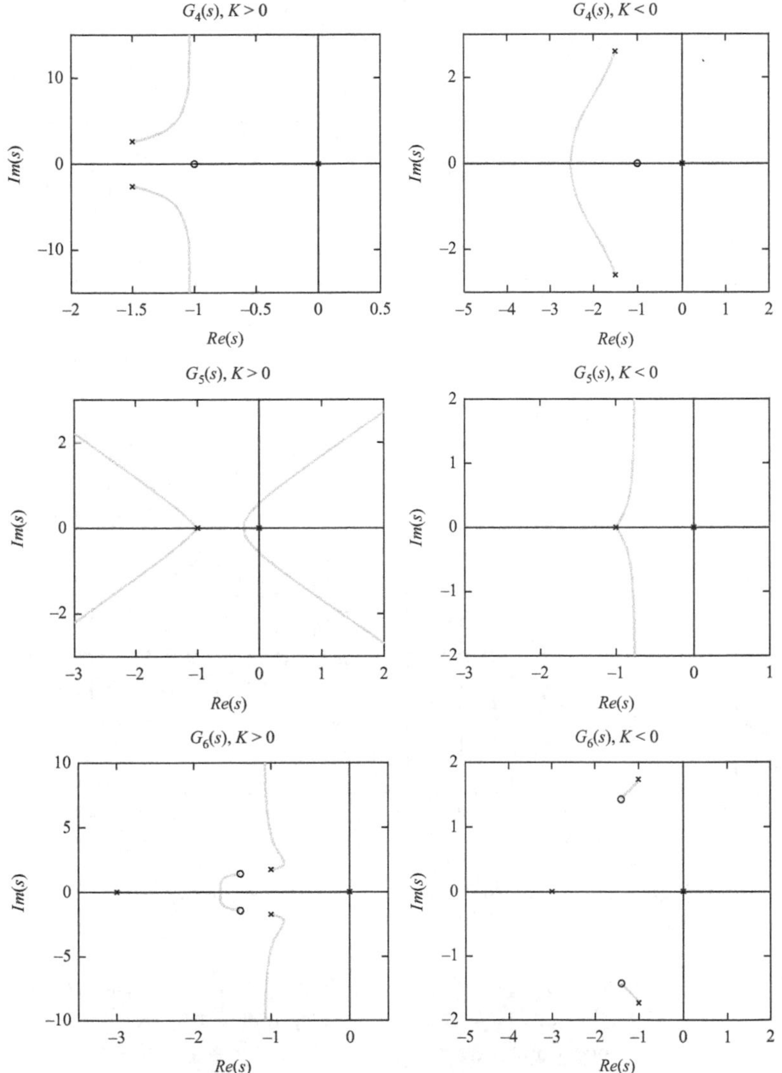

Figure 4.3 Root-locus plots of Problem 4.1 (continued)

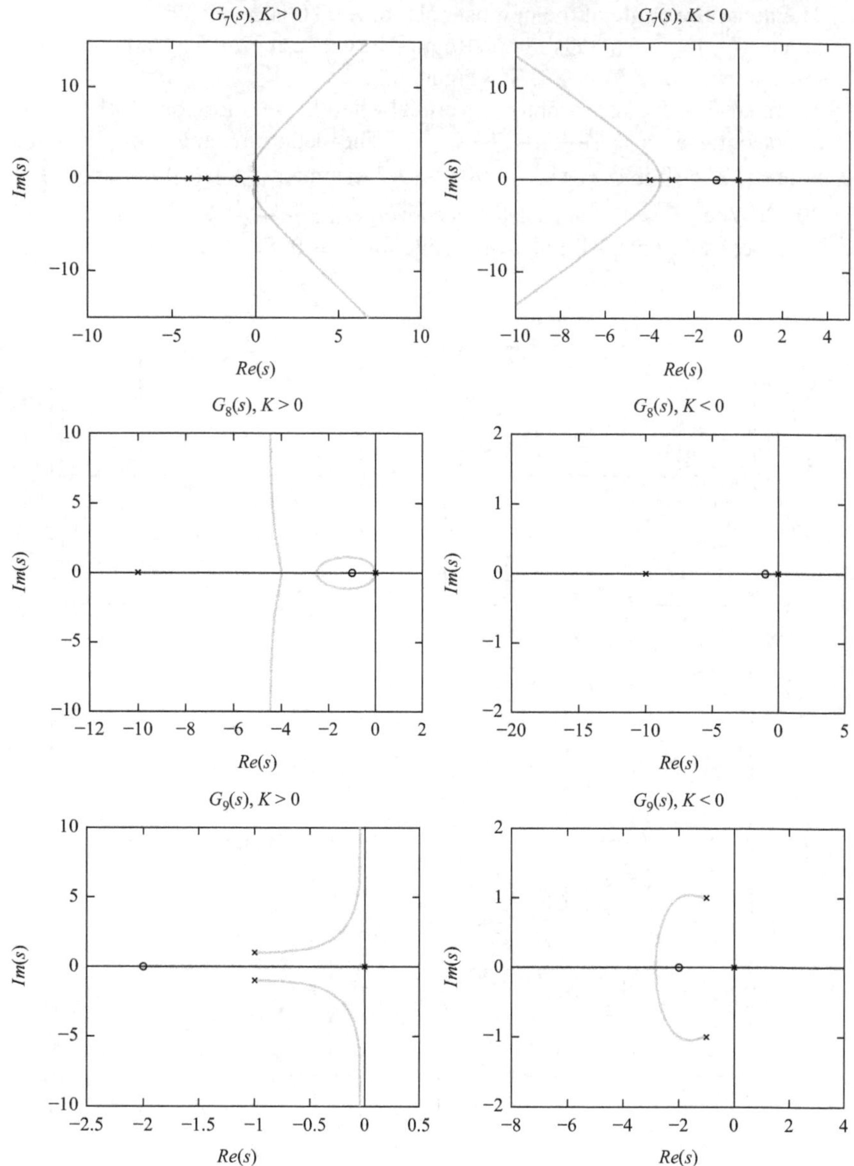

Figure 4.4 Root-locus plots of Problem 4.1 (continued)

5. There are four asymptotes, with $\theta = \pm 45°$ and $\theta = 135°$ when $K > 0$, and $\theta = \pm 90°$, $\theta = 180°$ and $\theta = 0°$ when $K < 0$, intersecting at $\sigma = \frac{0-1-1-1}{4} = -0.75$. Divergence points are at $\frac{d}{ds} s^4 + 3s^3 + 3s^2 + s = 0 \Leftrightarrow s = -1 \vee s = -1 \vee s = -0.25$. Using the Routh–Hurwitz criterion we find that all poles are stable for $0 < K < \frac{8}{9}$. See Figure 4.3.
6. There are $4 - 2 = 2$ asymptotes, vertical when $K > 0$, horizontal when $K < 0$, intersecting at $\sigma = \frac{0-3-1-1+1.4+1.4}{2} = -1.1$. The departure angle

of the upmost pole for $K > 0$ is $\phi = 180° - 90° - (90° + \arctan \frac{1}{1.73}) - \arctan \frac{1.73}{2} + \arctan \frac{0.3}{0.4} + \arctan \frac{3.16}{0.4} = 48.8°$; the arrival angle of the upmost zero for $K > 0$ is $\psi = 180° - 90° + \arctan \frac{1.43}{1.6} + (90° + \arctan \frac{1.43}{1.4}) + (90° + \arctan \frac{0.4}{3.16}) + (180° + \arctan \frac{0.3}{0.4}) = 221.5°$. When $K < 0$, the root-locus diverges at $\frac{d}{ds} \frac{s^4+5s^3+10s^2+12s}{s^2+2.8s+4} = 0 \Leftrightarrow s = -1.66$ (only real solution). All poles are stable for $K > 0$. See Figure 4.3.

7. There are $4 - 1 = 3$ asymptotes, with $\theta = \pm 60°$ and $\theta = 180°$ when $K > 0$, and $\theta = \pm 120°$ and $\theta = 0°$ when $K < 0$, intersecting at $\sigma = \frac{0+0-3-4+1}{3} = -2$. When $K < 0$, the root-locus converges at $\frac{d}{ds} \frac{s^3+7s^2+12s}{s+1} = 0 \Leftrightarrow s = -3.5$ (only real solution). Using the Routh–Hurwitz criterion we find that all poles are stable for $0 < K < \frac{3}{5}$. See Figure 4.4.
8. There are $3 - 1 = 2$ asymptotes, vertical when $K > 0$, horizontal when $K < 0$, intersecting at $\sigma = \frac{0+0-10+1}{2} = -4.5$. Convergence/divergence points are at $\frac{d}{ds} \frac{s^3+10s^2}{s+1} = 0 \Leftrightarrow s = 0 \vee s = -4 \vee s = -2.5$. All poles are stable for $K > 0$. See Figure 4.4.
9. There are $3 - 1 = 2$ asymptotes, vertical when $K > 0$, horizontal when $K < 0$, intersecting at $\sigma = \frac{0-1-1+2}{2} = 0$. The departure angle of the upmost pole for $K > 0$ is $\phi = 180° - 90° - 135° + 45° = 0°$. When $K < 0$, the root-locus converges at $\frac{d}{ds} \frac{s^3+2s^2+2s}{s+2} = 0 \Leftrightarrow s = -2.8$ (only real solution). Using the Routh–Hurwitz criterion we find that all poles are stable for $K > 0$, which means that the root-locus never crosses the imaginary axis. See Figure 4.4.
10. Notice that the open-loop already includes a minus sign, leading to the rules for case (4.10) when $K > 0$. There are $3 - 1 = 2$ asymptotes, horizontal when $K > 0$, vertical when $K < 0$, intersecting at $\sigma = \frac{0-1.5-1.5+1}{2} = -1$. Convergence/divergence points are at $\frac{d}{ds} \frac{s^3+40s^2+1025s}{-10s+40} = 0 \Leftrightarrow s = 12.4 \vee s = -10.4 \vee s = -16.0$. The departure angle of the upmost pole for $K > 0$ is $\phi = -90° - (90° + \arctan \frac{20}{25}) + (90° + \arctan \frac{24}{25}) = -84°$. Using the Routh–Hurwitz criterion we find that all poles are stable for $0 < K < 93.2$. See Figure 4.5.

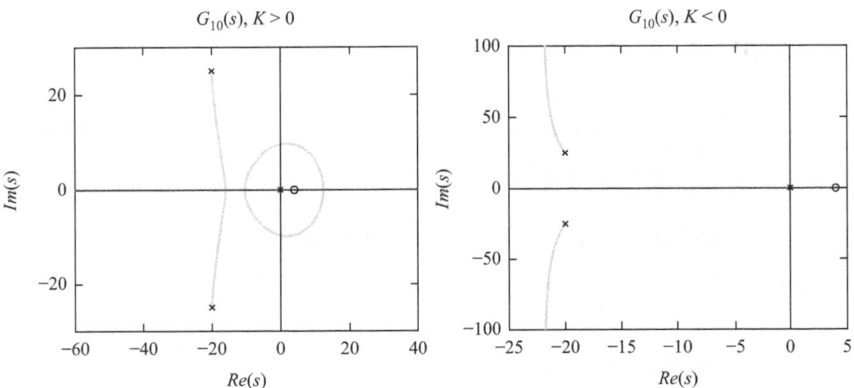

Figure 4.5 Root-locus plots of Problem 4.1 (continued)

Problem 4.2 Figure 4.6 shows the root-locus plot (for $K > 0$) of a plant with the open-loop transfer function $GH(s)$. Then:

A) $GH(s) = \dfrac{K(s+z)}{s(s+p)^2}$

B) $GH(s) = \dfrac{K(s+z)}{s(s+p)}$

C) $GH(s) = \dfrac{K(s+z)}{s^2(s+p)^2}$

D) $GH(s) = \dfrac{K(s+z)}{s^2(s+p)}$.

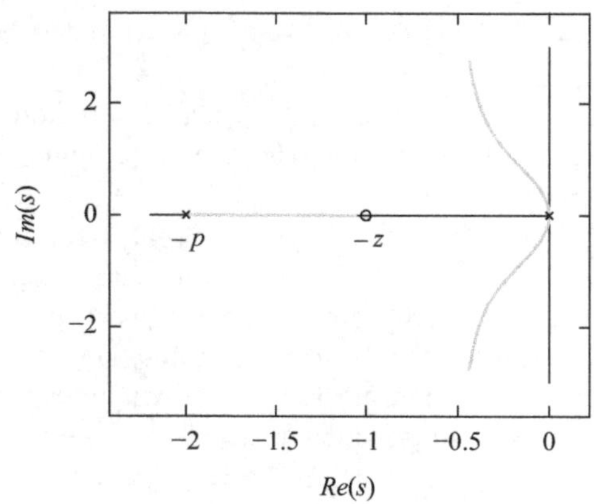

Figure 4.6 Root-locus plot of Problem 4.2

Resolution Two poles are leaving the origin, so the transfer function must have s^2 in the denominator. Only one pole leaves $-p$ for zero $-z$, so the correct answer is option **D)**.

4.3 Proposed problems

Exercise 4.1 Consider the root-locus of a plant with the open-loop transfer function $G(s) = \dfrac{K(s+3)}{s(s+1)(s+2)}$, $K > 0$.

1. The point on the real axis σ where the asymptotes intersect is:
   A) $\sigma = 0$
   B) $\sigma = -\frac{1}{2}$
   C) $\sigma = -1$
   D) None of the above.

2. The following intervals of real numbers belong to the root-locus plot:
   E) $\sigma \in]-\infty, -3] \cup [-2, 0]$
   F) $\sigma \in]-3, -2] \cup [-1, 0]$
   G) $\sigma \in]-\infty, -3] \cup [-2, -1]$
   H) None of the above.
3. The root-locus plot crosses the imaginary axis for $K = 0,\ s = 0$ and also for:
   I) $K = 1,\ s = \pm j2$
   J) $K = 2,\ s = \pm j3$
   K) $K - 2,\ s = \pm j2$
   L) None of the above.

Exercise 4.2 Consider the root-locus of a plant with the open-loop transfer function $G(s) = \dfrac{K(s+2)}{s(s+1)^2(s+3)},\ K > 0.$

1. The point on the real axis σ where the asymptotes intersect is:
   A) $\sigma = -\frac{3}{2}$
   B) $\sigma = -\frac{1}{2}$
   C) $\sigma = -2$
   D) $\sigma = -1.$
2. The following intervals of real numbers belong to the root-locus plot:
   E) $\sigma \in]-\infty, -3] \cup [-2, 0]$
   F) $\sigma \in]-\infty, -2] \cup [-1, 0]$
   G) $\sigma \in]-\infty, -1]$
   H) $\sigma \in]-\infty, -2]$.

Exercise 4.3 Consider the root-locus of a plant with the open-loop transfer function $G(s) = \dfrac{K}{s(s+1)(s+2)(s+3)},\ K > 0.$

1. The point on the real axis σ where the asymptotes intersect is:
   A) $\sigma = -1$
   B) $\sigma = -\frac{1}{2}$
   C) $\sigma = -2$
   D) $\sigma = -\frac{3}{2}.$
2. The following intervals of real numbers belong to the root-locus plot:
   E) $\sigma \in]-\infty, -3] \cup [-1, 0]$
   F) $\sigma \in]-\infty, -3] \cup [-2, -1]$
   G) $\sigma \in]-3, -2] \cup [-1, 0]$
   H) $\sigma \in]-3, -2] \cup [-2, -1]$.

Exercise 4.4 Consider a plant with the open-loop transfer function $GH(s) = \dfrac{K(s+1)(s+3)}{s^2(s+2)}$, $K > 0$. In the root-locus of Figure 4.7, point A where the plot joins the real axis is given by:

A) $\sigma = -5.531, K = 9.643$
B) $\sigma = -5.153, K = 9.364$
C) $\sigma = -3.315, K = 3.463$
D) None of the above.

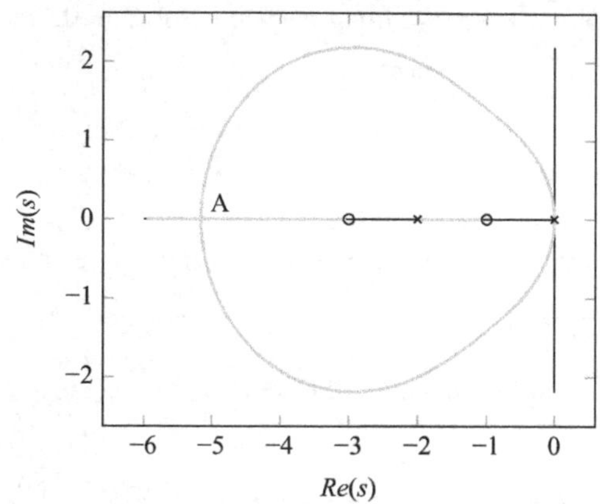

Figure 4.7 Root-locus plot of Problem 4.4

Exercise 4.5 Consider the root-locus of a plant with the open-loop transfer function $GH(s) = \dfrac{K(s+3)}{s(s^2+2s+2)}, K > 0$, shown in Figure 4.8. The point on the real axis σ where the asymptotes intersect is:

A) $\sigma = -1$
B) $\sigma = +1$
C) $\sigma = -1/2$
D) $\sigma = +1/2$.

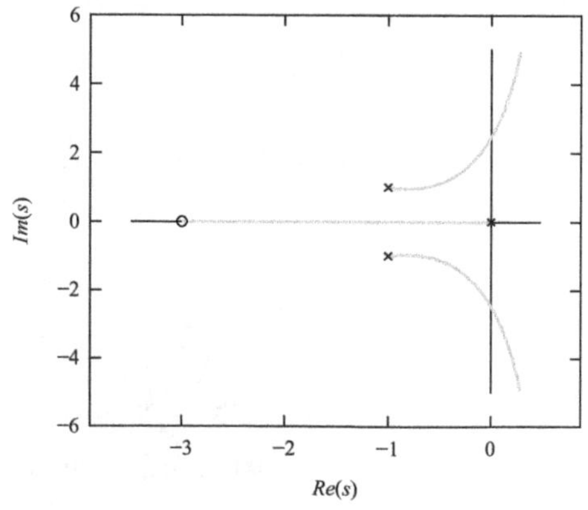

Figure 4.8 Root-locus plot of Exercise 4.5

Exercise 4.6 Consider the root-locus of a plant with the open-loop transfer function $GH(s) = K\dfrac{s+1}{s^2(s+2)(s+5)}$, $K > 0$. The point on the real axis σ where the three asymptotes intersect is:

A) $\sigma = -1$
B) $\sigma = -2$
C) $\sigma = -3$
D) None of the above.

Exercise 4.7 Consider the root-locus of a plant with the open-loop transfer function $GH(s) = K\dfrac{s+2}{s^3(s+4)}$, $K > 0$. The asymptotes are as follows:

A) There is an asymptote at an angle $\theta = 180°$ with the positive real axis.
B) There are two asymptotes at angles $\theta = \pm 90°$ with the positive real axis.
C) There are three asymptotes at angles $\theta = \pm 60°$ and $\theta = 180°$ with the positive real axis.
D) None of the above.

Exercise 4.8 Consider the root-locus of a plant with the open-loop transfer function $GH(s) = \dfrac{K(s+1)}{s^2(s+10)}$, $K > 0$.

1. The point on the real axis σ where the asymptotes intersect is:
   A) $\sigma = -3.5$
   B) $\sigma = -2.5$
   C) $\sigma = -5.5$
   D) None of the above.
2. The root-locus converges or diverges on the real axis for the following values of gain K:
   E) $K = -4$ and $K = -2.5$
   F) $K = 31.25$, $K = 0$ and $K = 32$
   G) $K = 1$ and $K = -2.5$
   H) $K = 21$, $K = 30$ and $K = 1$.
3. The root-locus plot crosses the imaginary axis for the following frequencies:
   I) $\omega = \pm 1$
   J) $\omega = \pm 1.213$
   K) $\omega = \pm 0.735$
   L) None of the above.

Exercise 4.9 Figure 4.9 shows the root-locus of a plant with the open-loop transfer function $G(s)$, $K > 0$. Then:

A) $G(s) = \dfrac{K(s+1)}{s(s+2)^2}$

B) $G(s) = \dfrac{K(s+1)}{s^2(s+2)}$

C) $G(s) = \dfrac{K(s+1)}{s(s+2)}$

D) $G(s) = \dfrac{K(s+1)^2}{s(s+2)}$.

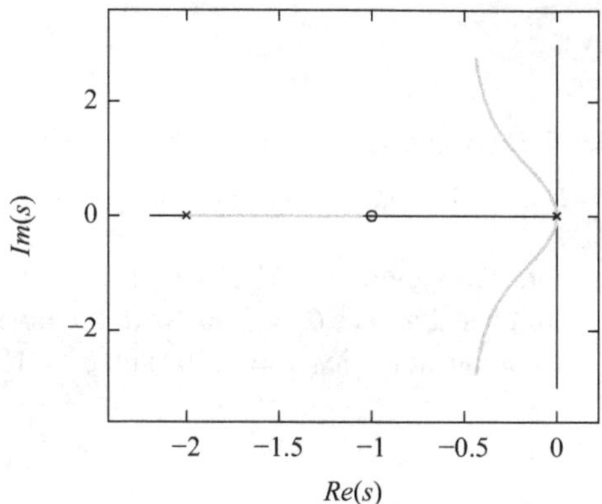

Figure 4.9 Root-locus plot of Exercise 4.9

Exercise 4.10 Figure 4.10 shows the root-locus of a plant with the open-loop transfer function $GH(s)$, $K > 0$. Then:

A) $GH(s) = \dfrac{K(s+z)}{(s+p_1)(s+p_2)}$

B) $GH(s) = \dfrac{K(s+z)}{s^2(s+p_1)(s+p_2)}$

C) $GH(s) = \dfrac{K(s+z)}{s(s+p_1)(s+p_2)}$

D) $GH(s) = \dfrac{K}{s(s+p_1)(s+p_2)}$.

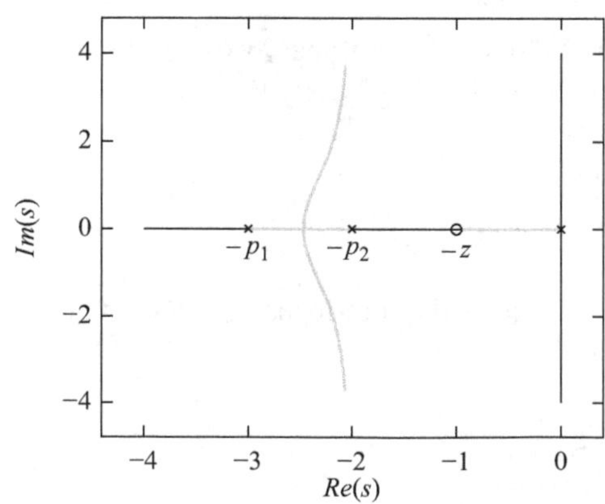

Figure 4.10 Root-locus plot of Exercise 4.10

Exercise 4.11 Figure 4.11 shows the root-locus of a plant with the open-loop transfer function $GH(s)$, $K > 0$. Then:

A) $GH(s) = \dfrac{K(s+z)}{s(s+p)^2}$

B) $GH(s) = \dfrac{K(s+z)}{s(s+p)}$

C) $GH(s) = \dfrac{K(s+z)}{s^2(s+p)^2}$

D) $GH(s) = \dfrac{K(s+z)}{s^2(s+p)}$.

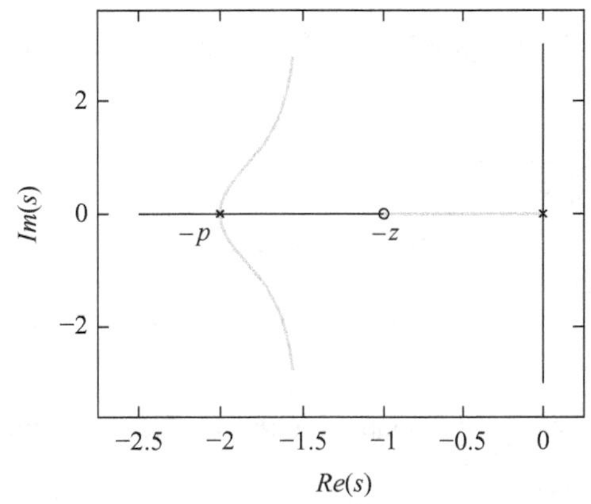

Figure 4.11 Root-locus plot of Exercise 4.11

Exercise 4.12 Figure 4.12 shows the root-locus plot of a plant with the open-loop transfer function $G(s)$. There are poles or zeros (simple or multiple) at $\sigma = 0, \sigma = -1$, $\sigma = -2$, and $\sigma = -3$. Real intervals $\sigma \in]-\infty, -3] \cup [-2, 0]$ belong to the root-locus plot. Hence:

A) $G(s) = \dfrac{K(s+2)}{s^2(s+1)(s+3)}$

B) $G(s) = \dfrac{K(s+3)}{s(s+1)^2(s+2)}$

C) $G(s) = \dfrac{K(s+2)}{s(s+1)^2(s+3)}$

D) $G(s) = \dfrac{K(s+3)}{s(s+1)(s+2)^2}$.

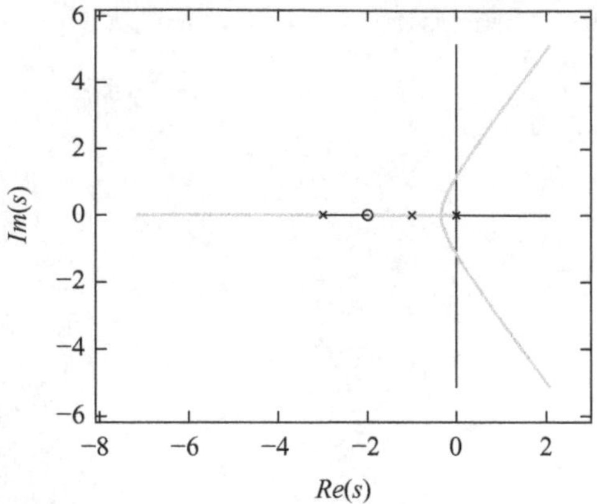

Figure 4.12 Root-locus plot of Exercise 4.12

Exercise 4.13 Consider a closed loop with unit feedback and the open-loop transfer function

$$GH(s) = \frac{K}{s(s^2 + 3s + 10)}, \quad K > 0$$

1. Plot the root-locus, determining all the relevant points, the asymptotes, and the departure and arrival angles.
2. Find the range of values of gain K for which the closed loop is stable.

Exercise 4.14 Consider a closed loop with unit feedback and the open-loop transfer function $GH(s) = K\frac{(s+5)^2}{s(s+2)}$, $K > 0$.

1. Plot the root-locus, determining all the relevant points, the asymptotes, and the departure and arrival angles.
2. Find the range of values of gain K for which the closed loop is stable.

Exercise 4.15 Consider a closed loop with unit feedback and the open-loop transfer function $G(s) = \frac{K(s+1)}{s(s+2)^2}$, $K > 0$.

1. Plot the root-locus, determining all the relevant points, the asymptotes, and the departure and arrival angles.
2. Find the range of values of gain K for which the closed loop is stable.

Exercise 4.16 Consider a closed loop with unit feedback and the open-loop transfer function $GH(s) = \dfrac{K}{(s+1)^2 + 2^2}$, $K > 0$.

1. Plot the root-locus, determining all the relevant points, the asymptotes, and the departure and arrival angles.
2. Find the range of values of gain K for which the closed loop is stable.

Exercise 4.17 Plot the root-locus of the closed loop shown in Figure 4.13 with the open-loop transfer function $GH(s) = \dfrac{K(s+1)(s+2)}{s^2(s+3)}$, $K > 0$.

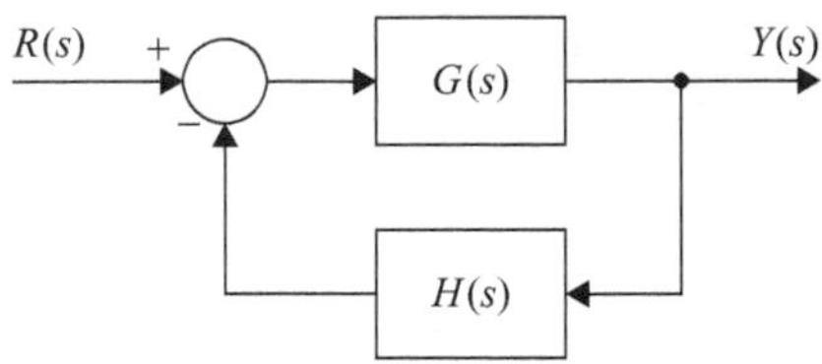

Figure 4.13 Closed loop of Exercise 4.17

4.4 Root-locus analysis using computer packages

This section presents several commands for the root-locus plot using the computer packages MATLAB, SCILAB and OCTAVE. Transfer function $G(s) = \frac{s+3}{s(s+1)(s+2)}$ from Exercise 4.1 is considered in what follows.

4.4.1 MATLAB

This subsection describes some basic commands that can be adopted with the package MATLAB.

The transfer function $G(s) = \frac{s+3}{s(s+1)(s+2)}$ is represented by two arrays num and den. The root-locus plot is accomplished by means of the command `rlocus`.

The complete code is as follows.

```
%%%% Root-locus plot %%%%
% Numerator and denominator of transfer function
num = [0 0 1 3];
den = [1 3 2 0];

% Root-locus
rlocus(num,den)

% Formatting the chart
grid
title('Root-locus plot of G(s) = (s+3)/[s(s+1)(s+2)]')
```

MATLAB creates the figure window in Figure 4.14.

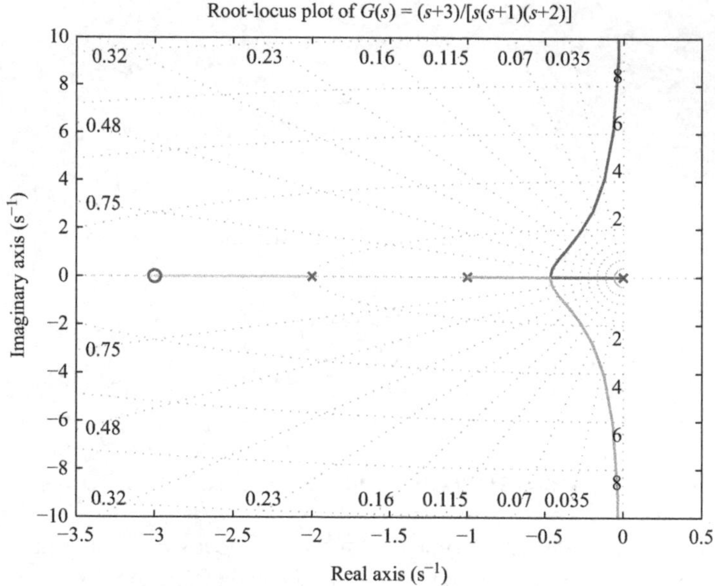

Figure 4.14 Root-locus plot of $G(s) = \frac{s+3}{s(s+1)(s+2)}$ *using MATLAB*

4.4.2 SCILAB

This subsection describes some basic commands that can be adopted with the package SCILAB.

The transfer function $G(s) = \frac{s+3}{s(s+1)(s+2)}$ is represented by two arrays `num` and `den`. The root-locus plot is accomplished by means of the command `evans`.

The complete code is as follows.

```
//// Root-locus plot ////
// Numerator and denominator of transfer function
num = poly([3 1 0 0],'s','coeff');
den = poly([0 2 3 1],'s','coeff');

// Transfer function
G = syslin('c',num/den);

// Root locus
clf();
evans(G);

// Formatting the chart
xgrid
```

SCILAB creates the figure window represented in Figure 4.15.

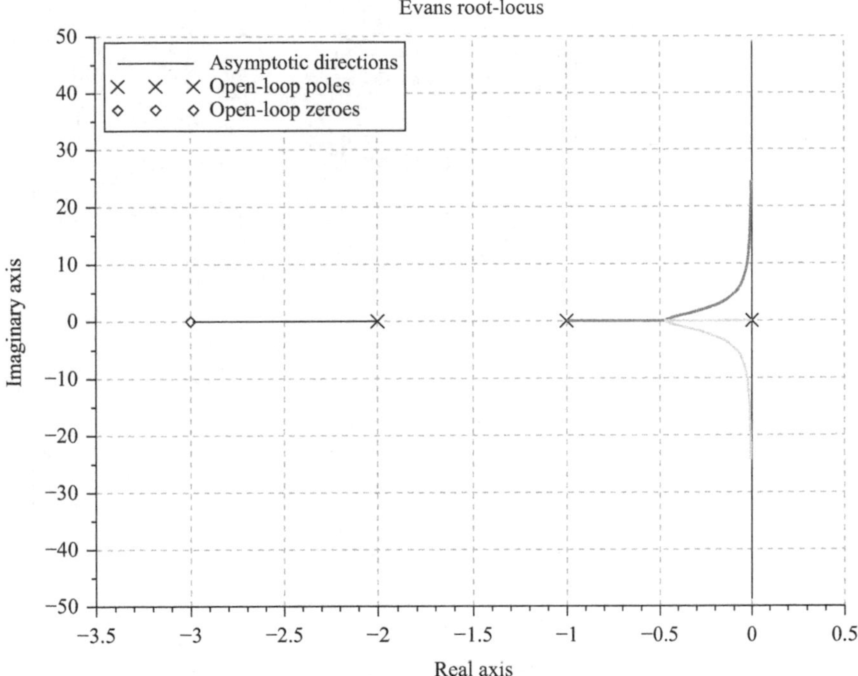

Figure 4.15 Root-locus plot of $G(s) = \frac{s+3}{s(s+1)(s+2)}$ *using SCILAB*

4.4.3 OCTAVE

This subsection describes some basic commands that can be adopted with the package OCTAVE. It is required to load package `control`.

The transfer function $G(s) = \frac{s+3}{s(s+1)(s+2)}$ is represented by two arrays `num`, `den` and `tf`. The root-locus plot is accomplished by means of the command `rlocus`.

The complete code is as follows.

```
%%%% Root-locus plot %%%%
% Numerator and denominator of transfer function
num = [0 0 1 3];
den = [1 3 2 0];

% Transfer function
G = tf(num,den);

% Root-locus
rlocus(G)
```

OCTAVE creates the figure window in Figure 4.16.

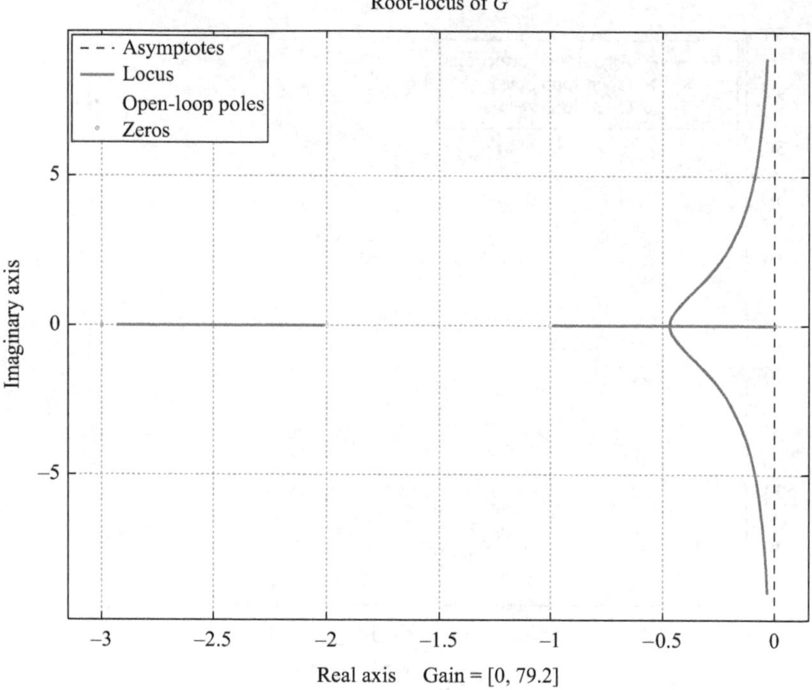

Figure 4.16 Root-locus plot of $G(s) = \frac{s+3}{s(s+1)(s+2)}$ *using OCTAVE*

Chapter 5
Frequency domain analysis

5.1 Fundamentals

Frequency response means system steady-state response to a sinusoidal input. We present three alternatives for representing graphically the frequency response of a dynamical system, namely Bode, Nyquist and Nichols plots [2–5,9]. Afterward, we address closed-loop stability and conditional stability criteria.

5.1.1 List of symbols

A	amplitude
d	number of open-loop poles
$G(s)$, $H(s)$	transfer function
G_{OL}	open-loop transfer function
K	gain
LM	log-magnitude
$M(\omega)$	magnitude function
M_r	resonant peak
n	number of open-loop zeros
p	multiplicity of a pole or zero
$r(t)$	time-domain input
s	Laplace variable
t	time
T	time constant
$y(t)$	time-domain output
y_{ss}	steady-state output
ζ	damping coefficient
$\phi(\omega)$	phase angle function
ω	frequency
ω_1	gain crossover frequency
ω_q	corner frequency
ω_r	resonant frequency
ω_π	phase crossover frequency

5.1.2 Frequency response preliminaries

Given an LTI system, its steady-state response to a sinusoidal input is also sinusoidal of the same frequency, but different amplitude and phase.

If the system transfer function is $G(s) = \frac{Y(s)}{R(s)}$, then the corresponding sinusoidal transfer function, $G(j\omega) = \frac{Y(j\omega)}{R(j\omega)}$, can be calculated by directly replacing the variable s by $j\omega$, where ω is the angular frequency. Thus, we have:

$$M(\omega) = |G(j\omega)| = \left|\frac{Y(j\omega)}{R(j\omega)}\right| \tag{5.1a}$$

$$\phi(\omega) = \underline{/G(j\omega)} = \underline{\Big/ \frac{Y(j\omega)}{R(j\omega)}} \tag{5.1b}$$

where $M(\omega)$ is the magnitude of $G(j\omega)$, representing the amplitude ratio between the output and the input sinusoids, and $\phi(\omega)$ is the phase angle of $G(j\omega)$, denoting the phase shift of the output relative to the input. The sinusoidal transfer function $G(j\omega)$ is a complex function, where ω represents a parameter that can be varied by the user within an interval of interest.

In time-domain, giving the input $r(t) = A\sin(\omega t)$, then the system steady-state response is:

$$y_{ss}(t) = M(\omega)A\sin[\omega t + \phi(\omega)] \tag{5.2}$$

5.1.3 Bode diagram

A Bode diagram consists of two graphs, namely the magnitude and the phase angle of $G(j\omega)$ versus frequency. The magnitude is usually represented in decibels (dB), $|G(j\omega)|_{dB} = 20\log_{10} G(j\omega)$, where $\log_{10}(\cdot)$ denotes the base ten logarithm, and the phase angle, ϕ, is expressed in degrees. For the frequency we use a logarithmic scale, while for both the log-magnitude (LM) and the phase angle we adopt a linear scale.

For sketching Bode plots we can use several software packages (e.g., MATLAB®, OCTAVE©, SCILAB™) [16–18]. Nevertheless, using straight lines we can easily obtain by hand an asymptotic approximation of the plot. In a first step, we sketch the contribution of each basic factor that occurs in the system transfer function. We then add the individual contributions for both the magnitude and phase plots.

Given a generic open-loop transfer function of a feedback control system, $G_{OL} = K \cdot G(j\omega)H(j\omega)$, the basic factors to consider are:

- Gain K: $LM = 20\log_{10} K$; $\phi = 0°$. The LM is a horizontal line intersecting the vertical axis at $20\log_{10} K$.
- Pole or zero at the origin $(j\omega)^{\pm p}$: $LM = \pm 20p \cdot \log_{10}\omega$; $\phi = \pm p \cdot 90°$. The LM is a straight line with slope $\pm 20p$ dB per decade.
- First-order factor $(1 + j\omega T)^{\pm p}$: $LM = \pm 20p \cdot \log_{10}\sqrt{1 + (T\omega)^2}$; $\phi = \pm p \cdot \arctan T\omega$. The LM can be approximated by two straight line asymptotes that intersect at the corner frequency $\omega_q = \frac{1}{T}$:
  - ◇ $LM \approx 0, \quad \omega \ll \frac{1}{T}$
  - ◇ $LM \approx \pm 20p \cdot \log_{10} T\omega, \quad \omega \gg \frac{1}{T}$

  The phase angle, ϕ, approaches zero and $\pm p \cdot 90°$ as ω tends to zero and infinity, respectively:

◇ $\phi \approx 0, \quad \omega \leq \frac{0.1}{T}$
◇ $\phi \approx \pm p \cdot 90°, \quad \omega \geq \frac{10}{T}$

- Second-order factor $[1 + 2\zeta(j\omega/\omega_r) + (j\omega/\omega_n)^2]^{\pm p}$: $LM = \pm 20p \cdot \log_{10} \sqrt{\left(1 - \frac{\omega^2}{\omega_n^2}\right)^2 + \left(\frac{2\zeta\omega}{\omega_n}\right)^2}$; $\phi = \pm p \cdot \arctan \frac{2\zeta\omega/\omega_n}{1-\omega^2/\omega_n^2}$. The LM can be approximated by straight line asymptotes that intersect at the corner frequency $\omega_q = \omega_n$:
  ◇ $LM \approx 0, \quad \omega \ll \omega_n$
  ◇ $LM \approx \pm 40p \cdot \log_{10} \frac{\omega}{\omega_n}, \quad \omega \gg \omega_n$

  The phase angle, ϕ, approaches zero and $\pm p \cdot 180°$ as ω tends to zero and infinity, respectively:
  ◇ $\phi \approx 0, \quad \omega \leq 0.1\omega_n$
  ◇ $\phi \approx \pm p \cdot 180°, \quad \omega \geq 10\omega_n$

For $\zeta \leq 0.707$ the second-order factor has a resonant peak, M_r at the resonant frequency, ω_r, given by:

$$M_r = 20 \log_{10} \frac{1}{2\zeta\sqrt{1-\zeta^2}} \tag{5.3a}$$

$$\omega_r = \omega_n\sqrt{1-2\zeta^2} \tag{5.3b}$$

Certain open-loop transfer functions include a transport delay factor, $G(j\omega) = e^{-j\omega T}$. In this case, we have $LM = 0$ and phase angle $\phi = -\omega T$ (radians), or $\phi = -57.3\omega T$ (degrees). This means that the phase angle varies linearly with the frequency. The effect of $G(j\omega) = e^{-j\omega T}$ can be easily included in the Bode diagram.

In Appendix A, we depict the exact and asymptotic Bode diagrams of elemental factors.

5.1.4 Nyquist diagram

The Nyquist, or polar, plot of a transfer function, $G(j\omega)$, is the graph of the magnitude versus phase angle (or imaginary versus real part) of $G(j\omega)$, parametrized in the frequency, for ω varying in the interval $\omega \in [0, \infty[$.

Figure 5.1 depicts a typical Nyquist, or polar, plot.

For a generic feedback controlled system with open-loop transfer function $G_{OL} = K \cdot G(j\omega)H(j\omega)$, the polar plot is constructed by calculating successively the magnitude and phase values (or the imaginary and real parts) of G_{OL} and plotting the points obtained. This is usually accomplished by means of specific software packages, such as MATLAB, OCTAVE and SCILAB.

In certain conditions, namely for systems with more open-loop poles, d, than open-loop zeros, n, the polar curves at low ($\omega \to 0$) and high frequency ($\omega \to \infty$), present typical shapes that can help in the sketching of the diagrams by hand.

Figures 5.2 and 5.3 depict the shape of polar curves at low and high frequency, for type 0, 1 and 2 systems. By using these typical shapes and a few more auxiliary points calculated at specific frequencies, the user can obtain quickly an approximate polar plot of the system.

In Appendix A, we present the polar plots of some common transfer functions.

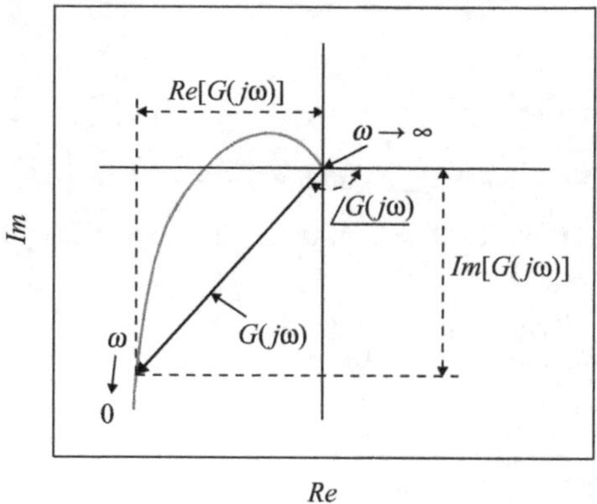

Figure 5.1 Representation of a typical Nyquist, or polar, plot

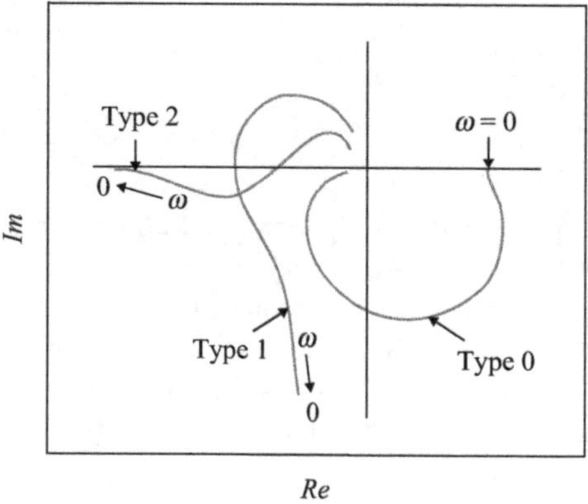

Figure 5.2 Polar plot of type 0, 1 and 2 systems, at low frequency ($\omega \to 0$)

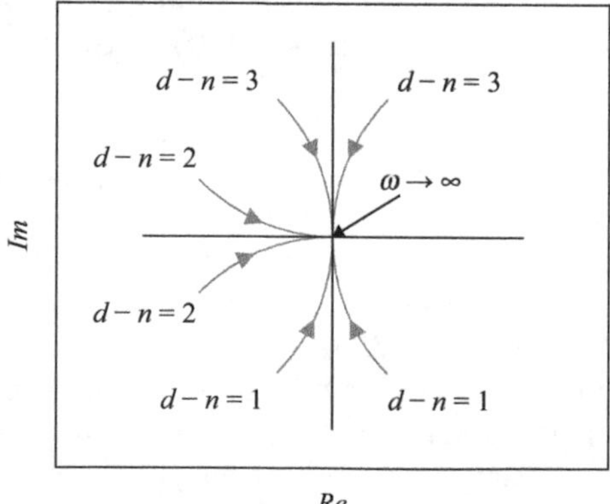

Figure 5.3 Polar plot of type 0, 1 and 2 systems, at high frequency ($\omega \to \infty$)

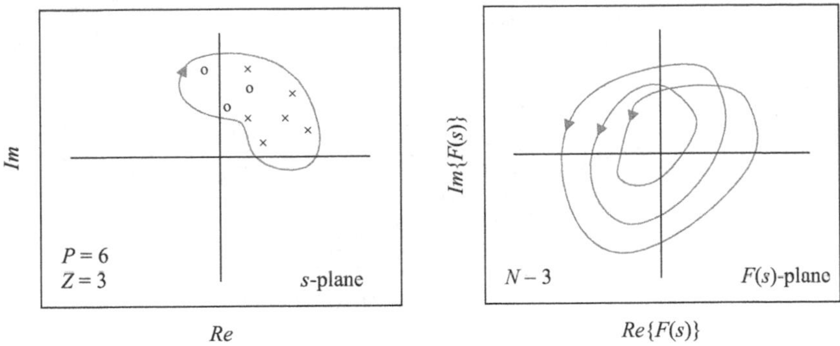

Figure 5.4 Illustration of the Cauchy's principle of argument

5.1.5 Nichols diagram

The Nichols diagram is the graph of the logarithmic magnitude versus the phase angle of $G(j\omega)$, parametrized in ω, for a frequency range of interest. The chart is constructed by calculating successively the LM and phase angle values of $G(j\omega)$ and plotting the points obtained. As said before, this is usually accomplished by means of specific software packages, namely MATLAB, OCTAVE and SCILAB.

In Appendix A, we present the Nichols plot of common transfer functions.

5.1.6 Nyquist stability

Given a generic open-loop transfer function, $G_{OL} = K \cdot G(j\omega)H(j\omega)$, the Nyquist stability criterion can be used to conclude about the stability of the closed-loop system.

The Nyquist stability criterion is based on the Cauchy's principle of argument, enunciated as follows. Let $F(s)$ be an analytic function in a closed region of the s plane, except at a finite number of points, namely its poles. Then, as s travels in the clockwise direction around a contour in the s plane, the function $F(s)$ encircles, in the same direction, the origin of the $F(s)$ plane, a number of times equal to the difference between the number of zeros and the number of poles of $F(s)$ that are inside the contour (Figure 5.4).

A contour covering the whole right-half s plane (Figure 5.5) maps on the locus of $G_{OL} = K \cdot G(j\omega)H(j\omega)$, as ω varies in the interval $]-\infty, +\infty[$. The loci of G_{OL} for $\omega \in]-\infty, 0]$ and $\omega \in [0, +\infty]$ are symmetrical about the real axis. Moreover, as we use $K \cdot G(j\omega)H(j\omega)$ to infer about the poles of $1 + K \cdot G(j\omega)H(j\omega)$, then the point of interest is $(-1, 0j)$ instead of the origin.

The Nyquist stability criterion states that the number of unstable closed-loop poles, Z, is equal to the number of unstable open-loop poles, P, plus the number of clockwise encirclements of the point $(-1, 0j)$, N, made by the locus of G_{OL}, as ω varies in the interval $]-\infty, +\infty[$:

$$Z = N + P \tag{5.4}$$

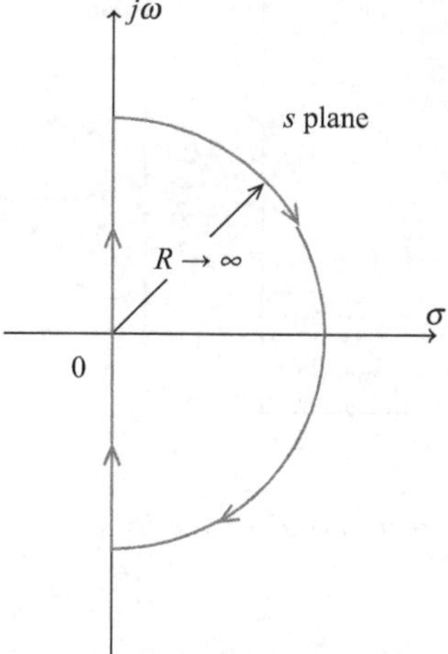

Figure 5.5 Nyquist contour

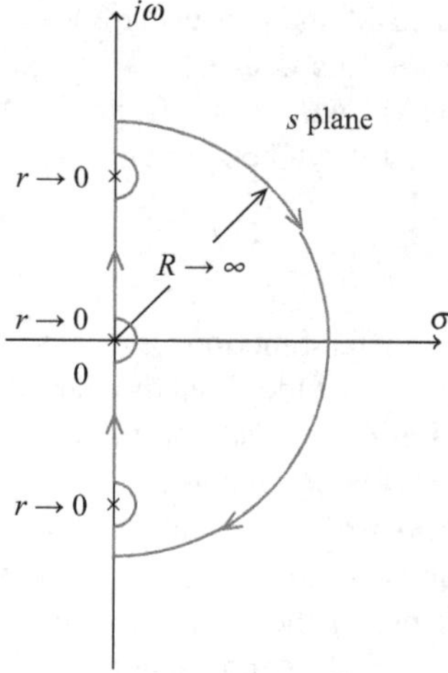

Figure 5.6 Modified Nyquist contour

If G_{OL} has poles or zeros on the imaginary axis, then the contour used for applying the Nyquist stability criterion must be modified. Usually we employ a semicircle with an infinitesimal radius, ϵ, in the right-half (or left-half) s plane, to deviate from those points (Figure 5.6).

5.1.7 Relative stability

In practical situations it is necessary not only that the system is stable, but also that it has adequate relative stability. The gain and phase margins (GM, PM) are two quantitative measures of how stable a system is.

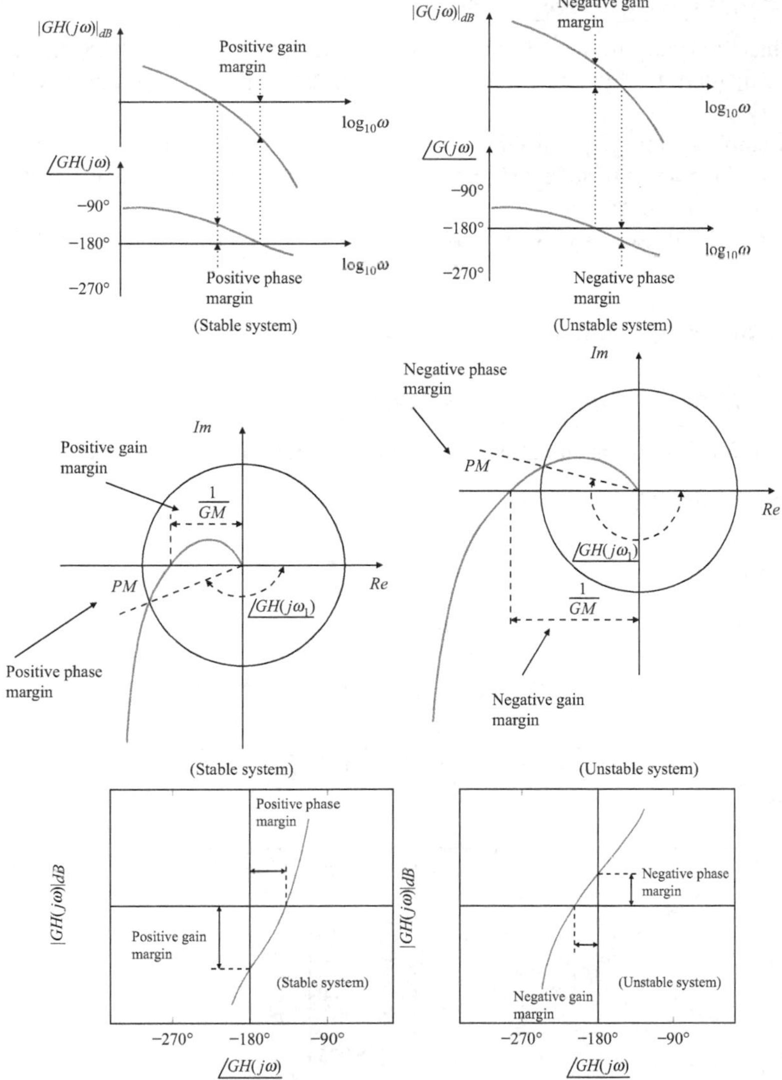

Figure 5.7 Gain and phase margins in Bode, Nyquist and Nichols plots, of stable and unstable systems

The GM is the reciprocal of the magnitude of $K \cdot G(j\omega)H(j\omega)$ at the phase crossover frequency, $\omega = \omega_\pi$. The PM is the amount of additional phase lag that is necessary to bring the system to instability, measured at the gain crossover frequency, $\omega = \omega_1$:

$$GM = \frac{1}{|K \cdot G(j\omega_\pi)H(j\omega_\pi)|} \tag{5.5a}$$

$$PM = \underline{/G(j\omega_1)H(j\omega_1)} + 180^\circ \tag{5.5b}$$

Phase and gain crossover frequencies are the frequencies at which the phase angle and the magnitude of the open-loop transfer function equals -180° and 1 (or 0 dB), respectively.

A stable system requires $GM > 1$ (0 dB) and $PM > 0$. When both margins are negative, the plant is unstable. If one margin is positive and the other is negative, no conclusions can be taken.

Figure 5.7 illustrates how to measure *GM* and *PM* in Bode, Nyquist and Nichols plots.

5.2 Solved problems

5.2.1 Bode diagram and phase margins

Problem 5.1 Plot the asymptotes of the Bode diagram of a plant with the open-loop transfer function $G(s) = \dfrac{e^{-\frac{\pi}{20}s}}{s(s+1)}$.

Resolution

$$G(j\omega) = \frac{1}{\omega\sqrt{\omega^2+1}} \arg\left[-\frac{\pi}{20}\omega - \frac{\pi}{2} - \arctan\omega\right] \tag{5.6}$$

The delay has no effect on the gain, and its phase does not have any asymptotes:

ω (rad/s)	0	0.1	1	10	20
$-\frac{\pi}{20}\omega$ (rad/s)	0	−0.0157	−0.157	−1.57	−3.14

See Figure 5.8.

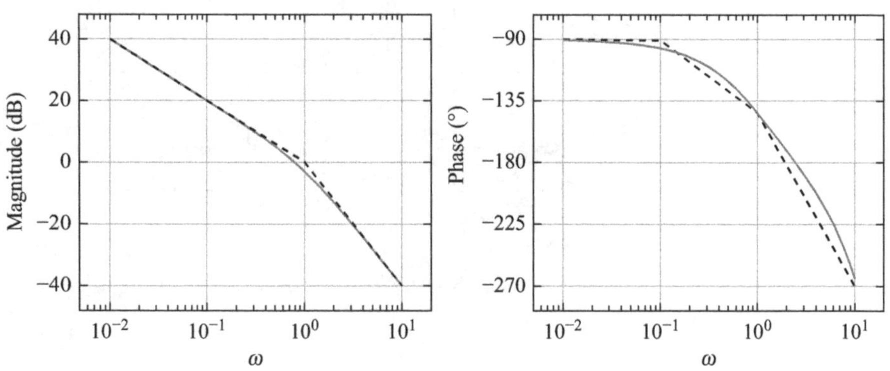

Figure 5.8 Resolution of Problem 5.1

Problem 5.2 Figure 5.9 shows the gain plot of the Bode diagram of a plant. Find the plant's transfer function.

Resolution At low frequencies the gain has a −20 dB/decade slope, so there is a pole at the origin. The slope becomes 0 dB/decade at 0.01 rad/s (20 dB/decade above

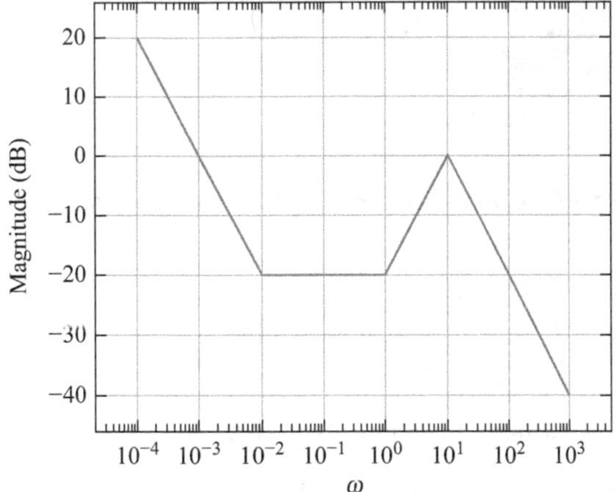

Figure 5.9 Gain plot of the Bode diagram of a plant in Problem 5.2

the previous value), so there is a zero at that frequency. It becomes 20 dB/decade at 1 rad/s (20 dB/decade above the previous value), so there is another zero at that frequency. And it becomes −20 dB/decade at 10 rad/s (40 dB/decade below the previous value), so there is a double pole at that frequency. Putting all this together we get $\frac{(s+0.01)(s+1)}{s(s+10)^2}$. At low frequencies, this transfer function is approximately $\frac{0.001}{s}$, and will have a unit gain (0 dB) at frequency 0.001 rad/s, as seen in the plot, so there is no need to change its gain.

Problem 5.3 Find analytically the phase margin of the system in Figure 5.10, where $G(s) = \frac{1}{(s+1)(s+3)}$ and $K = 10$.

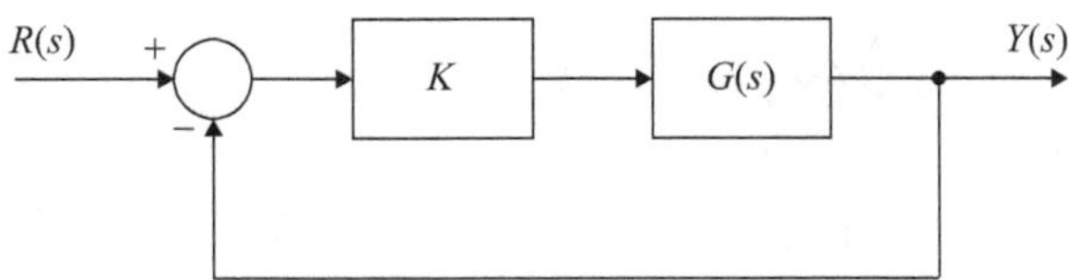

Figure 5.10 System of Problem 5.3

Resolution

$$|KG(s)| = 1 \Leftrightarrow \frac{10}{\sqrt{(\omega^2+1)(\omega^2+9)}} \Rightarrow \omega_1 = 2.402 \text{ rad/s}$$

$$PM = 180^\circ - \arctan\left(\frac{\omega_1}{1}\right) - \arctan\left(\frac{\omega_1}{3}\right) = 73.9^\circ \tag{5.7}$$

Problem 5.4 The block diagram in Figure 5.11 corresponds to the control system of a space vehicle.

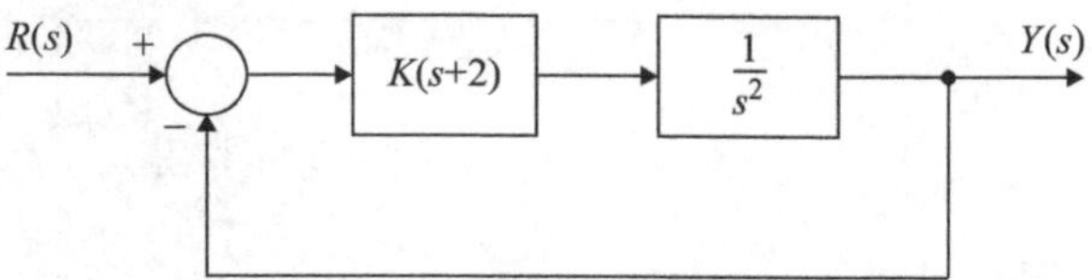

Figure 5.11 *Control system of a space vehicle in Problem 5.4*

1. Find gain K so that the phase margin is $PM = 50°$.
2. For that value of K, what is the gain margin?

Resolution First, determine the gain crossover frequency from the phase margin:

$$\begin{aligned} \arg\left[G\left(j\omega_1\right)H\left(j\omega_1\right)\right] &= PM - 180° \Leftrightarrow \\ \Leftrightarrow \arg\left[\frac{K\left(j\omega_1+2\right)}{\left(j\omega_1\right)^2}\right] &= -130° \Leftrightarrow \\ \Leftrightarrow \omega_1 &= 2.38 \text{ rad/s} \end{aligned}$$

Consequently:

$$\begin{aligned} \left|G\left(j\omega_1\right)H\left(j\omega_1\right)\right| &= 1 \Leftrightarrow \\ \Leftrightarrow \left|\frac{K\left(j\omega_1+2\right)}{\left(j\omega_1\right)^2}\right| &= 1 \Leftrightarrow \\ \Leftrightarrow K &= 1.81 \end{aligned}$$

To find the gain margin, first determine the phase crossover frequency:

$$\begin{aligned} \arg\left[G\left(j\omega_\pi\right)H\left(j\omega_\pi\right)\right] &= -180° \Leftrightarrow \\ \Leftrightarrow \omega_\pi &= 0 \text{ rad/s} \end{aligned}$$

Then

$$GM = -20\log_{10}\left|G\left(j\omega_\pi\right)H\left(j\omega_\pi\right)\right| = \infty$$

Problem 5.5 Consider a plant with transfer function $\dfrac{Y(s)}{X(s)} = \dfrac{Ms^2}{Ms^2+Bs+K}$, where $M = 1$ kg, $B = 10$ N · s · m^{-1} and $K = 12.5$ N/m.

1. Plot its Bode diagram.
2. What will the output $y(t)$ of the plant be, when the input is $x(t) = \sin(10\,t)$?

Resolution
See the Bode diagram in Figure 5.12. From the poles of the transfer function, we get

$$G(j\omega) = \frac{(j\omega)^2}{(j\omega+8.535)(j\omega+1.465)}$$

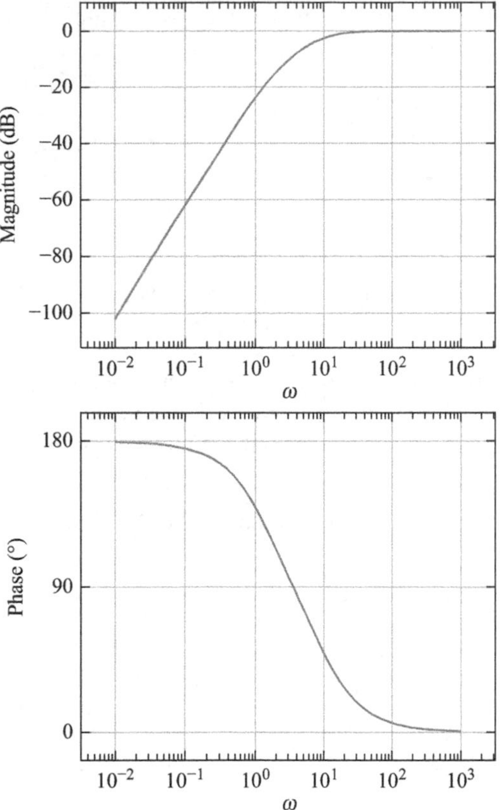

Figure 5.12 Resolution of Problem 5.5

When $\omega = 10$ rad/s,

$$\begin{aligned} |G(10j)| &= \frac{100}{\sqrt{100+72.65}\sqrt{100+2.15}} = 0.75 \\ \arg\left[G(10j)\right] &= \pi - \arctan\left(\frac{10}{8.535}\right) - \arctan\left(\frac{10}{1.465}\right) = 0.852 \text{ rad} \end{aligned} \tag{5.8}$$

Hence $y(t) = 0.75 \sin(10t + 0.852)$.

5.2.2 Nyquist and Nichols diagrams

Problem 5.6 Consider the control system in Figure 5.13, where the plant is $G(s) = \dfrac{s-1}{s(s+1)}$ and the controller is $K > 0$.

1. Plot the Bode diagram and the polar diagram of the plant.
2. For which values of $K > 0$ will the closed loop be stable?

3. Find a minimum phase transfer function $G_m(s)$ having the same gain plot of the Bode diagram as $G(s)$.

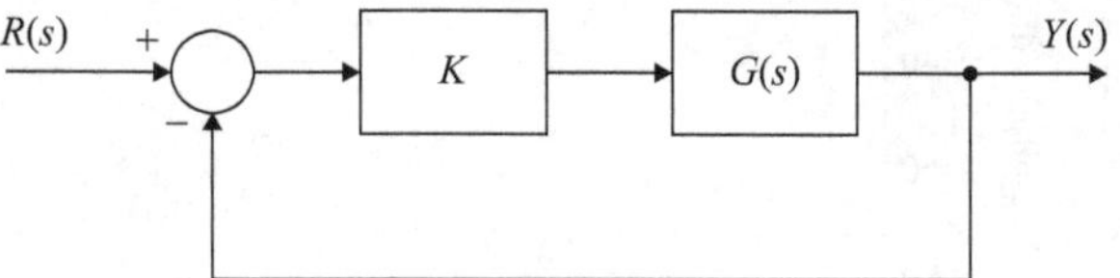

Figure 5.13 Control system of Problem 5.6

Resolution See Figure 5.14. Since $G(s)$ has no unstable poles and the Nyquist diagram encircles point -1 once, irrespective of the value of $K > 0$, the control system is always unstable. This can be seen from the root-locus as well, since the pole at the origin goes right to zero $+1$.

Plant $G_m(s) = \dfrac{1}{s}$ has the same gain plot of the Bode diagram as $G(s)$.

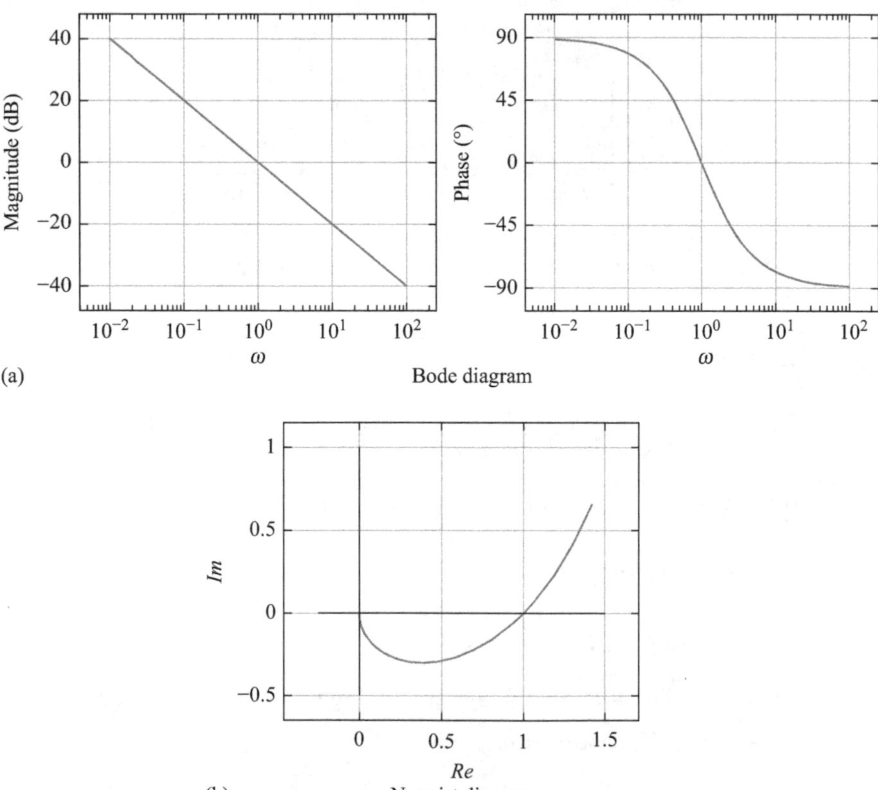

Figure 5.14 Resolution of Problem 5.6

Problem 5.7 Consider a plant with the open-loop transfer function $G(s) = \dfrac{Y(s)}{R(s)} = \dfrac{25}{s^2 + 4s + 25}$.

1. Find the damping factor ζ and the undamped natural frequency ω_n.
2. Find the rise time t_r, the peak of the response $y(t_p)$ and the peak time t_p, when the input is a unit step $r(t) = 1,\ t \geq 0$.
3. Find the maximum gain M_r of the frequency response $G(j\omega)$ and the frequency ω_r at which this is to be found.
4. Plot $G(j\omega)$ in a Bode diagram.

Resolution See Figure 5.15.

$$\begin{cases} \omega_n^2 = 25 \\ 2\zeta\omega_n = 4 \end{cases} \Rightarrow \begin{cases} \omega_n = 5 \text{ rad/s} \\ \zeta = 0.4 \end{cases}$$

$$t_p = \frac{\pi}{5\sqrt{1 - 0.4^2}} = 0.857 \text{ s}$$

$$y(t_p) = 1 + e^{-\frac{0.4\pi}{\sqrt{1-0.4^2}}} = 1.254$$

$$t_r = \frac{e^{\frac{\arccos 0.4}{\tan \arccos 0.4}}}{5} = 0.332 \text{ s}$$

$$\omega_r = 5\sqrt{1 - 2 \times 0.4^2} = 4.1 \text{ rad/s}$$

$$M_r = \frac{1}{2 \times 0.4\sqrt{1 - 0.4^2}} = 1.364 \tag{5.9}$$

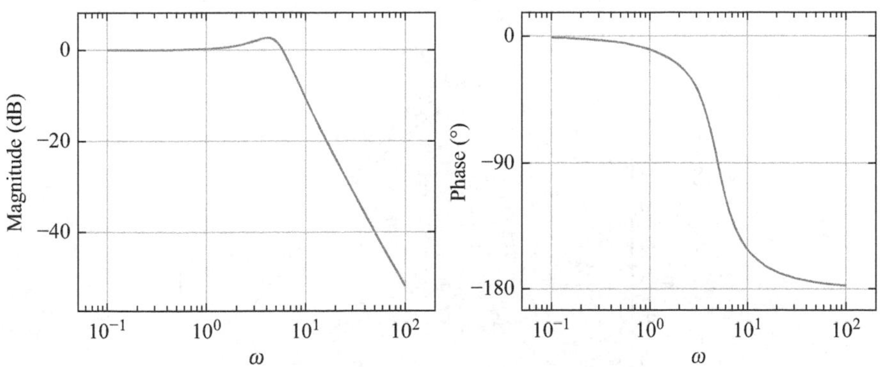

Figure 5.15 Resolution of Problem 5.7

5.3 Proposed problems

5.3.1 Bode diagram and phase margins

Exercise 5.1 Plot the asymptotes of the gain plot of the Bode diagram of transfer function $G(s) = \dfrac{120(s+1)}{s(s+2)^2(s+3)}$.

Exercise 5.2 Plot the asymptotes of the Bode diagrams of the plants with the open-loop transfer functions given below. Mark the gain and phase margins in your plots, and conclude about the stability of the plants.

1. $G_1(s) = \dfrac{s^2}{(s+0.5)\,(s+10)}$
2. $G_2(s) = \dfrac{10s}{(s+10)\,\left(s^2+s+2\right)}$
3. $G_3(s) = \dfrac{(s+4)\;(s+20)}{(s+1)\,(s+80)}$.

Exercise 5.3 Consider a plant with the open-loop transfer function $G(s) = \dfrac{K}{s(s+1)^2}$.

1. The phase margin is $PM = 60°$ for gain:
   A) $K = 0.287$
   B) $K = 2.941$
   C) $K = 0.562$
   D) $K = 5.670$.
2. The asymptotic gain plot of the Bode diagram in Figure 5.16 corresponds to gain:
   E) $K = 0.287$
   F) $K = 2.871$
   G) $K = 0.562$
   H) $K = 5.672$.

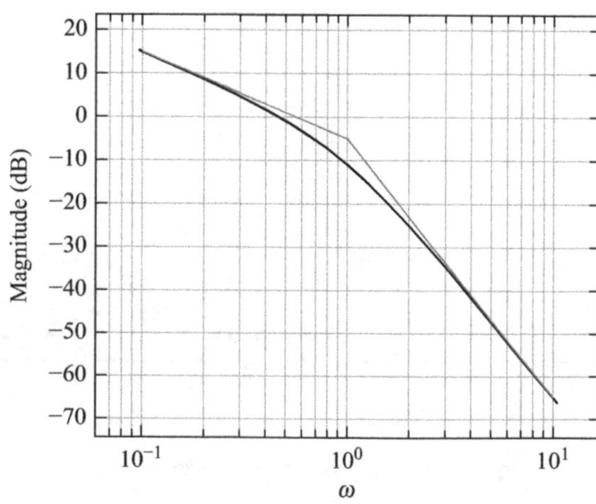

Figure 5.16 Asymptotic gain plot of the Bode diagram of Exercise 5.3

Exercise 5.4 Consider a plant with the open-loop transfer function $G(s) = \dfrac{2(s+3)}{s(s+1)(s+2)}$.

1. In Figure 5.17, the correct asymptotic gain plot of the Bode diagram is:
   A) plot A
   B) plot B
   C) plot C
   D) plot D.
2. The phase margin is:
   E) $PM = 15.8°$
   F) $PM = 32.8°$
   G) $PM = 23.5°$
   H) $PM = 49.8°$.

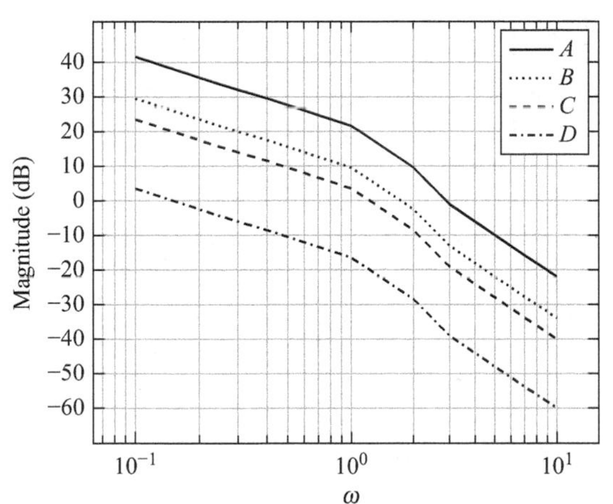

Figure 5.17 Asymptotic gain plot of Bode diagrams of Exercise 5.4

Exercise 5.5 Consider a plant with the open-loop transfer function $G(s) = \dfrac{5(s+2)}{(s+1)(s^2+2s+4)}$. The phase margin is:

A) $PM = 128.5°$
B) $PM = 45.5°$
C) $PM = \infty$
D) None of the above.

Exercise 5.6 Figure 5.18 shows the asymptotes of the Bode diagram of transfer function $G(s)$. The plant's transfer function is:

A) $G(s) = \dfrac{1}{(s+1)(s+2)}$

B) $G(s) = \dfrac{1}{s^2(s+1)^2}$

C) $G(s) = \dfrac{s+3}{s(s+1)(s+2)}$

D) $G(s) = \dfrac{(s+3)(s+4)}{s(s+1)(s+2)}.$

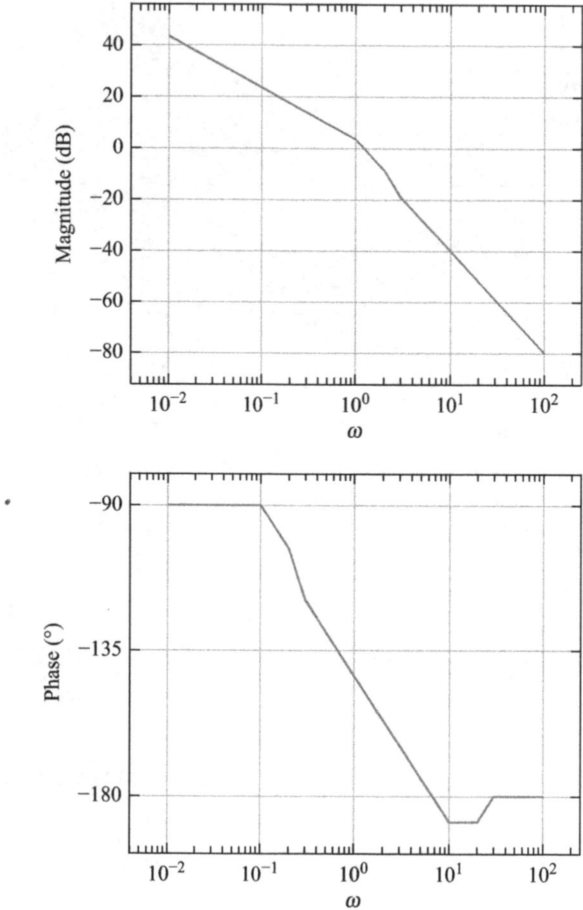

Figure 5.18 Bode diagram of Exercise 5.6

Exercise 5.7 Figure 5.19 shows the Bode diagram of transfer function $G(s)$. The plant's transfer function is:

A) $G(s) = \dfrac{1}{(s+1)(s+2)}$

B) $G(s) = \dfrac{s+1}{s^2(s+2)}$

C) $G(s) = \dfrac{s+3}{s(s+1)(s+2)}$

D) $G(s) = \dfrac{(s+3)(s+4)}{s(s+1)(s+2)}$.

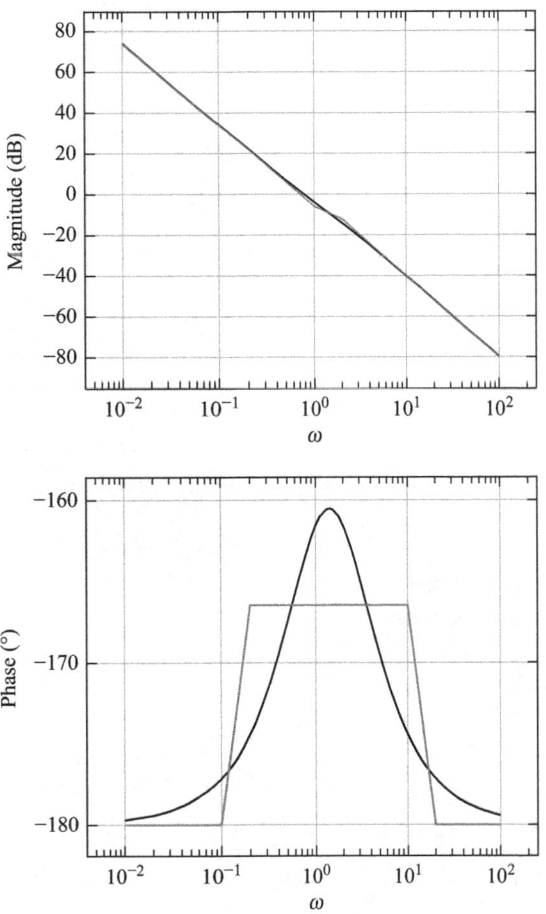

Figure 5.19 Bode diagram of Exercise 5.7

Exercise 5.8 Figure 5.20 shows the gain plot of the Bode diagram of plant $G(s)$, together with its asymptotes. The plant's transfer function is:

A) $G(s) = \dfrac{10\,(s+1)}{s\,(s+10)}$

B) $G(s) = \dfrac{10}{s(s+1)}$

C) $G(s) = \dfrac{10}{s^2\,(s+1)}$

D) $G(s) = \dfrac{10}{s}$.

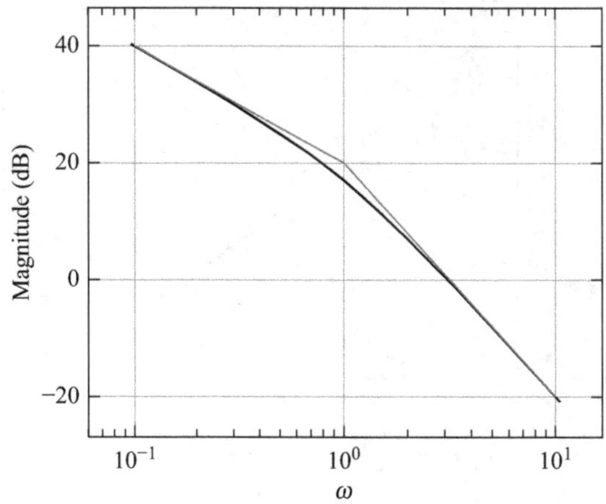

Figure 5.20 Gain plot of the Bode diagram of Exercise 5.8

Exercise 5.9 Figure 5.21 shows the asymptotes of the gain plot of the Bode diagram of plant $G(s)$. The plant's transfer function is:

A) $G(s) = \dfrac{1}{s^2+s+1}$

B) $G(s) = \dfrac{1}{s+1}$

C) $G(s) = \dfrac{1}{s^2+1}$

D) None of the above.

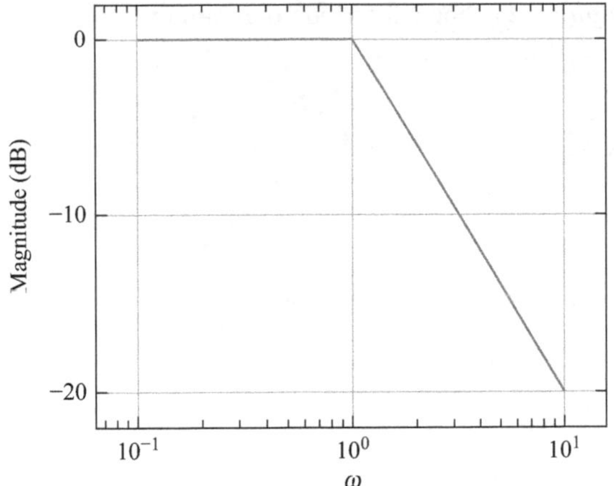

Figure 5.21 Asymptotes of the gain plot of the Bode diagram of Exercise 5.9

Exercise 5.10 Figure 5.22 shows the asymptotes of the gain plot of the Bode diagram of a plant. The plant's transfer function is:

A) $G(s) = \dfrac{2(s+1)}{s(s+2)}$

B) $G(s) = \dfrac{s}{(s+1)(s+2)}$

C) $G(s) = \dfrac{1}{(s+1)(s+2)}$

D) None of the above.

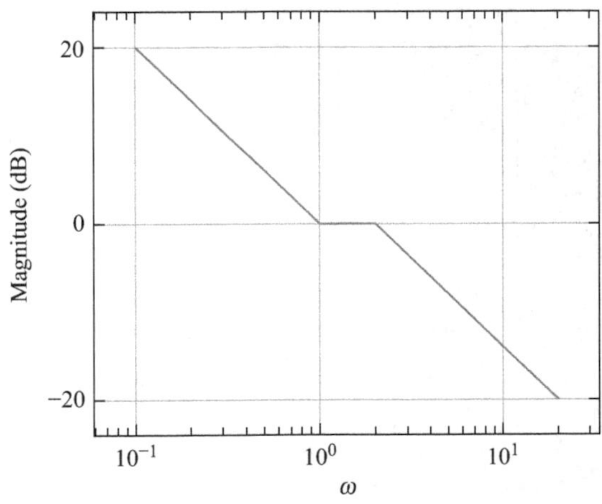

Figure 5.22 Asymptotes of the gain plot of the Bode diagram of Exercise 5.10

Exercise 5.11 Figure 5.23 shows the Bode diagram of a plant. The plant's transfer function is:

A) $G(s) = \dfrac{1}{s^2 + s + 1}$

B) $G(s) = \dfrac{1}{s + 1}$

C) $G(s) = \dfrac{1}{s^2 + 1}$

D) None of the above.

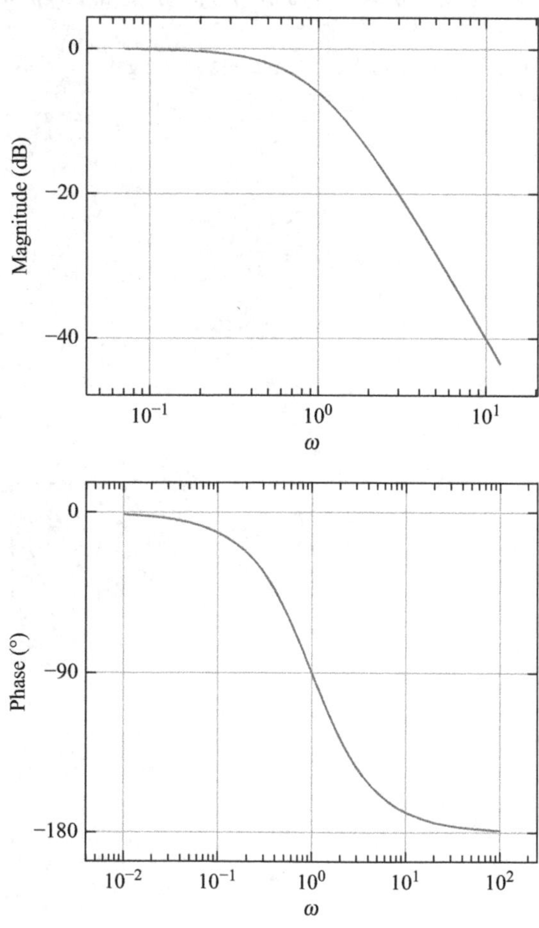

Figure 5.23 Bode diagram of Exercise 5.11

Exercise 5.12 Figure 5.24 shows the asymptotes of the gain plot of the Bode diagram of a plant. The plant's transfer function is:

A) $G(s) = \dfrac{3.43(s+5)}{(s+2)(s+6)}$

B) $G(s) = \dfrac{8.53(s+5)}{(s+2)^2(s+6)}$

C) $G(s) = \dfrac{7.51(s+5)}{(s+2)^2(s+6)}$

D) $G(s) = \dfrac{6.33(s+5)}{(s+2)(s+6)^2}$.

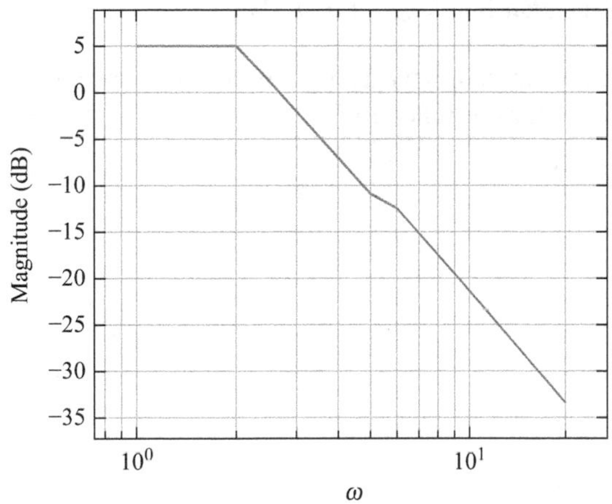

Figure 5.24 Asymptotes of the gain plot of the Bode diagram of Exercise 5.12

Exercise 5.13 Figure 5.25 shows the asymptotes of the gain plot of the Bode diagram of a plant. The plant's transfer function is:

A) $G(s) = \dfrac{2(s+3)}{s^2(s+2)}$

B) $G(s) = \dfrac{2(s+3)}{s(s+2)}$

C) $G(s) = \dfrac{4(s+3)}{s+2}$

D) $G(s) = \dfrac{4(s+3)}{3s(s+2)}$.

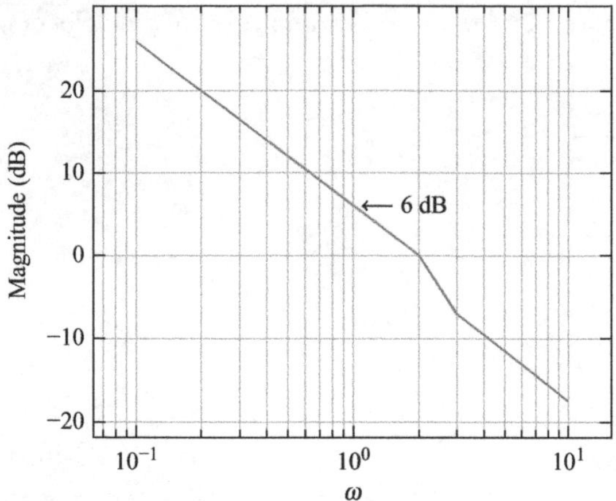

Figure 5.25 Asymptotes of the gain plot of the Bode diagram of Exercise 5.13

Exercise 5.14 Figure 5.26 shows the asymptotes of the gain plot of the Bode diagram of a plant. The plant's transfer function is:

A) $G(s) = \dfrac{10}{s(s+2)(s+5)}$

B) $G(s) = \dfrac{14}{s(s+2)(s+5)}$

C) $G(s) = \dfrac{1}{s(s+2)(s+5)}$

D) None of the above.

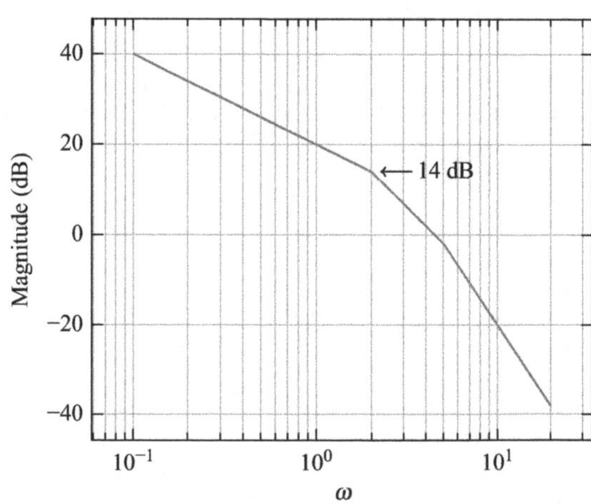

Figure 5.26 Asymptotes of the gain plot of the Bode diagram of Exercise 5.14

Exercise 5.15 Figure 5.27 shows the phase plot of the Bode diagram of plant $G(s)$, together with its asymptotes. $G(s)$ has a positive gain K. Its transfer function is:

A) $G(s) = \dfrac{K}{s^2(s+1)}$

B) $G(s) = \dfrac{K}{s+1}$

C) $G(s) = \dfrac{K}{s(s+1)^2}$

D) $G(s) = \dfrac{K}{s(s+1)}$.

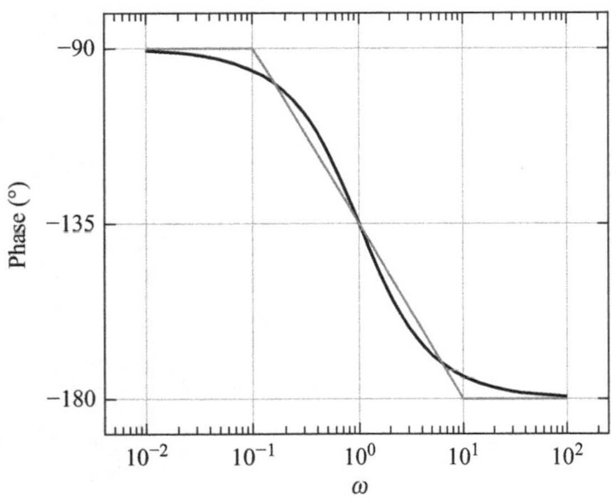

Figure 5.27 Phase plot of the Bode diagram of Exercise 5.15

Exercise 5.16 Consider the second-order plant $G(s) = \dfrac{\omega_n^2}{s^2 + 2\zeta\omega_n s + \omega_n^2}$, where ζ is the damping factor and ω_n the (undamped) natural frequency. Figure 5.28 shows the poles of $G(s)$ when $0 < \zeta < 1$ (under-damped system). In this case:

A) $\theta = \arccos(\zeta)$
B) $\theta = \arcsin(\zeta)$
C) $\theta = \arctan(\zeta)$
D) None of the above.

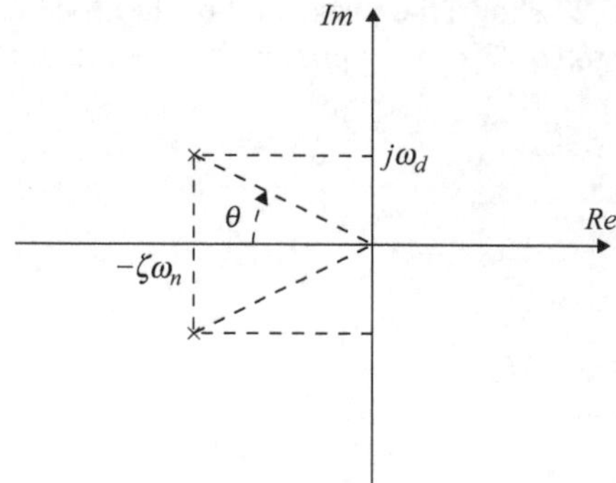

Figure 5.28 Poles of G(s) when $0 < \zeta < 1$, of Exercise 5.16

Exercise 5.17 Figure 5.29 shows the Bode diagram of a plant with transfer function $\frac{Y(s)}{R(s)} = \frac{\omega_n^2}{s^2 + 2\zeta\omega_n s + \omega_n^2}$. The damping coefficient ζ and the undamped natural frequency ω_n are:

A) $\zeta = 1.0,\ \omega_n = 10$ rad/s
B) $\zeta = 2.0,\ \omega_n = 1$ rad/s
C) $\zeta = 0.1,\ \omega_n = 2$ rad/s
D) $\zeta = 0.4,\ \omega_n = 5$ rad/s.

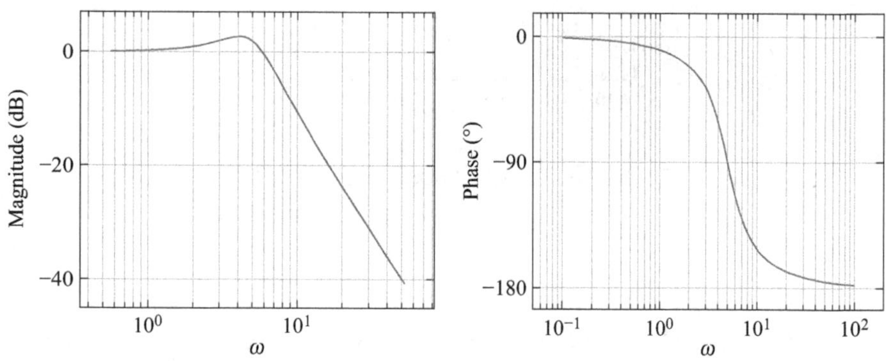

Figure 5.29 Bode diagram of Exercise 5.17

Exercise 5.18 Consider a plant with the transfer function $G(s) = \frac{0.5}{s(s+1)}e^{-2s}$. Find its gain and phase margins.

Exercise 5.19 Consider a plant with the open-loop transfer function $G(s) = \dfrac{Y(s)}{R(s)} = \dfrac{16}{s^2 + 4s + 16}$.

1. Find the damping factor ζ and the undamped natural frequency ω_n.
2. Find the rise time t_r, the peak of the response $y(t_p)$ and the peak time t_p, when the input is a unit step $r(t) = 1,\ t \geq 0$.
3. Find the maximum gain M_r of the frequency response $G(j\omega)$ and the frequency ω_r at which this is to be found.
4. Plot $G(j\omega)$ in a Bode diagram.

Exercise 5.20 Consider a plant with the open-loop transfer function $G(s) = \dfrac{10K}{s(s+1)^2}$, with $K = 0.1$. The phase margin is:

A) $PM = 11.4°$
B) $PM = 21.4°$
C) $PM = 31.4°$
D) $PM = 41.4°$.

Exercise 5.21 Consider a second-order system with transfer function $\dfrac{Y(s)}{R(s)} = \dfrac{\omega_n^2}{s^2 + 2\zeta\omega_n s + \omega_n^2}$. Figure 5.30 shows the gain of its frequency response $G(j\omega)$, with resonance frequency $\omega_r = 2.474$ rad/s and peak gain $M_r = 2.696$ dB. Find the values of the undamped natural frequency ω_n and the damping coefficient ζ.

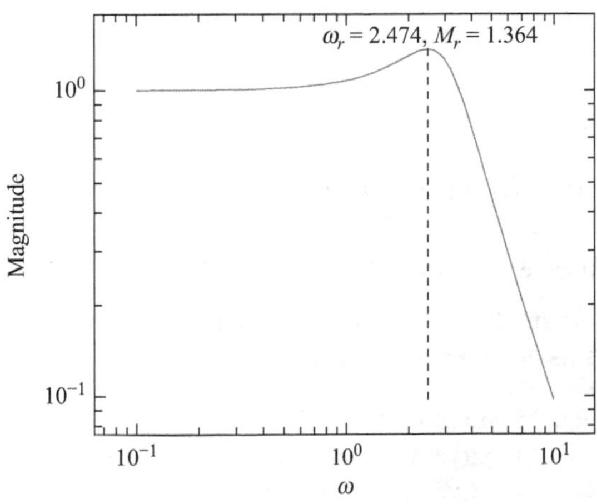

Figure 5.30 Gain of Exercise 5.21

Exercise 5.22 Consider the fluid level control system in Figure 5.31. The closed loop includes a proportional controller such that $q_i = K(r - h_2)$, where r is the reference. Let $h_2(t)$ be the output and $q_d(t)$ a disturbance.

1. Draw the closed-loop block diagram.
2. Determine the steady-state response $h_2(\infty)$ when the disturbance is a unit step, $q_d(t) = 1$, and $r(t) = 0$.
3. Find analytical expressions for the undamped natural frequency ω_n and the damping coefficient ζ of the closed-loop transfer function, when $q_d(t) = 0$.

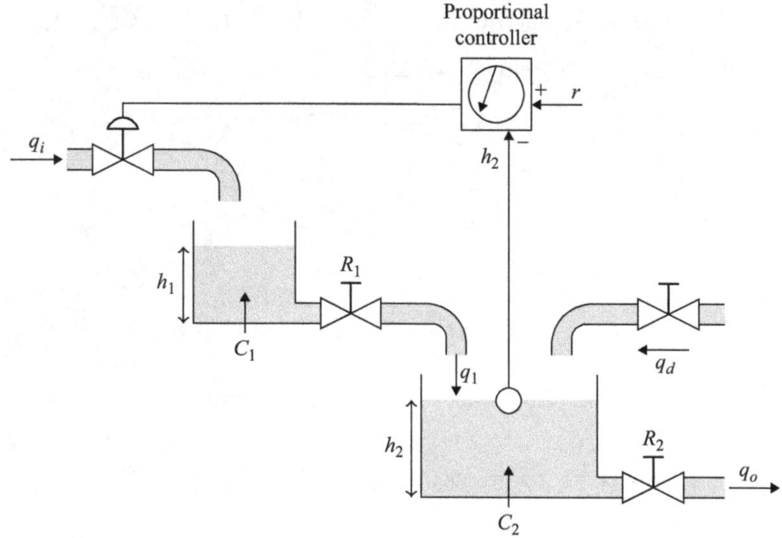

Figure 5.31 Fluid level control system of Exercise 5.22

5.3.2 Nyquist and Nichols diagrams

Exercise 5.23 Consider the second-order plant $G(s) = \dfrac{\omega_{n^2}}{s^2 + 2\zeta\omega_n s + \omega_{n^2}}$, where $\zeta \in \{0.25, 0.5, 1\}$ is the damping factor and ω_n the (undamped) natural frequency. Figure 5.32 shows its polar diagram. Then:

A) Curve A: $\zeta = 0.25$; curve B: $\zeta = 0.5$; curve C: $\zeta = 1$
B) Curve C: $\zeta = 0.25$; curve A: $\zeta = 0.5$; curve B: $\zeta = 1$
C) Curve B: $\zeta = 0.25$; curve C: $\zeta = 0.5$; curve A: $\zeta = 1$
D) None of the above.

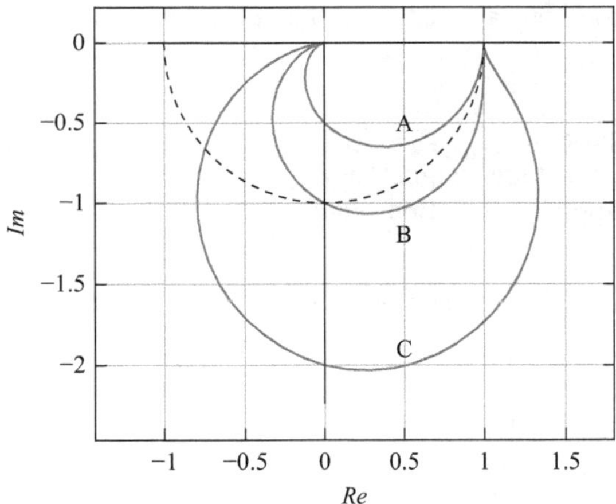

Figure 5.32 Polar diagram of Exercise 5.23

Exercise 5.24 Figure 5.33 shows the polar diagram of a plant with transfer function $G(s) = e^{-sT}$, $T > 0$. Then:

A) $\omega_0 = \pi$ rad/s

B) $\omega_0 = \dfrac{\pi}{T}$ rad/s

C) $\omega_0 = \dfrac{\pi}{2T}$ rad/s

D) None of the above.

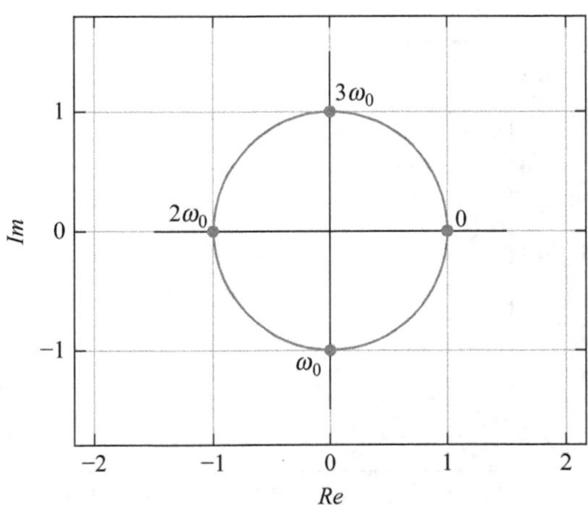

Figure 5.33 Polar diagram of Exercise 5.24

Exercise 5.25 Consider a plant with the open-loop transfer function $G(s) = \frac{30}{s(s+3)}e^{-0.1s}$.

1. Draw its Nyquist plot.
2. Find its gain and phase margins.
3. Draw the asymptotes of its Bode diagram.

Exercise 5.26 Consider a plant with the open-loop transfer function $G(s) = \frac{K(s+2)}{s(s+1)(s+3)(s+4)}$, with $K = 20$.

1. Draw its Nyquist plot.
2. Draw the asymptotes of the gain plot of its Bode diagram.

Exercise 5.27 Consider a plant with the open-loop transfer function $G(s) = \frac{K(s+1)}{s(s-1)}$, with $K > 0$.

1. Draw the plant's Nyquist plot.
2. Let $K = 2$. Draw the asymptotes of the open loop's Bode diagram, find the gain and phase margins, and comment on stability.

Exercise 5.28 Figure 5.34 sketches the Nyquist diagram of a plant with open-loop transfer function $G(s)H(s) = \frac{12}{s(s+2)(s+4)}$.

1. At frequency $\omega = \sqrt{8}$, the value of $\underline{/GH}$, is:
   A) $\underline{/GH} = -\pi/2$
   B) $\underline{/GH} = -\pi$

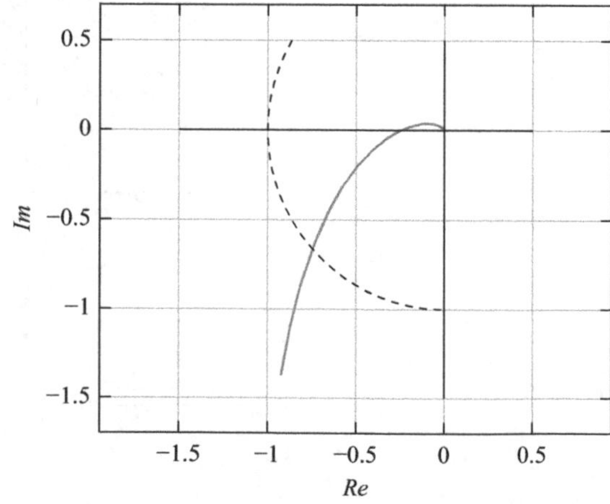

Figure 5.34 Nyquist diagram of Exercise 5.28

C) $\underline{/GH} = -3\pi/4$
D) $\underline{/GH} = +\pi/2$.

2. At frequency $\omega = \sqrt{8}$ the value of $|GH|$, is:
   E) $|GH| = 6$
   F) $|GH| = 1$
   G) $|GH| = 1/4$
   H) $|GH| = 1/12$.
3. At frequency $\omega = 1.22356$ rad/s we have $|GH| = 1$. Thus, the plant's phase margin is:
   I) 41.5°
   J) −30.0°
   K) 52.0°
   L) 19.3°.
4. The plant's gain margin is:
   M) −12 dB
   N) 12 dB
   O) −3 dB
   P) 6 dB.

Exercise 5.29 Consider a unit feedback control system with transfer function in the direct loop $G(s) = \dfrac{2}{s+2}e^{-Ts}$, $T > 0$ and the Nyquist diagram shown in Figure 5.35. We conclude that the time delay T is:

A) $T = 1$ s
B) $T = 2$ s
C) $T = 3$ s
D) $T = 4$ s.

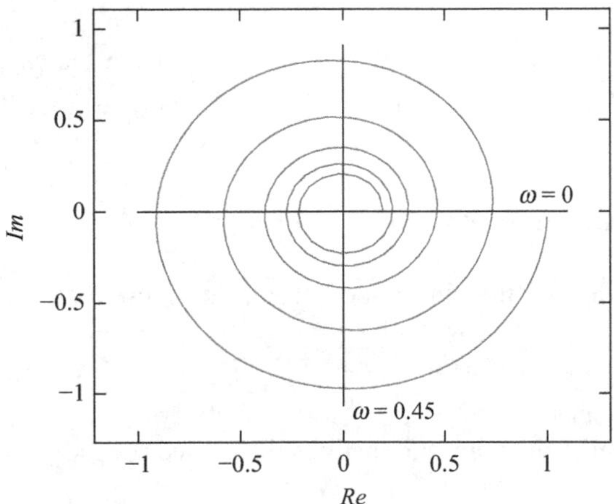

Figure 5.35 Nyquist diagram of Exercise 5.29

Exercise 5.30 Consider a plant with the open-loop transfer function $GH(s) = \dfrac{100(s+2)}{(s+1)(s+3)(s+4)(s+5)}$. Figure 5.36 shows the corresponding Nyquist diagram.

1. Point A, where $GH(j\omega)$ crosses the real axis, corresponds to frequency ω_π:
   A) $\omega_\pi = 2.039$ rad/s
   B) $\omega_\pi = 6.095$ rad/s
   C) $\omega_\pi = 2.818$ rad/s
   D) None of the above.
2. Point B, for which $|GH(j\omega)| = 1$, corresponds to frequency ω_1:
   E) $\omega_1 = 2.039$ rad/s
   F) $\omega_1 = 6.095$ rad/s
   G) $\omega_1 = 2.818$ rad/s
   H) None of the above.

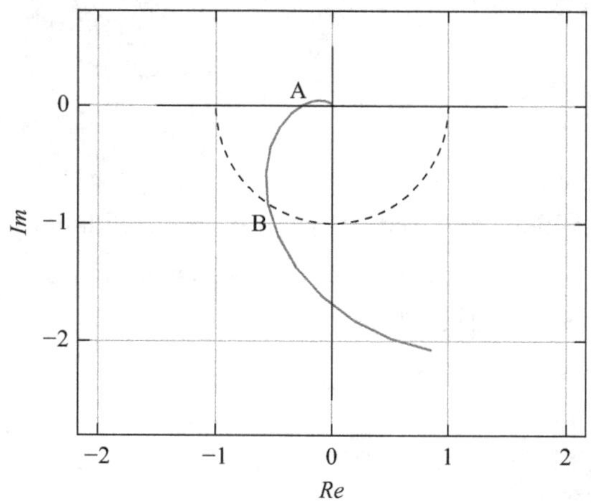

Figure 5.36 Nyquist diagram of Exercise 5.30

Exercise 5.31 The step responses of Figure 5.37, the Bode diagrams of Figure 5.38 and the Nichols diagrams of Figure 5.39 correspond to four different systems. Establish a correspondence between them.

5.3.3 Root-locus and frequency domain analysis

Exercise 5.32 Consider the closed loop shown in Figure 5.40 with the open-loop transfer function $GH(s) = \dfrac{K(s+2)(s+4)}{s(s+1)(s+3)}$, $K \geq 0$.

1. Plot its root-locus.
2. Plot the Nyquist diagram and analyze the closed loop's stability for different values of K.
3. Plot the Nichols diagram.

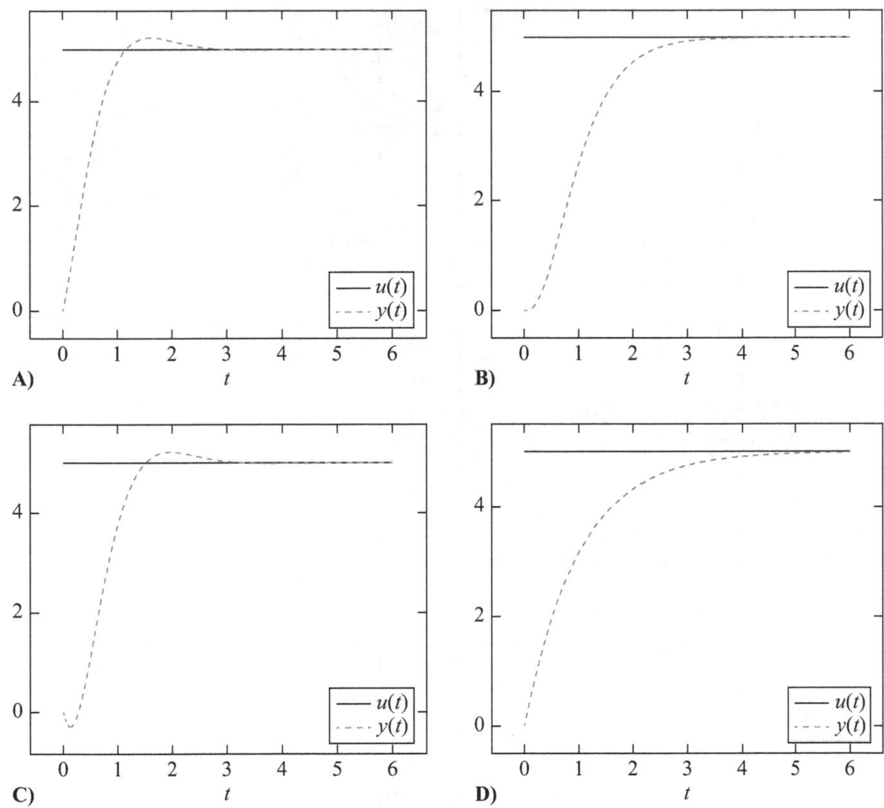

Figure 5.37 Unit step responses of Exercise 5.31

Exercise 5.33 Consider the closed loop shown in Figure 5.40 with the open-loop transfer function $GH(s) = \dfrac{K(s+1)}{s(s+2)(s+3)}$, $K \geq 0$.

1. Plot its root-locus.
2. Plot the Nyquist diagram and analyze the closed loop's stability for different values of K.
3. Plot the Nichols diagram.

Exercise 5.34 Consider a plant with the open-loop transfer function $GH(s) = \dfrac{K(s+0.5)}{s^2(s+2)(s+3)}$. Figure 5.41 shows the corresponding root-locus plot for $K > 0$ and the Nyquist diagram for $K = 10$. Then:

1. The point where the asymptotes meet is
   A) $\sigma = -1.5$
   B) $\sigma = -1.0$
   C) $\sigma = -2.0$
   D) None of the above.

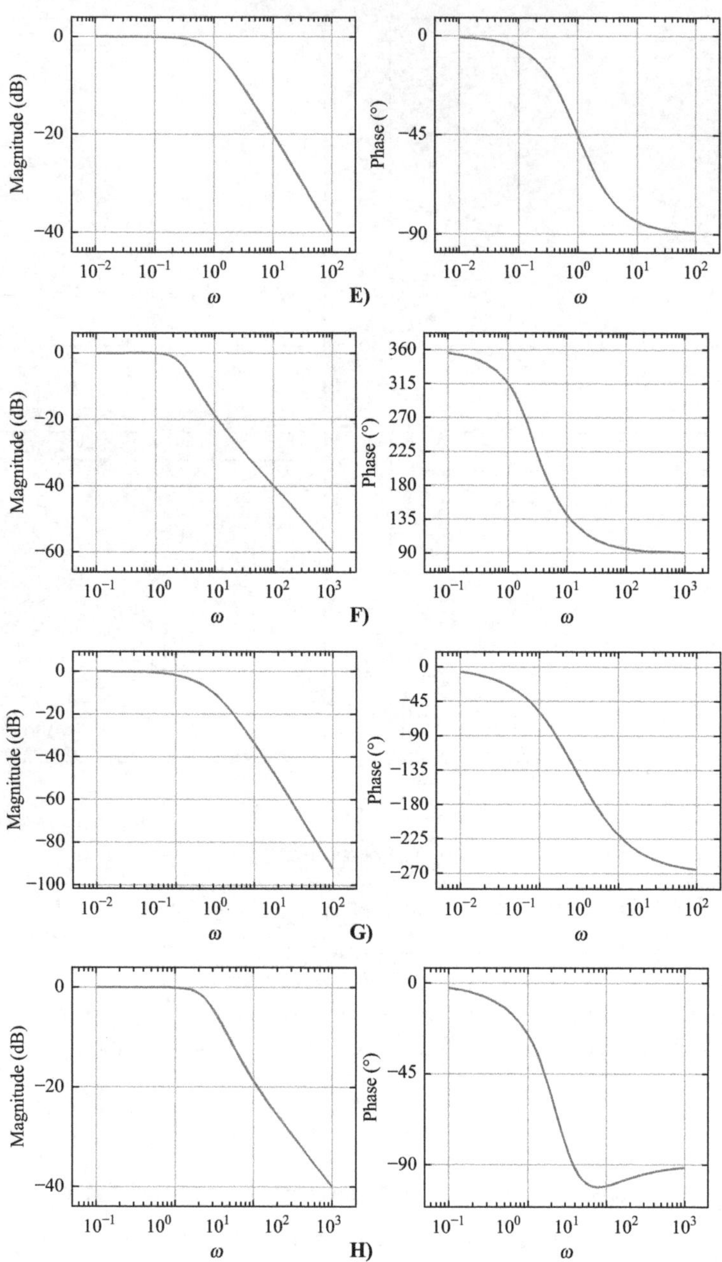

Figure 5.38 Bode diagrams of Exercise 5.31

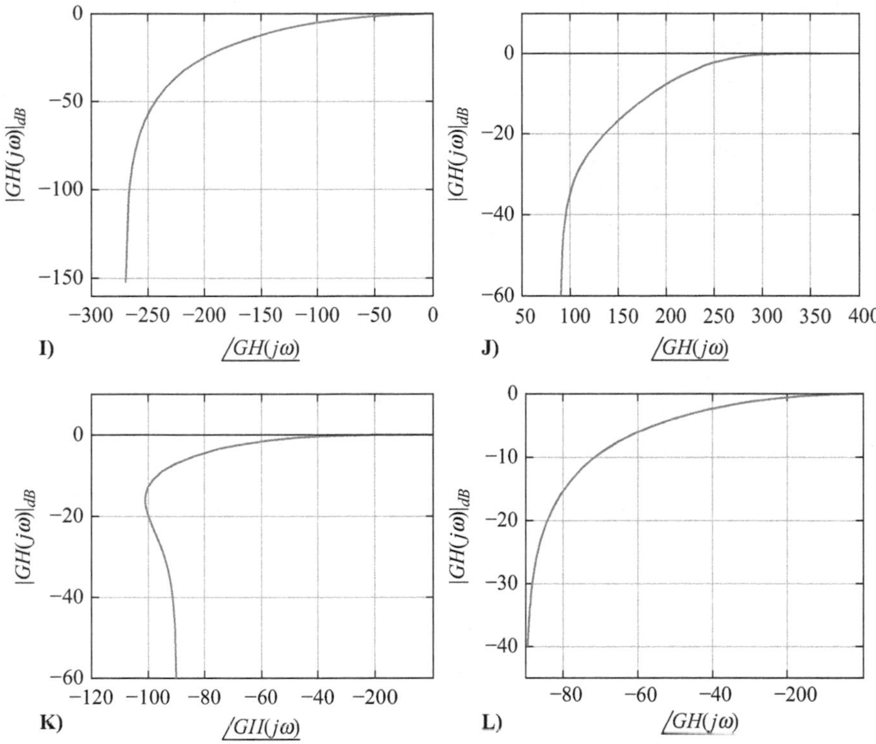

Figure 5.39 Nichols diagrams of Exercise 5.31

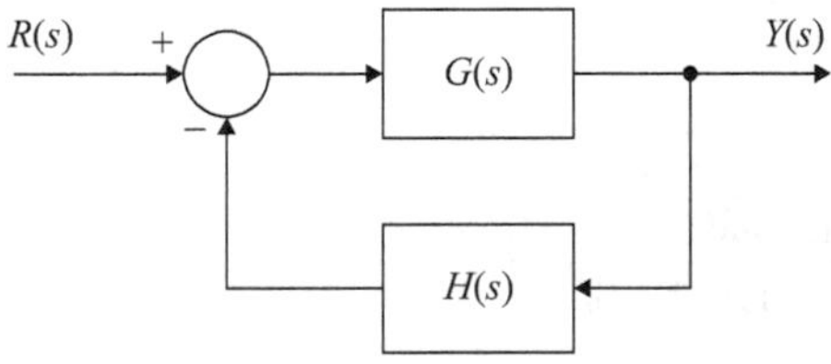

Figure 5.40 Closed loop of Exercises 5.32 and 5.33

2. The limit of stability occurs for
   E) $K = 10.5$, $s = \pm j\ 1.153$ rad/s
   F) $K = 15.5$, $s = \pm j\ 1.567$ rad/s
   G) $K = 17.5$, $s = \pm j\ 1.871$ rad/s
   H) None of the above.

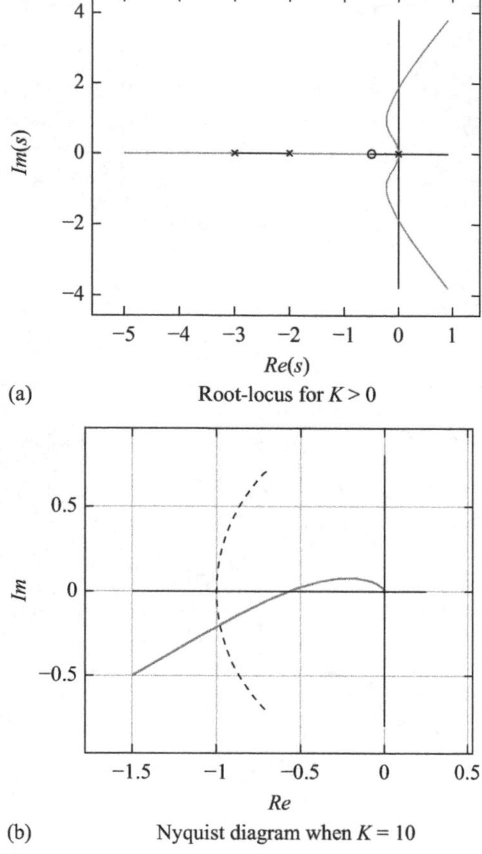

(a) Root-locus for $K > 0$

(b) Nyquist diagram when $K = 10$

Figure 5.41 Plots of Exercise 5.34

3. When $K = 10$, the phase margin in closed loop is
   I) $10.5°$ for $\omega = 1.153$ rad/s
   J) $11.5°$ for $\omega = 1.346$ rad/s
   K) $4.86°$ for $\omega = 1.871$ rad/s
   L) None of the above.
4. Again for $K = 10$, the gain margin in closed loop is
   M) 10.5 dB for $\omega = 1.153$ rad/s
   N) 11.5 dB for $\omega = 1.346$ rad/s
   O) 4.86 dB for $\omega = 1.871$ rad/s
   P) None of the above.
5. The asymptotic plot of the gain Bode diagram in open loop $GH(s)$ is (see Figure 5.42)

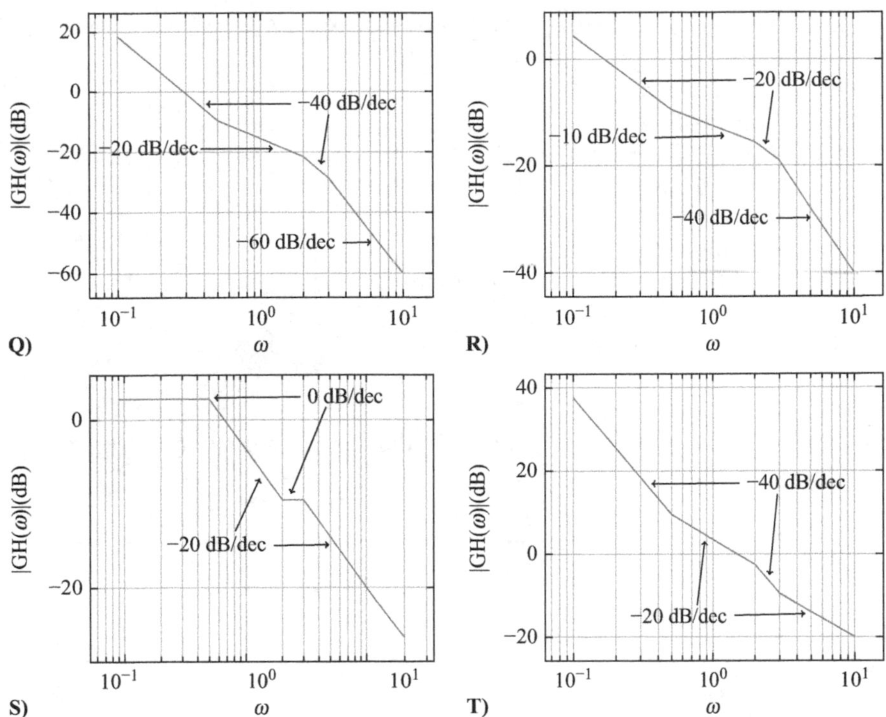

Figure 5.42 Gain plots of the Bode diagram of Exercise 5.34

Exercise 5.35 Consider the root-locus of a plant with the open-loop transfer function $GH(s) = \dfrac{K(s+2)}{s(s+1)(s+5)}$, $K > 0$.

1. The point on the real axis σ where the asymptotes intersect is:
   A) $\sigma = -1.5$
   B) $\sigma = -1.0$
   C) $\sigma = -2.0$
   D) None of the above.
2. The limit of stability (the value of K for which the root-locus crosses the imaginary axis) is:
   E) $K = 1$
   F) $K = 2$
   G) $K = 3$
   H) None of the above.
3. Asymptotes are as follows:
   I) There is an asymptote at an angle $\theta = 180°$ with the positive real axis.
   J) There are two asymptotes at angles $\theta = \pm 90°$ with the positive real axis.

K) There are two asymptotes at angles $\theta = \pm 60°$ and $\theta = 180°$ with the positive real axis.
L) None of the above.

4. The root-locus diverges:
M) At $\sigma = -0.128$ for $K = 0.16$.
N) At $\sigma = -0.558$ for $K = 0.76$.
O) Nowhere
P) None of the above.

5. The Nyquist diagram $GH(j\omega)$ verifies, for $\omega \to 0^+$:
Q) $\text{Re}[GH(j\omega)] = -(7/25)K$
R) $\text{Re}[GH(j\omega)] = -(2/5)K$
S) $\text{Re}[GH(j\omega)] = 0$
T) None of the above.

6. The asymptotic plot of the gain Bode diagram in open loop $GH(s)$ is (see Figure 5.43)

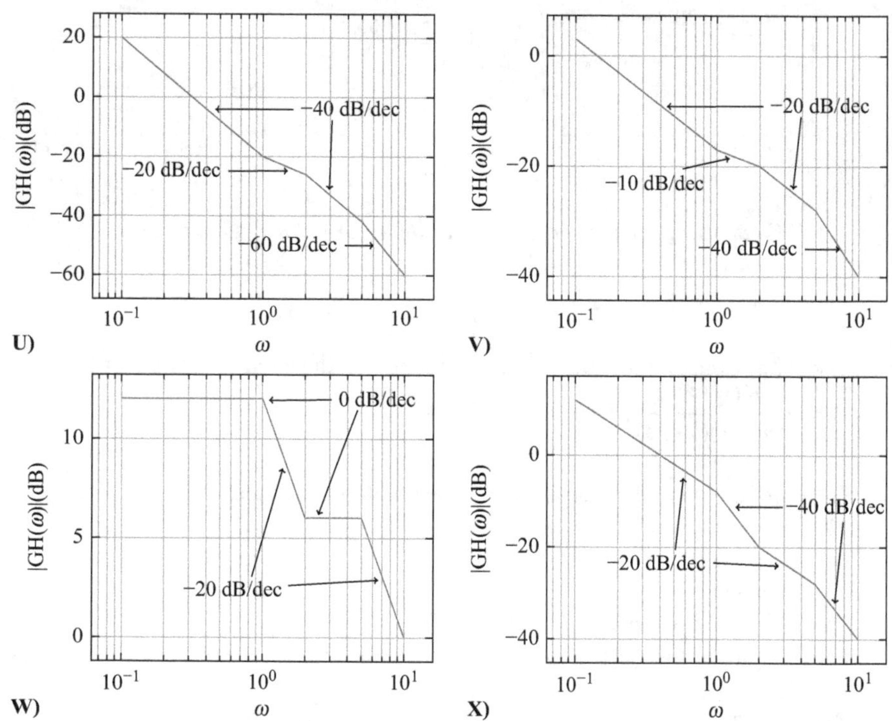

Figure 5.43 Gain plots of the Bode diagram of Exercise 5.35

Exercise 5.36 Consider the root-locus of a plant with the open-loop transfer function $GH(s) = \dfrac{K}{(s+1)(s^2+8s+20)}$, $K \geq 0$).

1. The point on the real axis σ where the asymptotes intersect is:
   A) $\sigma = -2.334$
   B) $\sigma = -3.000$
   C) $\sigma = -1.400$
   D) None of the above.
2. The root-locus plot crosses the imaginary axis for the following frequencies ω and corresponding values of gain K:
   E) $\omega = \pm 5.29$ and $K = 232$
   F) $\omega = \pm 3.30$ and $K = 40$
   G) $\omega = \pm 4.10$ and $K = 102$
   H) None of the above.
3. The departure angle α from pole $p_1 = -4 + j2$ is:
   I) $\alpha = -43.5°$
   J) $\alpha = -56.3°$
   K) $\alpha = -65.1°$
   L) None of the above.
4. Point $s = -3.8$ in the complex plane:
   M) belongs to the root-locus of $GH(s)$ for $K \geq 0$
   N) does not belong to the root-locus of $GH(s)$ for $K \geq 0$
5. When $K = 5$, the plant's gain margin is:
   O) $GM = -5.00$ dB
   P) $GM = 30.00$ dB
   Q) $GM = 33.33$ dB
   R) $GM = 6.00$ dB.

Exercise 5.37 Figure 5.44 shows the root-locus plot of a plant with the open-loop transfer function $GH(s) = \dfrac{K}{s(s+2)(s+1)}$, $K > 0$.

1. Point B, where asymptotes cross the real axis, is:
   A) $\sigma = -1.500$
   B) $\sigma = -1$
   C) $\sigma = -1.667$
   D) None of the above.
2. Point A, where the root-locus leaves the real axis, is:
   E) $\sigma = -1.500$, $K = 7.071$
   F) $\sigma = -0.785$, $K = 2.113$
   G) $\sigma = -0.423$, $K = 0.385$
   H) None of the above.
3. Point C, where the root-locus crosses the imaginary axis, is:
   I) $\omega = 0.785$ e, $K = 7.701$
   J) $\omega = 1.667$ e, $K = 2.113$
   K) $\omega = 1.41$, $K = 6$
   L) None of the above.

4. When $K = 1$, the plant's phase margin is:
   M) $PM = 53.4°$
   N) $PM = -3.0°$
   O) $PM = 2.1°$
   P) None of the above.
5. When $K = 1$, the plant's gain margin is:
   Q) $GM = 15.6$ dB
   R) $GM = 20.2$ dB
   S) $GM = 5.5$ dB
   T) None of the above.

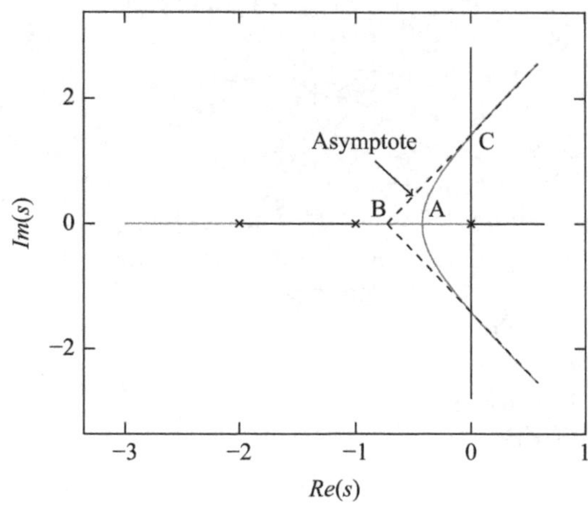

Figure 5.44 Root-locus of Exercise 5.37

Exercise 5.38 Consider the plant in Figure 5.45, where $G(s) = \dfrac{1}{(s+1)(s+3)(s+8)}$, $K > 0$.

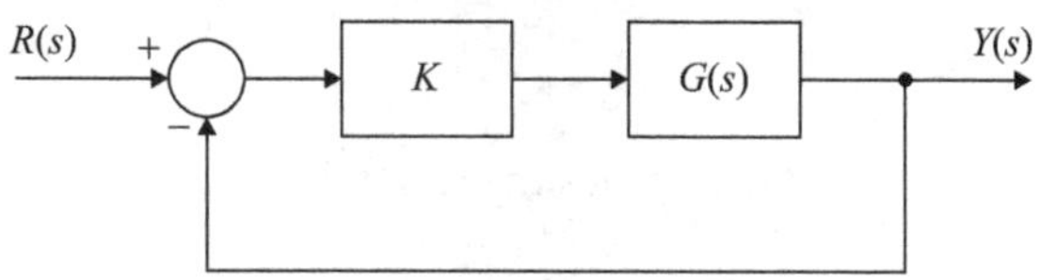

Figure 5.45 Closed loop of Exercise 5.38

1. Plot its root-locus.
2. For which values of K is the system stable?
3. Can you find a 5% steady-state error e_{ss} for a unit-step input, solely by changing gain K?
4. Plot the Bode diagram of $G(s)$ for $K = 240$.

5.4 Frequency domain analysis using computer packages

This section presents several commands for handling system frequency analysis using the computer packages MATLAB, SCILAB and OCTAVE.

5.4.1 MATLAB

This subsection describes some basic commands that can be adopted with the package MATLAB.

We consider Problem 5.7. The transfer function $G(s) = \dfrac{Y(s)}{R(s)} = \dfrac{25}{s^2 + 4s + 25}$ is represented by two arrays, num and den.

The Bode diagrams of amplitude and phase are obtained by means of the command bode.

The complete code is as follows.

```
%%%% Bode diagram %%%%
% Numerator and denominator of transfer function
num = [0 0 25];
den = [1 4 25];

% Bode plots
bode(num,den)

% Formatting the chart
grid
title('Bode diagrams of G(s) = 25/(s^2+4s+25)')
```

MATLAB creates the figure window represented in Figure 5.46.

The Nyquist or the Nichols diagrams can be obtained using the commands nyquist or nichols.

The complete code is as follows.

```
%%%% Nyquist diagram %%%%
% Numerator and denominator of transfer function
num = [0 0 25];
den = [1 4 25];

% Nyquist plot
nyquist(num,den)

% Formatting the chart
grid
title('Nyquist diagram of G(s) = 25/(s^2+4s+25)')
```

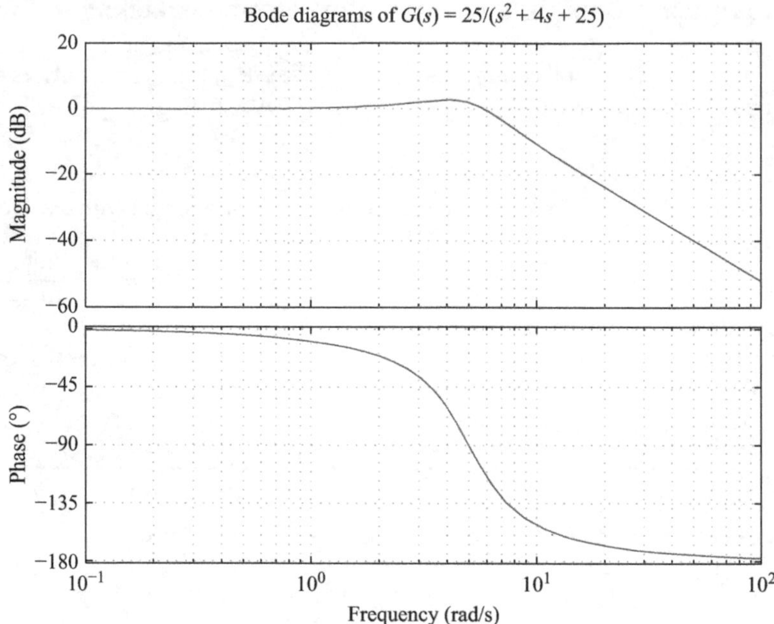

Figure 5.46 Bode diagrams of amplitude and phase for the transfer function $G(s) = \frac{Y(s)}{R(s)} = \frac{25}{s^2+4s+25}$ *using MATLAB*

```
%%%% Nichols diagram %%%%
% Numerator and denominator of transfer function
num = [0 0 25];
den = [1 4 25];

% Nichols  plot
nichols (num,den)

% Formatting the chart
grid
title('Nichols diagram of G(s)  = 25/(s^2+4s+25)')
```

MATLAB creates the figure windows represented in Figure 5.47 or Figure 5.48.

For obtaining the gain and phase margins and the crossover frequencies (in [rad]), we can use the command `margin`.

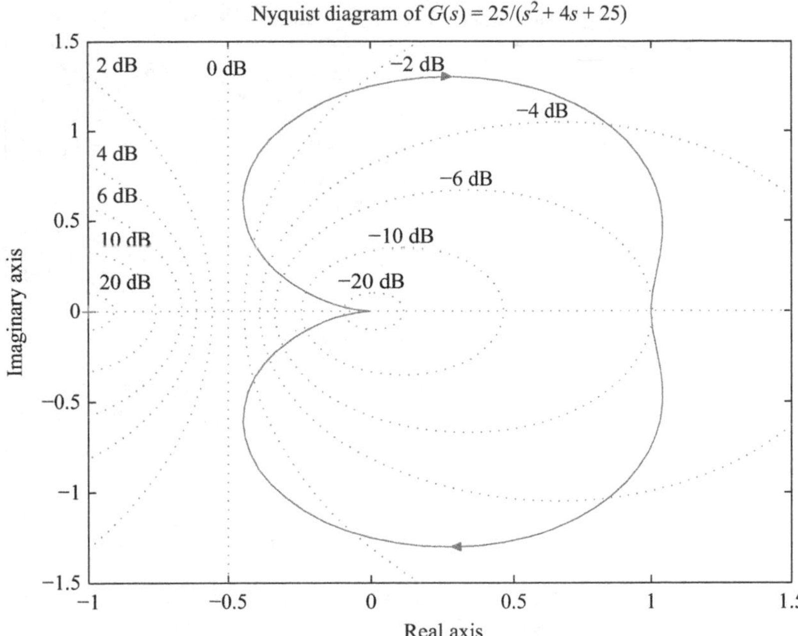

Figure 5.47 *Nyquist diagram for the transfer function* $G(s) = \frac{Y(s)}{R(s)} = \frac{25}{s^2+4s+25}$ *using MATLAB*

```
%%%% Gain and Phase margins %%%%
% Numerator and denominator of transfer function
num = [0 0 25];
den = [1 4 25];

% Create transfer function model, convert to transfer
% function model
G = tf(num,den);

% Gain margin, phase margin, and crossover frequencies
% [rad]
[Gm,Pm,Wgm,Wpm] = margin(G)
```

MATLAB creates the following Command window.

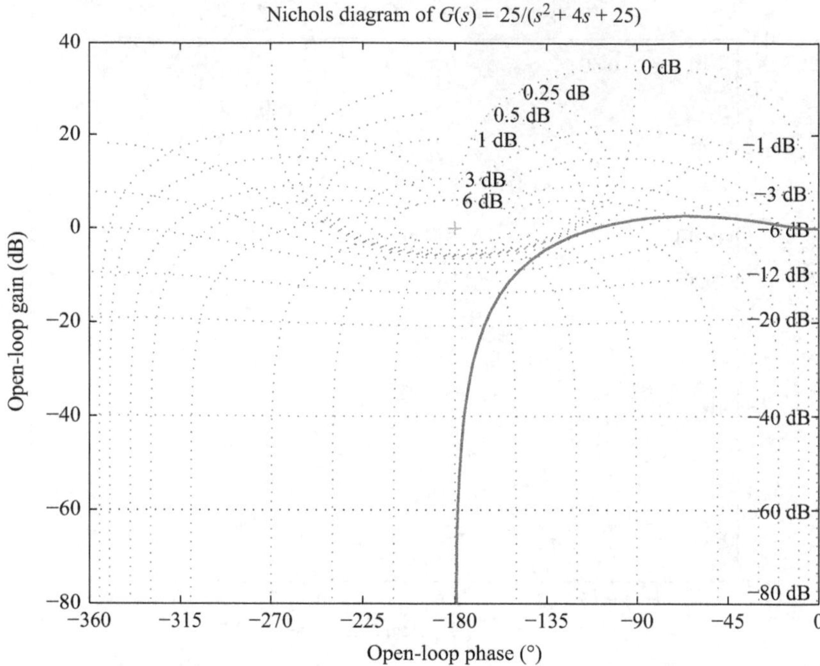

Figure 5.48 Nichols diagram for the transfer function $G(s) = \frac{Y(s)}{R(s)} = \frac{25}{s^2+4s+25}$ *using MATLAB*

```
Gm =

   Inf


Pm =

   68.9154


Wgm =

   Inf


Wpm =

    5.8302
```

5.4.2 SCILAB

This subsection describes some basic commands that can be adopted with the package SCILAB.

We consider Problem 5.7. The transfer function $G(s) = \dfrac{Y(s)}{R(s)} = \dfrac{25}{s^2 + 4s + 25}$ is represented by two arrays num and den.

The Bode diagrams of amplitude and phase are obtained by means of the command bode.

The complete code is as follows.

```
//// Bode diagram ////
// Numerator and denominator of transfer function
num = poly([25 0 0],'s','coeff');
den = poly([25 4 1],'s','coeff');

//Create a Scilab continuous system LTI object
G = syslin('c',num,den)

// Bode plots
bode(G)

// Formatting the chart
title('Bode diagrams of G(s) = 25/(s^2+4s+25)')
```

SCILAB creates the figure window represented in Figure 5.49.

The Nyquist or the Nichols diagrams can be obtained using the commands nyquist or nichols.

The complete code is as follows.

```
//// Nyquist diagram ////
// Numerator and denominator of transfer function
num = poly([25 0 0],'s','coeff');
den = poly([25 4 1],'s','coeff');

//Create a Scilab continuous system LTI object
G = syslin('c',num,den)

// Nyquist plot
nyquist(G)

// Formatting the chart
title('Nyquist diagram of G(s) = 25/(s^2+4s+25)')
```

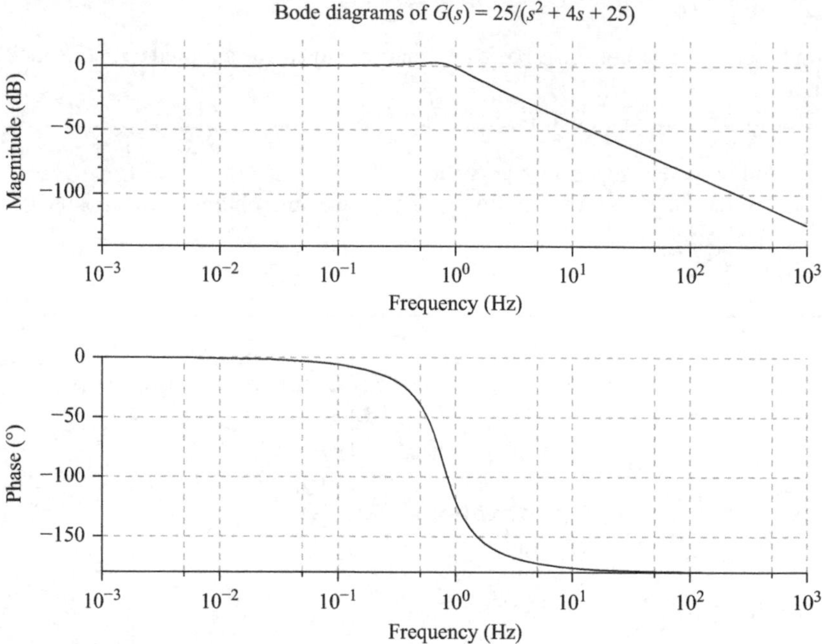

Figure 5.49 Bode diagrams of amplitude and phase for the transfer function $G(s) = \frac{Y(s)}{R(s)} = \frac{25}{s^2+4s+25}$ *using SCILAB*

```
//// Nichols diagram ////
// Numerator and denominator of transfer function
num = poly([25 0 0],'s','coeff');
den = poly([25 4 1],'s','coeff');

//Create a Scilab continuous system LTI object
G = syslin('c',num,den)

// Nichols  plot
black(G)

// Formatting the chart
title('Nichols diagram of G(s)  = 25/(s^2+4s+25)')
```

SCILAB creates the figure windows represented in Figure 5.50 or Figure 5.51.

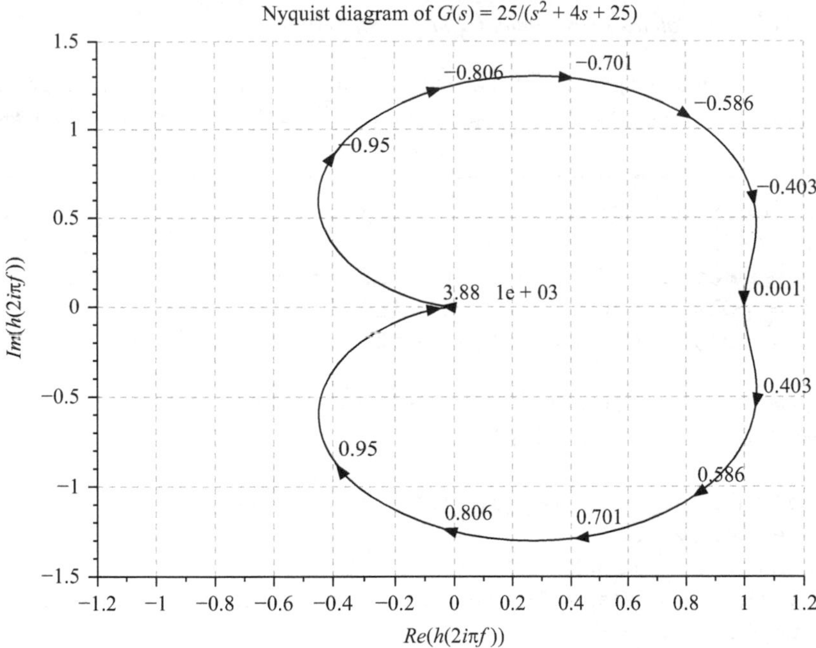

Figure 5.50 Nyquist diagram for the transfer function $G(s) = \frac{Y(s)}{R(s)} = \frac{25}{s^2+4s+25}$ *using SCILAB*

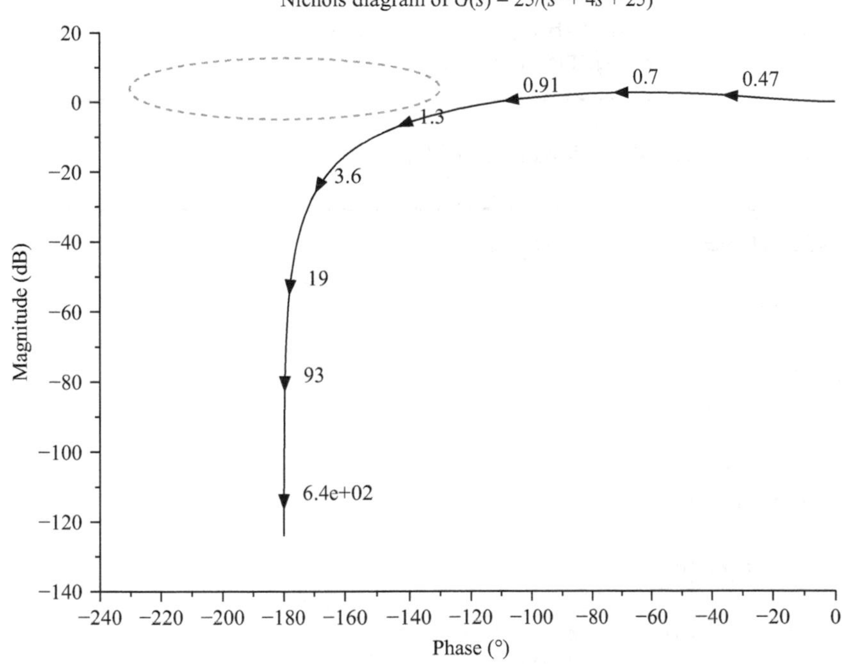

Figure 5.51 Nichols diagram for the transfer function $G(s) = \frac{Y(s)}{R(s)} = \frac{25}{s^2+4s+25}$ *using SCILAB*

For obtaining the gain and phase margins and the crossover frequencies, we can use the commands `g_margin` and `p_margin`. The complete code is as follows.

```
//// Gain and Phase margins ////

// Numerator and denominator of transfer function
num = poly([25 0 0],'s','coeff');
den = poly([25 4 1],'s','coeff');

//Create a Scilab continuous system LTI object
G = syslin('c',num,den)

//Calculates gain margin [dB] and corresponding
//frequency [Hz]
[GainMargin,freqGM] = g_margin(G)
//Calculates phase [deg] and corresponding freq [Hz]
//of phase margin
[PhaseMargin,freqPM] = p_margin(G)

disp(GainMargin,"Gain Margin [dB]")
disp(freqGM,"freq GM [Hz]")

disp(PhaseMargin,"Phase Margin [deg]")
disp(freqPM,"freq PM [Hz]")
```

SCILAB creates the following Console.

```
Gain Margin [dB]

   Inf

freq GM [Hz]

    []

Phase Margin [deg]

   68.899804

freq PM [Hz]

   0.9280248
```

5.4.3 OCTAVE

This subsection describes some basic commands that can be adopted with the package OCTAVE. It is required to load package `control`.

We consider Problem 5.7. The transfer function $G(s) = \dfrac{Y(s)}{R(s)} = \dfrac{25}{s^2+4s+25}$ is represented by two arrays `num` and `den`.

The Bode diagrams of amplitude and phase are obtained by means of the command `bode`.

The complete code is as follows.

```
%%%% Bode diagram %%%%
% Numerator and denominator of transfer function
num = [0 0 25];
den = [1 4 25];

% Transfer function
G = tf(num,den)

% Bode plots
bode(G)
```

OCTAVE creates the figure window represented in Figure 5.52.

The Nyquist or the Nichols diagrams can be obtained using the commands `nyquist` or `nichols`.

The complete code is as follows.

```
%%%% Nyquist diagram %%%%
% Numerator and denominator of transfer function
num = [0 0 25];
den = [1 4 25];

% Transfer function
G = tf(num,den)

% Nyquist plot
nyquist(G)
```

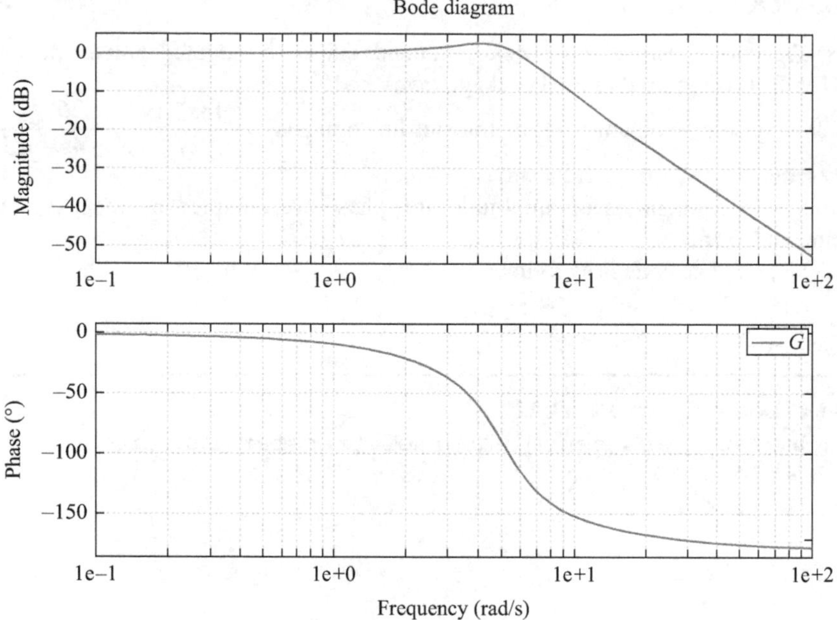

Figure 5.52 Bode diagrams of amplitude and phase for the transfer function $G(s) = \frac{Y(s)}{R(s)} = \frac{25}{s^2 + 4s + 25}$ *using OCTAVE*

```
%%%% Nichols diagram %%%%
% Numerator and denominator of transfer function
num = [0 0 25];
den = [1 4 25];

% Transfer function
G = tf(num,den)

% Nichols  plot
nichols (G)
```

OCTAVE creates the figure windows represented in Figure 5.53 or Figure 5.54.

For obtaining the gain and phase margins and the crossover frequencies (in [rad]), we can use the command `margin`.

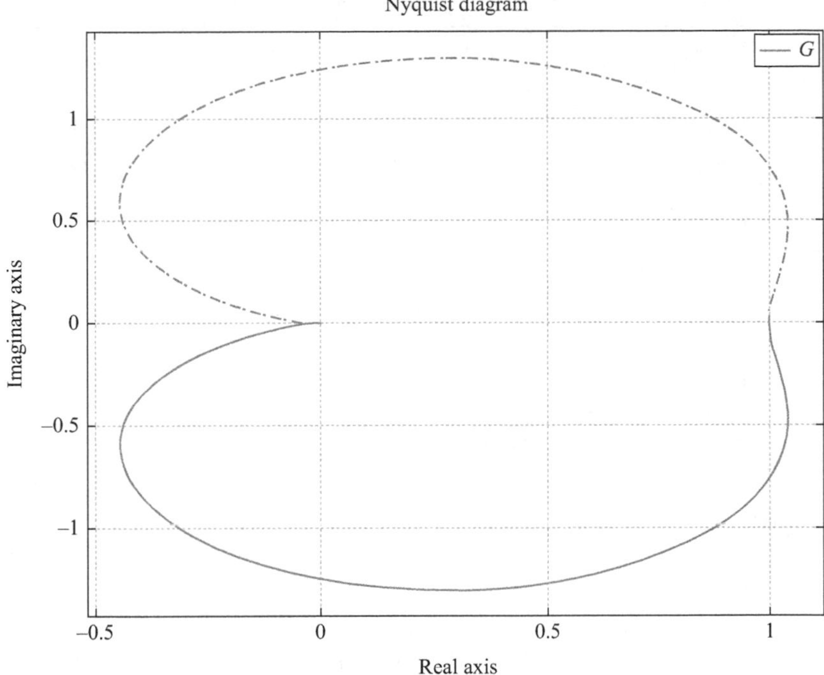

Figure 5.53 Nyquist diagram for the transfer function $G(s) = \frac{Y(s)}{R(s)} = \frac{25}{s^2 + 4s + 25}$ *using OCTAVE*

```
%%%% Gain and Phase margins %%%%
% Numerator and denominator of transfer function
num = [0 0 25];
den = [1 4 25];

% Create transfer function model, convert to transfer
% function model
G = tf(num,den);

% Gain margin, phase margin, and crossover frequencies
% [rad]
[Gm,Pm,Wgm,Wpm] = margin(G)
```

OCTAVE creates the following Command window.

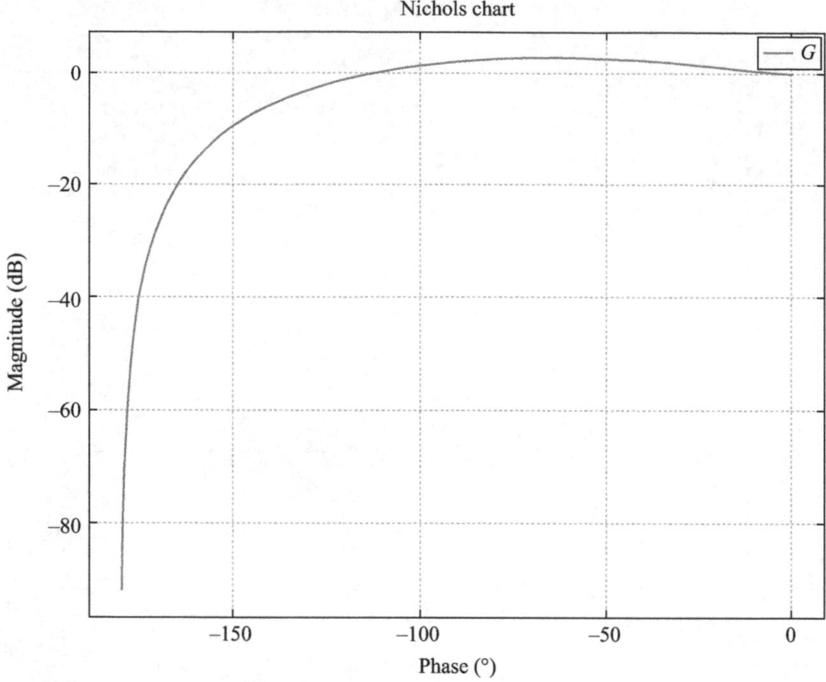

Figure 5.54 Nichols diagram for the transfer function $G(s) = \frac{Y(s)}{R(s)} = \frac{25}{s^2+4s+25}$ *using OCTAVE*

```
Gm = Inf
Pm =  68.900
Wgm = NaN
Wpm =  5.8310
```

Chapter 6
PID controller synthesis

6.1 Fundamentals

A proportional, integral and derivative (PID) controller is a simple yet versatile feedback compensator that is widely used in industrial control systems. We present the effect of each PID component on the closed-loop dynamics of a feedback-controlled system [2–5,9,19,20]. Afterward, we address different PID tuning methods.

6.1.1 List of symbols

A	amplitude
$e(t)$	tracking error
e_{ss}	steady-state error
$G(s)$	controlled system
K	gain
K_d	derivative constant
K_i	integral constant
K_p	proportional constant
K_u	ultimate gain
$m(t)$	control variable
P_u	ultimate oscillation period
$r(t)$	time-domain input function
s	Laplace variable
t	time
t_d	time delay
T	time constant
T_d	derivative time constant
T_i	integral time constant
$y(t)$	time-domain output function

6.1.2 The PID controller

A PID controller calculates the difference over time between the system output variable, $y(t)$, and the desired reference input, $r(t)$, and attempts to minimize this difference (or tracking error), $e(t)$, by adjusting the control variable, $m(t)$, that drives the system.

In time and Laplace domains, we have:

$$m(t) = K\left[e(t) + \frac{1}{T_i}\int_0^t e(\tau)d\tau + T_d\frac{de(t)}{dt}\right] = K_p e(t) + K_i\int_0^t e(\tau)d\tau + K_d\frac{de(t)}{dt} \tag{6.1a}$$

$$M(s) = K\left(1 + \frac{1}{T_i s} + T_d s\right)E(s) \tag{6.1b}$$

where K is a gain that affects all PID terms, T_i and T_d represent the integral and derivative time constants, and $K_p = K$, $K_i = \frac{K}{T_i}$ and $K_d = KT_d$ are the PID gains.

Figure 6.1 depicts a block diagram representative of the PID controller, where $G(s)$ represents the controlled system.

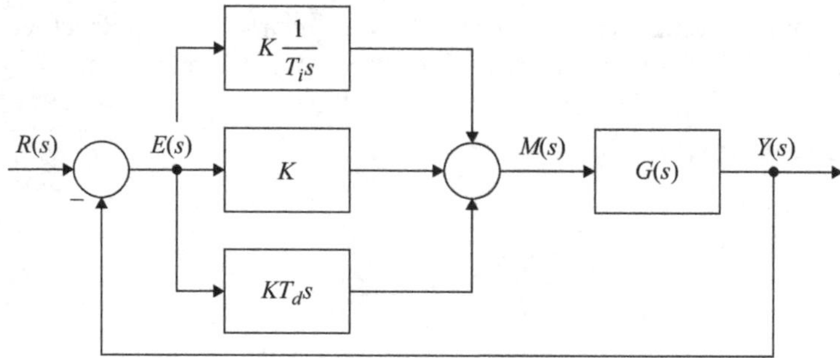

Figure 6.1 Block diagram of a PID controller

6.1.2.1 Proportional action

For $K_d = 0$, $K_i = 0$ and $K_p > 0$ we have a proportional (P) controller. Given a type zero system, the steady-state error, e_{ss}, to the unit-step input is:

$$e_{ss} = \lim_{t\to\infty} e(t) = \lim_{s\to 0} sE(s) = \frac{1}{1+K_p} \tag{6.2}$$

This means that e_{ss} can be made small if K_p is chosen high enough. However, augmenting the proportional gain increases the settling time. The system becomes more oscillatory and can reach instability.

6.1.2.2 Integral action

For $K_d = 0$, $K_i > 0$ and $K_p > 0$ we have a proportional plus integral (PI) controller. The integral action adds one pole to the system open-loop transfer function at the origin of the s plane. For a stable system, this yields zero steady-state error to the unit step. The Integral time, T_i, represents the time necessary for the integral and proportional actions to become equal, meaning that the integral action necessitates a certain time to produce a significant control output.

In practical cases, limitations of the actuators may lead to integral wind-up. This is particularly important when large changes occur in the set point, or disturbance, inputs. For both situations, the system output may exhibit large overshoot.

6.1.2.3 Derivative action

For $K_d > 0$, $K_i > 0$ and $K_p > 0$ we have a PID controller. The derivative contribution serves to overcome some limitations revealed by both the proportional and integral actions. The derivative part reacts faster than its proportional and integral counterparts, even when the actuating error is small. Parameter T_d gives a measure of the reaction time. When the input, $r(t)$, is constant, the derivative component is calculated based on the system output, $-\frac{dy(t)}{dt}$, instead on the error, with the advantage of having a smoother control action in response to step changes in the set point. For systems with poles and zeros in the left-hand side of the s plane, the derivative component tends to stabilize the controlled system.

The effects of increasing each of the controller parameters $\{K_p, K_i, K_d\}$ are illustrated in Table 6.1 [21].

6.1.3 PID tuning

Tuning a PID controller means choosing the values of the parameters K_p, T_i and T_d so that the controlled system exhibits a given set of specifications defined by the user [22–25].

To calculate the PID parameters we need to know the model of the system. Such model can be obtained by means of system identification techniques, but in most cases it is difficult for high order models.

In practical cases, the two following models, $G_1(s)$ and $G_2(s)$, are sufficient to describe most systems of interest:

$$G_1(s) = \frac{K_G \cdot e^{-t_d s}}{1 + Ts} \tag{6.3a}$$

$$G_2(s) = \frac{K_G \cdot e^{-t_d s}}{s} \tag{6.3b}$$

where K_G is the steady-state open-loop gain, T represents a time constant and t_d denotes time delay.

Table 6.1 Effects on the system response when the controller parameters $\{K_p, K_i, K_d\}$ increase

Parameter	**Rise time**	**Overshoot**	**Settling time**	**Steady-state error**	**Stability**
K_p	Decreases	Increases	Small effect	Decreases	Degrades
K_i	Decreases	Increases	Increases	Eliminates	Degrades
K_d	Minor effect	Decreases	Decreases	No effect	Improves (if K_d is small)

An approach to calculate $\{K_p, T_i, T_d\}$ consists of two steps:

1. The execution of a test for estimating the system model parameters,
2. The adoption of known expressions that give the controller parameters as a function of the system parameters.

6.1.3.1 Open-loop tuning

In open-loop tuning, the model parameters are estimated based on the system response to a step input of amplitude A.

For $G_1(s)$, the constant K_G is determined from the steady-state value of the step response, A', yielding $K_G = \frac{A'}{A}$. The time delay, t_d, is estimated from the intercept of the steepest tangent to the system response curve, as shown in Figure 6.2. The time constant, T, is given by $T = t' - t_d$, where $y(t') = 0.63A'$.

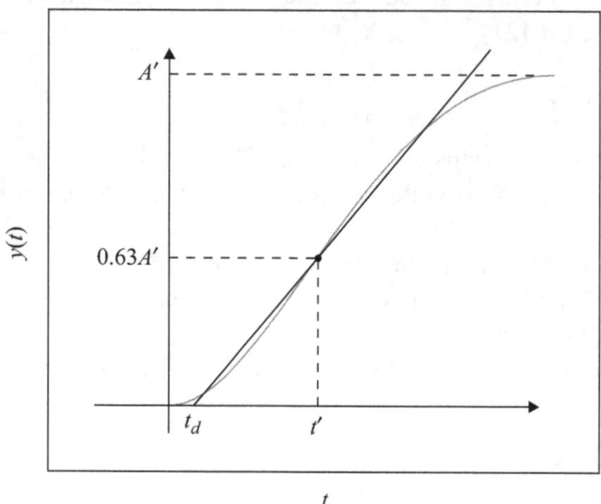

Figure 6.2 Approximation of the system response by the time response of $G_1(s)$ to a step input of amplitude A

For system $G_2(s)$, the constant K_G is determined from the slope of the line tangent to the steady-state step response, given by $A \cdot K_G$. The intercept of the tangent line corresponds to the time delay, t_d, (Figure 6.3).

The controller parameters $\{K_p, T_i, T_d\}$ are set based on the system parameters as shown in Table 6.2.

6.1.3.2 Closed-loop tuning

In closed-loop tuning, the K_i and K_d gains are first set to zero. The proportional gain, K_p, is increased until it reaches the ultimate gain, K_u, at which the output oscillates with constant amplitude. The controller parameters $\{K_p, T_i, T_d\}$ are

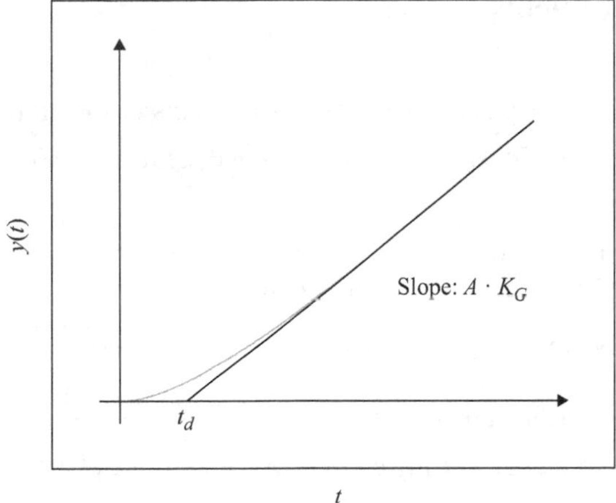

Figure 6.3 Approximation of the system response by the time response of $G_2(s)$ to a step input of amplitude A

Table 6.2 Formulas for PID controller tuning

Controller	Setting	Ziegler–Nichols (closed loop)	Shinskey (closed loop)	Ziegler–Nichols (open loop)	Cohen–Coon (open loop)
P	K_p	$0.5K_u$	$0.5K_u$	$\frac{T}{t_d K_G}$	$\frac{T}{t_d K_G}\left(1 + 0.33\frac{t_d}{T}\right)$
PI	K_p	$0.45K_u$	$0.5K_u$	$\frac{0.9T}{t_d K_G}$	$\frac{T}{t_d K_G}\left(0.9 + 0.082\frac{t_d}{T}\right)$
	T_i	$0.833P_u$	$0.43P_u$	$3.33t_d$	$t_d\left(\frac{3.33+0.3t_d/T}{1+2.2t_d/T}\right)$
PID	K_p	$0.6K_u$	$0.5K_u$	$\frac{1.2T}{t_d K_G}$	$\frac{T}{t_d K_G}\left(1.35 + 0.27\frac{t_d}{T}\right)$
	T_i	$0.5P_u$	$0.34P_u$	$2t_d$	$t_d\left(\frac{2.5+0.5t_d/T}{1+0.6t_d/T}\right)$
	T_d	$0.125P_u$	$0.08P_u$	$0.5t_d$	$t_d\left(\frac{0.37}{1+0.2t_d/T}\right)$

set based on K_u and the oscillation period P_u, according to different expressions (Table 6.2).

6.1.3.3 Tuning rules

We address the Ziegler–Nichols and Shinskey rules, that use model $G_2(s)$, and the Cohen–Coon expressions, that consider model $G_1(s)$. The values obtained for the parameters $\{K_p, T_i, T_d\}$ are distinct and, usually, should be tuned in subsequent iterations.

Table 6.2 summarizes the Ziegler–Nichols, Shinskey and Cohen–Coon rules.

6.2 Solved problems

Problem 6.1 Consider a plant with the open-loop transfer function $G(s) = \dfrac{0.5e^{-2s}}{3s+1}$. Obtain a PID using the Cohen–Coon method. Analyze its transfer function.

Resolution The dead time is $t_d = 2$, the response time is $T = 3$, and the gain is $K_G = 0.5$. From these values the controller parameters $K_P = 4.59$, $T_i = 4.05$ and $T_d = 0.653$ can be found. The controller is $C(s) = \dfrac{1.134 + 4.59s + 3.00s^2}{s}$ and the open-loop transfer function is $C(s)G(s) = \dfrac{(s+0.31)(s+1.22)}{s}\dfrac{0.5e^{-2s}}{s+0.33}$; so the controller has a zero that practically cancels the pole of the plant.

Problem 6.2 Consider a plant with the open-loop transfer function $G(s) = \dfrac{1}{s(s+1)(s+2)(s+3)}$. Obtain a PID using the closed loop (critical gain) Ziegler–Nichols method.

Resolution The closed-loop transfer function with variable gain K is $\dfrac{K}{s^4+6s^3+11s^2+6s+K}$. Using the Routh–Hurwitz criterion, we get

s^4	1	11	K
s^3	6	6	
s^2	10	K	
s^1	$\frac{60-6K}{10}$		
s^0	K		

From the two last lines, we see that the system is stable for $K \in]0, 10[$. For $K = K_u = 10$, we find from the s^2 line that $10s^2 + 10 = 0 \Rightarrow s = \pm j$, and so the period of the critical oscillations is $P_u = \dfrac{2\pi}{|\pm j|} = 6.28$. From these values the controller parameters $K_P = 6$, $T_i = 3.14$ and $T_d = 0.785$ can be found. Thus the desired PID is $6\left(1 + \dfrac{1}{3.14s} + 0.785s\right) = 6 + \dfrac{1.91}{s} + 4.71s$.

Problem 6.3 Consider a plant with the open-loop transfer function $G(s) = \dfrac{e^{-st_d}}{sT+1}$, where $t_d = 1.2$, and $T = 0.7$, controlled by a controller with transfer function $G_1(s) = K_1 + K_2\dfrac{1}{s}$.

1. Find the gain margin of the closed loop when the controller parameters are $K_1 = 0.1$ and $K_2 = 0.3$.
2. Find the controller parameters using the Cohen–Coon method.

Resolution From condition $\arg[GG_1] = -\pi$ we find that

$$-1.2\omega_\pi + \arctan\left(\frac{0.1\omega_\pi}{0.3}\right) - \frac{\pi}{2} - \arctan\left(\frac{0.7\omega_\pi}{1}\right) = -\pi \Rightarrow \omega_\pi = 1.06 \text{ rad}$$

$$|GG_1(\omega_\pi)| = 0.241$$

Consequently the $GM = \frac{1}{0.241} 4.15$ or $GM = -20\log_{10} 0.241 = 12.4$ dB.

As to the Cohen–Coon PI, the dead time is $t_d = 1.2$, the response time is $T = 0.7$, and the gain is $K_G = 1$. From these values the controller parameters $K_P = 0.607$ and $T_i = 0.967$ can be found; hence the controller is $0.607 + \frac{0.628}{s}$.

6.3 Proposed problems

Exercise 6.1 The transfer function of a PID controller with proportional gain K_p, integral gain K_i and differential gain K_d is:

A) $G(s) = (K_p + K_i + K_d)s$
B) $G(s) = K_p + K_i s + K_d s^2$
C) $G(s) = K_p + K_i/s + K_d s$
D) None of the above.

Exercise 6.2 Plot the Bode diagram of a PID controller given by $C(s) = 2.2 + \frac{2}{s} + 0.2s$.

Exercise 6.3 Consider a plant with the open-loop transfer function $G(s) = \frac{7e^{-3s}}{5s+1}$. A PID is obtained using the Cohen–Coon method. Parameters K_P (proportional gain), T_i (integral time constant) and T_d (differential time constant) are given by:

A) $K_P = 0.360$, $T_i = 6.176$, $T_d = 0.991$
B) $K_P = 0.630$, $T_i = 7.167$, $T_d = 0.199$
C) $K_P = 0.066$, $T_i = 1.176$, $T_d = 0.919$
D) None of the above.

Exercise 6.4 Consider a plant with the open-loop transfer function $G(s) = \frac{10e^{-3s}}{2s+1}$. A PID is obtained using the Cohen–Coon method. Parameters K_P (proportional gain), T_i (integral time constant) and T_d (differential time constant) are given by:

A) $K_P = 0.238$, $T_i = 6.223$, $T_d = 0.657$
B) $K_P = 0.571$, $T_i = 3.184$, $T_d = 0.501$
C) $K_P = 0.117$, $T_i = 5.132$, $T_d = 0.854$
D) $K_P = 0.973$, $T_i = 2.101$, $T_d = 0.325$.

Exercise 6.5 Consider a plant with the open-loop transfer function $G(s) = \dfrac{1}{(s+1)^4}$. A PID is obtained using the closed loop (critical gain) Ziegler–Nichols method. Parameters K_P (proportional gain), T_i (integral time constant) and T_d (differential time constant) are given by:

A) $K_P = 3.200$, $T_i = 2.176$, $T_d = 0.391$
B) $K_P = 1.600$, $T_i = 4.163$, $T_d = 0.509$
C) $K_P = 2.400$, $T_i = 3.142$, $T_d = 0.785$
D) None of the above.

Exercise 6.6 Consider a plant with the open-loop transfer function $G(s) = \dfrac{10}{s(s+1)^2}$. A PID is obtained using the closed loop (critical gain) Ziegler–Nichols method. Parameters K_P (proportional gain), T_i (integral time constant) and T_d (differential time constant) are given by:

A) $K_P = 0.238$, $T_i = 0.923$, $T_d = 0.257$
B) $K_P = 0.120$, $T_i = 3.14$, $T_d = 0.785$
C) $K_P = 0.571$, $T_i = 1.184$, $T_d = 0.501$
D) $K_P = 0.373$, $T_i = 2.101$, $T_d = 0.325$.

Exercise 6.7 Find a PI controller and a PID controller for plant $G(s) = \dfrac{-\frac{1}{2}s + 1}{s^2 + s + \frac{1}{3}}$, using the open-loop (reaction curve) Ziegler–Nichols method.

Exercise 6.8 Consider the control system in Figure 6.4, for the position of a satellite. Here θ represents the position, $2L$ is the length of the satellite, J its inertia, and $F/2$ the force exerted by each thrust.

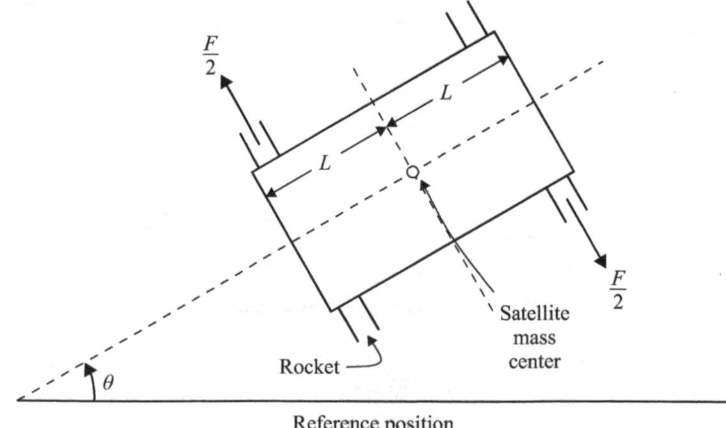

Figure 6.4 Satellite of Exercise 6.8

Also suppose that a PD controller is employed, as seen in Figure 6.5. Find the derivative time constant T_d so that $\zeta = 0.7$.

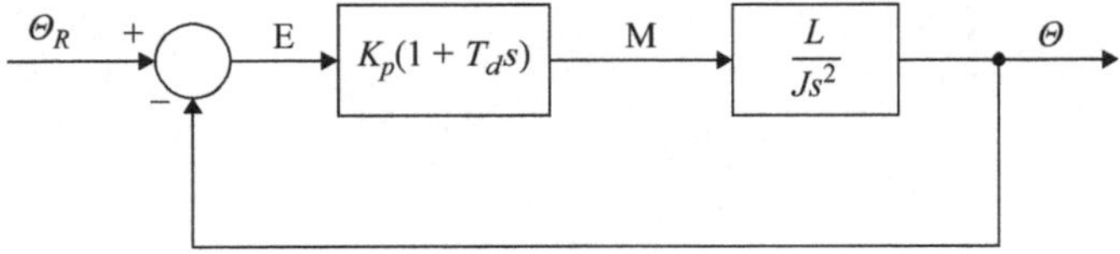

Figure 6.5 PID of Exercise 6.8

Chapter 7
State space analysis of continuous systems

7.1 Fundamentals

Modern control theory represents the system dynamics as a set of coupled first-order differential equations in a set of internal variables, known as state variables, together with a set of algebraic equations that combine the state into physical output variables [2–6]. The state-space representation of LTI systems surpasses several limitations of the classical methods that are mostly based on input–output descriptions. Moreover, the increase in the number of inputs, or outputs, does not affect the complexity of the state-space representations.

7.1.1 List of symbols

a_i, b_i, c_i, k_i, α_i	constant coefficients
$\mathbf{A}$	$n \times n$ dimensional state (or system) matrix
$\mathbf{B}$	$n \times q$ dimensional input matrix
C	capacitance, thermal capacitance
$\mathbf{C}$	$m \times n$ dimensional output matrix
$\mathbf{D}$	$m \times q$ dimensional feedthrough (or feedforward) matrix
i	current
$\mathbf{I}_n$	$n \times n$ dimensional identity matrix
J	inertia
$\mathbf{J}$, $\mathbf{M}$, $\mathbf{P}$, $\mathbf{V}$, $\mathbf{T}$, $\mathbf{W}$	$n \times n$ dimensional matrices
$\mathbf{J}_k$	Jordan block
K	stiffness
L	inductance
$\mathscr{L}$	Laplace operator
M	mass
n	number of state variables (system order)
p_i	poles of the transfer function
$\mathbf{Q}$	state controllability matrix
$\mathbf{R}$	observability matrix
s	Laplace variable

$\mathbf{S}$	output controllability matrix
t	time
T	temperature
t_0	initial time
$u(t)$	time-domain input function
$\mathbf{u}$	q dimensional input (or control) vector
v	voltage
$\mathbf{v}_i$	eigenvector
$\mathbf{w}_i$	left eigenvector
x	displacement
$\dot{x}$	linear velocity
$\mathbf{x}$	n dimensional state vector
$\mathbf{x}_0$	initial state vector
$\mathbf{y}$	m dimensional output vector
$y(t)$	time-domain output function
λ_i	eigenvalue
$\boldsymbol{\Lambda}$	$n \times n$ dimensional diagonal matrix of eigenvalues
$\boldsymbol{\Phi}(t)$	state transition matrix
ω	angular velocity

7.1.2 State space representation

Given the single-input single-output LTI system:

$$\frac{d^n y}{dt^n} + a_1 \frac{d^{n-1} y}{dt^{n-1}} + \cdots + a_n y = b_0 \frac{d^n u}{dt^n} + b_1 \frac{d^{n-1} u}{dt^{n-1}} + \cdots + b_n u \tag{7.1}$$

where u and y represent the input and the output, respectively, then the system transfer function is given by:

$$\frac{Y(s)}{U(s)} = \frac{b_0 s^n + b_1 s^{n-1} + \cdots + b_n}{s^n + a_1 s^{n-1} + \cdots + a_n} \tag{7.2}$$

Considering the intermediate variable x_1 such that:

$$\frac{X_1(s)}{U(s)} = \frac{1}{s^n + a_1 s^{n-1} + \cdots + a_n} \tag{7.3}$$

we obtain:

$$Y(s) = b_0 s^n X_1(s) + b_1 s^{n-1} X_1(s) + \cdots + b_{n-1} s X_1(s) + b_n X_1(s) \tag{7.4}$$

Choosing $X_2(s) = sX_1(s)$, $X_3(s) = s^2X_1(s)$, …, $X_n(s) = s^{n-1}X_1(s)$, the following equation results:

$$\begin{cases} \dot{x}_1 = x_2 \\ \dot{x}_2 = x_3 \\ \cdots \\ \dot{x}_{n-1} = x_n \\ \dot{x}_n = -a_1x_n - a_2x_{n-1} - \cdots - a_nx_1 + u \\ y = b_nx_1 + b_{n-1}x_2 + \cdots + b_1x_n + b_0(u - a_1x_n - a_2x_{n-1} - \cdots - a_nx_1) \end{cases} \tag{7.5}$$

Using matrix notation we may write the system state-space representation as:

$$\begin{cases} \dot{\mathbf{x}} = \mathbf{A}\mathbf{x} + \mathbf{B}u \\ y = \mathbf{C}\mathbf{x} + \mathbf{D}u \end{cases} \tag{7.6}$$

where

$$\mathbf{A} = \begin{bmatrix} 0 & 1 & 0 & \cdots & 0 \\ 0 & 0 & 1 & \cdots & 0 \\ \vdots & \vdots & \vdots & & \vdots \\ 0 & 0 & 0 & & 1 \\ -a_n & -a_{n-1} & -a_{n-2} & \cdots & -a_1 \end{bmatrix}, \quad \mathbf{B} = \begin{bmatrix} 0 \\ 0 \\ \vdots \\ 0 \\ 1 \end{bmatrix},$$

$$\mathbf{C}^T = \begin{bmatrix} b_n - a_nb_0 \\ b_{n-1} - a_{n-1}b_0 \\ \vdots \\ b_2 - a_2b_0 \\ b_1 - a_1b_0 \end{bmatrix}, \quad \mathbf{D} = b_0$$

$$\mathbf{x} = \begin{bmatrix} x_1 \\ x_2 \\ \vdots \\ x_{n-1} \\ x_n \end{bmatrix}, \quad \dot{\mathbf{x}} = \begin{bmatrix} \dot{x}_1 \\ \dot{x}_2 \\ \vdots \\ \dot{x}_{n-1} \\ \dot{x}_n \end{bmatrix}$$

This corresponds to the controllable canonical form representation of the system, as depicted in the block diagram of Figure 7.1.

Alternatively, by integrating n times (7.1) we obtain:

$$y + a_1\int ydt + \cdots + a_n\underbrace{\int\cdots\int}_{n} ydt = b_0u + \cdots + b_n\underbrace{\int\cdots\int}_{n} udt \tag{7.7}$$

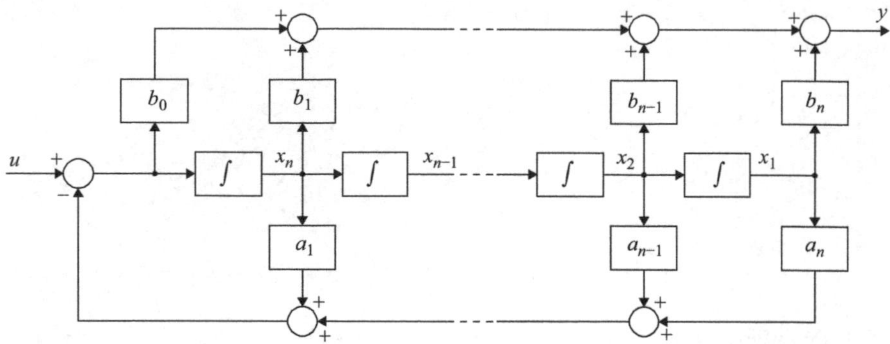

Figure 7.1 Block diagram of a system represented in the controllable canonical form

or,

$$y = b_0 u + \int (b_1 u - a_1 y)dt + \cdots + \underbrace{\int \cdots \int}_{n} (b_n u - a_n y)dt \tag{7.8}$$

yielding

$$\mathbf{A} = \begin{bmatrix} -a_1 & 1 & 0 & \cdots & 0 \\ -a_2 & 0 & 1 & \cdots & 0 \\ \vdots & \vdots & \vdots & & \vdots \\ -a_{n-1} & 0 & 0 & & 1 \\ -a_n & 0 & 0 & \cdots & 0 \end{bmatrix}, \quad \mathbf{B} = \begin{bmatrix} b_1 - a_1 b_0 \\ b_2 - a_2 b_0 \\ \vdots \\ b_{n-1} - a_{n-1} b_0 \\ b_n - a_n b_0 \end{bmatrix},$$

$$\mathbf{C} = \begin{bmatrix} 1 & 0 & \cdots & 0 & 0 \end{bmatrix}, \quad \mathbf{D} = b_0$$

This leads to the system represented in the observable canonical form, as shown in the block diagram of Figure 7.2.

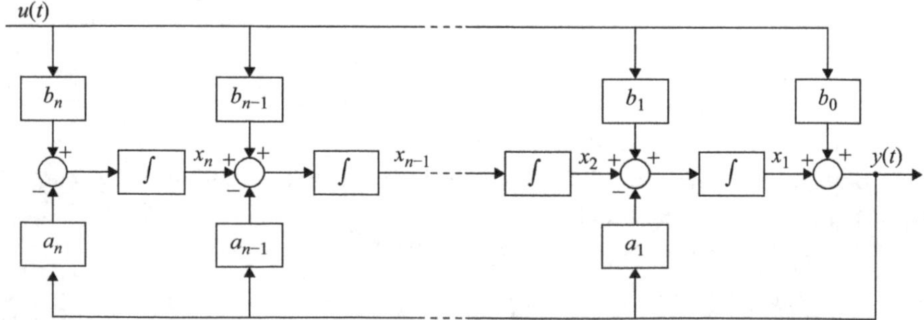

Figure 7.2 Block diagram of a system represented in the observable canonical form

Another canonical representation is the diagonal canonical form. This can be easily derived from the transfer function for the particular case where the denominator has distinct real roots:

$$\frac{Y(s)}{U(s)} = \frac{b_0 s^n + b_1 s^{n-1} + \cdots + b_n}{(s+p_1)(s+p_2)\cdots(s+p_n)} = b_0 + \frac{k_1}{s+p_1} + \frac{k_2}{s+p_2} + \cdots + \frac{k_n}{s+p_n} \quad (7.9)$$

that leads to:

$$\mathbf{A} = \mathrm{diag}(-p_1, -p_2, \ldots, -p_n), \quad \mathbf{B} = \begin{bmatrix} 1 \\ 1 \\ \vdots \\ 1 \end{bmatrix}, \quad \mathbf{C} = [\, k_1 \quad k_2 \quad \cdots \quad k_n \,], \quad \mathbf{D} = b_0$$

Figure 7.3 depicts the block diagram of a system represented in the diagonal canonical form.

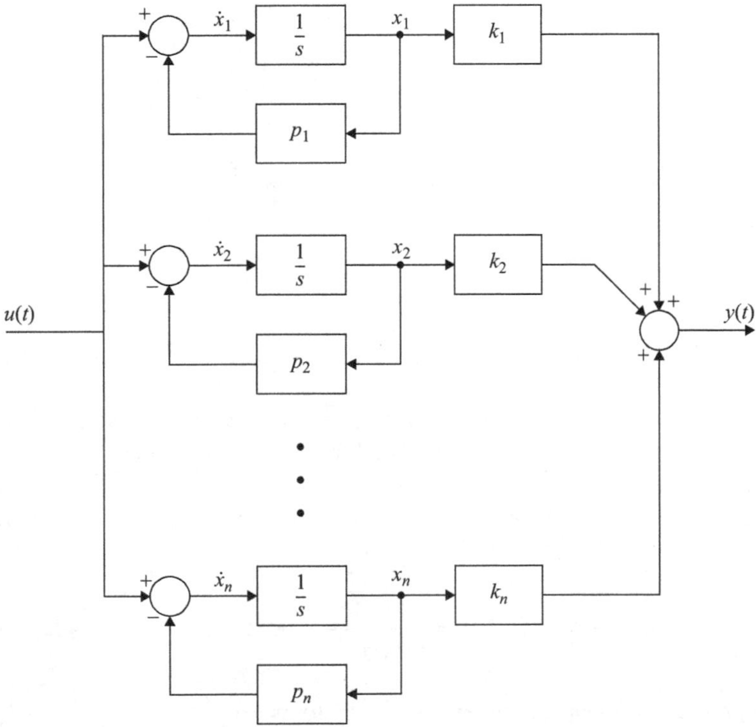

Figure 7.3 Block diagram of a system represented in the diagonal canonical form

For multiple real roots the previous diagonal canonical form can be generalized, yielding the Jordan canonical form. In this case we have:

$$\begin{aligned}\frac{Y(s)}{U(s)} &= \frac{b_0 s^n + b_1 s^{n-1} + \cdots + b_n}{(s+p_1)^r (s+p_{r+1}) \cdots (s+p_n)} \\ &= b_0 + \frac{k_{11}}{(s+p_1)} + \cdots + \frac{k_{1r}}{(s+p_1)^r} + \frac{k_{r+1}}{(s+p_{r+1})} \cdots + \frac{k_n}{(s+p_n)}\end{aligned} \tag{7.10}$$

that leads to:

$$\mathbf{A} = \begin{bmatrix} -p_1 & 1 & 0 & \cdots & \cdots & 0 & 0 & \cdots & \cdots & 0 \\ 0 & -p_1 & 1 & 0 & \cdots & 0 & 0 & \cdots & \cdots & 0 \\ \vdots & \ddots & \ddots & \ddots & \ddots & \vdots & \vdots & \vdots & \vdots & \vdots \\ 0 & \cdots & 0 & -p_1 & 1 & 0 & \vdots & \vdots & \vdots & \vdots \\ 0 & \cdots & \cdots & 0 & -p_1 & 1 & 0 & \cdots & \cdots & 0 \\ 0 & \cdots & \cdots & \cdots & 0 & -p_1 & 0 & \cdots & \cdots & 0 \\ 0 & 0 & \cdots & \cdots & \cdots & 0 & -p_{r+1} & 0 & \cdots & 0 \\ 0 & 0 & \cdots & \cdots & \cdots & 0 & \ddots & -p_{r+2} & \ddots & \vdots \\ \vdots & \vdots & \vdots & \vdots & \vdots & \vdots & \vdots & \ddots & \ddots & 0 \\ 0 & 0 & \cdots & \cdots & \cdots & 0 & 0 & \cdots & 0 & -p_n \end{bmatrix}$$

$$\mathbf{B} = [0 \;\; 0 \;\; \cdots \;\; \cdots \;\; 0 \;\; 1 \;\; 1 \;\; \cdots \;\; \cdots \;\; 1]^T$$

$$\mathbf{C} = [k_{1r} \;\; k_{1r-1} \;\; \cdots \;\; \cdots \;\; k_{12} \;\; k_{11} \;\; k_{r+1} \;\; k_{r+2} \;\; \cdots \;\; k_n], \quad \mathbf{D} = b_0$$

Figure 7.4 depicts the block diagram of a system represented in the Jordan canonical form.

All alternatives constitute state-space representations of the system (7.1). The output, $y(t)$, for every $t \geq t_0$, can be obtained from the system input, $u(t)$, $t \geq t_0$, and the state $\mathbf{x}(t)$.

Applying the Laplace transform to (7.6), we can obtain the system transfer function:

$$\frac{Y(s)}{U(s)} = \mathbf{C}(s\mathbf{I}_n - \mathbf{A})^{-1}\mathbf{B} + \mathbf{D} = \frac{1}{\det(s\mathbf{I}_n - \mathbf{A})}\mathbf{C}[\mathrm{adj}(s\mathbf{I}_n - \mathbf{A})]\mathbf{B} + \mathbf{D} \tag{7.11}$$

where $\mathbf{I}_n$ represents the $n \times n$ dimensional identity matrix.

The characteristic polynomial is given by $\det(s\mathbf{I}_n - \mathbf{A})$ and the poles of the transfer function correspond to the eigenvalues of matrix $\mathbf{A}$.

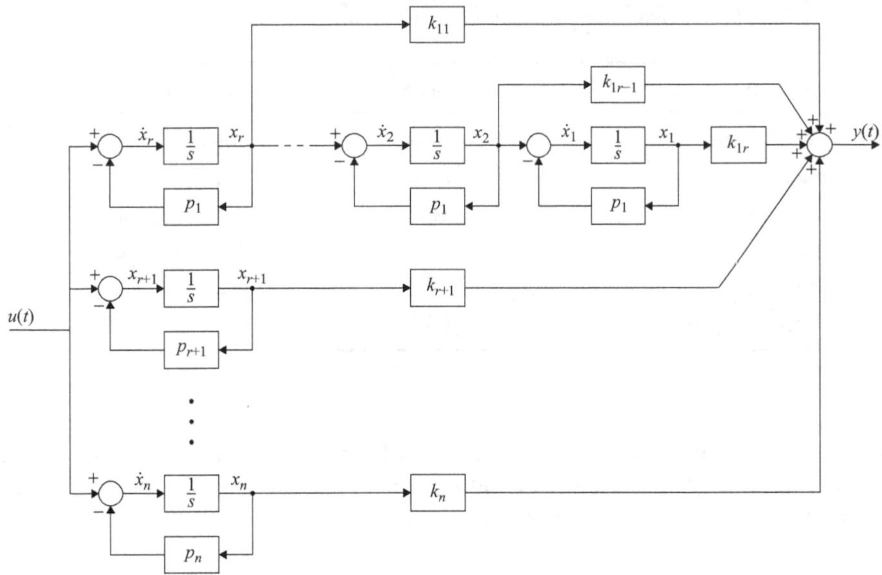

Figure 7.4 Block diagram of a system represented in the Jordan canonical form

For a multi-input multi-output LTI system the state-space representation has the general structure:

$$\begin{cases} \dot{\mathbf{x}} = \mathbf{Ax} + \mathbf{Bu} \\ \mathbf{y} = \mathbf{Cx} + \mathbf{Du} \end{cases} \tag{7.12}$$

where $\mathbf{x} \in \mathbb{R}^n$, $\mathbf{u} \in \mathbb{R}^q$, $\mathbf{y} \in \mathbb{R}^m$, $\mathbf{A} \in \mathbb{R}^{n\times n}$, $\mathbf{B} \in \mathbb{R}^{n\times q}$, $\mathbf{C} \in \mathbb{R}^{m\times n}$ and $\mathbf{D} \in \mathbb{R}^{m\times q}$.

7.1.2.1 State variables

The set of state variables $x_1, x_2, \ldots, x_n$ is not unique. In practical cases we choose variables associated to energy storage elements. Examples are shown in Table 7.1. Nevertheless, it should be noted that:

1. We must choose only linearly independent variables,
2. For certain physical systems we often need additional variables.

7.1.2.2 Similarity transformation

Consider the system:

$$\begin{cases} \dot{\mathbf{x}} = \mathbf{Ax} + \mathbf{Bu} \\ \mathbf{y} = \mathbf{Cx} + \mathbf{Du} \\ \mathbf{x}(0) = \mathbf{x}_0 \end{cases} \tag{7.13}$$

Table 7.1 Energy storage elements

Element	Energy	Physical variable
Capacitance, C	$\frac{1}{2}Cv^2$	Voltage, v
Inductance, L	$\frac{1}{2}Li^2$	Current, i
Mass, M	$\frac{1}{2}M\dot{x}^2$	Linear velocity, $\dot{x}$
Inertia, J	$\frac{1}{2}J\omega^2$	Angular velocity, ω
Stiffness, K	$\frac{1}{2}Kx^2$	Displacement, x
Thermal capacitance, C	$\frac{1}{2}CT^2$	Temperature, T

As the state variables are not unique, if we apply the linear transformation $\mathbf{x} = \mathbf{P}\bar{\mathbf{x}}$, where $\bar{\mathbf{x}}$ is a new state-space vector and $\mathbf{P}$ is an $n \times n$ non-singular matrix, then we obtain the equivalent state space representation:

$$\begin{cases} \dot{\bar{\mathbf{x}}} = \bar{\mathbf{A}}\bar{\mathbf{x}} + \bar{\mathbf{B}}\mathbf{u} \\ \mathbf{y} = \bar{\mathbf{C}}\bar{\mathbf{x}} + \bar{\mathbf{D}}\mathbf{u} \\ \bar{\mathbf{x}}(0) = \bar{\mathbf{x}}_0 \end{cases} \tag{7.14}$$

with the new matrices:

$$\begin{cases} \bar{\mathbf{A}} = \mathbf{P}^{-1}\mathbf{A}\mathbf{P} \\ \bar{\mathbf{B}} = \mathbf{P}^{-1}\mathbf{B} \\ \bar{\mathbf{C}} = \mathbf{C}\mathbf{P} \\ \bar{\mathbf{D}} = \mathbf{D} \\ \bar{\mathbf{x}}(0) = \mathbf{P}^{-1}\mathbf{x}(0) \end{cases} \tag{7.15}$$

7.1.2.3 Modal transformation

The modal transformation is a linear transformation that diagonalizes matrix $\mathbf{A}$.

For the eigenvalues λ_i and eigenvectors $\mathbf{v}_i$, $i = 1, \ldots, n$, such that $\mathbf{A}\mathbf{v}_i = \lambda_i \mathbf{v}_i$, if $\mathbf{P} = \mathbf{V} = \left[\, \mathbf{v}_1 \middle| \mathbf{v}_2 \middle| \cdots \middle| \mathbf{v}_n \,\right]$, then it results $\bar{\mathbf{A}} = \mathbf{V}^{-1}\mathbf{A}\mathbf{V} = \boldsymbol{\Lambda} = \operatorname{diag}(\lambda_1, \lambda_2, \ldots, \lambda_n)$.

Alternatively, we can write $\mathbf{V}^{-1}\mathbf{A} = \boldsymbol{\Lambda}\mathbf{V}^{-1}$, or $\mathbf{W}^T\mathbf{A} = \boldsymbol{\Lambda}\mathbf{W}^T$, where $\mathbf{W}^T = \mathbf{V}^{-1}$ and $\mathbf{W}^T\mathbf{V} = \mathbf{I}_n$.

The left eigenvectors $\mathbf{w}_i$, $i = 1, \ldots, n$, are given by:

$$\begin{cases} \mathbf{w}_i^T\mathbf{A} = \lambda_i \mathbf{w}_i^T \\ \mathbf{A}^T\mathbf{w}_i = \lambda_i \mathbf{w}_i \end{cases} \tag{7.16}$$

where $\mathbf{W} = [\mathbf{w}_1 | \mathbf{w}_2 | \cdots | \mathbf{w}_n]$.

7.1.3 The Cayley–Hamilton theorem

Given a matrix $\mathbf{A} \in \mathbb{R}^{n \times n}$, its characteristic polynomial is:

$$p(\lambda) = \det(\lambda \mathbf{I}_n - \mathbf{A}) = \prod_{i=1}^{d} (\lambda - \lambda_i)^{n_i} \tag{7.17}$$

where $n_1 + n_2 + \cdots + n_d = n$. The roots of the characteristic equation are the eigenvalues of $\mathbf{A}$.

The Cayley–Hamilton theorem states that matrix $\mathbf{A}$ satisfies its own characteristic polynomial, such that:

$$p(\mathbf{A}) = \prod_{i=1}^{d} (\mathbf{A} - \lambda_i)^{n_i} = \mathbf{A}^n + c_{n-1}\mathbf{A}^{n-1} + \cdots + c_1\mathbf{A} + c_0\mathbf{I}_n = 0 \tag{7.18}$$

This means that for all $p \geq 0$, $\mathbf{A}^p$ can be expressed as a linear combination of the first $n - 1$ powers of $\mathbf{A}$, namely $\mathbf{I}_n, \mathbf{A}^1, \mathbf{A}^2, \ldots, \mathbf{A}^{n-1}$.

7.1.4 Matrix exponential

For $\mathbf{A} \in \mathbb{R}^{n \times n}$ and $t \in \mathbb{R}$, the matrix exponential is:

$$\exp(\mathbf{A}t) = \mathbf{I}_n + \mathbf{A}t + \frac{\mathbf{A}^2 t^2}{2!} + \cdots = \sum_{k=0}^{\infty} \frac{\mathbf{A}^k t^k}{k!} \tag{7.19}$$

The matrix exponential has the following properties:

- $\exp(\mathbf{0}) = \mathbf{I}_n$
- $\exp[(\mathbf{A}_1 + \mathbf{A}_2)t] = \exp(\mathbf{A}_1 t) \cdot \exp(\mathbf{A}_2 t), \ \mathbf{A}_1\mathbf{A}_2 = \mathbf{A}_2\mathbf{A}_1$
- $\exp[\mathbf{A}(t_1 + t_2)] = \exp(\mathbf{A}t_1) \cdot \exp(\mathbf{A}t_2)$
- $[\exp(\mathbf{A}t)]^{-1} = \exp(-\mathbf{A}t)$
- $\frac{d}{dt}[\exp(\mathbf{A}t)] = \mathbf{A} \cdot \exp(\mathbf{A}t) = \exp(\mathbf{A}t) \cdot \mathbf{A}$
- If $\mathbf{v}$ is an eigenvector of $\mathbf{A}$ associated to the eigenvalue λ, then $\mathbf{v}$ is also an eigenvector of $\exp(\mathbf{A}t)$ associated to the eigenvalue $\exp(\lambda t)$
- $\mathcal{L}[\exp(\mathbf{A}t)] = (s\mathbf{I}_n - \mathbf{A})^{-1}$

7.1.5 Computation of the matrix exponential

7.1.5.1 Taylor series expansion

For a given $\varepsilon \in \mathbb{R}^+$, and $l \in \mathbb{N}$, if $\|\frac{\mathbf{A}^l t^l}{l!}\| < \varepsilon$, then the matrix exponential is approximated by the truncated series:

$$\exp(\mathbf{A}t) \approx \sum_{k=0}^{l} \frac{\mathbf{A}^k t^k}{k!} \tag{7.20}$$

7.1.5.2 Diagonalization

Distinct real eigenvalues

If $\mathbf{A}$ has distinct real eigenvalues, λ_i, with associated eigenvectors, $\mathbf{v}_i$, $i = 1, \ldots, n$, then $\mathbf{A} = \mathbf{V}\boldsymbol{\Lambda}\mathbf{V}^{-1}$, where

$$\boldsymbol{\Lambda} = \text{diag}(\lambda_1, \lambda_2, \ldots, \lambda_n) \tag{7.21a}$$

$$\mathbf{V} = [\mathbf{v}_1 | \mathbf{v}_2 | \cdots | \mathbf{v}_n] \tag{7.21b}$$

The matrix exponential is written in the form:

$$\exp(\mathbf{A}t) = \mathbf{V} \cdot \exp(\boldsymbol{\Lambda}t) \cdot \mathbf{V}^{-1} \tag{7.22}$$

with

$$\exp(\boldsymbol{\Lambda}t) = \begin{bmatrix} e^{\lambda_1 t} & 0 & \cdots & 0 \\ 0 & e^{\lambda_2 t} & \cdots & 0 \\ \vdots & \vdots & & \vdots \\ 0 & 0 & \cdots & e^{\lambda_n t} \end{bmatrix} \tag{7.23}$$

Distinct complex eigenvalues

If all eigenvalues are distinct, but k are complex numbers, $\lambda_i = \sigma_i \pm j\omega_i$, $i = 1, 3, \ldots, k-1$, then $\mathbf{A} = \mathbf{PMP}^{-1}$, where

$$\mathbf{M} = \begin{bmatrix} \begin{bmatrix} \sigma_1 & \omega_1 \\ -\omega_1 & \sigma_1 \end{bmatrix} & & & & & & 0 \\ & \begin{bmatrix} \sigma_3 & \omega_3 \\ -\omega_3 & \sigma_3 \end{bmatrix} & & & & & \\ & & \ddots & & & & \\ & & & \begin{bmatrix} \sigma_{k-1} & \omega_{k-1} \\ -\omega_{k-1} & \sigma_{k-1} \end{bmatrix} & & & \\ & & & & \lambda_{k+1} & & \\ 0 & & & & & \ddots & \\ & & & & & & \lambda_n \end{bmatrix} \tag{7.24a}$$

$$\mathbf{P} = [\text{Re}(\mathbf{v}_1) | \text{Im}(\mathbf{v}_1) | \cdots | \text{Re}(\mathbf{v}_{k-1}) | \text{Im}(\mathbf{v}_{k-1}) | \mathbf{v}_{k+1} | \cdots | \mathbf{v}_n] \tag{7.24b}$$

The matrix exponential is written in the form:

$$\exp(\mathbf{A}t) = \mathbf{P} \cdot \exp(\mathbf{M}t) \cdot \mathbf{P}^{-1} \tag{7.25}$$

with

$$\exp(\mathbf{M}t) = \begin{bmatrix} \mathbf{M}_1 & & & & & & 0 \\ & \mathbf{M}_3 & & & & & \\ & & \ddots & & & & \\ & & & \mathbf{M}_{k-1} & & & \\ & & & & e^{\lambda_{k+1}t} & & \\ 0 & & & & & \ddots & \\ & & & & & & e^{\lambda_n t} \end{bmatrix} \tag{7.26a}$$

$$\mathbf{M}_i = e^{\sigma_i t} \begin{bmatrix} \cos\omega_i t & \sin\omega_i t \\ -\sin\omega_i t & \cos\omega_i t \end{bmatrix} \tag{7.26b}$$

Multiple real eigenvalues
If $\mathbf{A}$ has multiple real eigenvalues, then $\mathbf{A} = \mathbf{TJT}^{-1}$, where the n columns of $\mathbf{T}$ are generalized eigenvectors of matrix $\mathbf{A}$. Matrix $\mathbf{T}$ transforms $\mathbf{A}$ into the Jordan canonical form, $\mathbf{J}$, that consists of bidiagonal Jordan blocks $\mathbf{J}_k$, $k = 1, \ldots, q$, of variable dimensions:

$$\mathbf{J} = \begin{bmatrix} \mathbf{J}_1 & & & \\ & \mathbf{J}_2 & & \\ & & \ddots & \\ & & & \mathbf{J}_q \end{bmatrix} \tag{7.27a}$$

$$\mathbf{J}_k = \begin{bmatrix} \lambda_k & 1 & \cdots & 0 \\ 0 & \lambda_k & \ddots & \vdots \\ \vdots & \ddots & \ddots & 1 \\ 0 & \cdots & 0 & \lambda_k \end{bmatrix} \tag{7.27b}$$

The matrix exponential is written in the following form:

$$\exp(\mathbf{A}t) = \mathbf{T} \cdot \exp(\mathbf{J}t) \cdot \mathbf{T}^{-1} \tag{7.28}$$

with

$$\exp(\mathbf{J}t) = \begin{bmatrix} \exp(\mathbf{J}_1 t) & 0 & \cdots & 0 \\ 0 & \exp(\mathbf{J}_2 t) & \cdots & 0 \\ \vdots & \vdots & & \vdots \\ 0 & 0 & \cdots & \exp(\mathbf{J}_q t) \end{bmatrix} \tag{7.29}$$

Let $\mathbf{J}_k$ be a $n_k \times n_k$ Jordan block, $k = 1, \ldots, q$, and $\sum_{k=1}^{q} n_k = n$. Then we have:

$$\exp(\mathbf{J}_k t) = e^{\lambda_k t} \begin{bmatrix} 1 & t & \frac{t^2}{2} & \cdots & \frac{t^{n_k-1}}{(n_k-1)!} \\ 0 & 1 & t & \cdots & \frac{t^{n_k-2}}{(n_k-2)!} \\ 0 & 0 & 1 & \cdots & \frac{t^{n_k-3}}{(n_k-3)!} \\ \vdots & \vdots & \vdots & \ddots & \vdots \\ 0 & 0 & 0 & \cdots & 1 \end{bmatrix} \tag{7.30}$$

7.1.5.3 Approximation based on the Cayley–Hamilton theorem

Based on the Cayley–Hamilton theorem we may write:

$$\exp(\mathbf{A}t) = \sum_{i=0}^{n-1} \alpha_i(t)\mathbf{A}^i \tag{7.31}$$

As the eigenvalues λ_j also satisfy the characteristic equation, then:

$$\exp(\lambda_j t) = \sum_{i=0}^{n-1} \alpha_i(t)\lambda_j^i \tag{7.32}$$

When $\mathbf{A}$ has n distinct eigenvalues, this yields a system of n equations and n variables.

7.1.5.4 Laplace transform

By applying the Laplace transform we have:

$$\exp(\mathbf{A}t) = \mathscr{L}^{-1}[(s\mathbf{I}_n - \mathbf{A})^{-1}] \tag{7.33}$$

7.1.6 Solution of the state-space equation

Given the state-space equation $\dot{\mathbf{x}}(t) = \mathbf{A}\mathbf{x}(t) + \mathbf{B}\mathbf{u}(t)$, its solution is:

$$\mathbf{x}(t) = \exp(\mathbf{A}t)\mathbf{x}(0) + \int_0^t \exp[\mathbf{A}(t-\tau)]\mathbf{B}\mathbf{u}(\tau)d\tau \tag{7.34}$$

where the first term is the solution of the homogeneous state equation, or the system response to the initial conditions, and the second term corresponds to the system response to the input $\mathbf{u}(t)$, without initial conditions. The system output is:

$$\mathbf{y}(t) = \mathbf{C}\exp(\mathbf{A}t)\mathbf{x}(0) + \mathbf{C}\int_0^t \exp[\mathbf{A}(t-\tau)]\mathbf{B}\mathbf{u}(\tau)d\tau + \mathbf{D}\mathbf{u}(t) \tag{7.35}$$

The solution of the homogeneous state equation $\dot{\mathbf{x}}(t) = \mathbf{A}\mathbf{x}(t)$ is often written as:

$$\mathbf{x}(t) = \boldsymbol{\Phi}(t)\mathbf{x}(0) \tag{7.36}$$

where $\boldsymbol{\Phi}(t) = \exp(\mathbf{A}t)$ is the state transition matrix.

7.1.7 Controllability

7.1.7.1 State controllability

A system is state controllable at $t = t_0$ if there exists an unconstrained input signal able to move the internal state from any initial state to any other final state in a finite time interval. If every state is controllable, then the system is completely state controllable.

Consider the state-space model of a LTI system:

$$\begin{cases} \dot{\mathbf{x}} = \mathbf{A}\mathbf{x} + \mathbf{B}\mathbf{u} \\ \mathbf{y} = \mathbf{C}\mathbf{x} + \mathbf{D}\mathbf{u} \end{cases} \tag{7.37}$$

where $\mathbf{x} \in \mathbb{R}^n$, $\mathbf{u} \in \mathbb{R}^q$ and $\mathbf{y} \in \mathbb{R}^m$.

If we construct the $n \times n \cdot q$ dimensional controllability matrix:

$$\mathbf{Q} = \left[\mathbf{B}\middle|\mathbf{AB}\middle|\cdots\middle|\mathbf{A}^{n-1}\mathbf{B}\right] \tag{7.38}$$

then the system is completely state controllable if and only if rank $(\mathbf{Q}) = n$.

7.1.7.2 Output controllability

A system is completely output controllable at $t = t_0$ if there exists an unconstrained input signal able to move any initial output to any final output in a finite time interval.

If we construct the $m \times (n + 1) \cdot q$ dimensional matrix:

$$\mathbf{S} = \left[\mathbf{CB}\middle|\mathbf{CAB}\middle|\cdots\middle|\mathbf{CA}^{n-1}\mathbf{B}\middle|\mathbf{D}\right] \tag{7.39}$$

then the system is completely output controllable if and only if rank $(\mathbf{S}) = m$.

7.1.8 Observability

A system is completely observable if every state $\mathbf{x}(t_0)$ can be determined from the observation of the system output over a finite time interval.

If we construct the $n \times m \cdot n$ dimensional observability matrix:

$$\mathbf{R} = \begin{bmatrix} \mathbf{C} \\ \hline \mathbf{CA} \\ \hline \vdots \\ \hline \mathbf{CA}^{n-1} \end{bmatrix} \tag{7.40}$$

then the system is completely observable if and only if rank $(\mathbf{R}) = n$.

7.2 Solved problems

Problem 7.1 Figure 7.5 shows a control system for the speed of a DC motor. Let $e_b = k_b\, d\theta/dt$ and $T = k_i i$. Then its transfer function is $\frac{\Theta(s)}{V(s)} = \frac{k_i}{s(sL+R)(sJ+B)+sk_bk_i}$, where $R = 8\Omega$, $L = 0.8$ H, $J = 1$ kg m^2, $B = 1$ N m rad^{-1} s, $k_i = 0.5$ N m A^{-1} and $k_b = 0.5$ V rad^{-1} s.

1. Find the system's observable canonical form.
2. Draw the corresponding block diagram.
3. Find matrices **Q** and **R** and check whether the system is controllable and observable.

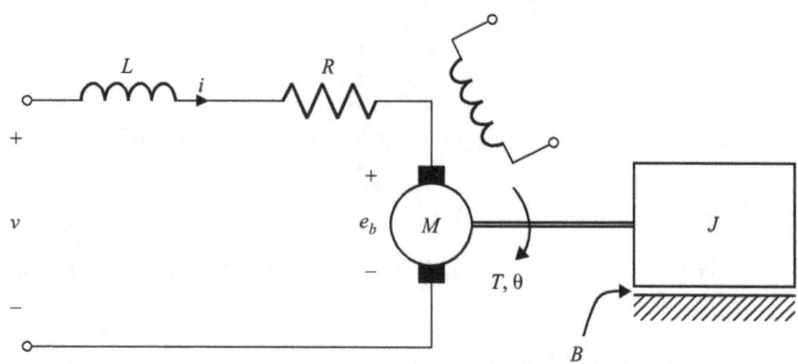

Figure 7.5 DC motor control system in Problem 7.1

Resolution From $\dfrac{\Theta(s)}{V(s)} = \dfrac{b_0 s^3 + b_1 s^2 + b_2 s + b_3}{s^3 + a_1 s^2 + a_2 s + a_3} = \dfrac{0.625}{s^3 + 11s^2 + 10.3125s}$ we get

$$\mathbf{A} = \begin{bmatrix} -a_1 & 1 & 0 \\ -a_2 & 0 & 1 \\ -a_3 & 0 & 0 \end{bmatrix} = \begin{bmatrix} -11 & 1 & 0 \\ -10.3125 & 0 & 1 \\ 0 & 0 & 0 \end{bmatrix}$$

$$\mathbf{B} = \begin{bmatrix} b_1 - a_1 b_0 \\ b_2 - a_2 b_0 \\ b_3 - a_3 b_0 \end{bmatrix} = \begin{bmatrix} 0 \\ 0 \\ 0.625 \end{bmatrix}$$

$$\mathbf{C} = [1 \quad 0 \quad 0] \tag{7.41}$$

$$\mathbf{Q} = \begin{bmatrix} \mathbf{B} & \mathbf{AB} & \mathbf{A}^2\mathbf{B} \end{bmatrix} = \begin{bmatrix} 0 & 0 & 0.625 \\ 0 & 0.625 & 0 \\ 0.625 & 0 & 0 \end{bmatrix}$$

$$\mathbf{R} = \begin{bmatrix} \mathbf{C} \\ \mathbf{CA} \\ \mathbf{CA}^2 \end{bmatrix} = \begin{bmatrix} 1 & 0 & 0 \\ -11 & 1 & 0 \\ 110.6875 & -11 & 0 \end{bmatrix}$$

As the rank of **Q** is 3, and the rank of **R** is 3, the system is both observable and controllable. Figure 7.6 shows the block diagram of the system.

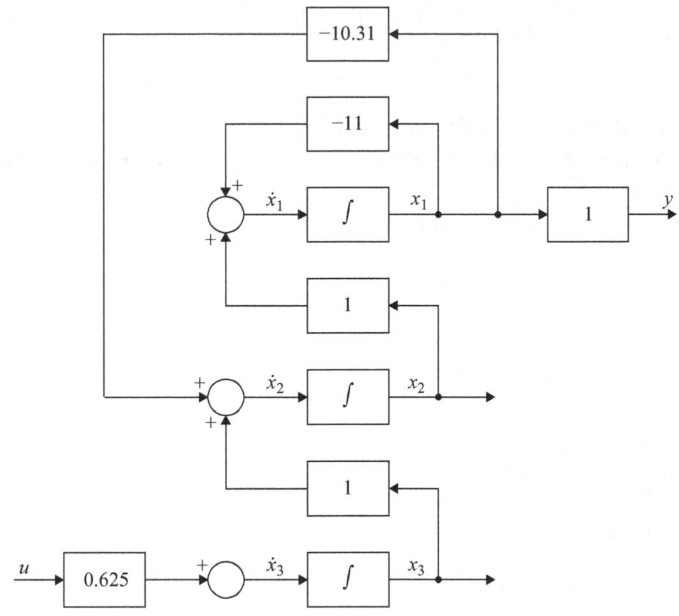

Figure 7.6 Block diagram of Problem 7.1

Problem 7.2 Consider the plant $\ddot{y} + 6\dot{y} + 5y = 2u$.

1. Find the system's controllable canonical form.
2. Find the state transition matrix $\boldsymbol{\Phi}(t) = e^{\mathbf{A}t}$.
3. Find transfer function $\dfrac{Y(s)}{U(s)}$ from the state-space representation.

Resolution

$$\begin{bmatrix} \dot{x}_1 \\ \dot{x}_2 \end{bmatrix} = \begin{bmatrix} 0 & 1 \\ -5 & -6 \end{bmatrix} \begin{bmatrix} x_1 \\ x_2 \end{bmatrix} + \begin{bmatrix} 0 \\ 1 \end{bmatrix} u$$

$$y = [2 \quad 0] \begin{bmatrix} x_1 \\ x_2 \end{bmatrix}$$

$$\boldsymbol{\Phi}(t) = e^{\mathbf{A}t} = \mathscr{L}^{-1}\left[(s\mathbf{I} - \mathbf{A})^{-1}\right] = \mathscr{L}^{-1}\left[\begin{bmatrix} s & -1 \\ 5 & s+6 \end{bmatrix}^{-1}\right]$$

$$= \mathscr{L}^{-1}\left[\begin{bmatrix} \frac{s+6}{s^2+6s+5} & \frac{1}{s^2+6s+5} \\ \frac{-5}{s^2+6s+5} & \frac{s}{s^2+6s+5} \end{bmatrix}\right] = \mathscr{L}^{-1}\left[\begin{bmatrix} \frac{-\frac{1}{4}}{s+5} + \frac{\frac{5}{4}}{s+1} & \frac{-\frac{1}{4}}{s+5} + \frac{\frac{1}{4}}{s+1} \\ \frac{\frac{5}{4}}{s+5} + \frac{-\frac{5}{4}}{s+1} & \frac{\frac{5}{4}}{s+5} + \frac{-\frac{1}{4}}{s+1} \end{bmatrix}\right]$$

$$= \begin{bmatrix} -\frac{1}{4}e^{-5t} + \frac{5}{4}e^{-t} & -\frac{1}{4}e^{-5t} + \frac{1}{4}e^{-t} \\ \frac{5}{4}e^{-5t} - \frac{5}{4}e^{-t} & \frac{5}{4}e^{-5t} - \frac{1}{4}e^{-t} \end{bmatrix}$$

$$\frac{Y(s)}{U(s)} = \mathbf{C}(s\mathbf{I} - \mathbf{A})^{-1}\mathbf{B} = \frac{2}{(s+5)(s+1)}$$

Problem 7.3 Consider the plant $\dfrac{Y(s)}{U(s)} = \dfrac{2(s+2)}{(s+1)(s+3)}$.

1. Find the system's observable canonical form.
2. Find matrices **Q** and **R** and check whether the system is controllable and observable.
3. Find the system's diagonal canonical form.

Resolution

$$\mathbf{A} = \begin{bmatrix} -a_1 & 1 \\ -a_2 & 0 \end{bmatrix} = \begin{bmatrix} -4 & 1 \\ -3 & 0 \end{bmatrix}$$

$$\mathbf{B} = \begin{bmatrix} b_1 - a_1 b_0 \\ b_2 - a_2 b_0 \end{bmatrix} = \begin{bmatrix} 2 \\ 4 \end{bmatrix}$$

$$\mathbf{C} = [1 \quad 0]$$

$$\mathbf{Q} = [\mathbf{B} \quad \mathbf{AB}] = \begin{bmatrix} 2 & -4 \\ 4 & -6 \end{bmatrix}$$

$$\mathbf{R} = \begin{bmatrix} \mathbf{C} \\ \mathbf{CA} \end{bmatrix} = \begin{bmatrix} 1 & 0 \\ -4 & 1 \end{bmatrix}$$

As the rank of **Q** is 2, the plant is controllable. As the rank of **R** is 2, the plant is observable.

For the diagonal canonical form, we need the following eigenvalues and eigenvectors:

$$\det[\lambda\mathbf{I} - \mathbf{A}] = \det\begin{bmatrix} \lambda+4 & -1 \\ 3 & \lambda \end{bmatrix} = (\lambda+4)\lambda + 3 = 0 \Leftrightarrow \lambda = -1 \vee \lambda = -3$$

$$\mathbf{A}\mathbf{v}_1 = \lambda_1 \mathbf{v}_1 \Leftrightarrow \begin{cases} -4e_1 + e_2 = -e_1 \\ -3e_1 = -e_2 \end{cases} \Rightarrow \mathbf{v}_1 = \begin{bmatrix} 1 \\ 3 \end{bmatrix}$$

$$\mathbf{A}\mathbf{v}_2 = \lambda_2 \mathbf{v}_2 \Leftrightarrow \begin{cases} -4e_1 + e_2 = -3e_1 \\ -3e_1 = -3e_2 \end{cases} \Rightarrow \mathbf{v}_2 = \begin{bmatrix} 1 \\ 1 \end{bmatrix}$$

The diagonal canonical form is given by

$$\dot{\mathbf{d}} = \mathbf{V}^{-1}\mathbf{A}\mathbf{V}\mathbf{d} + \mathbf{V}^{-1}\mathbf{B}u$$

$$y = \mathbf{C}\mathbf{V}\mathbf{d}$$

$$\mathbf{V} = [\mathbf{v}_1 \quad \mathbf{v}_2] = \begin{bmatrix} 1 & 1 \\ 3 & 1 \end{bmatrix}$$

$$\mathbf{V}^{-1} = \begin{bmatrix} -\frac{1}{2} & \frac{1}{2} \\ \frac{3}{2} & -\frac{1}{2} \end{bmatrix}$$

$$\mathbf{V}^{-1}\mathbf{AV} = \begin{bmatrix} -1 & 0 \\ 0 & -3 \end{bmatrix}$$

$$\mathbf{V}^{-1}\mathbf{B} = \begin{bmatrix} 1 \\ 1 \end{bmatrix}$$

$$\mathbf{CV} = [1 \quad 1]$$

Problem 7.4 Consider the plant $\dot{\mathbf{x}} = \begin{bmatrix} 0 & 1 & 0 \\ 0 & 0 & 1 \\ -15 & -11 & -5 \end{bmatrix}\mathbf{x} + \begin{bmatrix} 0 \\ 0 \\ 1 \end{bmatrix}u$, $y = [1 \quad 0 \quad 0]\mathbf{x}$, where $\mathbf{x} = [x_1 \quad x_2 \quad x_3]^T$, $\dot{\mathbf{x}} = [\dot{x}_1 \quad \dot{x}_2 \quad \dot{x}_3]^T$.

1. Find its diagonal canonical form.
2. Find its transfer function $\dfrac{Y(s)}{U(s)}$.

Resolution

$$s\mathbf{I} - \mathbf{A} = \begin{bmatrix} s & -1 & 0 \\ 0 & s & -1 \\ 15 & 11 & s+5 \end{bmatrix}$$

$$\det[s\mathbf{I} - \mathbf{A}] = (s+3)(s+1+2j)(s+1-2j)$$

For eigenvalue -3,

$$\begin{bmatrix} -3 & -1 & 0 \\ 0 & -3 & -1 \\ 15 & 11 & 2 \end{bmatrix}\begin{bmatrix} x_1 \\ x_2 \\ x_3 \end{bmatrix} = \begin{bmatrix} 0 \\ 0 \\ 0 \end{bmatrix} \Rightarrow \begin{cases} -3x_1 = x_2 \\ -3x_2 = x_3 \end{cases} \Rightarrow \mathbf{v}_1 = \begin{bmatrix} 1 \\ -3 \\ 9 \end{bmatrix}$$

For eigenvalues $-1 \pm 2j$,

$$\begin{bmatrix} -1+2j & -1 & 0 \\ 0 & -1+2j & -1 \\ 15 & 11 & 4+2j \end{bmatrix}\begin{bmatrix} x_1 \\ x_2 \\ x_3 \end{bmatrix} = \begin{bmatrix} 0 \\ 0 \\ 0 \end{bmatrix} \Rightarrow \begin{cases} (-1+2j)x_1 = x_2 \\ (-1+2j)x_2 = x_3 \end{cases}$$

$$\Rightarrow \mathbf{v}_2 = \begin{bmatrix} 1 \\ -1+2j \\ -3-4j \end{bmatrix} = \begin{bmatrix} 1 \\ -1 \\ -3 \end{bmatrix} + \begin{bmatrix} 0 \\ 2 \\ -4 \end{bmatrix}j$$

Thus,

$$\mathbf{P} = [\mathrm{Re}[\mathbf{v}_2] \quad \mathrm{Im}[\mathbf{v}_2] \quad \mathbf{v}_1] = \begin{bmatrix} 1 & 0 & 1 \\ -1 & 2 & -3 \\ -3 & -4 & 9 \end{bmatrix}$$

$$\mathbf{P}^{-1} = \frac{1}{8}\begin{bmatrix} 3 & -2 & -1 \\ 9 & 6 & 1 \\ 5 & 2 & 1 \end{bmatrix}$$

$$\mathbf{d} = \mathbf{P}\mathbf{x}$$

$$\dot{\mathbf{d}} = \mathbf{P}^{-1}\mathbf{A}\mathbf{P}\mathbf{d} + \mathbf{P}^{-1}\mathbf{B}u = \begin{bmatrix} -1 & 2 & 0 \\ -2 & -1 & 0 \\ 0 & 0 & -3 \end{bmatrix}\mathbf{d} + \begin{bmatrix} -\frac{1}{8} \\ \frac{1}{8} \\ \frac{1}{8} \end{bmatrix} u$$

$$y = \mathbf{C}\mathbf{P}\mathbf{d} = [\,1 \quad 0 \quad 1\,]\mathbf{d}$$

$$\frac{Y(s)}{U(s)} = \mathbf{C}(s\mathbf{I} - \mathbf{A})^{-1}\mathbf{B} = \frac{1}{(s+3)(s^2+2s+5)} \tag{7.42}$$

Problem 7.5 Consider the plant $\dot{\mathbf{x}} = \begin{bmatrix} 0 & 1 & 0 \\ 0 & 0 & 1 \\ -8 & -14 & -7 \end{bmatrix}\mathbf{x} + \begin{bmatrix} 0 \\ 0 \\ 1 \end{bmatrix} u$, $y = [\,8 \quad 8 \quad 0\,]\mathbf{x}$, where $\mathbf{x} = [\,x_1 \quad x_2 \quad x_3\,]^T$, $\dot{\mathbf{x}} = [\,\dot{x}_1 \quad \dot{x}_2 \quad \dot{x}_3\,]^T$.

1. Draw the system's block diagram.
2. Find the system's transfer function $\frac{Y(s)}{U(s)}$.
3. Is the system controllable?
4. Is the system observable?

Resolution

$$(s\mathbf{I} - \mathbf{A})^{-1} = \begin{bmatrix} s & -1 & 0 \\ 0 & s & -1 \\ 8 & 14 & s+7 \end{bmatrix}^{-1}$$

$$= \frac{1}{(s+1)(s+2)(s+4)}\begin{bmatrix} s^2+7s+14 & s+7 & 1 \\ -8 & s(s+7) & s \\ -8s & -2(7s+4) & s^2 \end{bmatrix}$$

$$\mathbf{C}(s\mathbf{I}-\mathbf{A})^{-1}\mathbf{B} = \frac{8}{(s+2)(s+4)}$$

$$\mathbf{Q} = [\mathbf{B} \quad \mathbf{AB} \quad \mathbf{A}^2\mathbf{B}] = \begin{bmatrix} 0 & 0 & 1 \\ 0 & 1 & -7 \\ 1 & -7 & 35 \end{bmatrix},$$

$$\mathbf{R} = \begin{bmatrix} \mathbf{C} \\ \mathbf{CA} \\ \mathbf{CA}^2 \end{bmatrix} = \begin{bmatrix} 8 & 8 & 0 \\ 0 & 8 & 8 \\ -64 & -112 & -48 \end{bmatrix}$$

Figure 7.7 shows the block diagram of the system. As the rank of **Q** is 3, the plant is controllable. As the rank of **R** is 2, the plant is not observable.

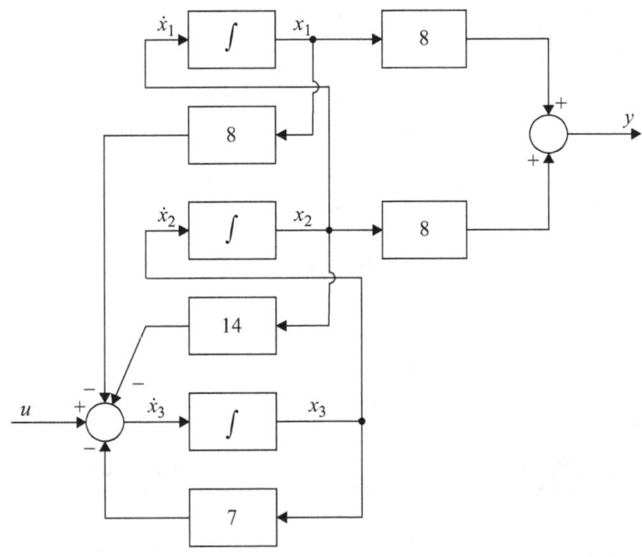

Figure 7.7 Block diagram of Problem 7.5

Problem 7.6 Consider the circuit in Figure 7.8.

1. Find a state-space representation for this system, using x_1 and x_2 as state variables.
2. Draw the corresponding block diagram.
3. Find the corresponding transfer function matrix $\mathbf{G}(s) = [Y_1(s)/V(s), Y_2(s)/V(s)]^T$.
4. Let $C_1 = C_2 = 1/2$ and $R_1 = R_2 = R_3 = 1$. Is the system controllable?

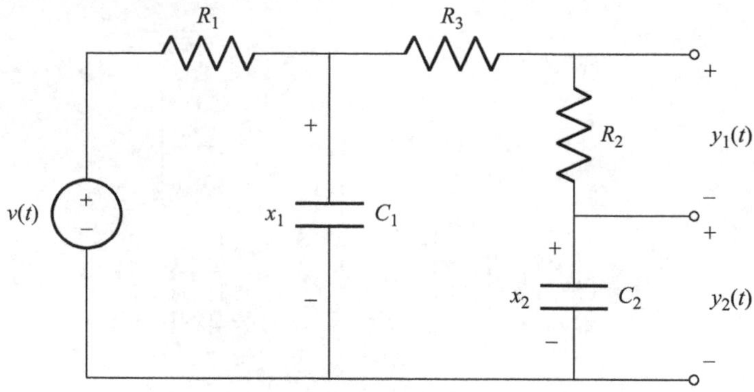

Figure 7.8 Circuit of Problem 7.6

Resolution

$$\begin{bmatrix} \dot{x}_1 \\ \dot{x}_2 \end{bmatrix} = \begin{bmatrix} -\frac{R_1+R_2+R_3}{R_1C_1(R_2+R_3)} & \frac{R_1}{R_1C_1(R_2+R_3)} \\ \frac{1}{C_2(R_2+R_3)} & -\frac{1}{C_2(R_2+R_3)} \end{bmatrix} \begin{bmatrix} x_1 \\ x_2 \end{bmatrix} + \begin{bmatrix} \frac{1}{R_1C_1} \\ 0 \end{bmatrix} u$$

$$\begin{bmatrix} y_1 \\ y_2 \end{bmatrix} = \begin{bmatrix} \frac{1}{C_2(R_2+R_3)} & -\frac{1}{C_2(R_2+R_3)} \\ 0 & 1 \end{bmatrix} \begin{bmatrix} x_1 \\ x_2 \end{bmatrix}$$

Figure 7.9 shows the block diagram of the system.

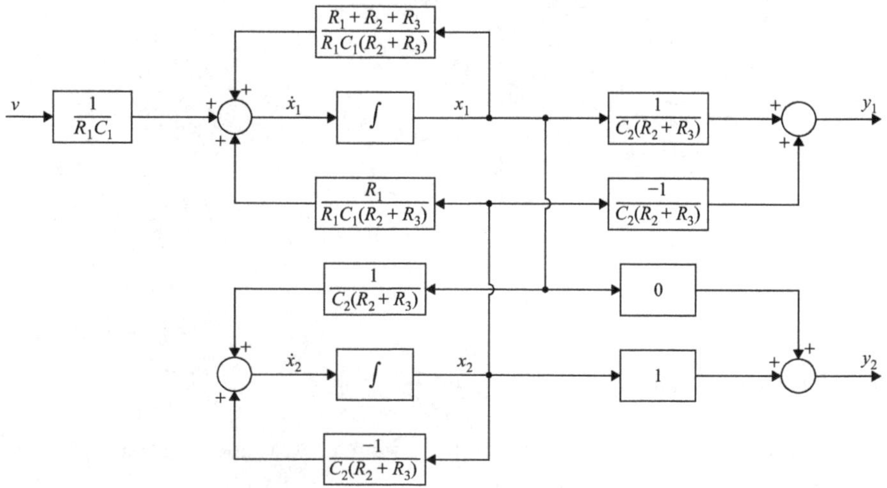

Figure 7.9 Block diagram of Problem 7.6

$$(s\mathbf{I}-\mathbf{A})^{-1} = \begin{bmatrix} s+3 & -1 \\ -1 & s+1 \end{bmatrix}^{-1} = \frac{1}{(s+1)(s+3)-1}\begin{bmatrix} s+1 & 1 \\ 1 & s+3 \end{bmatrix}$$

$$\mathbf{C}(s\mathbf{I}-\mathbf{A})^{-1}\mathbf{B} = \begin{bmatrix} \frac{2s}{s^2+4s+2} \\ \frac{2}{s^2+4s+2} \end{bmatrix}$$

$$\mathbf{Q} - [\mathbf{B} \quad \mathbf{AB}] = \begin{bmatrix} 2 & -6 \\ 0 & 2 \end{bmatrix},$$

As the rank of **Q** is 2, the plant is controllable.

7.3 Proposed problems

Exercise 7.1 Given a system's state-space representation $\begin{cases} \dot{\mathbf{x}} = \mathbf{Ax} + \mathbf{B}u \\ y = \mathbf{Cx} + \mathbf{D}u \end{cases}$, the corresponding transfer function $\frac{Y(s)}{U(s)}$ is:

A) $\frac{Y(s)}{U(s)} = [\mathbf{C}(s\mathbf{I}-\mathbf{A})]^{-1}\mathbf{B}+\mathbf{D}$

B) $\frac{Y(s)}{U(s)} = \mathbf{C}(s\mathbf{I}-\mathbf{A})^{-1}\mathbf{B}+\mathbf{D}$

C) $\frac{Y(s)}{U(s)} = \mathbf{C}(s\mathbf{I}-\mathbf{A})\mathbf{B}+\mathbf{D}$

D) $\frac{Y(s)}{U(s)} = \mathbf{C}(s\mathbf{I}-\mathbf{A}^{-1})\mathbf{B}+\mathbf{D}.$

Exercise 7.2 Consider the plant represented in Figure 7.10 in state space.

1. Find its model of the form $\dot{\mathbf{x}}(t) = \mathbf{Ax}(t) + \mathbf{B}u(t)$, $y(t) = \mathbf{Cx}(t)$, where $\mathbf{x}(t) = [x_1 \quad x_2 \quad x_3]^T$.
2. Analyze whether the plant is stable, controllable, and observable.
3. Find its transfer function $Y(s)/U(s)$.
4. Find its output $y(t)$ for a unit step input $u(t)$, $t \geq 0$, when the initial conditions are zero.
5. Find a discrete equivalent for the plant $\mathbf{x}(k+1) = \mathbf{G}(h)\mathbf{x}(k) + \mathbf{H}(h)u(t)$, $y(k) = \mathbf{Cx}(k)$, when signal $u(t)$ passes through a zero order hold and the sample time is $h = 1$.

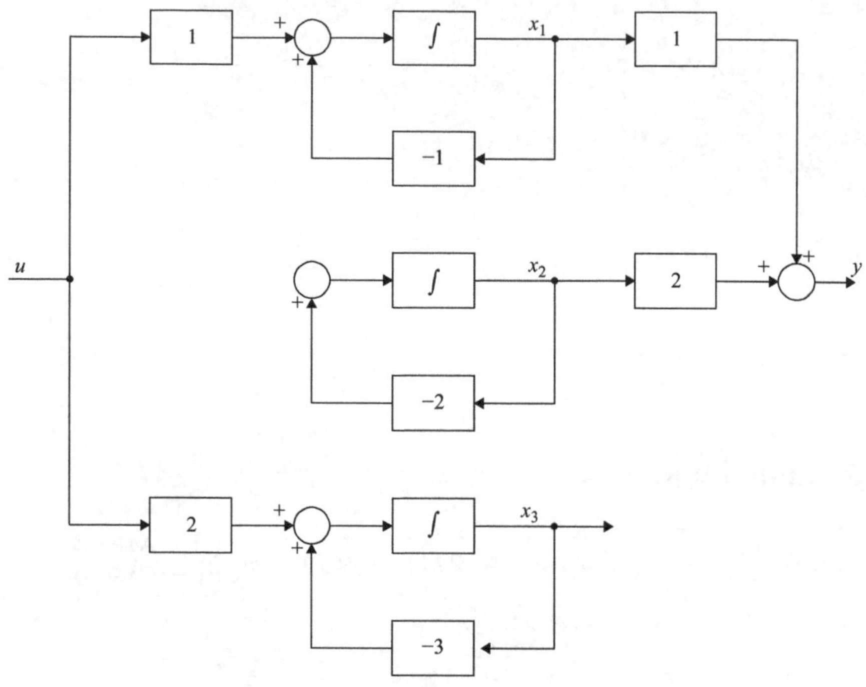

Figure 7.10 Block diagram of Exercise 7.2

Exercise 7.3 Consider the plant $\dot{\mathbf{x}} = \begin{bmatrix} 1 & -1 \\ 6 & -4 \end{bmatrix}\mathbf{x} + \begin{bmatrix} 1 \\ 3 \end{bmatrix}u$, $y = [-2 \quad 1]\mathbf{x}$, $\mathbf{x} = [x_1 \quad x_2]^T$. Is it observable? Is it controllable?

Exercise 7.4 Consider the plant $\dot{\mathbf{x}} = \mathbf{A}\mathbf{x} + \mathbf{B}u, y = \mathbf{C}\mathbf{x}$, where $\mathbf{A} = \begin{bmatrix} 2 & 0 & 0 \\ 1 & 3 & 0 \\ -1 & 2 & -5 \end{bmatrix}$, $\mathbf{B} = \begin{bmatrix} 1 \\ 0 \\ 0 \end{bmatrix}$, $\mathbf{C} = \begin{bmatrix} 1 & 0 & 0 \end{bmatrix}$ and $\mathbf{x} = \begin{bmatrix} x_1 & x_2 & x_3 \end{bmatrix}^T$.

1. The plant's transfer function $\dfrac{Y(s)}{U(s)}$ is:

   A) $\dfrac{Y(s)}{U(s)} = \dfrac{1}{s-2}$

   B) $\dfrac{Y(s)}{U(s)} = \dfrac{1}{(s-3)(s+5)}$

   C) $\dfrac{Y(s)}{U(s)} = \dfrac{1}{(s-2)(s+5)}$

   D) $\dfrac{Y(s)}{U(s)} = \dfrac{1}{(s-2)(s-3)(s+5)}$.

2. The plant is:
   E) observable and controllable
   F) non-observable and controllable
   G) observable and non-controllable
   H) non-observable and non-controllable.

Exercise 7.5 Consider the plant $\dot{\mathbf{x}} = \begin{bmatrix} -3 & 0 \\ -1 & -2 \end{bmatrix}\mathbf{x} + \begin{bmatrix} 1 \\ 0 \end{bmatrix}u$, where $\mathbf{x} = [\,x_1 \quad x_2\,]^T$.

The exponential matrix $e^{\mathbf{A}t}$ is:

A) $e^{\mathbf{A}t} = \begin{bmatrix} 3e^{-3t} & 2e^{-2t} \\ e^{-2t} + e^{-3t} & e^{-2t} \end{bmatrix}$

B) $e^{\mathbf{A}t} = \begin{bmatrix} e^{-2t} & 0 \\ 0 & e^{-3t} \end{bmatrix}$

C) $e^{\mathbf{A}t} = \begin{bmatrix} e^{-2t} + e^{-3t} & -1 \\ e^{-2t} - e^{-3t} & e^{-3t} \end{bmatrix}$

D) $e^{\mathbf{A}t} = \begin{bmatrix} e^{-3t} & 0 \\ -e^{-2t} + e^{-3t} & e^{-2t} \end{bmatrix}$.

Exercise 7.6 Consider the plant $\dot{\mathbf{x}} = \mathbf{A}\mathbf{x} + \mathbf{B}u$, $y = \mathbf{C}\mathbf{x}$, where $\mathbf{A} = \begin{bmatrix} 0 & 1 & 0 \\ 0 & 0 & 1 \\ -15 & -11 & -5 \end{bmatrix}$, $\mathbf{B} = \begin{bmatrix} 0 \\ 0 \\ 1 \end{bmatrix}$, $\mathbf{C} = [\,1 \quad 0 \quad 0\,]$ and $\mathbf{x} = [\,x_1 \quad x_2 \quad x_3\,]^T$.
The eigenvalues of $\mathbf{A}$ are $\lambda = -1 + j2$, $\lambda = -1 - j2$ and $\lambda = -3$. The corresponding eigenvectors are $\mathbf{v}_1 = [\,1 \quad -1+2j \quad -3-4j\,]^T$, $\mathbf{v}_2 = [\,1 \quad -1-2j \quad -3+4j\,]^T$ and $\mathbf{v}_3 = [\,1 \quad -3 \quad 9\,]^T$.

1. The plant's transfer function $\dfrac{Y(s)}{U(s)}$ is:

   A) $\dfrac{Y(s)}{U(s)} = \dfrac{s+1}{(s+3)\left(s^2+2s+5\right)}$

   B) $\dfrac{Y(s)}{U(s)} = \dfrac{s^2+s+1}{(s+3)\left(s^2+2s+5\right)}$

   C) $\dfrac{Y(s)}{U(s)} = \dfrac{1}{(s+3)\left(s^2+2s+5\right)}$

   D) $\dfrac{Y(s)}{U(s)} = \dfrac{s^2+1}{(s+3)\left(s^2+2s+5\right)}$.

2. The plant's Jordan canonical state-space representation of the plant is:

E) $\dot{\mathbf{d}} = \begin{bmatrix} -1 & 0 & 0 \\ 0 & -1 & 0 \\ 0 & 0 & -3 \end{bmatrix}\mathbf{d} + \begin{bmatrix} 1 \\ 1 \\ 1 \end{bmatrix}u,\ y = [\,1 \quad 1 \quad 1\,]\mathbf{x}$

F) $\dot{\mathbf{d}} = \begin{bmatrix} -1 & 2 & 0 \\ -2 & -1 & 0 \\ 0 & 0 & -3 \end{bmatrix}\mathbf{d} + \begin{bmatrix} -\frac{1}{8} \\ \frac{1}{8} \\ \frac{1}{8} \end{bmatrix}u,\ y = [\,1 \quad 0 \quad 1\,]\mathbf{x}$

G) $\dot{\mathbf{d}} = \begin{bmatrix} 1 & -2 & 0 \\ -2 & 1 & 0 \\ 0 & 0 & 3 \end{bmatrix}\mathbf{d} + \begin{bmatrix} \frac{1}{8} \\ \frac{1}{8} \\ \frac{1}{8} \end{bmatrix}u,\ y = [\,0 \quad 1 \quad 0\,]\mathbf{x}$

H) $\dot{\mathbf{d}} = \begin{bmatrix} 1 & 2 & 0 \\ 2 & 1 & 0 \\ 0 & 0 & 3 \end{bmatrix}\mathbf{d} + \begin{bmatrix} -1 \\ 1 \\ 1 \end{bmatrix}u,\ y = \left[-\frac{1}{8} \quad \frac{1}{8} \quad \frac{1}{8}\right]\mathbf{x}.$

Exercise 7.7 Consider the linear plant $\mathbf{x} = [\,x_1 \quad x_2\,]^T$, $\dot{\mathbf{x}} = \begin{bmatrix} -2 & 0 \\ 0 & -4 \end{bmatrix}\mathbf{x} + \begin{bmatrix} 1 \\ -3 \end{bmatrix}u,\ y = [\,6 \quad 5\,]\mathbf{x}.$

1. The eigenvalues are:
   A) $\lambda_1 = 1,\ \lambda_2 = -3$
   B) $\lambda_1 = 6,\ \lambda_2 = 5$
   C) $\lambda_1 = -2,\ \lambda_2 = -4$
   D) None of the above.

2. The plant is:
   E) non-controllable
   F) non-observable
   G) controllable
   H) unstable.

Exercise 7.8 Consider the plant $\begin{cases} \dot{\mathbf{x}} = \mathbf{A}\mathbf{x} + \mathbf{B}u \\ y = \mathbf{C}\mathbf{x} \end{cases}$, where $\mathbf{A} = \begin{bmatrix} -1 & 1 & 1 \\ 0 & 0 & 1 \\ -6 & 0 & -5 \end{bmatrix}$, $\mathbf{B} = \begin{bmatrix} 0 \\ 0 \\ 1 \end{bmatrix}$, $\mathbf{C} = [\,1 \quad 0 \quad 0\,]$.

1. Draw its block diagram.
2. Find its transfer function $\dfrac{Y(s)}{U(s)}$.
3. Find matrix $\mathbf{Q}$ and see if the plant is controllable.

Exercise 7.9 Consider the block diagram in Figure 7.11.

1. **A)** The plant is controllable and observable.
   B) The plant is non-controllable and observable.
   C) The plant is controllable and non-observable.
   D) The plant is non-controllable and non-observable.

2. The plant's transfer function is:

   E) $\dfrac{Y(s)}{U(s)} = \dfrac{7}{s-2}$

   F) $\dfrac{Y(s)}{U(s)} = \dfrac{2}{s+4}$

   G) $\dfrac{Y(s)}{U(s)} = \dfrac{12}{s+3}$

   H) None of the above.

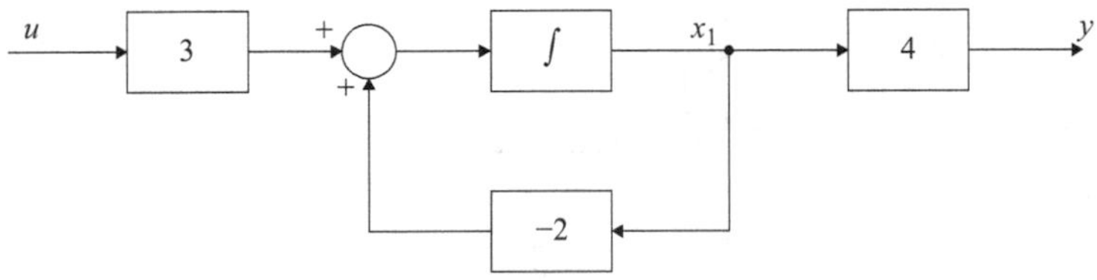

Figure 7.11 Block diagram of Exercise 7.9

Exercise 7.10 Consider the circuit in Figure 7.12. Let x be the current in the inductance. The state-space representation of the plant is:

A) $\dot{x} = -\dfrac{R_1}{LR_2}x + \dfrac{1}{LR_2}u$

B) $\dot{x} = -\dfrac{R_1}{L(R_1+R_2)}x + \dfrac{1}{L(R_1+R_2)}u$

C) $\dot{x} = -\dfrac{R_1}{L(\frac{R_1}{R_2}+1)}x + \dfrac{1}{L(\frac{R_1}{R_2}+1)}u$

D) None of the above.

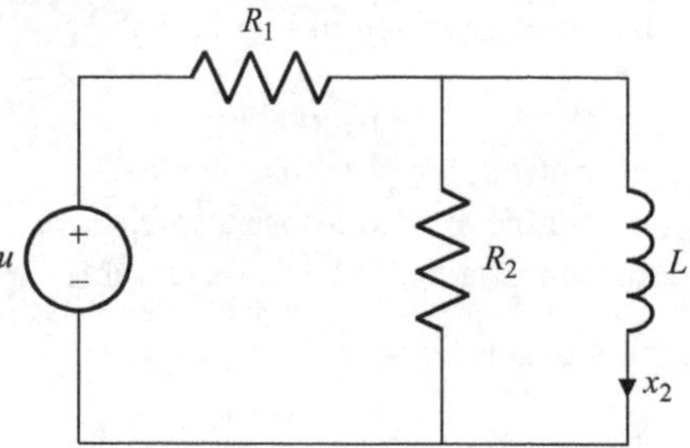

Figure 7.12 Circuit of Exercise 7.10

Exercise 7.11 Consider the block diagram of Figure 7.13. Let $\mathbf{x} = [x_1 \quad x_2]^T$. The plant's state-space representation is:

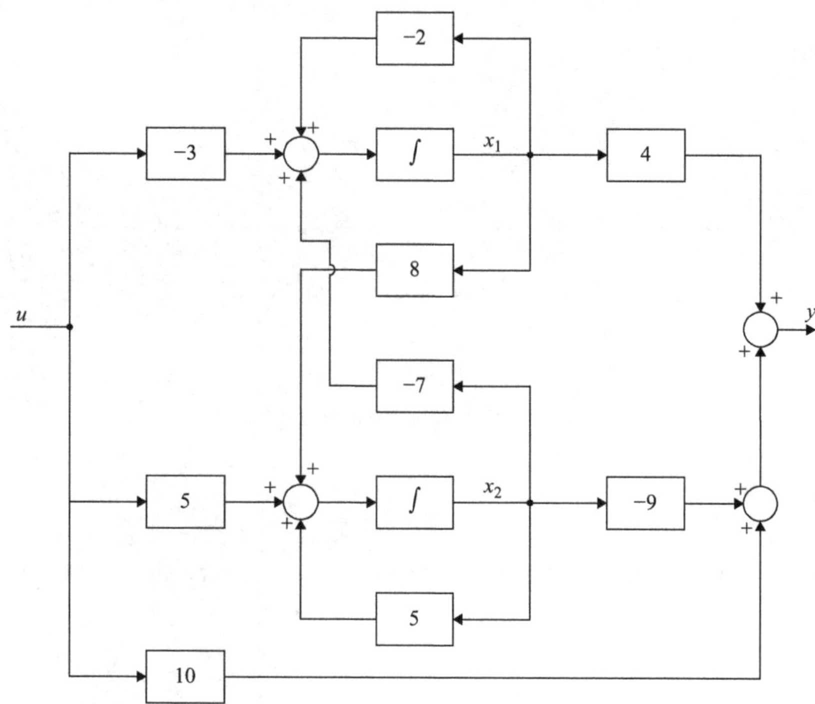

Figure 7.13 Block diagram of Exercise 7.11

A) $\begin{cases} \dot{\mathbf{x}} = \begin{bmatrix} 2 & 7 \\ 5 & 8 \end{bmatrix} \mathbf{x} + \begin{bmatrix} 3 \\ 5 \end{bmatrix} u \\ y = \begin{bmatrix} 4 & 9 \end{bmatrix} \mathbf{x} \end{cases}$

B) $\begin{cases} \dot{\mathbf{x}} = \begin{bmatrix} -2 & -7 \\ 8 & 5 \end{bmatrix}\mathbf{x} + \begin{bmatrix} -3 \\ 5 \end{bmatrix}u \\ y = \begin{bmatrix} 4 & -9 \end{bmatrix}\mathbf{x} + 10u \end{cases}$

C) $\begin{cases} \dot{\mathbf{x}} = \begin{bmatrix} 2 & -7 \\ 8 & 5 \end{bmatrix}\mathbf{x} + \begin{bmatrix} 3 \\ -5 \end{bmatrix}u \\ y = \begin{bmatrix} 4 & 9 \end{bmatrix}\mathbf{x} - 10u \end{cases}$

D) None of the above.

Exercise 7.12 Consider the plant $\dot{\mathbf{x}} = \begin{bmatrix} -4 & 0 \\ 0 & -5 \end{bmatrix}\mathbf{x} + \begin{bmatrix} 3 \\ 0 \end{bmatrix}u$, $y = [\,2 \quad 0\,]\mathbf{x}$ with $\mathbf{x} = [\,x_1 \quad x_2\,]^T$.

1. Is the system controllable?
   A) Yes
   B) No.
2. Is the system observable?
   C) Yes
   D) No.
3. Is the system stable?
   E) Yes
   F) No.
4. The system's transfer function $\dfrac{Y(s)}{U(s)}$ is:
   G) $\dfrac{Y(s)}{U(s)} = \dfrac{6}{s+4}$
   H) $\dfrac{Y(s)}{U(s)} = \dfrac{5}{(s+4)(s+5)}$
   I) $\dfrac{Y(s)}{U(s)} = \dfrac{6}{s+5}$
   J) None of the above.

Exercise 7.13 Consider system $\dddot{y} + 9\ddot{y} + 26\dot{y} + 24y = \dot{u} + 4u$. Let $\mathbf{x} = [\,x_1 \quad x_2 \quad x_3\,]^T$ and $\dot{\mathbf{x}} = [\,\dot{x}_1 \quad \dot{x}_2 \quad \dot{x}_3\,]^T$.

1. The system's controllable canonical form is:

   A) $\begin{cases} \dot{\mathbf{x}} = \begin{bmatrix} 0 & 1 & 0 \\ 0 & 0 & 1 \\ -24 & -26 & -9 \end{bmatrix}\mathbf{x} + \begin{bmatrix} 0 \\ 0 \\ 1 \end{bmatrix}u \\ y = [\,4 \quad 1 \quad 0\,]\mathbf{x} \end{cases}$

B)
$$\begin{cases} \dot{\mathbf{x}} = \begin{bmatrix} -9 & 1 & 0 \\ -26 & 0 & 1 \\ -24 & 0 & 0 \end{bmatrix} \mathbf{x} + \begin{bmatrix} 0 \\ 1 \\ 4 \end{bmatrix} u \\ y = [1 \quad 0 \quad 0]\mathbf{x} \end{cases}$$

C)
$$\begin{cases} \dot{\mathbf{x}} = \begin{bmatrix} -2 & 0 & 0 \\ 0 & -3 & 0 \\ 0 & 0 & -4 \end{bmatrix} \mathbf{x} + \begin{bmatrix} 0 \\ 0 \\ 1 \end{bmatrix} u \\ y = [1 \quad 0 \quad 0]\mathbf{x} \end{cases}$$

D) None of the above.

2. The system's observable canonical form is:

E)
$$\begin{cases} \dot{\mathbf{x}} = \begin{bmatrix} 0 & 1 & 0 \\ 0 & 0 & 1 \\ -24 & -26 & -9 \end{bmatrix} \mathbf{x} + \begin{bmatrix} 0 \\ 0 \\ 1 \end{bmatrix} u \\ y = [4 \quad 1 \quad 0]\mathbf{x} \end{cases}$$

F)
$$\begin{cases} \dot{\mathbf{x}} = \begin{bmatrix} -9 & 1 & 0 \\ -26 & 0 & 1 \\ -24 & 0 & 0 \end{bmatrix} \mathbf{x} + \begin{bmatrix} 0 \\ 1 \\ 4 \end{bmatrix} u \\ y = [1 \quad 0 \quad 0]\mathbf{x} \end{cases}$$

G)
$$\begin{cases} \dot{\mathbf{x}} = \begin{bmatrix} -2 & 0 & 0 \\ 0 & -3 & 0 \\ 0 & 0 & -4 \end{bmatrix} \mathbf{x} + \begin{bmatrix} 0 \\ 0 \\ 1 \end{bmatrix} u \\ y = [1 \quad 0 \quad 0]\mathbf{x} \end{cases}$$

H) None of the above.

Exercise 7.14 Consider the block diagram in Figure 7.14.

1. Let $\mathbf{x} = [x_1 \quad x_2]^T$ and $\dot{\mathbf{x}} = [\dot{x}_1 \quad \dot{x}_2]^T$. The system's state-space representation is:

A)
$$\begin{cases} \dot{\mathbf{x}} = \begin{bmatrix} -3 & 0 \\ -1 & 2 \end{bmatrix} \mathbf{x} + \begin{bmatrix} 1 \\ -1 \end{bmatrix} u \\ y = [3 \quad 2]\mathbf{x} \end{cases}$$

B)
$$\begin{cases} \dot{\mathbf{x}} = \begin{bmatrix} -3 & 2 \\ -1 & 0 \end{bmatrix} \mathbf{x} + \begin{bmatrix} 2 \\ 3 \end{bmatrix} u \\ y = [-1 \quad 1]\mathbf{x} \end{cases}$$

C) $\begin{cases} \dot{\mathbf{x}} = \begin{bmatrix} -3 & -1 \\ 2 & 0 \end{bmatrix}\mathbf{x} + \begin{bmatrix} -1 \\ 1 \end{bmatrix}u \\ y = [\,2 \quad 3\,]\mathbf{x} \end{cases}$

D) None of the above.

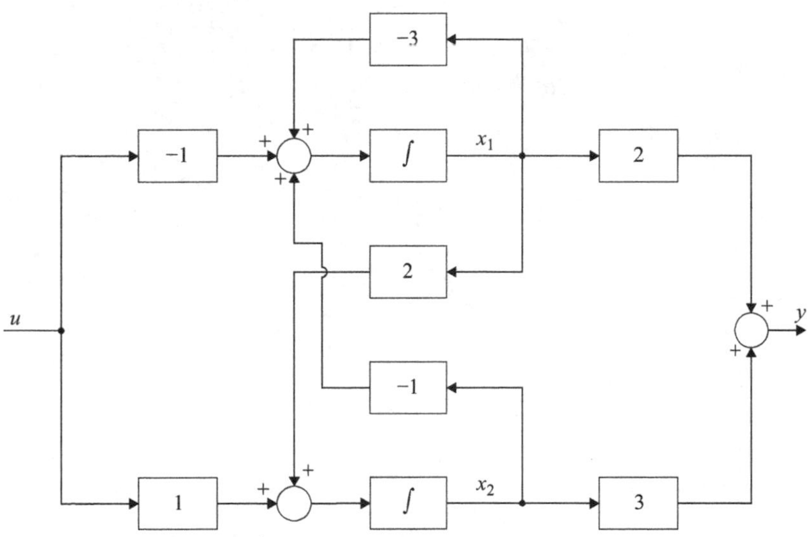

Figure 7.14 Block diagram of Exercise 7.14

2. The system's diagonal canonical form is:

E) $\begin{cases} \dot{\mathbf{x}} = \begin{bmatrix} 1 & 0 \\ 0 & 2 \end{bmatrix}\mathbf{x} + \begin{bmatrix} -1 \\ -1 \end{bmatrix}u \\ y = [\,2 \quad 3\,]\mathbf{x} \end{cases}$

F) $\begin{cases} \dot{\mathbf{x}} = \begin{bmatrix} -1 & 0 \\ 0 & -2 \end{bmatrix}\mathbf{x} + \begin{bmatrix} 0 \\ -1 \end{bmatrix}u \\ y = [\,-4 \quad -1\,]\mathbf{x} \end{cases}$

G) $\begin{cases} \dot{\mathbf{x}} = \begin{bmatrix} 1 & 0 \\ 0 & 2 \end{bmatrix}\mathbf{x} + \begin{bmatrix} -1 \\ -2 \end{bmatrix}u \\ y = [-1 \quad -2]\mathbf{x} \end{cases}$

H) None of the above.

3. The exponential matrix $\boldsymbol{\Phi}(t) = e^{\mathbf{A}t}$ is:

I) $\boldsymbol{\Phi}(t) = \begin{bmatrix} e^{-2t} - 2e^{-t} & 2e^{-2t} - 2e^{-t} \\ e^{-t} - e^{-2t} & e^{-t} - 2e^{-2t} \end{bmatrix}$

J) $\boldsymbol{\Phi}(t) = \begin{bmatrix} e^{-2t} + e^{-t} & e^{-2t} - e^{-t} \\ e^{-2t} - 2e^{-t} & e^{-t} + 2e^{-2t} \end{bmatrix}$

K) $\boldsymbol{\Phi}(t) = \begin{bmatrix} 2e^{-2t} - e^{-t} & e^{-2t} - e^{-t} \\ 2e^{-t} - 2e^{-2t} & 2e^{-t} - e^{-2t} \end{bmatrix}$

L) None of the above.

4. Transfer function $\dfrac{Y(s)}{U(s)}$ is:

M) $\dfrac{Y(s)}{U(s)} = \dfrac{1}{s+2}$

N) $\dfrac{Y(s)}{U(s)} = \dfrac{1}{(s+1)(s+2)}$

O) $\dfrac{Y(s)}{U(s)} = \dfrac{1}{s+1}$

P) None of the above.

5. The plant is:

Q) Controllable and observable
R) Non-controllable and observable
S) Controllable and non-observable
T) Non-controllable and non-observable.

Exercise 7.15 Consider the circuit in Figure 7.15.

1. Let $\mathbf{x} = [\,x_1 \quad x_2\,]^T$ and $\dot{\mathbf{x}} = [\,\dot{x}_1 \quad \dot{x}_2\,]^T$. Then the system's state-space representation is:

A) $\begin{cases} \dot{\mathbf{x}} = \begin{bmatrix} 0 & -\frac{1}{L} \\ \frac{1}{C} & -\frac{L}{R} \end{bmatrix} \mathbf{x} + \begin{bmatrix} \frac{1}{L} \\ 0 \end{bmatrix} i(t) \\ y(t) = [\,0 \quad -R\,]\mathbf{x} \end{cases}$

B) $\begin{cases} \dot{\mathbf{x}} = \begin{bmatrix} 0 & -1 \\ \frac{1}{R} & -\frac{C}{R} \end{bmatrix} \mathbf{x} + \begin{bmatrix} 0 \\ \frac{1}{R} \end{bmatrix} i(t) \\ y(t) = [\,1 \quad 0\,]\mathbf{x} \end{cases}$

C) $\begin{cases} \dot{\mathbf{x}} = \begin{bmatrix} 0 & -\frac{1}{C} \\ \frac{1}{L} & -\frac{R}{L} \end{bmatrix} \mathbf{x} + \begin{bmatrix} \frac{1}{C} \\ 0 \end{bmatrix} i(t) \\ y(t) = [\,0 \quad R\,]\mathbf{x} \end{cases}$

D) None of the above.

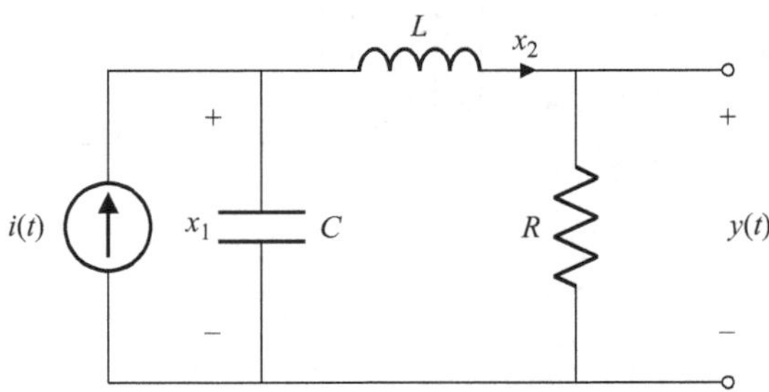

Figure 7.15 Circuit of Exercise 7.15

2. The system's transfer function $\dfrac{Y(s)}{I(s)}$ is:

E) $\dfrac{Y(s)}{I(s)} = \dfrac{R}{s^2LC + sRC + 1}$

F) $\dfrac{Y(s)}{I(s)} = \dfrac{1}{s^2 + sLC + RC}$

G) $\dfrac{Y(s)}{I(s)} = \dfrac{sL}{s^2L + sC + R}$

H) None of the above.

Exercise 7.16 Consider the block diagrams in Figure 7.16. Let $\mathbf{x} = [\,x_1 \quad x_2\,]^T$ and $\dot{\mathbf{x}} = [\,\dot{x}_1 \quad \dot{x}_2\,]^T$.

1. The state-space representation of system 1 is:

A) $\begin{cases} \dot{\mathbf{x}} = \begin{bmatrix} -\lambda & 1 \\ 1 & -\lambda \end{bmatrix} \mathbf{x} + \begin{bmatrix} 1 \\ 1 \end{bmatrix} u \\ y = [1 \quad 1]\mathbf{x} \end{cases}$

B) $\begin{cases} \dot{\mathbf{x}} = \begin{bmatrix} -\lambda & 1 \\ 0 & -\lambda \end{bmatrix} \mathbf{x} + \begin{bmatrix} 0 \\ 1 \end{bmatrix} u \\ y = [1 \quad 0]\mathbf{x} \end{cases}$

C) $\begin{cases} \dot{\mathbf{x}} = \begin{bmatrix} -\lambda & 1 \\ 0 & -\lambda \end{bmatrix} \mathbf{x} + \begin{bmatrix} 1 \\ 0 \end{bmatrix} u \\ y = [0 \quad 1]\mathbf{x} \end{cases}$

D) None of the above.

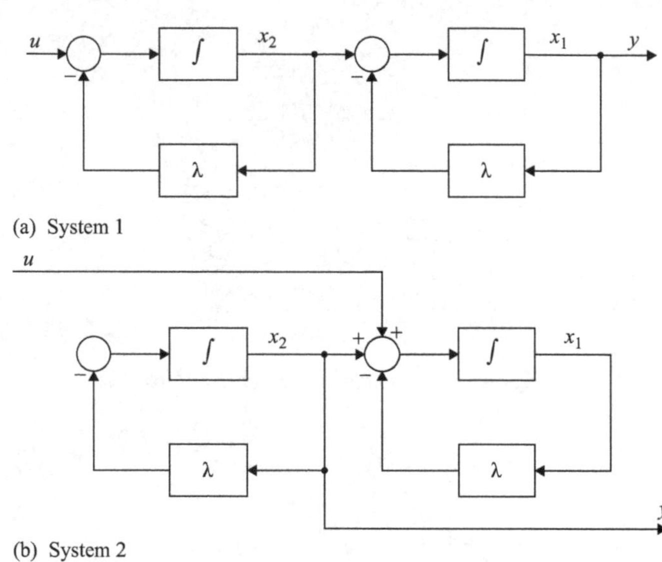

Figure 7.16 Block diagrams of Exercise 7.16

2. The state-space representation of system 2 is:

E) $\begin{cases} \dot{\mathbf{x}} = \begin{bmatrix} -\lambda & 1 \\ 1 & -\lambda \end{bmatrix} \mathbf{x} + \begin{bmatrix} 1 \\ 1 \end{bmatrix} u \\ y = [1 \quad 1]\mathbf{x} \end{cases}$

F) $\begin{cases} \dot{\mathbf{x}} = \begin{bmatrix} -\lambda & 1 \\ 0 & -\lambda \end{bmatrix} \mathbf{x} + \begin{bmatrix} 0 \\ 1 \end{bmatrix} u \\ y = [1 \quad 0]\mathbf{x} \end{cases}$

G) $\begin{cases} \dot{\mathbf{x}} = \begin{bmatrix} -\lambda & 1 \\ 0 & -\lambda \end{bmatrix} \mathbf{x} + \begin{bmatrix} 1 \\ 0 \end{bmatrix} u \\ y = [0 \quad 1]\mathbf{x} \end{cases}$

H) None of the above.

3. System 1 is:
   I) Controllable and observable
   J) Non-controllable and observable
   K) Controllable and non-observable
   L) Non-controllable and non-observable.

4. System 2 is:
   M) Controllable and observable
   N) Non-controllable and observable
   O) Controllable and non-observable
   P) Non-controllable and non-observable.

Exercise 7.17 Consider the circuit in Figure 7.17. Let $\mathbf{x}(t) = [x_1(t) \quad x_2(t)]^T$, $\dot{\mathbf{x}}(t) = [\dot{x}_1(t) \quad \dot{x}_2(t)]^T$ and $\mathbf{y}(t) = [y_1(t) \quad y_1(t)]^T$. Then the system's state-space representation is:

A)
$$\begin{cases} \dot{\mathbf{x}} = \begin{bmatrix} 0 & \frac{1}{L} \\ -\frac{1}{C} & -\frac{R_1}{C(R_1+R_2+R_3)} \end{bmatrix}\mathbf{x} + \begin{bmatrix} 0 \\ \frac{R_1}{C(R_1+R_2+R_3)} \end{bmatrix}u(t) \\ \mathbf{y} = \begin{bmatrix} 1 & 0 \\ 0 & -\frac{R_1}{R_1+R_2+R_3} \end{bmatrix}\mathbf{x} + \begin{bmatrix} 0 \\ \frac{1}{R_1+R_2+R_3} \end{bmatrix}u(t) \end{cases}$$

B)
$$\begin{cases} \dot{\mathbf{x}} = \begin{bmatrix} 0 & \frac{1}{C} \\ -\frac{1}{L} & -\frac{R_1(R_2+R_3)}{L(R_1+R_2+R_3)} \end{bmatrix}\mathbf{x} + \begin{bmatrix} 0 \\ \frac{R_2+R_3}{L(R_1+R_2+R_3)} \end{bmatrix}u(t) \\ \mathbf{y} = \begin{bmatrix} 1 & 0 \\ 0 & -\frac{R_1R_2}{R_1+R_2+R_3} \end{bmatrix}\mathbf{x} + \begin{bmatrix} 0 \\ \frac{R_2}{R_1+R_2+R_3} \end{bmatrix}u(t) \end{cases}$$

C)
$$\begin{cases} \dot{\mathbf{x}} = \begin{bmatrix} 0 & C \\ -\frac{1}{L} & -\frac{R_1(R_2+R_3)}{L(R_1+R_2+R_3)} \end{bmatrix}\mathbf{x} + \begin{bmatrix} 0 \\ \frac{R_2+R_3}{L(R_1+R_2+R_3)} \end{bmatrix}u(t) \\ \mathbf{y} = \begin{bmatrix} 1 & 0 \\ 0 & -R_1R_2 \end{bmatrix}\mathbf{x} + \begin{bmatrix} 0 \\ R_2 \end{bmatrix}u(t) \end{cases}$$

D) None of the above.

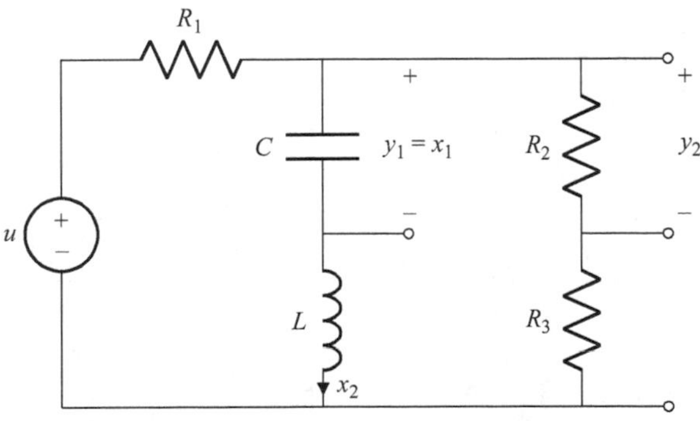

Figure 7.17 Circuit of Exercise 7.17

Exercise 7.18 Consider the block diagram in Figure 7.18. Let $\mathbf{x} = [x_1 \quad x_2]^T$ and $\dot{\mathbf{x}} = [\dot{x}_1 \quad \dot{x}_2]^T$.

1. The system's state-space representation is:

A) $$\begin{cases} \dot{\mathbf{x}} = \begin{bmatrix} 3 & -1 \\ 1 & 9 \end{bmatrix} \mathbf{x} + \begin{bmatrix} 2 \\ 3 \end{bmatrix} u \\ y = [7 \quad 5]\,\mathbf{x} \end{cases}$$

B) $$\begin{cases} \dot{\mathbf{x}} = \begin{bmatrix} 3 & -1 \\ -1 & 9 \end{bmatrix} \mathbf{x} + \begin{bmatrix} 2 \\ -3 \end{bmatrix} u \\ y = [7 \quad -5]\,\mathbf{x} \end{cases}$$

C) $$\begin{cases} \dot{\mathbf{x}} = \begin{bmatrix} -3 & -1 \\ 1 & -9 \end{bmatrix} \mathbf{x} + \begin{bmatrix} 2 \\ 3 \end{bmatrix} u \\ y = [5 \quad 7]\,\mathbf{x} \end{cases}$$

D) None of the above.

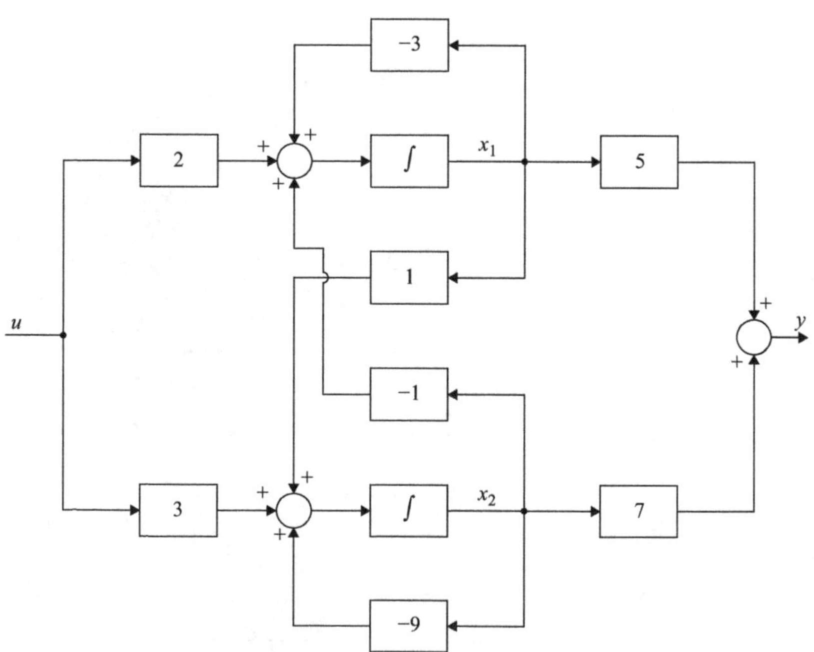

Figure 7.18 Block diagram of Exercise 7.18

2. The system's transfer function $\dfrac{Y(s)}{U(s)}$ is:

E) $\dfrac{Y(s)}{U(s)} = \dfrac{s+15}{s^2+2s+8}$

F) $\dfrac{Y(s)}{U(s)} = \dfrac{3s+12}{s^2+12s+8}$

G) $\dfrac{Y(s)}{U(s)} = \dfrac{31s+152}{s^2+12s+28}$

H) None of the above.

Exercise 7.19 Consider system $\dot{\mathbf{x}} = \mathbf{A}\mathbf{x} + \mathbf{B}u$, where the state vector is $\mathbf{x} = \begin{bmatrix} x_1 \\ x_2 \end{bmatrix}$, u is the input, $\mathbf{A} = \begin{bmatrix} 0 & 1 \\ -2 & -3 \end{bmatrix}$ and $\mathbf{B} = \begin{bmatrix} 1 \\ -1 \end{bmatrix}$. The exponential matrix $e^{\mathbf{A}t}$ is:

A) $e^{\mathbf{A}t} = \begin{bmatrix} 2e^{-t} - e^{-2t} & e^{-t} - e^{-2t} \\ -2e^{-t} + 2e^{-2t} & -e^{-t} + 2e^{-2t} \end{bmatrix}$

B) $e^{\mathbf{A}t} = \begin{bmatrix} 2e^{-t} + e^{-2t} & -e^{-t} + e^{-2t} \\ 2e^{-t} - 2e^{-2t} & -e^{-t} - 2e^{-2t} \end{bmatrix}$

C) $e^{\mathbf{A}t} = \begin{bmatrix} e^{-t} - 2e^{-2t} & e^{-2t} - e^{-t} \\ 2e^{-t} + e^{-2t} & e^{-t} + e^{-2t} \end{bmatrix}$

D) None of the above.

Exercise 7.20 Consider the system $\dot{\mathbf{x}} = \mathbf{A}\mathbf{x}$, where $\mathbf{x} = \begin{bmatrix} x_1 \\ x_2 \end{bmatrix}$ is the 2×1 state vector, with initial conditions $x(0) = \begin{bmatrix} 4 \\ -2 \end{bmatrix}$, and $\mathbf{A} = \begin{bmatrix} -3 & -1 \\ 2 & 0 \end{bmatrix}$. The evolution of the states with time $x(t) = e^{\mathbf{A}t}x(0)$ is:

A) $x(t) = \begin{bmatrix} e^{-5t} - e^{-3t} \\ 6e^{-5t} + e^{-t} \end{bmatrix}$

B) $x(t) = \begin{bmatrix} 5e^{-3t} \\ 6e^{-2t} \end{bmatrix}$

C) $x(t) = \begin{bmatrix} e^{-t} + 3e^{-2t} \\ 5e^{-t} + 2e^{-2t} \end{bmatrix}$

D) $x(t) = \begin{bmatrix} -2e^{-t} + 6e^{-2t} \\ 4e^{-t} - 6e^{-2t} \end{bmatrix}$.

Exercise 7.21 Consider the plant $\dot{\mathbf{x}} = \mathbf{A}\mathbf{x} + \mathbf{B}u$, where $\mathbf{x} = \begin{bmatrix} x_1 \\ x_2 \end{bmatrix}$ is the 2×1 state vector, u is the input, and $\mathbf{A} = \begin{bmatrix} 1 & 4 \\ 2 & 3 \end{bmatrix}$, $\mathbf{B} = \begin{bmatrix} 2 \\ -1 \end{bmatrix}$. The exponential matrix $e^{\mathbf{A}t}$ is:

A) $e^{\mathbf{A}t} = \begin{bmatrix} \frac{1}{3}(2e^{5t} - e^{-t}) & \frac{1}{3}(e^{5t} + e^{-t}) \\ \frac{2}{3}(e^{5t} + e^{-t}) & \frac{1}{3}(e^{5t} - e^{-t}) \end{bmatrix}$

B) $e^{\mathbf{A}t} = \begin{bmatrix} \frac{1}{3}(e^{5t} + 2e^{-t}) & \frac{2}{3}(e^{5t} - e^{-t}) \\ \frac{1}{3}(e^{5t} - e^{-t}) & \frac{1}{3}(2e^{5t} + e^{-t}) \end{bmatrix}$

C) $e^{\mathbf{A}t} = \begin{bmatrix} \frac{2}{3}(e^{5t} + 2e^{-t}) & \frac{1}{3}(2e^{5t} - e^{-t}) \\ \frac{1}{3}(e^{5t} + e^{-t}) & \frac{1}{3}(2e^{5t} - e^{-t}) \end{bmatrix}$

D) None of the above.

Exercise 7.22 Consider the block diagram in Figure 7.19.

1. Find its state-space representation $\dot{\mathbf{x}} = \mathbf{A}\mathbf{x} + \mathbf{B}u,\ y = \mathbf{C}\mathbf{x}$, when $\mathbf{x} = [x_1 \quad x_2]^T$.
2. Is the system observable?
3. Is the system controllable?

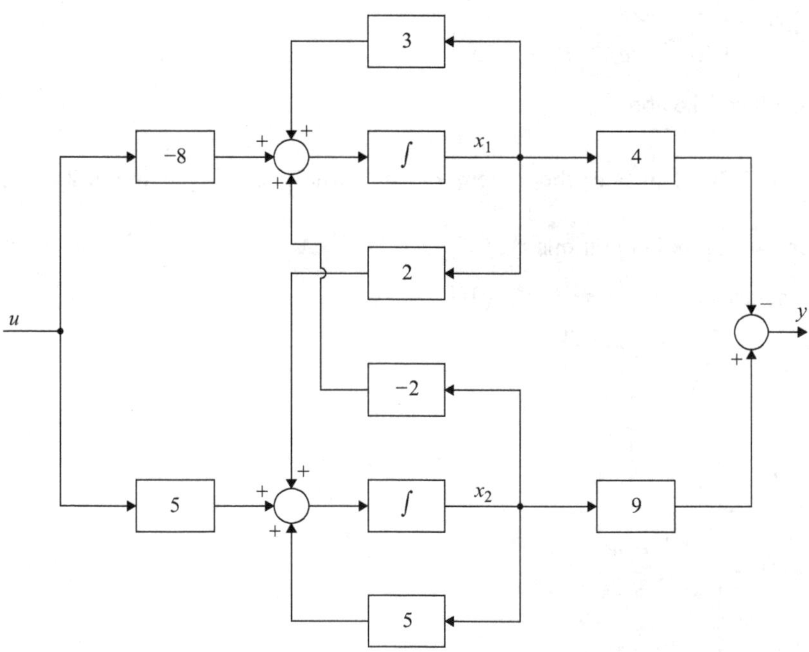

Figure 7.19 Block diagram of Exercise 7.22

Exercise 7.23 Consider the plant $\ddot{y} + 6\dot{y} + 8y = \ddot{u} + 4\dot{u} + 3u$.

1. Find the system's controllable canonical form.
2. Find the system's observable canonical form.
3. Find the system's transfer function $\dfrac{Y(s)}{U(s)}$.

Exercise 7.24 Consider the circuit in Figure 7.20.

1. Find a state-space representation for this system, using x_1, x_2 and x_3 as state variables.
2. Draw the corresponding block diagram.
3. Let $C_1 = C_2 = \frac{1}{2}$, $L = \frac{1}{3}$ and $R_1 = R_2 = 1$. Find the corresponding transfer function $\dfrac{Y(s)}{V(s)}$.

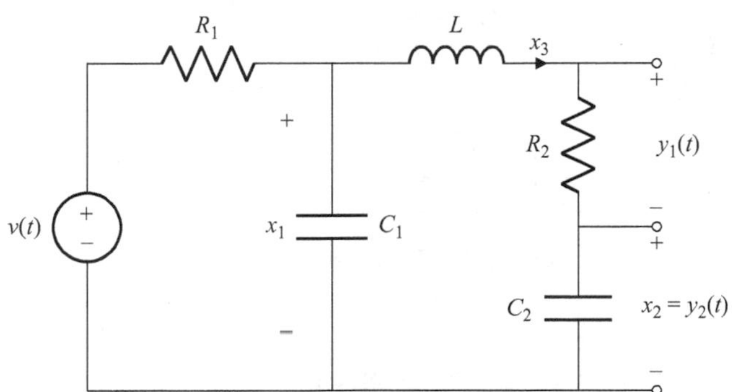

Figure 7.20 Circuit of Exercise 7.24

Exercise 7.25 Consider the block diagram in Figure 7.21. Let $\mathbf{x} = [x_1 \quad x_2]^T$ and $\lambda_1, \lambda_2 \in \mathbb{R} \setminus \{0\}$.

1. Find the state-space representation for this system.
2. Is the system controllable?
3. Is the system observable?
4. Find the range of values of λ_1, λ_2 for which the system is stable.
5. Prove that the system's transfer function is $G(s) = \dfrac{1}{(s + \lambda_1)(s + \lambda_2)}$.
6. Find the diagonal state-space representation of the plant and draw the corresponding block diagram.

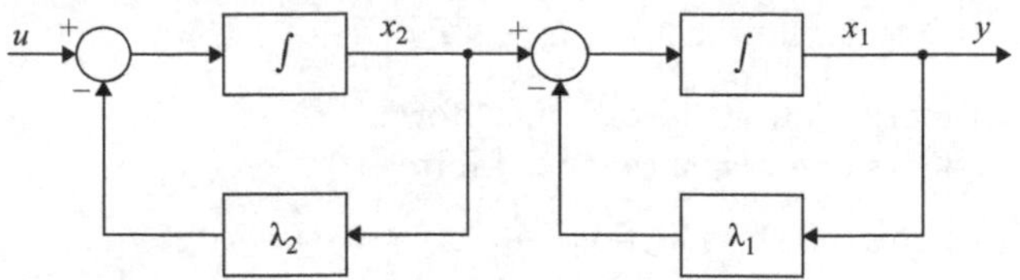

Figure 7.21 Block diagram of Exercise 7.25

Exercise 7.26 Consider the circuit in Figure 7.22. Find its state-space model $\dot{\mathbf{x}} = \mathbf{A}\mathbf{x} + \mathbf{B}u$, $y = \mathbf{C}\mathbf{x}$.

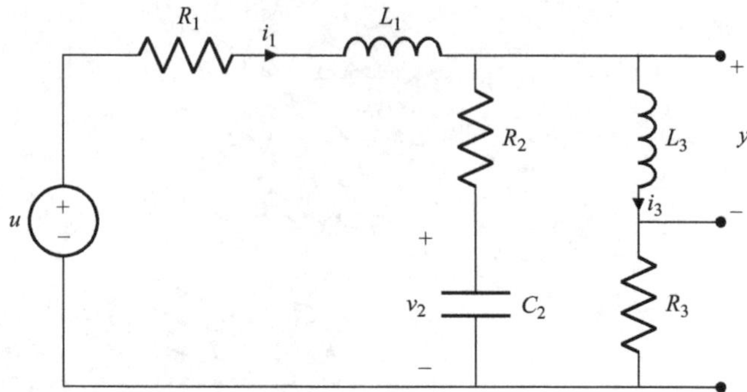

Figure 7.22 Circuit of Exercise 7.26

Exercise 7.27 Consider the plant represented in Figure 7.23.

1. Find its model of the form $\dot{\mathbf{x}} = \mathbf{A}\mathbf{x} + \mathbf{B}u$, $y = \mathbf{C}\mathbf{x}$, with $\mathbf{x} = [\,x_1 \quad x_2 \quad x_3\,]^T$.
2. Analyze whether the plant is stable and controllable.
3. Find its transfer function $Y(s)/U(s)$.

Exercise 7.28 Let $\mathbf{x} = [\,x_1 \quad x_2\,]^T$ and $\dot{\mathbf{x}} = [\,\dot{x}_1 \quad \dot{x}_2\,]^T$. Then the state-space representation of the block diagram in Figure 7.24 is:

A) $\begin{cases} \dot{\mathbf{x}} = \begin{bmatrix} 2 & -1 \\ 4 & 0 \end{bmatrix}\mathbf{x} + \begin{bmatrix} 3 \\ 5 \end{bmatrix}u \\ y = [\,2 \quad 6\,]\mathbf{x} \end{cases}$

B) $\begin{cases} \dot{\mathbf{x}} = \begin{bmatrix} 0 & 4 \\ -1 & 2 \end{bmatrix}\mathbf{x} + \begin{bmatrix} 3 \\ 5 \end{bmatrix}u \\ y = [\,6 \quad 2\,]\mathbf{x} \end{cases}$

C) $\begin{cases} \dot{\mathbf{x}} = \begin{bmatrix} 2 & 4 \\ -1 & 0 \end{bmatrix}\mathbf{x} + \begin{bmatrix} 2 \\ 6 \end{bmatrix}u \\ y = [\,3 \quad 5\,]\mathbf{x} \end{cases}$

D) None of the above.

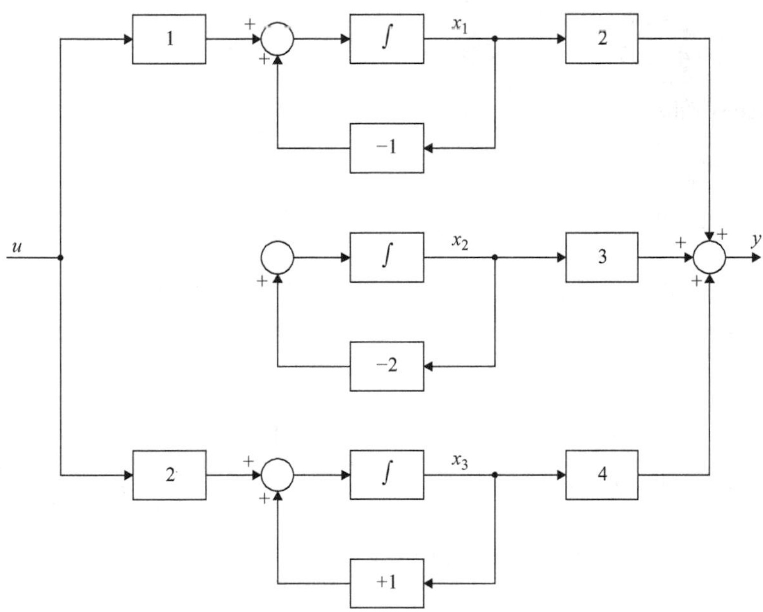

Figure 7.23 Block diagram of Exercise 7.27

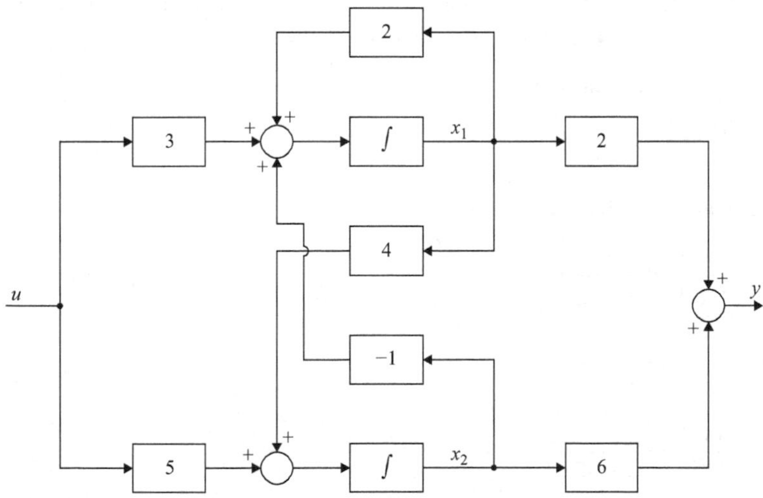

Figure 7.24 Block diagram of Exercise 7.28

Exercise 7.29 The state-space representation of the plant in Figure 7.25 is:

A) $\begin{cases} \dot{x} = 5x + 2u \\ y = 4x - 3u \end{cases}$

B) $\begin{cases} \dot{x} = 3x - 2u \\ y = 5x + 4u \end{cases}$

C) $\begin{cases} \dot{x} = -2x + 3u \\ y = -4x + 5u \end{cases}$

D) None of the above.

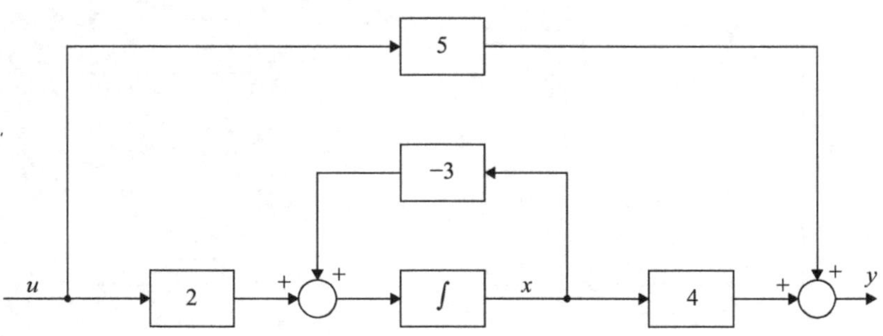

Figure 7.25 Block diagram of Exercise 7.29

Exercise 7.30 Consider the block diagram of a system, as shown in Figure 7.26, with $b_i, c_i, \lambda_i \in \mathbb{R}$ $(i = 1, 2, 3)$, $b_1 = -b_2$, $\lambda_1 = \lambda_2$, and $c_1 = c_2$.

1. Is the plant controllable?

   A) Yes
   B) No

2. The plant's transfer function $\dfrac{Y(s)}{U(s)}$ is:

   C) $\dfrac{Y(s)}{U(s)} = \dfrac{b_3 c_3}{s - \lambda_3}$

   D) $\dfrac{Y(s)}{U(s)} = \dfrac{c_3}{s - \lambda_3}$

   E) $\dfrac{Y(s)}{U(s)} = \dfrac{1}{s + \lambda_3}$

   F) None of the above.

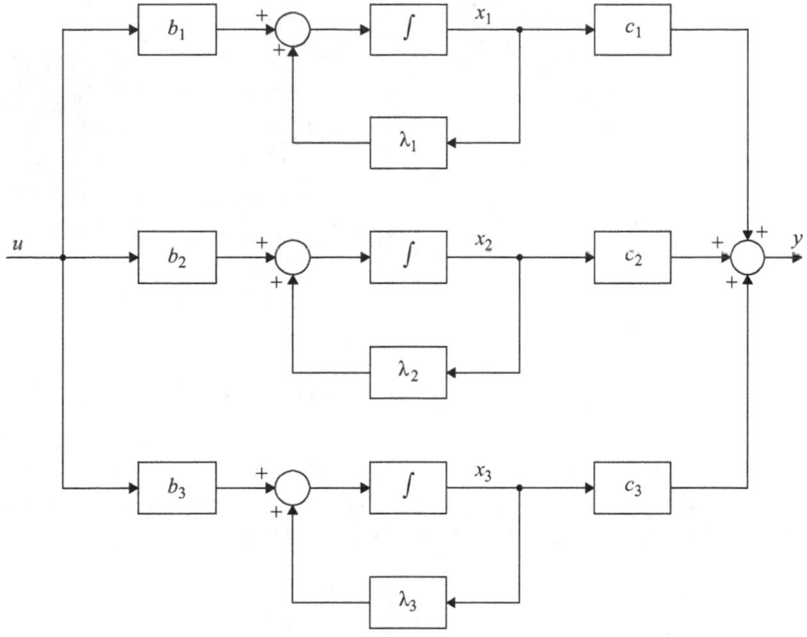

Figure 7.26 Block diagram of Exercise 7.30

Exercise 7.31 Find a model of the form $\dot{\mathbf{x}} = \mathbf{A}\mathbf{x} + \mathbf{B}u$, $y = \mathbf{C}\mathbf{x}$ for the plant in Figure 7.27.

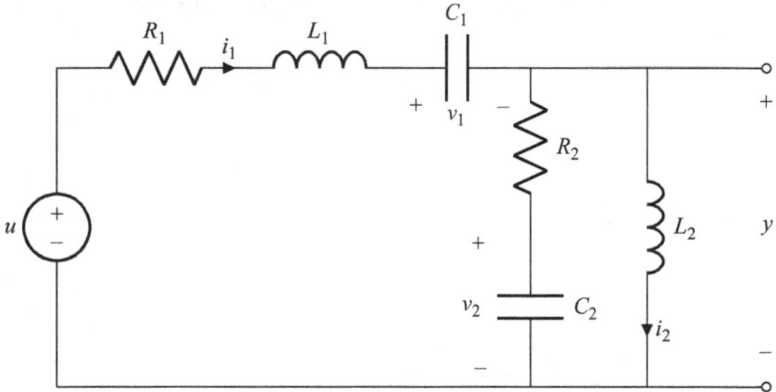

Figure 7.27 Circuit of Exercise 7.31

Exercise 7.32 Consider the circuit in Figure 7.28.

1. Find its state-space representation when the state vector is $[x_1 \quad x_2]^T$.
2. Find the exponential matrix $e^{\mathbf{A}t}$.
3. Find transfer function $Y_1(s)/U(s)$.
4. Is the plant is controllable and observable?

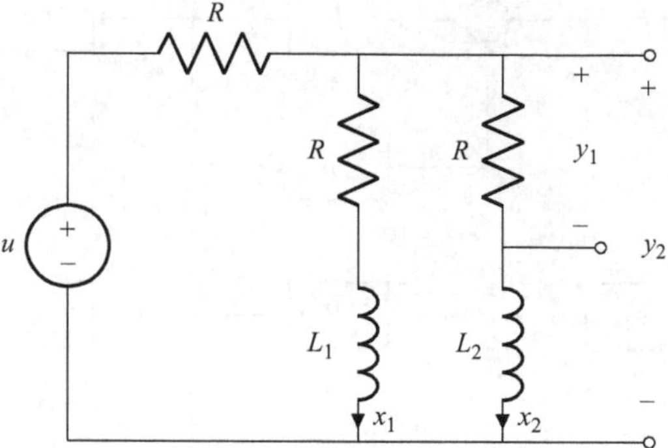

Figure 7.28 Circuit of Exercise 7.32

Exercise 7.33 Consider the hydraulic circuit in Figure 7.29. The input is flow $q_i = u$; the output is flow $q_o = y$. Assume laminar flow, for which:

$$\begin{cases} q_{out} = \frac{h}{R} \\ A \cdot \frac{dh}{dt} = q_{in} - q_{out} \end{cases}$$

1. Find a state-space representation for this plant, using the states $x_1 = h_1,\ x_2 = h_2$.
2. Let $R_1 = \frac{1}{2},\ R_2 = 1,\ A_1 = 1,\ A_2 = 2$. Find the transfer function $\dfrac{Q_o(s)}{Q_i(s)}$ from the state-space representation.
3. Find matrices **Q** and **R** and check whether the plant's representation is controllable and observable.
4. Find the diagonal state-space representation of the plant.

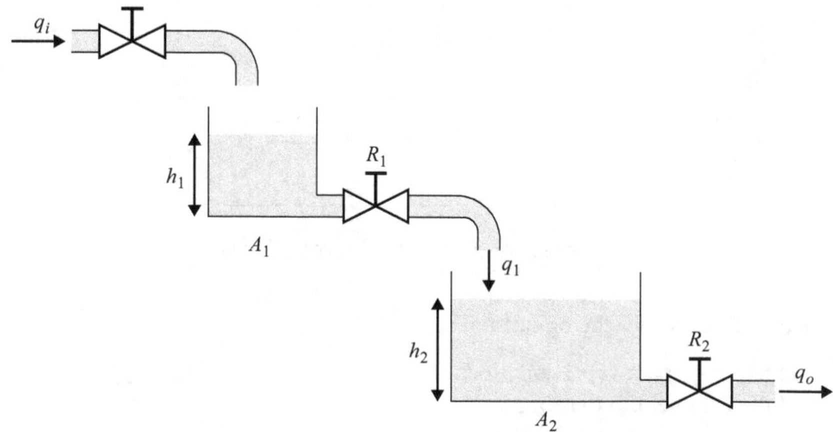

Figure 7.29 Hydraulic circuit of Exercise 7.33

Exercise 7.34 Consider the circuit in Figure 7.30.

1. Let $x_1 = v_1$, $x_2 = v_2$, $x_3 = i$, $y = v_2$ and $u = v_i$. Find the system's state-space representation.
2. Let $\frac{1}{RC} = 3$, $\frac{1}{C} = 1$ and $\frac{1}{L} = 2$. Find from the state-space representation the plant's transfer function $\frac{V_2(s)}{V_i(s)}$.

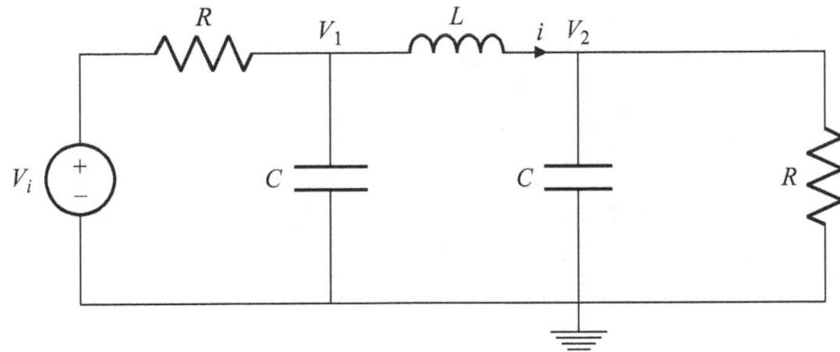

Figure 7.30 Circuit of Exercise 7.34

7.4 State space analysis of continuous systems using computer packages

In this section we consider the transfer function $G(s) = \frac{s+3}{s(s+1)(s+2)}$ and the state-space representation using the computer packages MATLAB®, SCILAB™ and OCTAVE©.

7.4.1 MATLAB

This subsection describes some basic commands that can be adopted with the package MATLAB.

The transfer function $G(s) = \frac{s+3}{s(s+1)(s+2)}$ is represented by two arrays `num` and `den`. The conversion from transfer function to state-space representation is accomplished by means of the command `tf2ss`. Note that there is an infinite number of possible state-space representations for the system.

The complete code is as follows.

```
% Conversion: Transfer function to state-space
% form

% Numerator and denominator of transfer function G(s)
num = [0 0 1 3];
```

```
den = [1 3 2 0];

% Convert G(s) to state-space form
[A,B,C,D] = tf2ss(num,den)
```

MATLAB outputs the following information at the Command Window.

```
A =

    -3    -2     0
     1     0     0
     0     1     0


B =

     1
     0
     0


C =

     0     0     1


D =

     0
```

It is also possible to perform the inverse conversion, that is, from the state-space representation to the transfer function. The conversion from state-space representation to transfer function is accomplished by means of the command `ss2tf`.

The complete code is as follows.

```
% Conversion: State-space form to transfer
% function

% State-space matrices
A = [-3 -2 0 ; 1 0 0 ; 0 1 0];
```

```
B = [1 ; 0 ; 0];
C = [0 0 1];
D = [0];

% Convert state-space representation to transfer
% function
[num,den] = ss2tf(A,B,C,D)
```

MATLAB outputs the following information at the Command Window.

```
num =

     0     0     0     1


den =

     1     3     2     0
```

The state-space canonical realization of the transfer function is accomplished by means of the command `canon`. The complete code is as follows.

```
%%%% State-space canonical realization %%%%

% Numerator and denominator of transfer function G(s)
num = [0 0 1 3];
den = [1 3 2 0];

G = tf(num,den);

% State-space canonical realization
Gc = canon(G,'modal')
```

MATLAB outputs the following information at the Command Window.

```
Gc =

  a =
         x1   x2   x3
```

```
   x1   0   0   0
   x2   0  -1   0
   x3   0   0  -2

 b =

       u1
   x1   1
   x2   3
   x3   3

 c =

             x1         x2         x3
   y1       1.5   -0.6667     0.1667

 d =

       u1
   y1   0

Continuous-time state-space model.
```

7.4.2 SCILAB

This subsection describes some basic commands that can be adopted with the package SCILAB.

The conversion from transfer function to state-space representation can be accomplished by means of the command abcd. Note that there is an infinite number of possible state-space representations for the system.

The complete code is as follows.

```
// Conversion: Transfer function to state-space
// form

// Numerator and denominator of transfer function G(s)
num = poly([3 1 0 0],'s','coeff');
den = poly([0 2 3 1],'s','coeff');

// Transfer function
G = syslin('c',num/den);

// Convert G(s) to state-space matrices
[A B C D] = abcd(G)
```

```
// Displays matrices A, B, C, D
disp(A,"A")
disp(B,"B")
disp(C,"C")
disp(D,"D")
```

SCILAB outputs the following information at the Console.

```
 A

      0.3            0.9539392     1.110D-16
   - 0.3040026   - 0.8934066   - 0.2085018
      0.5635445      2.6375481   - 2.4065934

 B

      0.
   - 0.6629935
      1.8869127

 C

   - 1.5811388     5.551D-17     0.

 D

      0.
```

It is also possible to perform the inverse conversion, that is, from the state-space representation to the transfer function. The conversion from state-space representation to transfer function is accomplished by means of the command `ss2tf`. The complete code is as follows.

```
// Conversion: State-space form to transfer
// function

// State-space matrices
A = [-3 -2 0 ; 1 0 0 ; 0 1 0];
B = [1 ; 0 ; 0];
C = [0 0 1];
```

```
D = [0];

//Linear system definition
S1 = syslin('c',A,B,C,D)

// Convert state-space representation to transfer
// function
[G] = ss2tf(S1)

// Displays transfer function
disp(G,"G(s)")
```

SCILAB outputs the following information at the Console.

```
 G(s)

                                   2
    1 - 2.665D-15s - 2.220D-15s
    ---------------------------
                      2   3
            2s + 3s + s
```

7.4.3 OCTAVE

This subsection describes some basic commands that can be adopted with the package OCTAVE. It is required to load packages `control` and `signal`.

The conversion from transfer function to state-space representation is accomplished by means of the command `tf2ss`. The complete code is as follows.

```
% Conversion: Transfer function to state-space
% form

% Numerator and denominator of transfer function G(s)
num = [0 0 1 3];
den = [1 3 2 0];

% Convert G(s) to state-space form
[A,B,C,D] = tf2ss(num,den)
```

OCTAVE creates the following Command Window.

```
A =

   0.00000    0.00000    0.00000
   1.00000   -0.00000    2.00000
   0.00000   -1.00000   -3.00000

B =

   3.00000
   1.00000
   0.00000

C =

   0   0  -1

D = 0
```

It is also possible to perform the inverse conversion, that is, from the state-space representation to the transfer function. The conversion from state-space representation to transfer function is accomplished by means of the command `ss2tf`.

The complete code is as follows.

```
% Conversion: State-space form to transfer
% function

% State-space matrices
A = [-3 -2 0 ; 1 0 0 ; 0 1 0];
B = [1 ; 0 ; 0];
C = [0 0 1];
D = [0];

% Convert state-space representation to transfer
% function
[num,den] = ss2tf(A,B,C,D)
```

OCTAVE creates the following Command Window.

```
num =   1.00000
den =

   1.00000   3.00000   2.00000   0.00000
```

Chapter 8
Controller synthesis by pole placement

8.1 Fundamentals

The pole placement synthesis technique allows placing all closed-loop poles at desired locations, so that the system closed-loop specifications can be met. Thus, the main advantage of pole placement over other classical synthesis techniques is that we can force both the dominant and the non-dominant poles to lie at arbitrary locations [2,4–6].

8.1.1 List of symbols

a_i, b_i, c_i, k_i, α_i	constant coefficients
$\mathbf{A}$	$n \times n$ dimensional state (or system) matrix
$\mathbf{B}$	$n \times q$ dimensional input matrix
C	capacitance, thermal capacitance
$\mathbf{C}$	$n \times m$ dimensional output matrix
$\mathbf{D}$	$m \times q$ dimensional feedthrough (or feedforward) matrix
e_{ss}	steady-state error
i	current
$\mathbf{I}_n$	$n \times n$ dimensional identity matrix
J	inertia
$\mathbf{J}$, $\mathbf{M}$, $\mathbf{P}$, $\mathbf{V}$, $\mathbf{T}$, $\mathbf{W}$	$n \times n$ dimensional matrices
$\mathbf{J}_k$	Jordan block
K	stiffness
L	inductance
$\mathscr{L}$	Laplace operator
M	mass
n	number of state variables (system order)
p_i	poles of the transfer function
$\mathbf{Q}$	state controllability matrix
$\mathbf{R}$	observability matrix
s	Laplace variable
$\mathbf{S}$	output controllability matrix
t	time
t_0	initial time

$u(t)$	time-domain input function
$\mathbf{u}$	q dimensional input (or control) vector
v	voltage
$\mathbf{v}_i$	eigenvector
$\mathbf{w}_i$	left eigenvector
x	displacement
$\dot{x}$	linear velocity
$\mathbf{x}$	n dimensional state vector
$\mathbf{x}_0$	initial state vector
$\mathbf{y}$	m dimensional output vector
$y(t)$	time-domain output function
ΔPM	phase margin variation
θ	temperature
λ_i	eigenvalue
$\mathbf{\Lambda}$	$n \times n$ dimensional diagonal matrix of eigenvalues
$\boldsymbol{\Phi}(t)$	state transition matrix
ω	angular velocity
ω^*	frequency at which the gain of the plant compensates that of a lead compensator
ω_1	gain crossover frequency

8.1.2 Pole placement using an input–output representation

When using an input–output representation (transfer function) of a plant, it is usually possible to place only the dominant pole, or the dominant pair of complex conjugate poles, of the closed loop. They are placed in a location that satisfies the requirements to fulfill, according to the results of Chapter 3, and then the performance is checked, to see if the influence of other poles and zeros is sufficiently small so that requirements are still verified.

The controller structure offered can be a PI, a PD, or a PID. Another usual controller structure employed in these cases is

$$C(s) = K\frac{s+a}{s+b} \tag{8.1}$$

- When the pole is to the left of the zero ($-b < -a$) and $\frac{Ka}{b} = 1$ (unitary low frequency gain), (8.1) is called a lead controller (or lead compensator).
- When the zero is to the left of the pole ($-a < -b$) and $K = 1$ (unitary high frequency gain), (8.1) is called a lag controller (or lag compensator).

Figure 8.1 shows the location of the pole and the zero for both cases. Figure 8.2 shows the corresponding Bode diagrams (from which the reason of the names of lead and lag compensators are obvious).

The lag controller is used to decrease a steady-state error, when eliminating it by adding a pole at the origin is not necessary or cannot be done:

- Coefficients K_p, K_v, K_a and K_j (3.35)–(3.38) get multiplied by a/b; hence, that ratio is found from the desired value of e_{ss}.

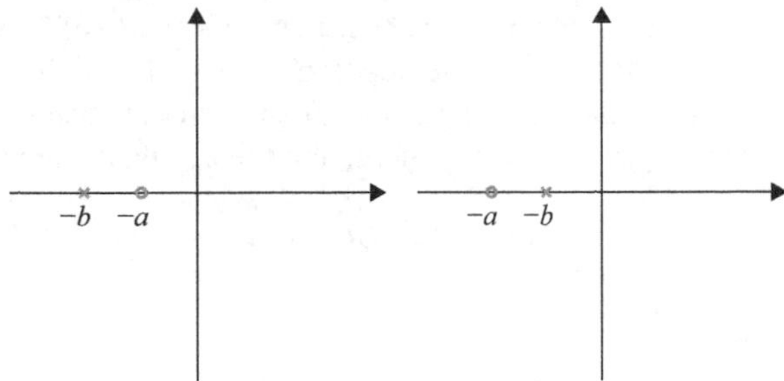

Figure 8.1 Location of the pole and the zero of lead (left) and lag (right) controllers

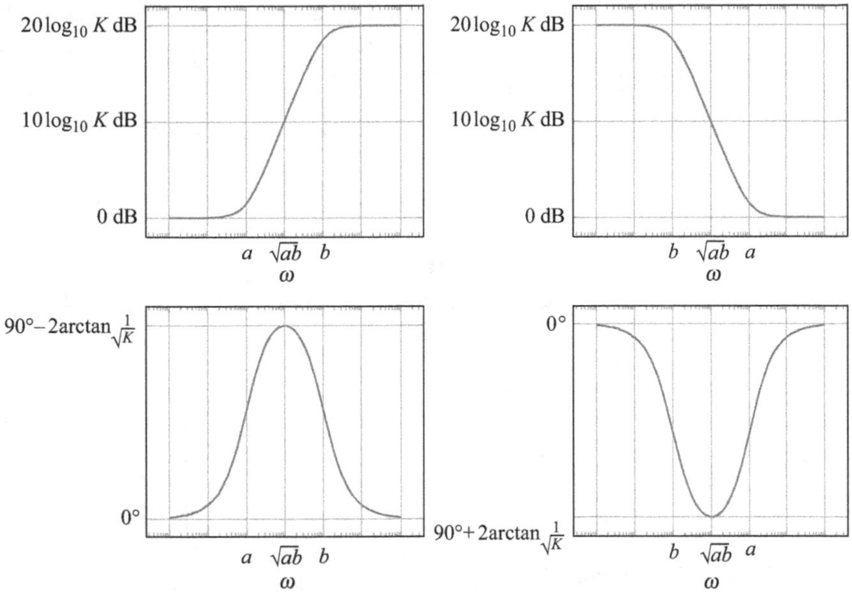

Figure 8.2 Bode diagrams of lead (left) and lag (right) controllers

- The zero and the pole ought to be at high frequencies: having the zero one decade below the gain-cross frequency ω_1 is a usual choice ($a = \frac{\omega_1}{10}$); if there are poles at the origin, a decade and a half, or more.
- $K = 1$.

The lead controller is used to increase a plant's phase margin. If the value of the desired increase ΔPM is known, from Figure 8.1, it is clear that the gain K must be

$$K = \left(\frac{1}{\tan \frac{90^\circ - \Delta PM}{2}} \right)^2 \tag{8.2}$$

but it is a good policy to allow for an extra phase margin increase of 5° or 10°, because of possible inaccuracies determining the position of the pole and the zero. Find the frequency $\omega^\star$ at which the gain of the plant $G(s)$ compensates that of the controller where the phase margin increase is maximum, and then a and b from that value:

$$10\log_{10} K = -20\log_{10}|G(j\omega^\star)| \Leftrightarrow |G(j\omega^\star)| = \frac{1}{\sqrt{K}} \tag{8.3}$$

$$a\sqrt{K} = \omega^\star \Leftrightarrow a = \frac{\omega^\star}{\sqrt{K}} \tag{8.4}$$

$$b = aK \tag{8.5}$$

From the above it is clear that a lag controller improves the steady-state response, while a lead controller improves the transient response. If both responses are unsatisfactory, a product of both controllers may be used.

8.1.3 Preliminaries of pole placement in state space

We now consider a single-input single-output LTI system represented in state space by:

$$\begin{cases} \dot{\mathbf{x}} = \mathbf{A}\mathbf{x} + \mathbf{B}u \\ y = \mathbf{C}\mathbf{x} \end{cases} \tag{8.6}$$

where $\mathbf{x} \in \mathbb{R}^n$, $u \in \mathbb{R}$, $y \in \mathbb{R}$, $\mathbf{A} \in \mathbb{R}^{n\times n}$, $\mathbf{B} \in \mathbb{R}^{n\times 1}$ and $\mathbf{C} \in \mathbb{R}^{1\times n}$.

If the system is represented in the controllable canonical form, then we have:

$$\mathbf{A} = \begin{bmatrix} 0 & 1 & 0 & \cdots & 0 \\ 0 & 0 & 1 & \cdots & 0 \\ \vdots & \vdots & \vdots & & \vdots \\ 0 & 0 & 0 & & 1 \\ -a_n & -a_{n-1} & -a_{n-2} & \cdots & -a_1 \end{bmatrix}, \quad \mathbf{B} = \begin{bmatrix} 0 \\ 0 \\ \vdots \\ 0 \\ 1 \end{bmatrix}, \quad \mathbf{C}^T = \begin{bmatrix} b_n - a_n b_0 \\ b_{n-1} - a_{n-1} b_0 \\ \vdots \\ b_2 - a_2 b_0 \\ b_1 - a_1 b_0 \end{bmatrix}$$

The characteristic polynomial is given by:

$$\det(s\mathbf{I}_n - \mathbf{A}) = a(s) = s^n + a_1 s^{n-1} + \cdots + a_n \tag{8.7}$$

and the eigenvalues of $\mathbf{A}$ (open-loop poles) are λ_i, $i = 1, \ldots, n$.

Suppose that for meeting some set of specifications, we need a new characteristic polynomial expressed by:

$$\alpha(s) = s^n + \alpha_1 s^{n-1} + \cdots + \alpha_n \tag{8.8}$$

with roots μ_i, $i = 1, \ldots, n$.

This polynomial can be obtained by defining a new external input, ν, such that $u = \nu - \mathbf{K}\mathbf{x}$ and $\mathbf{K} = [K_1, K_2, \ldots, K_n]$. The system state-space closed-loop model results in:

$$\begin{cases} \dot{\mathbf{x}} = (\mathbf{A} - \mathbf{B}\mathbf{K})\mathbf{x} + \mathbf{B}\nu \\ y = \mathbf{C}\mathbf{x} \end{cases} \tag{8.9}$$

Choosing the feedback gain as $\mathbf{K} = [\alpha_n - a_n, \alpha_{n-1} - a_{n-1}, \ldots, \alpha_1 - a_1]$ and knowing that

$$\mathbf{BK} = \begin{bmatrix} 0 & 0 & \cdots & 0 \\ 0 & 0 & \cdots & 0 \\ \vdots & \vdots & \ddots & \vdots \\ K_1 & K_2 & \cdots & K_n \end{bmatrix} \tag{8.10}$$

then $\alpha(s)$ is obtained, with:

$$\det[s\mathbf{I}_n - (\mathbf{A} - \mathbf{BK})] = \alpha(s) \tag{8.11}$$

Figure 8.3 depicts a block diagram of a system with pole assignment by state feedback.

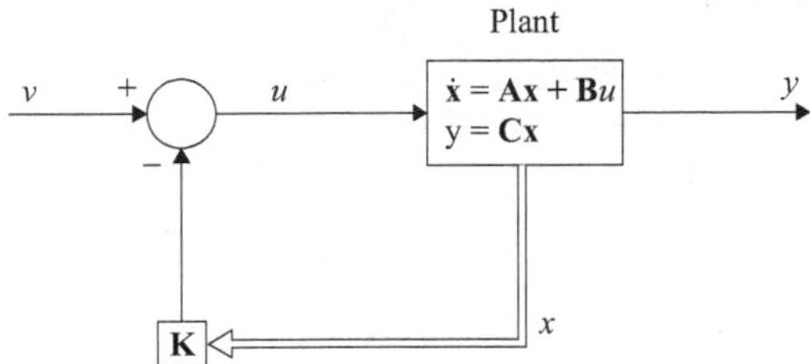

Figure 8.3 Block diagram of system with pole placement

It should be noted that:

1. The necessary and sufficient condition for arbitrary pole placement is that the system is completely state controllable.
2. The state variables are measurable and are available for feedback (otherwise they have to be estimated).
3. The state feedback **K** does not change the location of the open-loop zeros.

If the system is controllable, then it can always be represented in the controllable canonical form. In fact, adopting the similarity transformation $\mathbf{x} = \mathbf{T}_c\bar{\mathbf{x}}$, we may write:

$$\begin{cases} \dot{\bar{\mathbf{x}}} = \mathbf{T}_c^{-1}\mathbf{A}\mathbf{T}_c\bar{\mathbf{x}} + \mathbf{T}_c^{-1}\mathbf{B}u = \bar{\mathbf{A}}\bar{\mathbf{x}} + \bar{\mathbf{B}}u \\ y = \mathbf{C}\mathbf{T}_c\bar{\mathbf{x}} = \bar{\mathbf{C}}\bar{\mathbf{x}} \end{cases} \tag{8.12}$$

where

$$\mathbf{T}_c = \mathbf{Q}\bar{\mathbf{Q}}^{-1} \tag{8.13a}$$

$$\mathbf{Q} = \left[\,\mathbf{B}\,|\,\mathbf{AB}\,|\cdots|\,\mathbf{A}^{n-1}\mathbf{B}\,\right] \tag{8.13b}$$

$$\bar{\mathbf{Q}} = \left[\,\bar{\mathbf{B}}\,|\,\bar{\mathbf{A}}\bar{\mathbf{B}}\,|\cdots|\,\bar{\mathbf{A}}^{n-1}\bar{\mathbf{B}}\,\right] \tag{8.13c}$$

8.1.4 Calculation of the feedback gain

8.1.4.1 Formula 1

Let $\mathbf{K}$ and $\bar{\mathbf{K}}$ be the feedback gains for the systems (8.6) and (8.12), respectively.

If $u = \nu - \bar{\mathbf{K}}\bar{\mathbf{x}}$, then

$$\dot{\bar{\mathbf{x}}} = (\bar{\mathbf{A}} - \bar{\mathbf{B}}\bar{\mathbf{K}})\bar{\mathbf{x}} + \bar{\mathbf{B}}\nu \tag{8.14a}$$

$$\dot{\mathbf{x}} = [\mathbf{T}_c\bar{\mathbf{A}}\mathbf{T}_c^{-1} - (\mathbf{T}_c\bar{\mathbf{B}})(\bar{\mathbf{K}}\mathbf{T}_c^{-1})]\mathbf{x} + \mathbf{T}_c\bar{\mathbf{B}}\nu \tag{8.14b}$$

The feedback gain is calculated by $\mathbf{K} = \bar{\mathbf{K}}\mathbf{T}_c^{-1}$.

8.1.4.2 Formula 2 — Ackermann formula

Let

$$\bar{\mathbf{K}} = [\alpha_n - a_n, \alpha_{n-1} - a_{n-1}, \ldots, \alpha_1 - a_1] \tag{8.15a}$$

$$\det(s\mathbf{I}_n - \bar{\mathbf{A}}) = a(s) = s^n + a_1 s^{n-1} + \cdots + a_n \tag{8.15b}$$

$$\det[s\mathbf{I}_n - (\bar{\mathbf{A}} - \bar{\mathbf{B}}\bar{\mathbf{K}})] = s^n + \alpha_1 s^{n-1} + \cdots + \alpha_n \tag{8.15c}$$

From the Cayley–Hamilton theorem, we have

$$a(\bar{\mathbf{A}}) = \bar{\mathbf{A}}^n + a_1\bar{\mathbf{A}}^{n-1} + \cdots + a_n\mathbf{I}_n = 0 \tag{8.16}$$

$$\bar{\mathbf{A}}^n = -a_1\bar{\mathbf{A}}^{n-1} - \cdots - a_n\mathbf{I}_n \tag{8.17}$$

$$\alpha(\bar{\mathbf{A}}) = \bar{\mathbf{A}}^n + \alpha_1\bar{\mathbf{A}}^{n-1} + \cdots + \alpha_n\mathbf{I}_n = (\alpha_1 - a_1)\bar{\mathbf{A}}^{n-1} + \cdots + (\alpha_n - a_n)\mathbf{I}_n \tag{8.18}$$

Defining $\mathbf{v}_1^T = \begin{bmatrix} 1 & 0 & \cdots & 0 \end{bmatrix}$, then we have $\mathbf{v}_1^T\bar{\mathbf{A}}^{n-1} = \mathbf{v}_n^T = \begin{bmatrix} 0 & 0 & \cdots & 1 \end{bmatrix}$ and, therefore:

$$\mathbf{v}_1^T\alpha(\bar{\mathbf{A}}) = (\alpha_1 - a_1)\mathbf{v}_n^T + \cdots + (\alpha_n - a_n)\mathbf{v}_1^T = \bar{\mathbf{K}} \tag{8.19}$$

As

$$\mathbf{K} = \bar{\mathbf{K}}\mathbf{T}_c^{-1} = \mathbf{v}_1^T\mathbf{T}_c^{-1}\alpha(\mathbf{A}) = \mathbf{v}_1^T\bar{\mathbf{Q}}\mathbf{Q}^{-1}\alpha(\mathbf{A}) \tag{8.20}$$

we finally have

$$\mathbf{K} = \begin{bmatrix} 0 & 0 & \cdots & 1 \end{bmatrix}\mathbf{Q}^{-1}\alpha(\mathbf{A}) \tag{8.21}$$

8.1.5 Estimating the system state

If a system is completely observable, then the state, $\mathbf{x}$, can be calculated by using the following expression:

$$\mathbf{x} = \mathbf{R}^{-1}[\mathbf{Y} - \mathbf{T}\mathbf{U}] \tag{8.22}$$

where

$$\mathbf{R} = \begin{bmatrix} \mathbf{C} \\ \hline \mathbf{CA} \\ \hline \vdots \\ \hline \mathbf{CA}^{n-1} \end{bmatrix} \tag{8.23a}$$

$$\mathbf{T} = \left[\begin{array}{c|c|c|c|c} 0 & 0 & & 0 & 0 \\ \mathbf{CB} & 0 & & 0 & 0 \\ \mathbf{CAB} & \mathbf{CB} & \cdots & \vdots & \vdots \\ \vdots & \vdots & & \mathbf{CB} & 0 \end{array}\right] \tag{8.23b}$$

$$\mathbf{Y} = \left[\, y \,|\, \dot{y} \,|\cdots|\, y^{(n-1)} \,\right]^T \tag{8.23c}$$

$$\mathbf{U} = \left[\, u \,|\, \dot{u} \,|\cdots|\, u^{(n-1)} \,\right]^T \tag{8.23d}$$

Alternatively, if the system is represented in the observable canonical form, then we have:

$$\dot{\mathbf{x}} = \begin{bmatrix} -a_1 & 1 & 0 & \cdots & 0 \\ -a_2 & 0 & 1 & \cdots & 0 \\ \vdots & \vdots & \vdots & & \vdots \\ -a_{n-1} & 0 & 0 & & 1 \\ -a_n & 0 & 0 & \cdots & 0 \end{bmatrix} \mathbf{x} + \begin{bmatrix} b_1 - a_1 b_0 \\ b_2 - a_2 b_0 \\ \vdots \\ b_{n-1} - a_{n-1} b_0 \\ b_n - a_n b_0 \end{bmatrix} u \tag{8.24a}$$

$$\mathbf{y} = \begin{bmatrix} 1 & 0 & \cdots & 0 & 0 \end{bmatrix} \mathbf{x} \tag{8.24b}$$

The state can be estimated as shown in Figure 8.4, where the inputs of the estimator subsystem are the system input and output variables.

Considering that $\hat{\mathbf{x}}$ is the estimated state vector, then the estimated error is given by:

$$\tilde{\mathbf{x}} = \mathbf{x} - \hat{\mathbf{x}} \tag{8.25}$$

The error dynamics is:

$$\dot{\tilde{\mathbf{x}}} = (\mathbf{A} - \mathbf{LC})\tilde{\mathbf{x}} \tag{8.26}$$

where $\mathbf{L}$ represents the estimator feedback gain. The eigenvalues of $(\mathbf{A} - \mathbf{LC})$ can be arbitrarily chosen by $\mathbf{L}$, namely they can be made with negative real parts, so that $\hat{\mathbf{x}}$ converges to $\mathbf{x}$.

Considering the system open-loop characteristic polynomial as:

$$\det(s\mathbf{I}_n - \mathbf{A}) = a(s) = s^n + a_1 s^{n-1} + \cdots + a_n \tag{8.27}$$

and the estimator closed-loop characteristic polynomial as:

$$\det[s\mathbf{I}_n - (\mathbf{A} - \mathbf{LC})] = \beta(s) = s^n + \beta_1 s^{n-1} + \cdots + \beta_n \tag{8.28}$$

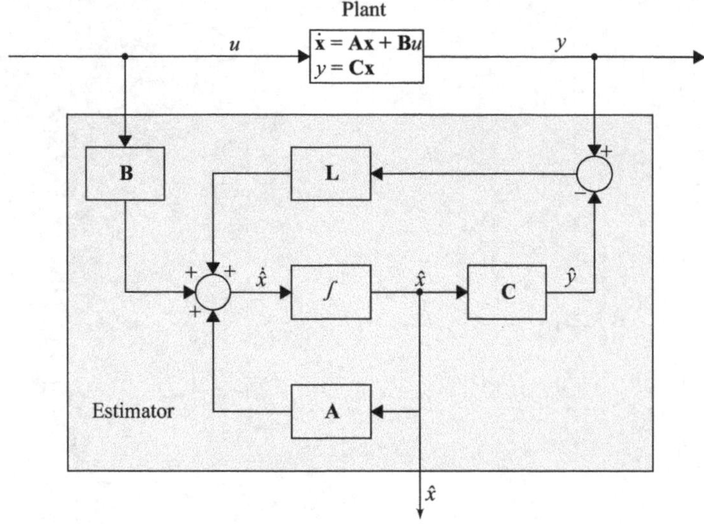

Figure 8.4 Block diagram of a state estimator

then the estimator feedback gain is given by:

$$\mathbf{L} = [\beta_1 - a_1, \beta_2 - a_2, \ldots, \beta_n - a_n] \tag{8.29}$$

If the system is observable, then it can always be represented in the observable canonical form, by adopting the similarity transformation $\mathbf{x} = \mathbf{T}_o\bar{\mathbf{x}}$, given by:

$$\mathbf{T}_o = \mathbf{R}^{-1}\bar{\mathbf{R}} \tag{8.30}$$

where

$$\mathbf{R} = \begin{bmatrix} \mathbf{C} \\ \hline \mathbf{CA} \\ \hline \vdots \\ \hline \mathbf{CA}^{n-1} \end{bmatrix} \tag{8.31a}$$

$$\bar{\mathbf{R}} = \begin{bmatrix} \bar{\mathbf{C}} \\ \hline \bar{\mathbf{C}}\bar{\mathbf{A}} \\ \hline \vdots \\ \hline \bar{\mathbf{C}}\bar{\mathbf{A}}^{n-1} \end{bmatrix} \tag{8.31b}$$

8.1.6 Calculation of the state estimator gain

Similarly to the formulas given in Section 8.1.4, the state estimator gain, **L**, can be calculated by:

$$\mathbf{L} = \mathbf{T}_o\bar{\mathbf{L}} \tag{8.32}$$

or, by using the Ackermann formula:

$$\mathbf{L} = \beta(\mathbf{A})\mathbf{R}^{-1}\begin{bmatrix} 0 \\ 0 \\ \vdots \\ 1 \end{bmatrix} \tag{8.33}$$

8.1.7 Simultaneous pole placement and state estimation

The estimated state can be used for pole assignment in a state feedback system, as shown in Figure 8.5.

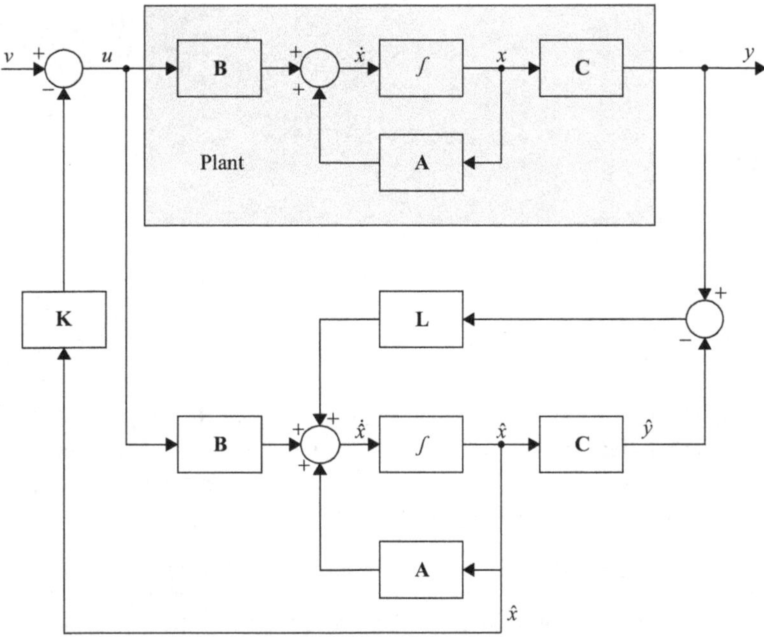

Figure 8.5 Block diagram of pole placement and state estimation

The global system has order $2n$ and can be represented by:

$$\begin{bmatrix} \dot{\mathbf{x}} \\ \dot{\hat{\mathbf{x}}} \end{bmatrix} = \left[\begin{array}{c|c} \mathbf{A} & -\mathbf{BK} \\ \hline \mathbf{LC} & \mathbf{A}-\mathbf{BK}-\mathbf{LC} \end{array}\right]\begin{bmatrix} \mathbf{x} \\ \hat{\mathbf{x}} \end{bmatrix} + \begin{bmatrix} \mathbf{B} \\ \mathbf{B} \end{bmatrix} v \tag{8.34a}$$

$$y = \left[\,\mathbf{C}\,|\,\mathbf{0}\,\right]\begin{bmatrix} \mathbf{x} \\ \hat{\mathbf{x}} \end{bmatrix} \tag{8.34b}$$

It should be noted that:

- The system transfer function remains the same no matter we use $\mathbf{x}$ or $\hat{\mathbf{x}}$ in the feedback $u = v - \mathbf{Kx}$.
- The error $\tilde{\mathbf{x}} = \mathbf{x} - \hat{\mathbf{x}}$ is uncontrollable from the control v.

- The system zeros are the same of the compensated system.
- If estimated states are used for state feedback, then the slowest eigenvalues of $\mathbf{A} - \mathbf{LC}$ should be faster than the eigenvalues of the state feedback system $\mathbf{A} - \mathbf{BK}$.

8.2 Solved problems

8.2.1 Pole placement using an input–output representation

Problem 8.1 To control plant $G_p(s) = \frac{20}{(s+0.5)(s+4)}$ in closed loop so as to obtain a step response with steady-state error equal to $\frac{1}{20}$, we can use a lag compensator given by

A) $C(s) = \dfrac{s + 0.366}{s + \dfrac{0.366}{2}}$

B) $C(s) = \dfrac{s + 0.366}{s + 1.9}$

C) $C(s) = \dfrac{s + 0.366}{s + \dfrac{0.366}{1.9}}$

D) None of the above.

Resolution For this plant, without any controller, $K_p = 10$, and thus $e_{ss} = \frac{1}{11}$. K_p should be 1.9 times larger. As $\omega_1 = 3.66$ rad/s, we will put the zero one decade below, so $a = 0.366$, and b must be 1.9 times smaller. Thus the correct answer is option **C)**. The first two answers lead to incorrect values of e_{ss}.

Problem 8.2 Figure 8.6 shows a system to position an antenna.

1. Plot the root-locus, determining all the relevant points, the asymptotes, and the departure and arrival angles.
2. This plant should stabilize in 4 s or less (according to the 2% criterion) with an overshoot of 16% or less. Verify that you cannot satisfy this just by tuning the gain $K > 0$.
3. Find a controller to satisfy the specifications.

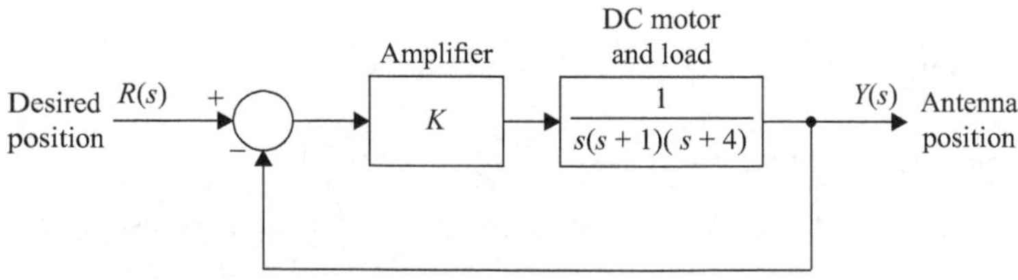

Figure 8.6 Block diagram of Problem 8.2

Resolution The specifications correspond to the zone of the complex plane shown in gray in Figure 8.7:

$$\begin{cases} -\omega_n\zeta \leq -1 \\ \zeta \geq 0.5 \end{cases} \tag{8.35}$$

The root-locus, shown in Figure 8.8, does not cross that area. Thus we need a lead controller to improve the transient response. The controller will be given by (8.1) and with three unknowns K, a and b there are infinite solutions. If we fix $b = 20$, the characteristic equation becomes $s^4 + 25s^3 + 104s^2 + (80 + K)s + Ka = 0$, and putting the poles at the corners of the admissible region of the complex plane, given by $-1 \pm \sqrt{3}j$, we get $(-1 + \sqrt{3}j)^4 + 25(-1 + \sqrt{3}j)^3 + 104(-1 + \sqrt{3}j)^2 + (80 + K)(-1 + \sqrt{3}j) + Ka = 0 \Rightarrow a = 1.8 \wedge K = 120$. The step response of the closed loop consisting of $G_p(s)$ and $G_c(s) = 120\frac{s+1.8}{s+20}$ has an acceptable settling time but the overshoot is 25%, because of the influence of the non-dominant pole of the open loop. We must thus increase the damping, i.e., bring the poles closer to the real axis. This must now be done with some trial and error. It can be seen, for instance, that $G_c(s) = 90\frac{s+1.5}{s+20}$ puts the dominant poles roughly at $-1 \pm j$ and has an acceptable time response.

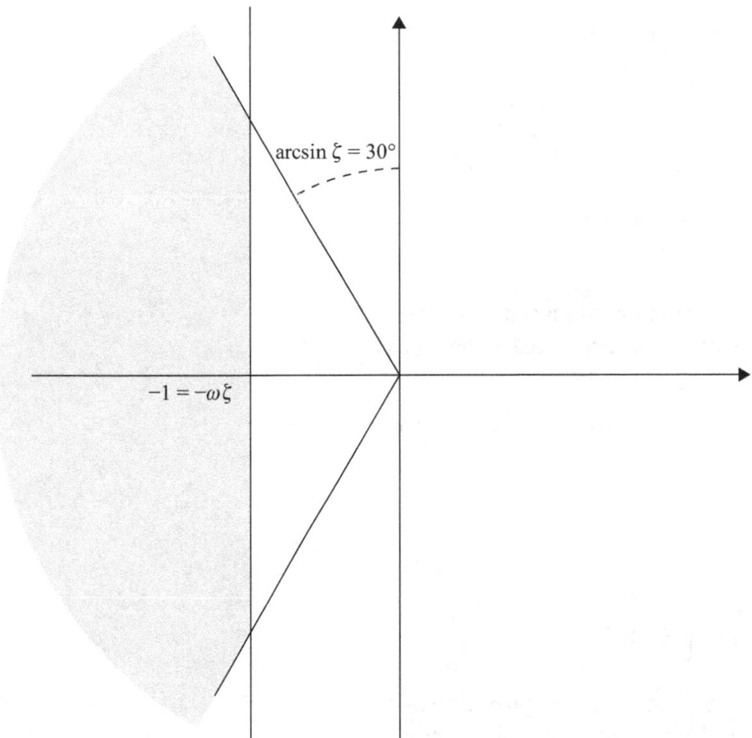

Figure 8.7 Specifications for Problem 8.2

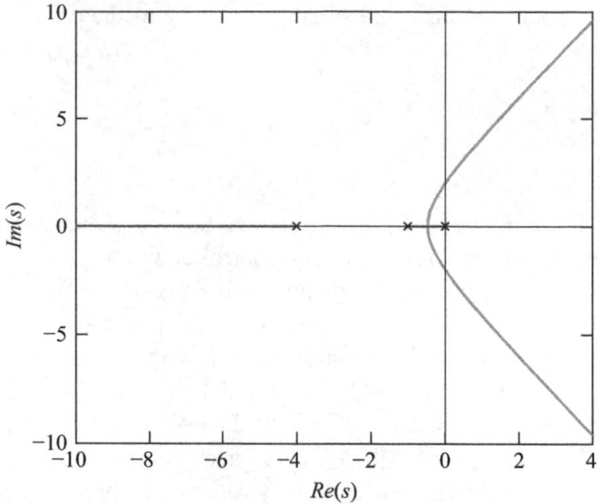

Figure 8.8 Root-locus of Problem 8.2

8.2.2 Pole placement in state space

Problem 8.3 Consider a plant given by

$$\dot{\mathbf{x}} = \begin{bmatrix} 0 & 1 \\ 2 & 3 \end{bmatrix}\mathbf{x} + \begin{bmatrix} 0 \\ 1 \end{bmatrix}u \tag{8.36}$$

$$y = [4 \quad 5]\mathbf{x} + 0u \tag{8.37}$$

1. Find the matrices of a state observer.
2. Use state feedback to place the poles at −5 and −2.

Resolution The matrices of the state observer are

$$\mathbf{R} = \begin{bmatrix} 4 & 5 \\ 10 & 19 \end{bmatrix} \tag{8.38}$$

$$\mathbf{T} = \begin{bmatrix} 0 & 0 \\ 5 & 0 \end{bmatrix} \tag{8.39}$$

As to state feedback, we have $a_1 = -3$, $a_2 = -2$, $b_0 = 0$, $b_1 = 5$, $b_2 = 4$, and, since the desired characteristic equation is $(s+5)(s+2) = s^2 + 7s + 10 = 0$, we also have $\alpha_1 = 7$ and $\alpha_2 = 10$. Thus we make $\mathbf{K} = [10 - (-2) \;\; 7 - (-3)] = [12 \;\; 10]$ and use (8.9).

8.3 Proposed problems

8.3.1 Pole placement using an input–output representation

Exercise 8.1 Plant $G_p(s) = \frac{(s+2)(s+30)}{s(s+3)(s+200)}$ is controlled in closed loop with a proportional controller $G_c(s) = 7$.

1. If the reference is a unit slope ramp, the steady-state error e_{ss} is:
   A) $e_{ss} = 0$
   B) $e_{ss} = \dfrac{7}{10}$
   C) $e_{ss} = \dfrac{10}{7}$
   D) $e_{ss} = \dfrac{10}{7+10}$.
2. To obtain the same steady-state error, the proportional controller could be replaced with:
   E) a lead compensator $G_c(s) = \dfrac{s+0.01}{s+0.07}$
   F) a lead compensator $G_c(s) = \dfrac{s+0.01}{s+\frac{0.01}{7}}$
   G) a lag compensator $G_c(s) = 7\dfrac{s+0.01}{s+0.07}$
   H) None of the above.

Exercise 8.2 The following lead controller can provide a phase margin increase of 45°:

A) $C(s) = 5.83\dfrac{s+10}{s+58.3}$

B) $C(s) = 5.83\dfrac{s+10}{s+1}$

C) $C(s) = \dfrac{s+5.83}{s+1}$

D) None of the above.

Exercise 8.3 Find a controller for the plant in Figure 8.9 that ensures two closed-loop poles at s = $-3 \pm j3$.

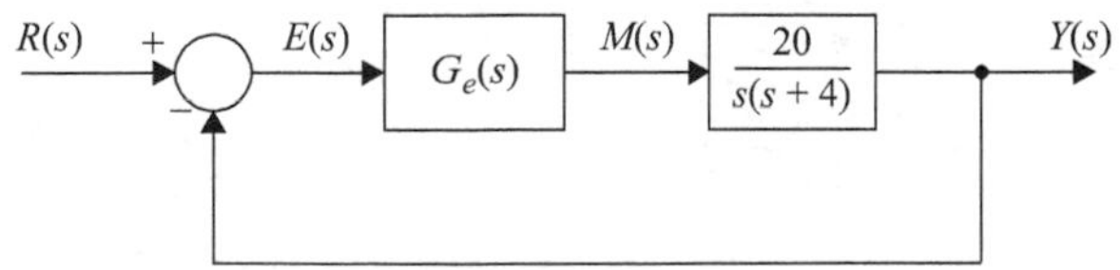

Figure 8.9 Block diagram of Exercise 8.3

Exercise 8.4 Consider the block diagram in Figure 8.10.

1. Plot the root-locus for $K > 0$, determining all the relevant points, the asymptotes, and the departure and arrival angles.
2. Is there any range of values of K for which this plant has an underdamped or oscillatory response?
3. Is there any range of values of K for which $p_{1,2} = -2 \pm 2j$ are poles of this system's closed-loop transfer function?

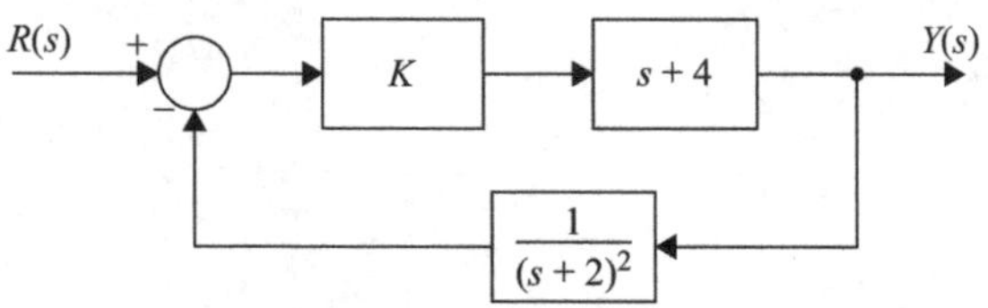

Figure 8.10 Block diagram of Exercise 8.4

Exercise 8.5 The force exerted by the position controller in Figure 8.11 is given by $F = A(x_R - x)$. Let $M = 1$ kg, $A = 5$ and $B = 6$ N·s·m^{-1}. Find the range of values of k_S for which the plant's output $x(t)$ has no oscillations.

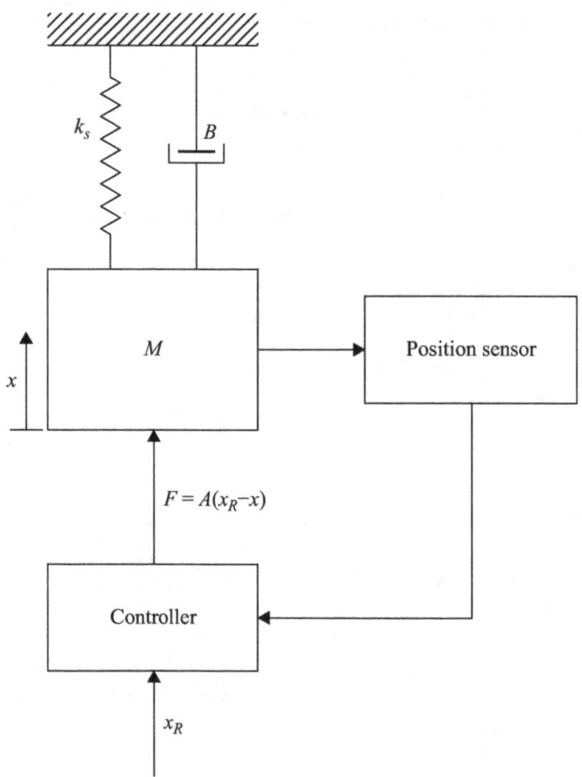

Figure 8.11 Mechanical system of Exercise 8.5

Exercise 8.6 Figure 8.12 shows a simplified diagram of the control system of a missile.

1. Plot this plant's root-locus.
2. Assuming that $H(s) = 1$, argue about this control system's viability from the root-locus.
3. Assume now that $H(s) = 1 + 0.5\,s$. Use the Routh–Hurwitz stability criterion to find the range of values of K for which the plant is stable.

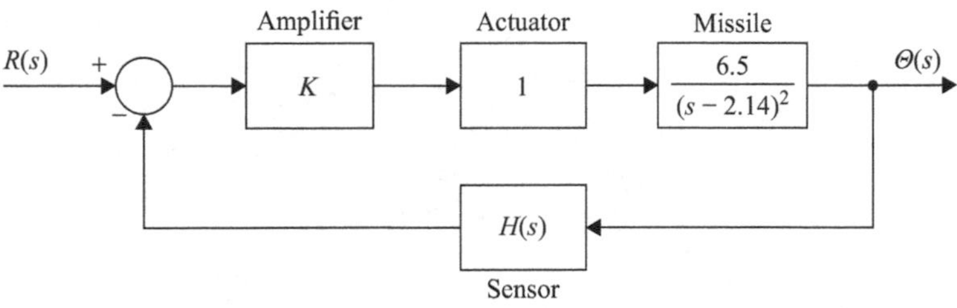

Figure 8.12 Block diagram of Exercise 8.6

Exercise 8.7 Figure 8.13 shows a plane, and Figure 8.14 gives the variation of its height $h(t)$ in metres when a step with an amplitude of $1°$ is applied to the angle of the elevators $\delta(t)$.

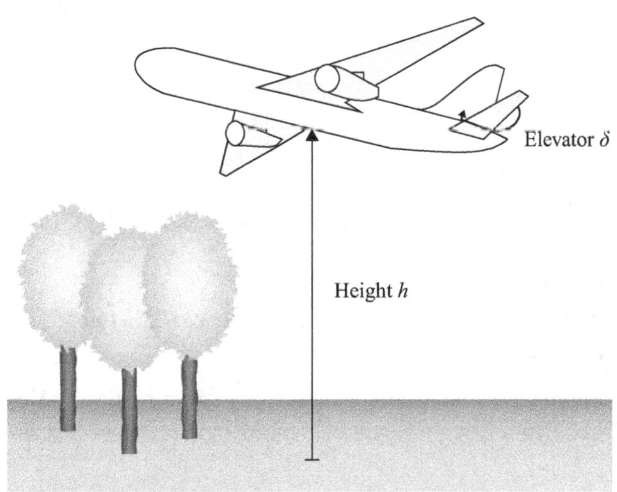

Figure 8.13 Plane of Exercise 8.7

1. Which of the plots A–H in Figure 8.15 corresponds to the root-locus of transfer function $\frac{H(s)}{\Delta(s)}$ when $K > 0$?
2. Which of the plots A–H in Figure 8.15 corresponds to the root-locus of transfer function $\frac{H(s)}{\Delta(s)}$ when $K < 0$?

3. Is a feedback loop with proportional control enough to ensure an accurate tracking of a constant vertical velocity $\dot{h}(t)$?
4. Knowing that a feedback loop with a proportional controller equal to 1 leads to a steady-state error of 0.9 when tracking a constant vertical velocity $\dot{h}(t) = 1$ m/s, design a controller reducing this steady-state error to 0.3, without significantly deteriorating the transient response.

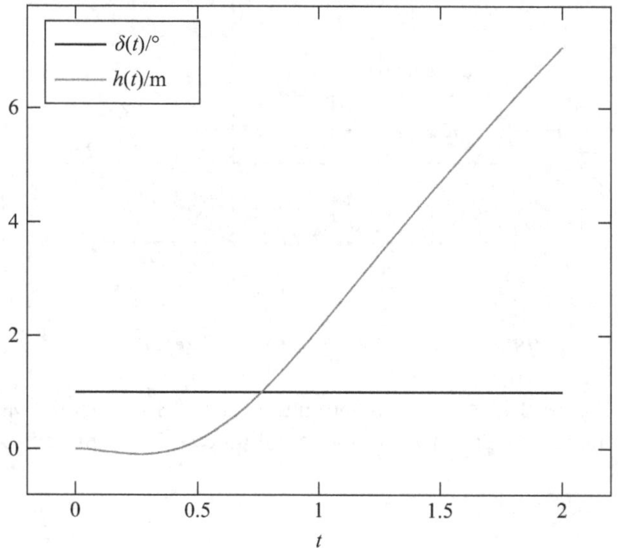

Figure 8.14 Unit step response of the plane of Exercise 8.7

Exercise 8.8 Consider plant $G(s) = \frac{s+5}{(s+0.5)(s^2+0.6s+1.09)}$.

1. Find a lead compensator $C(s)$ for this plant that fulfills the following specifications:
   - The gain margin must be infinite.
   - The phase margin must be $PM = 20°$ with a $\pm 10\%$ tolerance.
   - The steady-state error cannot be affected.
2. Find a lead compensator $C(s)$ for this plant that puts a pair of poles at $-4 + 10j$, and verify the 2%-settling time against the performance expected.

Exercise 8.9 A rotating robot can be described by transfer function $G(s) = \frac{\Theta(s)}{U(s)} = \frac{1}{s(12s+1)}$, where θ is the angle of the robot and u is the voltage applied to its motor. Performance specifications are as follows:

- No steady-state error for constant references of the rotation angle.
- 5%-settling time under 6 s.
- Step response overshoot under 10%.

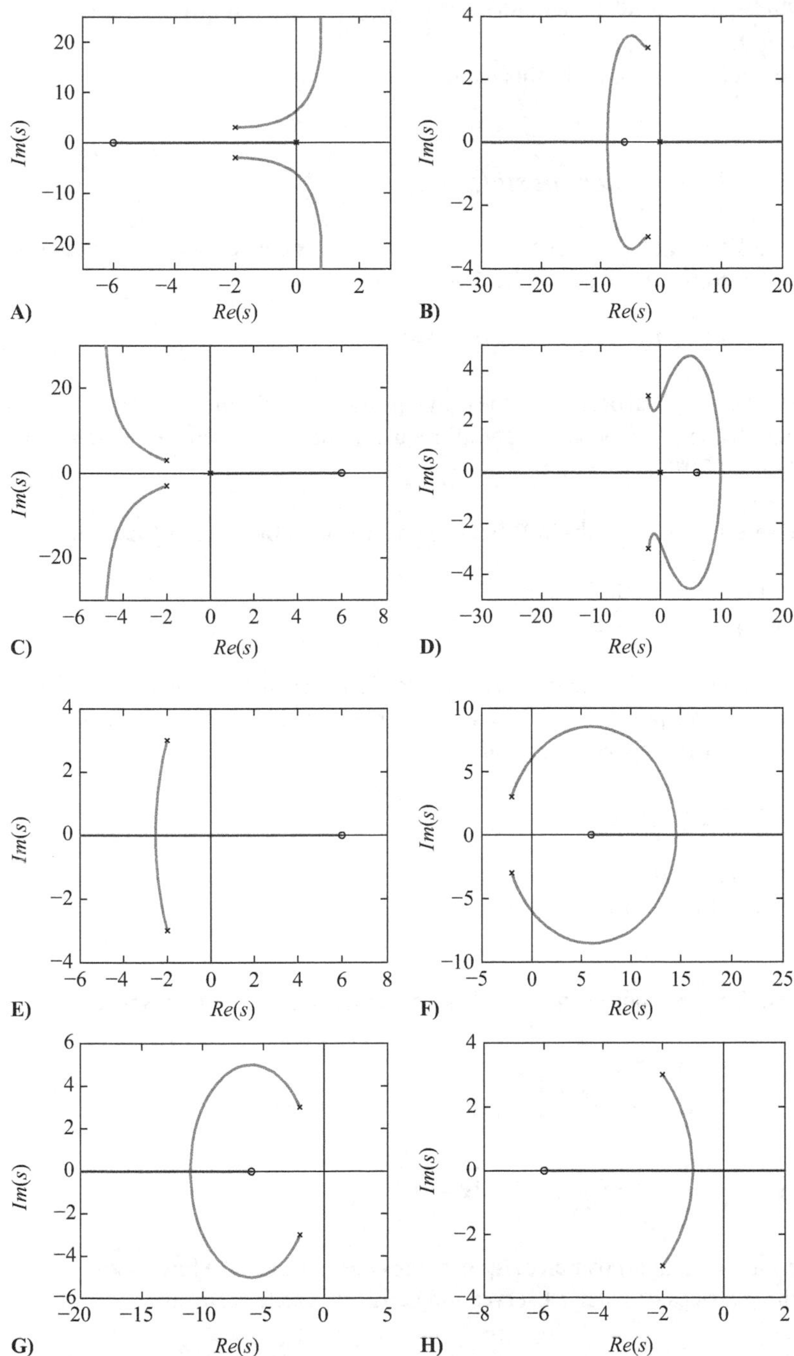

Figure 8.15 Root-locus plots of Exercise 8.7

1. Find the zone of the complex plane where closed-loop poles must fulfill the specifications.
2. Find a PD controller for this plant.

8.3.2 Pole placement in state space

Exercise 8.10 The behavior of a heaving wave energy converter (WEC), that produces electricity from sea waves, can be described by

$$y(t) = 2 \times 10^{-6} f(t) - 0.2\dot{y}(t) - 0.6\ddot{y}(t) \tag{8.40}$$

where y is the vertical position of the heaving element and f the force acting thereupon. Use state feedback to obtain a natural frequency of 6 rad/s and a unit step maximum overshoot of 30%.

Exercise 8.11 Suppose that a WEC can be described by transfer function

$$\frac{Y(s)}{F(s)} = \frac{2 \times 10^{-6}}{s^2 + 3s + 2} \tag{8.41}$$

where y is the vertical position of the heaving element and f the force acting thereupon. To have a double pole at −3, we can put the plant in the controllable canonical form, and then use the feedback gain matrix

A) $\mathbf{K} = [6 \quad 4]$
B) $\mathbf{K} = [1 \quad -2]$
C) $\mathbf{K} = [11 \quad 9]$
D) None of the above.

Exercise 8.12 Suppose that a WEC can be described by the state space

$$\dot{\mathbf{x}} = \begin{bmatrix} -\frac{1}{3} & -\frac{5}{3} \\ 1 & 0 \end{bmatrix} \mathbf{x} + \begin{bmatrix} 1 \\ 0 \end{bmatrix} f \tag{8.42}$$

$$y = \begin{bmatrix} -\frac{1}{6} \times 10^{-5} & \frac{1}{3} \times 10^{-5} \end{bmatrix} \mathbf{x} + 0f \tag{8.43}$$

where y is the vertical position of the heaving element and f the force acting thereupon. Find the matrices of a state observer and verify that states can be recovered when the input is

1. a unit step;
2. a unit slope ramp.

Exercise 8.13 Consider a plant described in state space by matrices

$$\mathbf{A} = \begin{bmatrix} -4 & 1 \\ -3 & 0 \end{bmatrix}$$
$$\mathbf{B} = \begin{bmatrix} 2 \\ 4 \end{bmatrix}$$
$$\mathbf{C} = \begin{bmatrix} 1 & 0 \end{bmatrix} \tag{8.44}$$

We can observe its states using matrices

A) $\mathbf{R} = \begin{bmatrix} 1 & 0 \\ -4 & 1 \end{bmatrix}, \quad \mathbf{T} = \begin{bmatrix} 0 & 0 \\ 2 & 0 \end{bmatrix}$

B) $\mathbf{R} = \begin{bmatrix} -4 & 1 \\ 1 & 0 \end{bmatrix}, \quad \mathbf{T} = \begin{bmatrix} 0 & 0 \\ 2 & 0 \end{bmatrix}$

C) $\mathbf{R} = \begin{bmatrix} 1 & 0 \\ -4 & 1 \end{bmatrix}, \quad \mathbf{T} = \begin{bmatrix} 2 & 0 \\ 0 & 0 \end{bmatrix}$

D) None of the above.

Exercise 8.14 Consider the plant $\dot{\mathbf{x}} = \begin{bmatrix} 0 & 1 & 0 \\ 0 & 0 & 1 \\ -8 & -14 & -7 \end{bmatrix}\mathbf{x} + \begin{bmatrix} 0 \\ 0 \\ 1 \end{bmatrix}u, y = \begin{bmatrix} 8 & 8 & 0 \end{bmatrix}\mathbf{x}$.

We can observe its states using matrices

A) $\mathbf{R} = \begin{bmatrix} 8 & 8 & 0 \\ 0 & 8 & 8 \\ -64 & -112 & -48 \end{bmatrix}, \quad \mathbf{T} = \begin{bmatrix} 0 & 0 & 0 \\ 0 & 0 & 0 \\ 8 & 0 & 0 \end{bmatrix}$

B) $\mathbf{R} = \begin{bmatrix} 8 & 8 & 0 \\ 0 & 8 & 8 \end{bmatrix}, \quad \mathbf{T} = \begin{bmatrix} 0 & 0 & 0 \\ 0 & 0 & 0 \end{bmatrix}$

C) $\mathbf{R} = \begin{bmatrix} 0 & 8 \\ -64 & -112 \end{bmatrix}, \quad \mathbf{T} = \begin{bmatrix} 0 & 0 \\ 8 & 0 \end{bmatrix}$

D) None of the above.

Chapter 9
Discrete-time systems and $\mathscr{Z}$-transform

9.1 Fundamentals

Nowadays, most control systems are based on microprocessors or microcontrollers. The structure of these computer-controlled or sampled-data systems is similar to the structure of continuous-time systems. However, as a digital computer is unable to monitor the process variables continuously, there are phenomena that occur in computer-controlled systems that have no correspondence in their analog counterparts. Computer-controlled systems contain both continuous and discrete-time or sampled signals. This mixture imposes additional difficulties that we usually avoid by considering the signals only at the sampling instants. The sampling interval is lower limited by the time involved in the calculation of the control law, being a multiple of the period of the computer internal clock. The presence of analog-to-digital (A/D) converters imposes a quantization interval that is responsible for a loss of accuracy that can result in limit cycles. Signals that are both discrete and quantized are digital signals, meaning that they are discrete both in time and in amplitude. Digital signals take part in the computation of the control action.

In this chapter, we introduce the main theory and tools necessary to deal with computer-controlled systems, namely the $\mathscr{Z}$-transform, discrete-time models, controllability and observability conditions, and stability criteria. Most concepts presented for continuous-time systems can be adapted to the discrete-time case [2,4,9,20,26].

9.1.1 List of symbols

$a_i, b_i, \alpha_i, \beta_i$	constant coefficients
$\mathbf{A}$	$n \times n$ dimensional state (or system) matrix
$\mathbf{B}$	$n \times q$ dimensional input matrix
$\mathbf{C}$	$m \times n$ dimensional output matrix
d_i	ith disturbance input
$\mathbf{G}$	$n \times n$ dimensional state matrix of a discrete-time system
$\mathbf{D}$	$m \times q$ dimensional feedthrough (or feedforward) matrix
$G(s)$, $W(s)$	continuous-time transfer function
h	sampling interval
$\mathbf{H}$	$m \times q$ dimensional input matrix of a discrete-time system
$\mathbf{I}_n$	$n \times n$ dimensional identity matrix

$\mathscr{L}$	Laplace operator
n	number of state variables (system order)
s	Laplace variable
t	time
$r(t)$	time-domain reference input function
$u(t)$	time-domain input (or control) function
$u^*(t)$	discrete-time version of $u(t)$
$\mathbf{u}$	q dimensional input (or control) vector
$\mathbf{x}$	n dimensional state vector
$\mathbf{x}_0$	initial state vector
$\mathbf{y}$	m dimensional output vector
$y(t)$	time-domain output function
z	z variable
$\mathscr{Z}$	z operator
$\delta(t)$	Dirac delta function
λ_i	eigenvalue, pole of continuous-time transfer function

9.1.2 Discrete-time systems preliminaries

Figure 9.1 depicts the structure of a typical computer-controlled system. The reference input, $r(t)$, and the output, $y(t)$, are continuous-time signals. These signals are converted into the digital form by the A/D converter, at the sampling instants $t = kh$, $k \in \mathbb{N}_0$, where h is the sampling interval. The computer interprets the converted signals as sequences of numbers and processes them according to a given control law. The controller output is then converted into an analog signal, $u(t)$, by a digital-to-analog (D/A) converter. This is normally done by keeping the control signal constant between the successive conversions, meaning that the system runs open loop in the time interval between the sampling instants, h. During the processing, all events are synchronized by the real-time clock of the computer. In Figure 9.1, the signals d_i, $i = 1, 2, 3$, denote disturbances that may affect the system performance, while $R(z)$ and $Y(z)$ are the $\mathscr{Z}$-transforms of $r(t)$ and $y(t)$, respectively, to be defined in the next subsection.

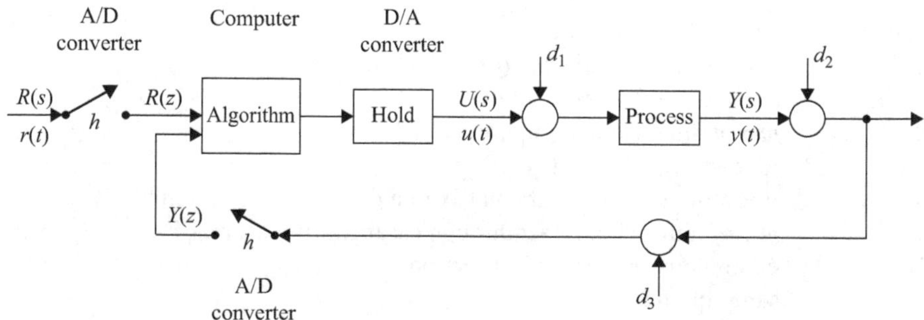

Figure 9.1 Block diagram of a computer-controlled system

The D/A converter can be modeled by the block $G(s) = \frac{1-e^{-hs}}{s}$. Its response to the Dirac impulse with area A, applied at time $t = 0$, is a rectangular pulse with height A and width h, starting at $t = 0$. This signal reconstruction is known as "zero-order hold," since during h the continuous-time signal is replaced by a constant. This process is time variant.

9.1.3 The $\mathscr{Z}$-transform

Let $u(t)$ be a continuous-time signal, with $u(t) = 0$ for $t < 0$. The sampled or discrete-time signal, $u^*(t)$, modeled as a sequence of equally spaced Dirac impulses is given by:

$$u^*(t) = \sum_{k=0}^{\infty} u(kh)\delta(t - kh) \tag{9.1}$$

Since $\mathscr{L}[\delta(t - a)] = e^{-as}$, we have:

$$\mathscr{L}[u^*(t)] = \sum_{k=0}^{\infty} u(kh)e^{-khs} \tag{9.2}$$

With the change of variable $z = e^{hs}$, we can write the $\mathscr{Z}$-transform of $u(t)$, to be denoted by $\mathscr{Z}[u(t)]$:

$$\mathscr{Z}[u(t)] = U(z) = \sum_{k=0}^{\infty} u(kh)z^{-k} \tag{9.3}$$

The $\mathscr{Z}$-transform is the discrete-time equivalent of the Laplace transform. In Appendix A, we present the $\mathscr{Z}$-transforms of some common functions and we summarize the main properties of the $\mathscr{Z}$-transform.

The inverse $\mathscr{Z}$-transform can be computed by means of the following expression:

$$\mathscr{Z}^{-1}[X(z)] = x(kh) = \frac{1}{2\pi j} \oint_{\Gamma} X(z)z^{k-1} dz \tag{9.4}$$

where Γ is a closed path enclosing all singularities of $X(z)$.

Usually, given a rational function, $X(z)$, we adopt the partial fraction decomposition or partial fraction expansion method for expressing $X(z)$ as the sum of a polynomial and one, or several, fractions with simpler denominators. We then use the tables of $\mathscr{Z}$-transforms to obtain $x(kh)$. Since the $\mathscr{Z}$-transform pairs appearing in the tables have z in the numerator, rather than $X(z)$, it is preferable to expand $\frac{X(z)}{z}$. Alternatively, we can divide the numerator and denominator polynomials of $X(z)$ to obtain an infinite series. This has the advantage of facilitating the conversion of a transfer function into its equivalent difference equation.

9.1.4 Discrete-time models

A linear discrete-time dynamical system can be described by a linear difference equation. Let $u(k)$ and $y(k)$, $k \in \mathbb{N}_0$, denote the input and output of the system, respectively, then the system discrete-time model is:

$$y(k+n) + a_{n-1}y(k+n-1) + \cdots + a_0 y(k) = b_n u(k+n) + \cdots + b_0 u(k) \tag{9.5}$$

Alternatively, we can describe the system by means of a state-space model analogous to that of a continuous-time system, that is:

$$\begin{cases} \mathbf{x}(k+1) = \mathbf{G}\mathbf{x}(k) + \mathbf{H}u(k) \\ y(k) = \mathbf{C}\mathbf{x}(k) + \mathbf{D}u(k) \end{cases} \tag{9.6}$$

where $\mathbf{x} \in \mathbb{R}^n$.

9.1.4.1 Discretization of state-space models

Consider the n-order continuous-time system of Figure 9.2, with q inputs and m outputs, and the state-space model given by:

$$\begin{cases} \dot{\mathbf{x}}(\mathbf{t}) = \mathbf{A}\mathbf{x}(t) + \mathbf{B}\mathbf{u}'(t) \\ \mathbf{y}(t) = \mathbf{C}\mathbf{x}(t) + \mathbf{D}\mathbf{u}'(t) \end{cases} \tag{9.7}$$

Suppose that the samplers are synchronized. Since the system is time-invariant and $\mathbf{u}'(t) = \mathbf{u}(kh)$, $kh \le t < (k+1)h$, we can write:

$$\mathbf{x}[(k+1)h] = \exp(\mathbf{A}h)\mathbf{x}(kh) + \int_0^h \exp[\mathbf{A}(h-t')]dt'\mathbf{B}\mathbf{u}(kh) \tag{9.8}$$

With the change of variable $t'' = h - t'$, we have:

$$\mathbf{x}[(k+1)h] = \exp(\mathbf{A}h)\mathbf{x}(kh) + \int_0^h \exp(\mathbf{A}t'')dt''\mathbf{B}\mathbf{u}(kh) \tag{9.9}$$

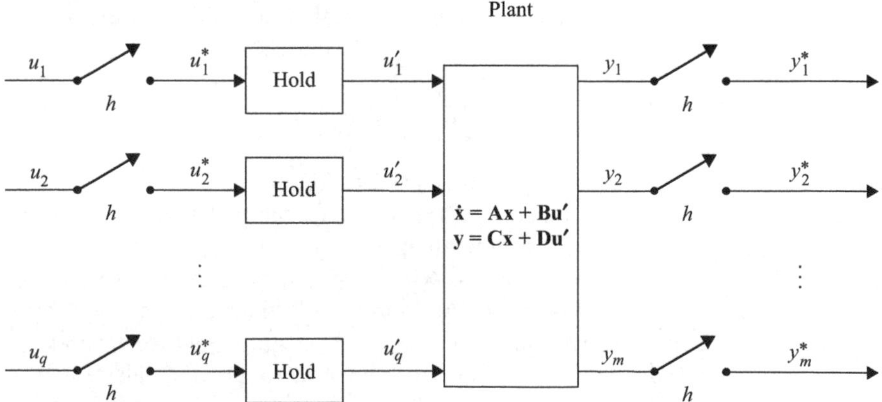

Figure 9.2 Sampling of a continuous-time system

that has the same structure of model (9.6), where:

$$\mathbf{G}(h) = \exp(\mathbf{A}h) \tag{9.10a}$$

$$\mathbf{H}(h) = \int_0^h \exp(\mathbf{A}t)dt\mathbf{B} \tag{9.10b}$$

Given the discrete-time state equation:

$$\mathbf{x}[(k+1)h] = \mathbf{G}\mathbf{x}(kh) + \mathbf{H}\mathbf{u}(kh) \tag{9.11}$$

its solution is given by:

$$\mathbf{x}(kh) = \mathbf{G}^k\mathbf{x}(0) + \sum_{j=0}^{k-1} \mathbf{G}^{k-j-1}\mathbf{H}\mathbf{u}(jh) \tag{9.12}$$

where the second term on the right-hand side is 0 for $k = 0$.

Alternatively, using the $\mathscr{Z}$-transform, the solution of (9.11) is:

$$\mathbf{x}(kh) = \mathscr{Z}^{-1}\left[(z\mathbf{I} - \mathbf{G})^{-1}z\right]\mathbf{x}(0) + \mathscr{Z}^{-1}\left[(z\mathbf{I} - \mathbf{G})^{-1}\mathbf{H}\mathbf{U}(z)\right] \tag{9.13}$$

We can note that:

- The solution of the discrete-time state equation includes one term corresponding to the system-free response, that depends only on the initial conditions, and one term corresponding to the system-forced response, that depends only on the system inputs.
- The second term on the right-hand side of (9.12) is the discrete version of the convolution integral.
- If $\mathbf{x}(0) = 0$, then $\mathbf{Y}(z) = [\mathbf{C}(z\mathbf{I} - \mathbf{G})^{-1}\mathbf{H}]\mathbf{U}(z)$, meaning that $\mathbf{C}(z\mathbf{I} - \mathbf{G})^{-1}\mathbf{H}$ is the discrete transfer matrix of the system.
- The equation

  $$\begin{cases} y = (0) \\ y(k) = \mathbf{C}\mathbf{G}^{k-1}\mathbf{H}u(0), \quad k \geq 1 \end{cases}$$

  is the sampled version of the continuous-time impulse response of the hold and system combination.

9.1.4.2 Discretization of transfer functions

Given a zero-order hold and a continuous-time system with transfer function $W(s)$, as shown in Figure 9.3, the discrete, or pulse, transfer function is the ratio $\frac{Y(z)}{X(z)}$, where $X(z)$ and $Y(z)$ are the $\mathscr{Z}$-transforms of the input, x, and output, y.

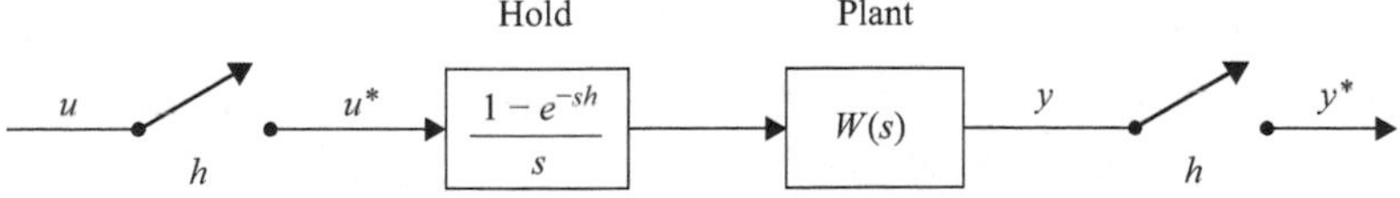

Figure 9.3 Zero-order hold sampling of a continuous-time system

The discrete, or pulse, transfer function can be obtained in three steps:

1. Compute the impulse response of $\frac{W(s)}{s}$.
2. Compute the $\mathscr{Z}$-transform of the impulse response.
3. Multiply the $\mathscr{Z}$-transform by $(1 - z^{-1})$.

The procedure is valid even when the system has a time delay that is multiple of the sampling interval, h.

9.1.4.3 Poles and zeros of sampled transfer functions

Given the transfer function of a continuous-time system:

$$W(s) = \frac{\beta_m s^m + \beta_{m-1}s^{m-1} + \cdots + \beta_1 s + \beta_0}{s^n + \alpha_{n-1}s^{n-1} + \cdots + \alpha_1 s + \alpha_0} \tag{9.14}$$

with poles $\{\lambda_1, \ldots, \lambda_n\}$, then the poles of the sampled transfer function, $W(z)$, are $\{e^{\lambda_1 h}, \ldots, e^{\lambda_n h}\}$.

The transformation $z = e^{hs}$ from the s-plane to the $\mathscr{Z}$-plane has the following properties:

- Two points s_1 and s_2 such that $s_1 - s_2 = \pm k\frac{2\pi}{h} \cdot j$, $k \in \mathbb{N}$, have the same image in $\mathscr{Z}$, since e^{hs} is periodic with period $\frac{2\pi}{h} \cdot j$.
- A line in the s-plane parallel to the imaginary axis and with abscissa $x \in \mathbb{R}$, is mapped into a circle in the $\mathscr{Z}$-plane with center at the origin and radius e^{hx}.
- A line in the left half s-plane, parallel to the real axis and with ordinate jy, $y \in \mathbb{R}$, is mapped into a line segment joining the origin of the $\mathscr{Z}$-plane with the point e^{jhy} on the unit circle.
- A line in the s-plane, corresponding to a constant damping coefficient ζ is mapped into a logarithmic spiral in the $\mathscr{Z}$-plane.
- A transfer function $W(s)$ with zeros in the left half s-plane may originate a transfer function $W(z)$ with zeros outside the unit circle in the $\mathscr{Z}$-plane.

9.1.4.4 State-space representation of difference equations

Let us consider the difference equation of order n:

$$y(k+n) + a_{n-1}y(k+n-1) + \cdots + a_0 y(k) = b_n u(k+n) + \cdots + b_0 u(k) \tag{9.15}$$

with $k \in \mathbb{N}_0$.

We define the delay operator, D, such that, $D[y(k)] = y(k-1)$. Knowing that the delay operator and the integrator play identical roles in discrete-time and in continuous-time systems, respectively, if we choose for state variables, $x_i(k)$, $i = 1, \ldots, n$, the outputs of the D blocks, then we can write:

$$\mathbf{x}(k+1) = \begin{bmatrix} 0 & 1 & 0 & \cdots & 0 \\ 0 & 0 & 1 & \cdots & 0 \\ \vdots & \vdots & \vdots & & \vdots \\ 0 & 0 & 0 & & 1 \\ -a_0 & -a_1 & -a_2 & \cdots & -a_{n-1} \end{bmatrix} \mathbf{x}(k) + \begin{bmatrix} 0 \\ 0 \\ \vdots \\ 0 \\ 1 \end{bmatrix} u(k) \tag{9.16a}$$

$$y(k) = \begin{bmatrix} (b_0 - a_0 b_n) \\ (b_1 - a_1 b_n) \\ \vdots \\ (b_{n-1} - a_{n-1} b_n) \end{bmatrix} \mathbf{x}(k) + b_n u(k) \tag{9.16b}$$

where $\mathbf{x}(k) = [x_1(k)\, x_2(k) \cdots x_n(k)]^T$ is the state vector.
This corresponds to the controllable canonical form representation of the system, as depicted in the block diagram of Figure 9.4. Different representations can be obtained by means of an approach similar to that used with continuous-time systems.

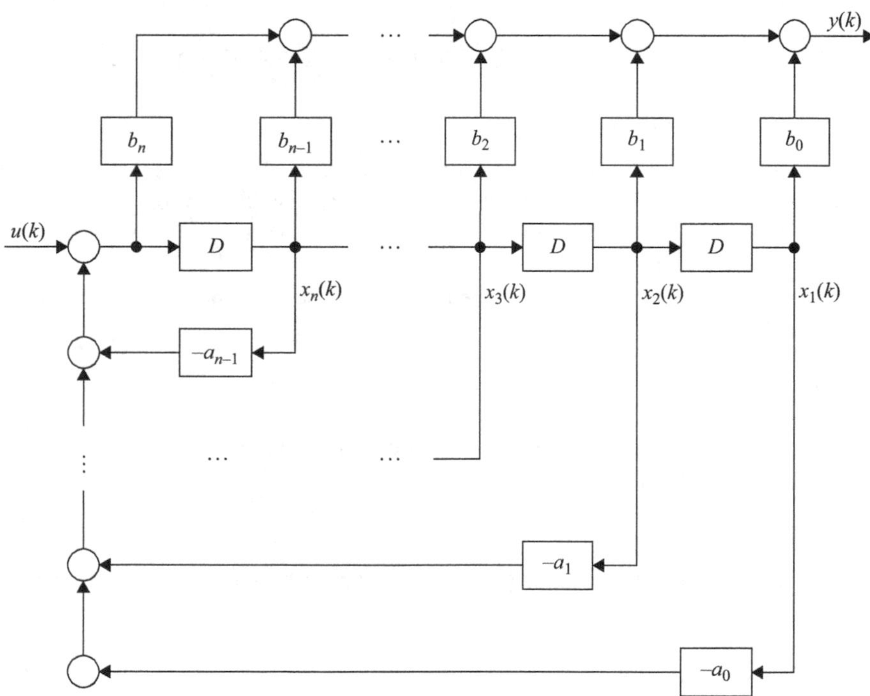

Figure 9.4 Controllable canonical form representation of a discrete-time system

9.1.5 Controllability and observability

The concepts of controllability and observability of continuous-time systems can be extended to the discrete-time case.

Given the n-order space-state model:

$$\begin{cases} \mathbf{x}[(k+1)h] = \mathbf{G}\mathbf{x}(kh) + \mathbf{H}u(kh) \\ y(kh) = \mathbf{C}\mathbf{x}(kh) \end{cases} \tag{9.17}$$

with matrix $\mathbf{G}$ non-singular. If we construct the matrix:

$$\left[\, \mathbf{H} \middle| \mathbf{GH} \middle| \cdots \middle| \mathbf{G}^{n-1}\mathbf{H} \,\right] \tag{9.18}$$

then the system is controllable if and only if the rank of the matrix is n.

If we construct the matrix:

$$\begin{bmatrix} \mathbf{C} \\ \hline \mathbf{CG} \\ \hline \vdots \\ \hline \mathbf{CG}^{n-1} \end{bmatrix} \tag{9.19}$$

then the system is observable if and only if the rank of the matrix is n.

9.1.6 Stability and the Routh–Hurwitz criterion

The stability analysis developed for continuous-time systems can be extended to the discrete-time case. However, the stability boundary is now the unit circle instead of the imaginary axis. This means that for a discrete-time system to be stable, all its poles must be inside the unit circle.

The Routh–Hurwitz criterion can be applied to discrete-time systems by means of the transformation:

$$w = \frac{z+1}{z-1} \tag{9.20}$$

that maps the $\mathscr{Z}$-plane unit circle into the left half w-plane.

For a given polynomial

$$P(z) = a_n z^n + a_{n-1} z^{n-1} + \cdots + a_1 z + a_0 \tag{9.21}$$

we transform $P(z)$ into a polynomial $Q(w)$ and then apply the Routh–Hurwitz criterion.

9.2 Solved problems

Problem 9.1 Consider a signal with $\mathscr{Z}$-transform given by $X(z) = \dfrac{1}{(z+1)(z-1)}$. Determine $x(k) = \mathscr{Z}^{-1}\{X(z)\}$, $k = 0, 1, 2, 3, 4, \ldots$.

Resolution From $X(z) = -1 + \dfrac{1}{2}\dfrac{z}{z+1} + \dfrac{1}{2}\dfrac{z}{z-1}$, we get $x(k) = -\delta(k) + \dfrac{1}{2}(-1)^k + \dfrac{1}{2}$, and thus $x(0) = 0$, $x(1) = 0$, $x(2) = 1$, $x(3) = 0$, $x(4) = 1$, $x(5) = 0$, and so on.

Problem 9.2 It is known that $\mathscr{Z}\{a^k\} = \frac{z}{z-a}$, with $a \in \mathbb{R}$ and $k = 0, 1, 2, \ldots$. Using this result, find the solution of the difference equation $x(k+2) = x(k)$, for the initial conditions $x(0) = 1$, $x(1) = 2$.

A) $x(k) = 1/4\left[1-(-2)^{k+1}\right], k = 0, 1, 2, \ldots$

B) $x(k) = 1/2\left[3-(-1)^{k}\right], k = 0, 1, 2, \ldots$

C) $x(k) = 3/3\left[3-(-2)^{k}\right], k = 0, 1, 2, \ldots$

D) None of the above.

Resolution Applying the $\mathscr{Z}$-transform to the equation, and taking into account initial conditions, we get:

$$z^2X(z) - z^2X(0) - zX(1) = X(z) \tag{9.22}$$

Replacing values and solving for $X(z)$,

$$X(z) = \frac{z(z+2)}{(z+1)(z-1)} = \frac{1}{2}\left(\frac{3z}{z-1} - \frac{z}{z+1}\right) \tag{9.23}$$

Applying the inverse $\mathscr{Z}$-transform, it can be seen that the correct answer is option **B)**.

Problem 9.3 Consider the system shown in Figure 9.5 with sampling period h. The impulsional transfer function is given by:

A) $\dfrac{Y(z)}{R(z)} = \dfrac{G_1(z)G_2(z)}{1+G_1(z)G_2(z)H(z)}$

B) $\dfrac{Y(z)}{R(z)} = \dfrac{G_1G_2(z)}{1+G_1G_2H(z)}$

C) $\dfrac{Y(z)}{R(z)} = \dfrac{G_1(z)G_2(z)}{1+G_1(z)G_2H(z)}$

D) None of the above.

Resolution

$$E(s) = R(s) - H(s)Y(s) = R(s) - H(s)G_2(s)M^*(s)$$

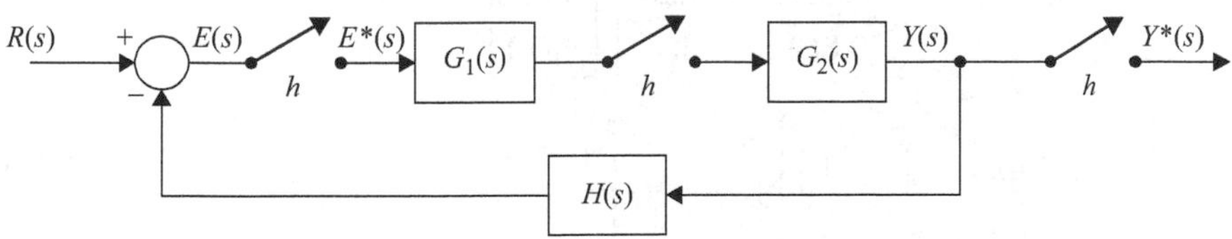

Figure 9.5 Block diagram of Problem 9.3

Because $M(s) = G_1(s)E^*(s) \Rightarrow M^*(s) = G_1^*(s)E^*(s)$, we get

$$E^*(s) = R^*(s) - [G_2H(s)]^* G_1^*(s)E^*(s) \Rightarrow$$

$$E^*(s) = \frac{R^*(s)}{1 + [G_2H(s)]^* G_1^*(s)} \Rightarrow$$

$$Y^*(s) = \frac{G_1^*(s)G_2^*(s)R^*(s)}{1 + [G_2H(s)]^* G_1^*(s)} \tag{9.24}$$

and thus the correct answer is option **C)**.

Problem 9.4 Consider the system in Figure 9.6. The sampling period is $h = 1$. Determine the discrete-time state-space representation of the system and sketch the corresponding block diagram using the $\mathscr{Z}$-transform.

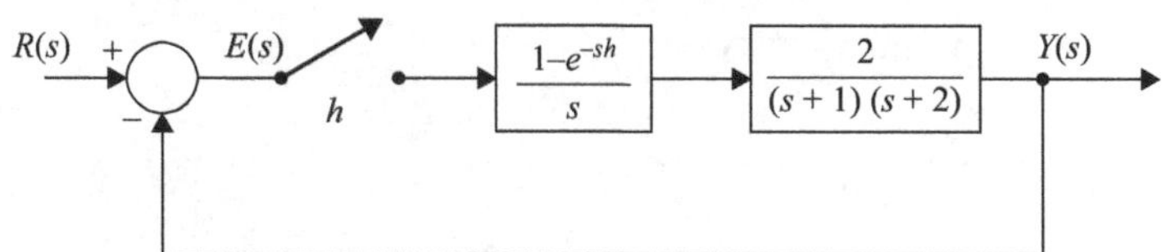

Figure 9.6 Block diagram of Problem 9.4

Resolution Let $G_1(s) = sG(s) = \dfrac{2}{s(s+1)(s+2)}$. Then $g(t) = \mathscr{L}^{-1}[G(s)] = 1 - 2e^{-t} + e^{-2t}$, and thus $G_1(z) = \dfrac{z}{z-1} - 2\dfrac{z}{z-e^{-1}} + \dfrac{z}{z-e^{-2}}$. We now need to multiply this by the $\mathscr{Z}$-transform of $1 - e^{-sh}$, the only that is still missing:

$$\begin{aligned} G(z) &= (1 - z^{-1})G_1(z) \\ &= \frac{z-1}{z} \frac{z0.4(z+0.368)}{(z-1)(z-0.368)(z-0.135)} \\ &= \frac{1.26}{z-0.368} - \frac{0.864}{z-0.135} \end{aligned}$$

This corresponds to the state space

$$\begin{bmatrix} x_1(k+1) \\ x_2(k+1) \end{bmatrix} = \begin{bmatrix} -0.892 & -1.26 \\ 0.864 & 1 \end{bmatrix} \begin{bmatrix} x_1(k) \\ x_2(k) \end{bmatrix} + \begin{bmatrix} 1.26 \\ -0.864 \end{bmatrix} u(k)$$

$$y(k) = \begin{bmatrix} 1 & 1 \end{bmatrix} \begin{bmatrix} x_1(k+1) \\ x_2(k+1) \end{bmatrix} \tag{9.25}$$

and to the block diagram in Figure 9.7.

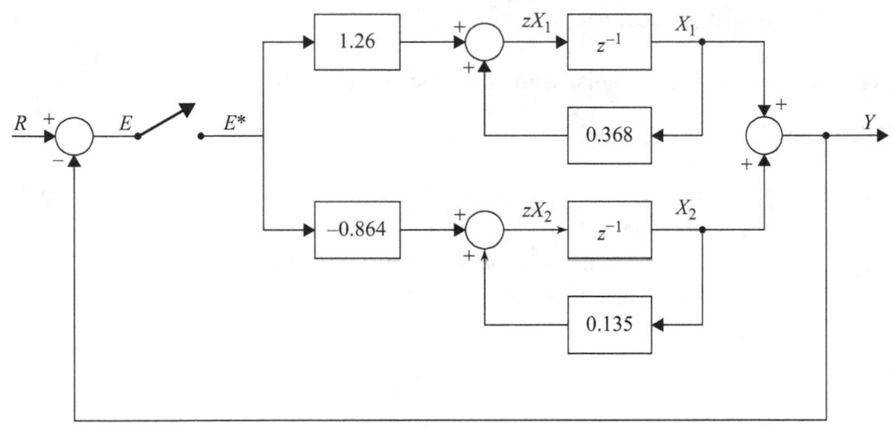

Figure 9.7 Resolution of Problem 9.4

Problem 9.5 Consider a system with state-space representation $\dot{\mathbf{x}} = \begin{bmatrix} 0 & 1 \\ 0 & -2 \end{bmatrix} \mathbf{x}$, $\mathbf{x} = [\, x_1 \quad x_2 \,]^T$. Knowing that $\mathbf{G}(h) = e^{\mathbf{A}h}$, and for a sampling period $h = \frac{1}{2}$, its representation in discrete time is:

A) $\mathbf{x}(k+1) = \begin{bmatrix} \frac{1}{2}\left(1 - e^{-1}\right) & 1 \\ e^{-1} & 1 \end{bmatrix} \mathbf{x}(k)$

B) $\mathbf{x}(k+1) = \begin{bmatrix} 1 & \frac{1}{2}\left(1 - e^{-1}\right) \\ 0 & e^{-1} \end{bmatrix} \mathbf{x}(k)$

C) $\mathbf{x}(k+1) = \begin{bmatrix} \frac{1}{2} & 1 - e^{-1} \\ 0 & \frac{1}{2}e^{-1} \end{bmatrix} \mathbf{x}(k)$

D) $\mathbf{x}(k+1) = \begin{bmatrix} 0 & e^{-1}\left(1 - e^{-1}\right) \\ e & 1 \end{bmatrix} \mathbf{x}(k).$

Resolution

$$(s\mathbf{I} - \mathbf{A}h)^{-1} = \begin{bmatrix} s & \frac{-1}{2} \\ 0 & s+1 \end{bmatrix}^{-1} = \frac{1}{s(s+1)} \begin{bmatrix} s+1 & \frac{1}{2} \\ 0 & s \end{bmatrix} = \begin{bmatrix} \frac{1}{s} & \frac{1}{2s} - \frac{\frac{1}{2}}{s+1} \\ 0 & \frac{1}{s+1} \end{bmatrix}$$

$$e^{\mathbf{A}h} = \begin{bmatrix} 1 & \frac{1}{2}\left(1 - e^{-1}\right) \\ 0 & e^{-1} \end{bmatrix}$$

Thus the correct answer is option **B)**.

9.3 Proposed problems

Exercise 9.1 Consider a signal with $\mathscr{Z}$-transform given by $X(z) = \dfrac{2z}{(z-1)\left(z-\frac{1}{2}\right)}$. Determine $x(t) = \mathscr{Z}^{-1}\{X(z)\}$.

Exercise 9.2 Let the signal $x(t)$ produce, by sampling, with sampling period h, the time-domain sequence $x^*(t)$. Therefore, the definition of the $\mathscr{Z}$-transform $X(z) = \mathscr{Z}\{x(t)\}$ is associated with the Laplace transform of the sampled signal $\mathscr{L}\{x^*(t)\}$ for the following change of variable:

A) $z = e^{-hs}$
B) $z = s$
C) $z = \dfrac{1-e^{-hs}}{s}$
D) None of the above.

Exercise 9.3 Consider the system shown in Figure 9.8 with sampling period h. The impulsional transfer function has an expression given by:

A) $\dfrac{Y(z)}{R(z)} = \dfrac{G(z)}{1+G(z)H(z)}$

B) $\dfrac{Y(z)}{R(z)} = \dfrac{G(z)}{1+GH(z)}$

C) None of the above.

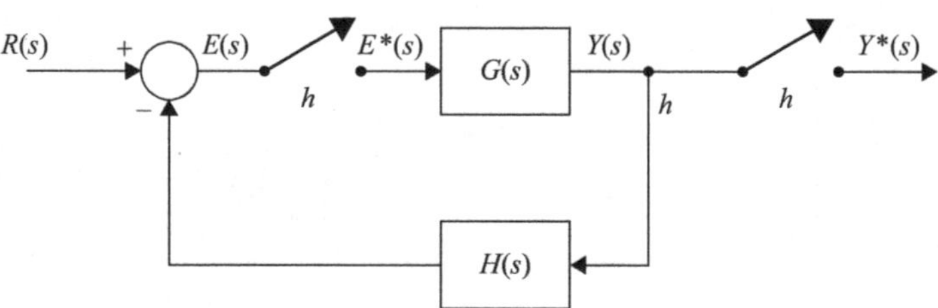

Figure 9.8 Block diagram of Exercise 9.3

Exercise 9.4 Consider the system shown in Figure 9.9 with sampling period h. The impulse transfer function has an expression given by:

A) $\dfrac{Y(z)}{R(z)} = \dfrac{G_1G_2(z)}{1+G_1G_2(z)}$

B) $\dfrac{Y(z)}{R(z)} = \dfrac{G_1(z)G_2(z)}{1+G_1G_2(z)}$

C) $\dfrac{Y(z)}{R(z)} = \dfrac{G_1G_2(z)}{1 + G_1(z)G_2(z)}$

D) $\dfrac{Y(z)}{R(z)} = \dfrac{G_1(z)G_2(z)}{1 + G_1(z)G_2(z)}$.

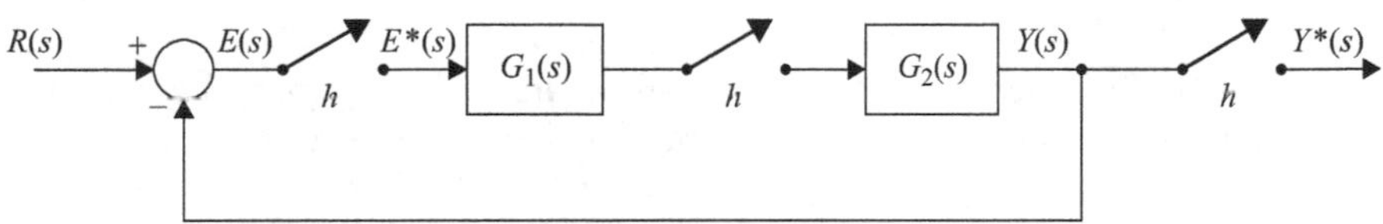

Figure 9.9 Block diagram of Exercise 9.4

Exercise 9.5 Consider signal $x(t)$, such that $X(s) = \mathscr{L}\{x(t)\} = \dfrac{s+2}{(s+1)^2(s+3)}$. Obtain $X(z) = \mathscr{Z}\{x(t)\}$.

Exercise 9.6 The $\mathscr{Z}$-transform $\mathscr{Z}\{e^{-at}\}$, with $a \in \mathbb{R}$, and h the sampling period, is:

A) $\mathscr{Z}\{e^{-at}\} = \dfrac{z}{z - e^{-ah}}$

B) $\mathscr{Z}\{e^{-at}\} = \dfrac{1}{z - e^{-ah}}$

C) $\mathscr{Z}\{e^{-at}\} = \dfrac{ze^{-ah}}{z - 1}$

D) None of the above.

Exercise 9.7 Knowing that $\mathscr{Z}\{a^k\} = \dfrac{z}{z-a}$, with $a \in \mathbb{R}$ and $k = 0, 1, 2, \ldots$, then:

A) $\mathscr{Z}^{-1}\left\{\dfrac{z}{z^2-1}\right\} = (-1)^k$

B) $\mathscr{Z}^{-1}\left\{\dfrac{z}{z^2-1}\right\} = \dfrac{1}{2}[1 - (-1)^k]$

C) $\mathscr{Z}^{-1}\left\{\dfrac{z}{z^2-1}\right\} = 1 - (-1)^k$

D) None of the above.

Exercise 9.8 Consider a signal $x(t)$ with transform $Z\{x(t)\} = X(z)$. If h is the sampling period, then:

A) $\mathscr{Z}\{x(t+h)\} = zX(z) - zx(0)$

B) $\mathscr{Z}\{x(t+h)\} = zX(z) + hx(0)$

C) $\mathscr{Z}\{x(t+h)\} = z^{-1}X(z) + z^{-1}x(0)$

D) None of the above.

Exercise 9.9 Consider a signal with $\mathscr{Z}$-transform given by $X(z) = \frac{1}{(1+\frac{1}{2}z^{-1})^2(1-2z^{-1})(1-3z^{-1})}$. Determine $x(t) = \mathscr{Z}^{-1}\{X(z)\}$.

Exercise 9.10 Consider the system shown in Figure 9.10 with sampling period h. The impulsional transfer function is given by:

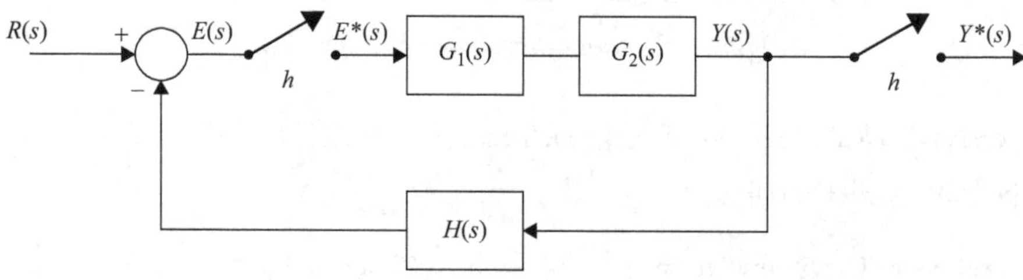

Figure 9.10 Block diagram of Exercise 9.10

A) $\dfrac{Y(z)}{R(z)} = \dfrac{G_1(z)G_2(z)}{1 + G_1(z)G_2H(z)}$

B) $\dfrac{Y(z)}{R(z)} = \dfrac{G_1(z)G_2(z)}{1 + G_1(z)G_2(z)H(z)}$

C) $\dfrac{Y(z)}{R(z)} = \dfrac{G_1G_2(z)}{1 + G_1G_2H(z)}$

D) None of the above.

Exercise 9.11 Given $X(z) = \dfrac{z^2-3}{(z-1)^2}$, determine $x(kh)$.

Exercise 9.12 It is known that $\mathscr{Z}\{e^{-at}\} = \dfrac{z}{z-e^{-ah}}$, for $a \in \mathbb{R}$, where h is the sampling period. So, taking into account other properties of the $\mathscr{Z}$-transform which you have learned, it can be concluded that:

A) $\mathscr{Z}\{\sin(\omega t)\} = \dfrac{z\cos(\omega h)}{z^2 + 2z\cos(\omega h) + 1}$

B) $\mathscr{Z}\{\sin(\omega t)\} = \dfrac{z\sin(\omega h)}{z^2 + 2z\sin(\omega h) + 1}$

C) $\mathscr{Z}\{\sin(\omega t)\} = \dfrac{z\sin(\omega h)}{z^2 - 2z\cos(\omega h) + 1}$

D) None of the above.

Exercise 9.13 Consider a signal $x(t)$ with transform $Z\{x(t)\} = X(z) = \sum_{k=0}^{\infty} x(kh)\, z^{-k}$, where h is the sampling period. Then the following property can be verified:

A) $x(0) = \lim_{z\to\infty} [(z-1)X(z)]$

B) $x(0) = \lim_{z\to\infty} [zX(z)]$

C) $x(0) = \lim_{z\to\infty} X(z)$

D) $x(0) = \lim_{z\to\infty} \left[\frac{dX(z)}{dz}\right].$

Exercise 9.14 Consider the following difference equation:

$$y\,(k+2) = u(k) - y\,(k+1) - \frac{1}{8}y(k), \quad k = 0, 1, 2, \ldots$$

where $u(t)$ is the unit step signal. What is the response of the system described by the equation above, considering zero initial conditions?

Exercise 9.15 Consider the system shown in Figure 9.11, a zero-order sample and hold with period h. The corresponding transfer function has an expression given by:

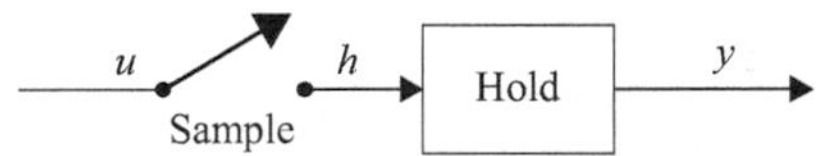

Figure 9.11 Block diagram of Exercise 9.15

A) $\dfrac{Y(s)}{U(s)} = \dfrac{1+e^{-hs}}{s}$

B) $\dfrac{Y(s)}{U(s)} = \dfrac{e^{-hs}}{1-s}$

C) $\dfrac{Y(s)}{U(s)} = \dfrac{s-e^{-hs}}{s}$

D) None of the above.

Exercise 9.16 Consider a signal with $\mathscr{Z}$-transform given by $\mathscr{Z}\{x(t)\} = X(z) = \dfrac{z}{(z+1)(z-1)}$. Therefore:

A) $x(t) = 1 - (-1)^n,\ n = 0, 1, 2, \ldots$

B) $x(t) = \frac{1}{2}[1 - (-1)^n],\ n = 0, 1, 2, \ldots$

C) $x(t) = 2[1 - (-1)^n],\ n = 0, 1, 2, \ldots$

D) None of the above.

Exercise 9.17 Consider the system shown in Figure 9.12 with sampling period h. The discrete-time equation that describes the output signal is:

A) $v[(k+1)h] = e^{-RCh}\, v(kh),\ k = 0, 1, \ldots$

B) $v[(k+1)h] = \dfrac{h}{RC}\, v(kh),\ k = 0, 1, \ldots$

C) $v[(k+1)h] = e^{-\frac{h}{RC}}\, v(kh),\ k = 0, 1, \ldots$

D) None of the above.

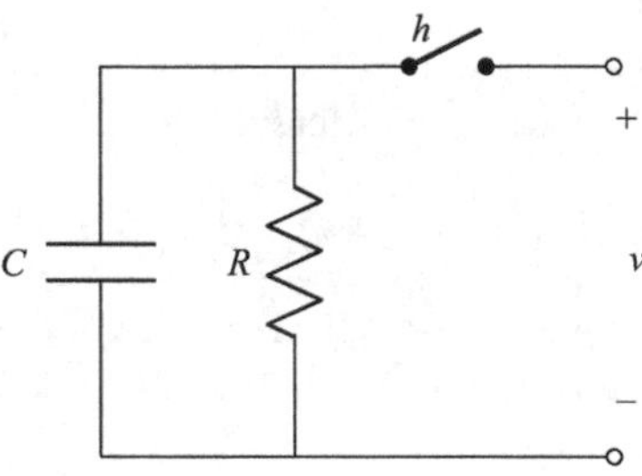

Figure 9.12 Circuit of Exercise 9.17

Exercise 9.18 Given $X(z) = \dfrac{z}{(z-1)^2\,(z-2)}$, determine $x(kh),\ k = 0, 1, 2, \ldots,$ where h is the sampling period.

Exercise 9.19 Consider a signal with $\mathscr{Z}$-transform given by $\mathscr{Z}\{x(t)\} = \frac{2z\left(z-\frac{5}{12}\right)}{\left(z-\frac{1}{2}\right)\left(z-\frac{1}{3}\right)}$.

1. Its inverse transform is:

   A) $x(t) = \left(\frac{1}{2}\right)^k + \left(\frac{1}{3}\right)^k,\ k = 0, 1, 2, \ldots$

   B) $x(t) = \left(\frac{1}{2}\right)^k + \left(\frac{1}{3}\right)^k - \left(\frac{5}{12}\right)^k,\ k = 0, 1, 2, \ldots$

   C) $x(t) = \left(\frac{5}{12}\right)^k - \left(\frac{1}{2}\right)^k - \left(\frac{1}{3}\right)^k,\ k = 0, 1, 2, \ldots$

   D) None of the above.
2. Let h be the sampling period. From the transform definition, it can be concluded that:

   E) $X(z) = 2 + \frac{5}{6}z^{-1} + \frac{13}{36}z^{-2} + \cdots \Rightarrow x(0) = 2,\ x(h) = \frac{5}{6},\ x(2h) = \frac{13}{36}, \ldots$

F) $X(z) = 1 + \frac{1}{2}z^{-1} + \frac{1}{3}z^{-2} + \cdots \Rightarrow x(0) = 1,\ x(h) = \frac{1}{2},\ x(2h) = \frac{1}{3}, \ldots$

G) $X(z) = \frac{5}{12} - \frac{1}{2}z^{-1} - \frac{1}{3}z^{-2} + \cdots \Rightarrow x(0) = \frac{5}{12},\ x(h) = -\frac{1}{2},\ x(2h) = -\frac{1}{3}, \ldots$

H) None of the above.

Exercise 9.20 Consider the systems shown in Figure 9.13. Let the sampling period be h.

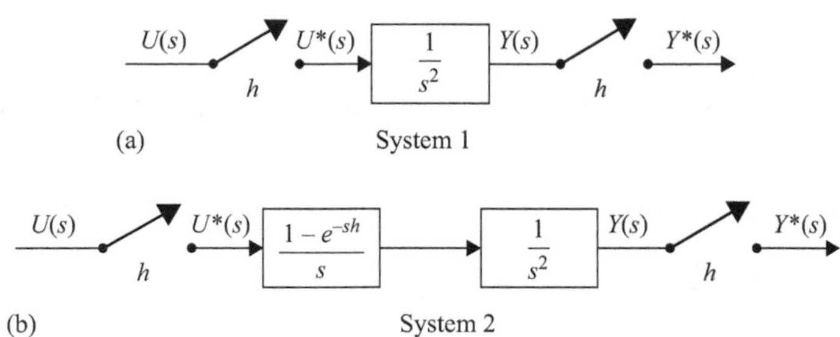

Figure 9.13 Block diagrams of Exercise 9.20

1. For system 1 the transfer function has an expression given by:

   A) $\frac{Y(z)}{U(z)} = \frac{hz}{(z-1)^2}$

   B) $\frac{Y(z)}{U(z)} = \frac{h^2(z+1)}{2(z-1)^2}$

   C) $\frac{Y(z)}{U(z)} = \frac{hz}{2(z-1)}$

   D) None of the above.

2. For system 2 the transfer function has an expression given by:

   E) $\frac{Y(z)}{U(z)} = \frac{hz}{(z-1)^2}$

   F) $\frac{Y(z)}{U(z)} = \frac{h^2(z+1)}{2\,(z-1)^2}$

   G) $\frac{Y(z)}{U(z)} = \frac{hz}{2(z-1)}$

   H) None of the above.

Exercise 9.21 The $\mathcal{Z}$-transform of the unit step signal, $\mathcal{Z}\{1(t)\}$, with sample period h, is:

A) $\mathcal{Z}\{1(t)\} = \dfrac{z}{z+1}$

B) $\mathcal{Z}\{1(t)\} = \dfrac{hz}{(z+1)^2}$

C) $\mathcal{Z}\{1(t)\} = \dfrac{z}{z-e^{-h}}$

D) $\mathcal{Z}\{1(t)\} = \dfrac{z}{z-1}$.

Exercise 9.22 The inverse $\mathcal{Z}$-transform of $X(z) = z^{-k}$, $\mathcal{Z}^{-1}\{z^{-k}\}$, with sample period h, is:

A) $\mathcal{Z}^{-1}\{z^{-k}\} = \delta(t-kh)$

B) $\mathcal{Z}^{-1}\{z^{-k}\} = 1$

C) $\mathcal{Z}^{-1}\{z^{-k}\} = 1(t-kh)$

D) $\mathcal{Z}^{-1}\{z^{-k}\} = t-kh$.

Exercise 9.23 Consider the system shown in Figure 9.14 with sampling period h.

1. The state-space representation, with $\mathbf{x}(k) = [x_1(k)\ x_2(k)]^T$, $k = 0, 1, 2, \ldots$, is:

A) $$\begin{cases} \mathbf{x}(k+1) = \begin{bmatrix} 1-e^{-h} & 1 \\ e^{-h} & 1 \end{bmatrix} \mathbf{x}(k) + \begin{bmatrix} 1-e^{-h} \\ h+e^{-h}-1 \end{bmatrix} u(k) \\ y(k) = \begin{bmatrix} 0 & 1 \end{bmatrix} \mathbf{x}(k) \end{cases}$$

B) $$\begin{cases} \mathbf{x}(k+1) = \begin{bmatrix} 1 & e^{-h} \\ 0 & 1-e^{-h} \end{bmatrix} \mathbf{x}(k) + \begin{bmatrix} e^{-h}-1 \\ h+1-e^{-h} \end{bmatrix} u(k) \\ y(k) = \begin{bmatrix} 1 & 1 \end{bmatrix} \mathbf{x}(k) \end{cases}$$

C) $$\begin{cases} \mathbf{x}(k+1) = \begin{bmatrix} 1 & 1-e^{-h} \\ 0 & e^{-h} \end{bmatrix} \mathbf{x}(k) + \begin{bmatrix} h+e^{-h}-1 \\ 1-e^{-h} \end{bmatrix} u(k) \\ y(k) = \begin{bmatrix} 1 & 0 \end{bmatrix} \mathbf{x}(k) \end{cases}$$

D) None of the above.

2. The impulsional transfer function, $Y(z)/U(z)$, is given by:

E) $\dfrac{Y(z)}{U(z)} = \dfrac{1}{z-1} + \dfrac{\left(e^{-h}-1\right)h}{z-e^{-h}}$

F) $\dfrac{Y(z)}{U(z)} = \dfrac{h}{z-1} + \dfrac{e^{-h}-1}{z-e^{-h}}$

G) $\dfrac{Y(z)}{U(z)} = \dfrac{h^2}{z-1} + \dfrac{\left(e^{-h}-1\right)h}{z-e^{-h}}$

H) None of the above.

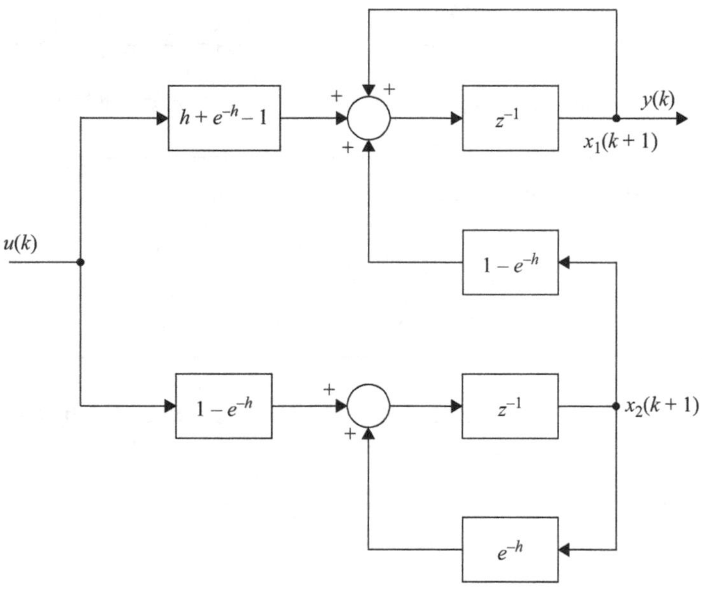

Figure 9.14 Block diagram of Exercise 9.23

9.4 Discrete-time systems and $\mathscr{Z}$-transform analysis using computer packages

This section presents several commands for handling discrete-time systems and $\mathscr{Z}$-transform analysis using the computer packages MATLAB®, SCILAB™ and OCTAVE©.

9.4.1 MATLAB

This subsection describes some basic commands that can be adopted with the package MATLAB.

We consider Problem 9.1 and a signal with $\mathscr{Z}$-transform given by $X(z) = \frac{1}{(z+1)(z-1)}$. We can determine $x(k)$ analytically or numerically.

The commands syms and iztrans produce the inverse $\mathscr{Z}$-transform in analytical form. The complete code is as follows.

```
%%%% Z-transform %%%%

% Creates symbolic variables z and n
syms z n

% Inverse Z-transform of X(z) = 1/((z+1)*(z-1))
disp('x(k) analytical expression')
iztrans(1/((z+1)*(z-1)))
```

MATLAB outputs the following information at the Command Window.

```
x(k) analytical expression

ans =

(-1)^n/2 - kroneckerDelta(n, 0) + 1/2
```

The command filter produces the inverse $\mathscr{Z}$-transform by means of a numerical series. The complete code is as follows.

```
%%%% Z-transform %%%%

% Inverse Z-transform of
% X(z) = 1/((z+1)*(z-1)) = z^(-2)/(1-z^(-2))
% using the Kronecker delta impulse

num = [0 0 1]; % numerator of X(z)
den = [1 0 -1];% denominator of X(z)

n = 6; % Number of terms
u = [1 zeros(1,n)]; % Kronecker delta input

disp('x(k) numerical series');
% filters the input data, u, using a rational transfer
% function
% defined by the numerator and denominator coefficients
x = filter(num,den,u)
```

MATLAB outputs the following information at the Command Window.

```
x(k) numerical series

x =

     0     0     1     0     1     0     1
```

It is also possible to perform the direct $\mathscr{Z}$-transform in analytical form by means of the command `ztrans`.

9.4.2 SCILAB

This subsection describes some basic commands that can be adopted with the package SCILAB.

The command `filter` produces the inverse $\mathscr{Z}$-transform by means of a numerical series. The complete code is as follows.

```
//// Z-transform ////

// Inverse Z-transform of
// X(z) = 1/((z+1)*(z-1)) = z^(-2)/(1-z^(-2))
// using the Kronecker delta impulse

num = poly([1 0 0],'z','coeff'); // numerator of X(z)
den = poly([-1 0 1],'z','coeff');// denominator of X(z)

n = 6; // Number of terms
u = [1 zeros(1,n)]; // Kronecker delta input

disp("x(k) numerical series");
// filters the input data, u, using a rational transfer
// function
// defined by the numerator and denominator
// coefficients
x = filter(num,den,u)

disp(x,"x(k)=")
```

SCILAB outputs the following information at the Console.

```
x(k) numerical series

x(k)=

   0.    0.    1.    0.    1.    0.    1.
```

9.4.3 OCTAVE

This subsection describes some basic commands that can be adopted with the package OCTAVE.

The command `filter` produces the inverse $\mathscr{Z}$-transform by means of a numerical series. The complete code is as follows.

```
%%%% Z-transform %%%%

% Inverse Z-transform of
% X(z) = 1/((z+1)*(z-1)) = z^(-2)/(1-z^(-2))
% using the Kronecker delta impulse

num = [0 0 1]; % numerator of X(z)
den = [1 0 -1];% denominator of X(z)

n = 6; % Number of terms
u = [1 zeros(1,n)]; % Kronecker delta input

disp('x(k) numerical series');
% filters the input data, u, using a rational transfer
% function
% defined by the numerator and denominator coefficients
x = filter(num,den,u)
```

OCTAVE outputs the following information at the Command Window.

```
x(k) numerical series
x =

   0   0   1   0   1   0   1
```

Chapter 10
Analysis of nonlinear systems with the describing function method

10.1 Fundamentals

The describing function (DF) is one method for the analysis of nonlinear systems. The main idea is to study the ratio between a sinusoidal input applied to the system and the fundamental harmonic component of the output. The DF allows the extension of the Nyquist stability criterion to nonlinear systems for detection of limit cycles, namely the prediction of limit cycle amplitude and frequency [27–29].

10.1.1 List of symbols

$c(t)$ time-domain system output signal
$e(t)$ time-domain error signal (input of the nonlinear element)
$G(s)$ linear transfer function
$m(t)$ output of the nonlinear element
N describing function
$r(t)$ time-domain system input signal
t time
X amplitude
ω angular frequency

10.1.2 The describing function

Suppose that the input to a nonlinear element is a sinusoidal signal $e(t) = X\sin(\omega t)$. In general, the output, $m(t)$, of the nonlinear element is not sinusoidal. Assume that the signal $m(t)$ is periodic, with the same period as the input, and containing higher harmonics in addition to the fundamental harmonic component. In the describing function method, we consider that only the fundamental harmonic component of the output is significant. Such assumption is often valid since the higher harmonics in the output of a nonlinear element are usually of smaller amplitude than the fundamental component. Moreover, most systems are "low-pass filters" with the result that the higher harmonics are further attenuated [30–32].

The DF of a nonlinear element, $N(X, \omega)$, is defined as the complex ratio of the fundamental harmonic component of the output and the input, that is:

$$N(X, \omega) = \frac{M_1}{X} e^{j\phi_1} \tag{10.1}$$

where the symbol N represents the DF, X is the amplitude of the input sinusoid, and M_1 and ϕ_1 are the amplitude and the phase shift of the fundamental harmonic component of the output, respectively.

If the system does not involve energy storage, then the DF depends only on the amplitude of the input signal, meaning that $N = N(X)$. However, if the system can store energy, then the DF depends both on the amplitude and frequency of the input signal, that is, $N = N(X, \omega)$.

10.1.3 Describing functions of common nonlinearities

The output, $m(t)$, can be expressed by means of the Fourier series expansion as:

$$m(t) = A_0 + \sum_{k=1}^{\infty} (A_k \cos k\omega t + B_k \sin k\omega t) = A_0 + \sum_{k=1}^{\infty} M_k \sin(k\omega t + \phi_k) \quad (10.2)$$

where

$$A_k = \frac{1}{\pi} \times \int_0^{2\pi} m(t) \cos k\omega t \cdot d(\omega t) \quad (10.3a)$$

$$B_k = \frac{1}{\pi} \times \int_0^{2\pi} m(t) \sin k\omega t \cdot d(\omega t) \quad (10.3b)$$

$$M_k = \sqrt{A_k^2 + B_k^2} \quad (10.3c)$$

$$\phi_k = \arctan\left(\frac{A_k}{B_k}\right) \quad (10.3d)$$

where M_k and ϕ_k are the amplitude and the phase shift of the kth harmonic component of the output $m(t)$, respectively.

If we assume that the nonlinearity is symmetrical with respect to the variation around zero, then $A_0 = 0$ and we have:

$$N = \frac{M_1}{X} \angle \phi_1 = \frac{\sqrt{A_1^2 + B_1^2}}{X} \angle \arctan\left(\frac{A_1}{B_1}\right) \quad (10.4)$$

The DF is a complex-valued function when $\phi_1 \neq 0$. The calculation of arbitrary DF is generally difficult and usually we have to adopt computer numerical approaches, or symbolic packages, since it is impossible to find a closed-form analytical solution. Nevertheless, in certain cases, we can find easily the relationship between DF of different nonlinear elements. For example, let N_1 and N_2 denote the DF of the dead zone and saturation nonlinear elements, respectively, then it is straightforward to conclude that $N_1 = k - N_2$ (please see table in Appendix A).

10.1.4 Nonlinear systems analysis

Let us consider the nonlinear feedback control system of Figure 10.1 with one nonlinear element N and one linear block $G(s)$ in the forward path.

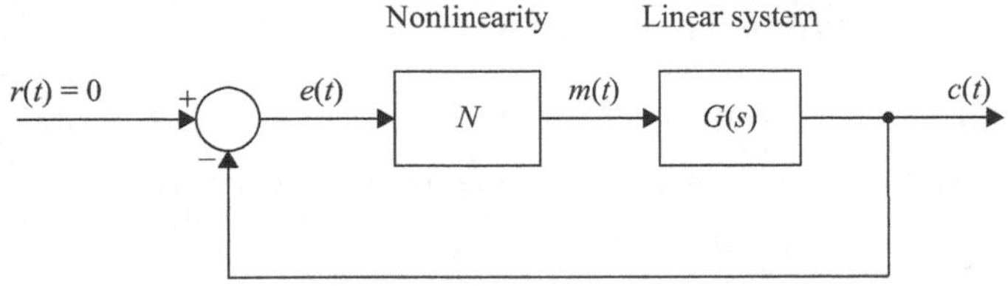

Figure 10.1 Nonlinear control system with one nonlinear element N and one linear block G(s) in the forward path

If the higher harmonics are sufficiently attenuated by $G(j\omega)$, then N can be treated as a real or complex variable gain and the closed-loop frequency response becomes:

$$\frac{C(j\omega)}{R(j\omega)} = \frac{N(X,\omega)G(j\omega)}{1 + N(X,\omega)G(j\omega)} \tag{10.5}$$

The characteristic equation is:

$$1 + N(X,\omega)G(j\omega) = 0 \tag{10.6}$$

If this equation can be satisfied for some values of X and ω, then the nonlinear system exhibits a limit cycle. Thus, in the describing function method, $-\frac{1}{N}$ is the geometric location of the critical points to be considered for stability analysis.

In practical terms, we plot the Nyquist diagram of the linear portion of the system and, on the same graph, we superimpose the locus of $-\frac{1}{N}$. If the two curves intersect, then a limit cycle is predicted by the DF. The amplitude and frequency of the oscillations can be estimated by means of the N and $G(j\omega)$ curves.

For determining the stability of a limit cycle, we can apply the Nyquist stability criterion. A limit cycle (or intersection point) with amplitude X_0 is stable if to a neighbor stable working point X_0' corresponds an increasing oscillation amplitude $X_0' > X_0$ and to a neighbor unstable working point X_0'' corresponds a decreasing oscillation amplitude $X_0'' < X_0$.

It should be noted that the DF analysis method has several limitations:

- it may predict non-existing limit cycles;
- it may fail in predicting an existing limit cycle;
- the predicted amplitude and frequency are merely approximations and can be far from the true values;
- it can be used only if the higher harmonics are sufficiently attenuated by $G(j\omega)$;
- it is difficult to determine, unless it can be related to, and calculated from, already known DF;
- in general, for two nonlinear elements connected in series, the overall DF is not the product of the two individual DF.

10.2 Solved problems

Problem 10.1 The control system in Figure 10.2 comprises a relay with dead zone, with $A = 1$ and $B = 2$, and a linear plant with transfer function $G(s) = \dfrac{K_G}{(s+1)^3}$. Use the describing function method to find:

1. approximate values for the amplitude X and the frequency ω of the limit cycle when $K_G = 10$;
2. the properties of the limit cycle (as seen in the Nyquist diagram);
3. the smallest K_G for which there no longer is a limit cycle.

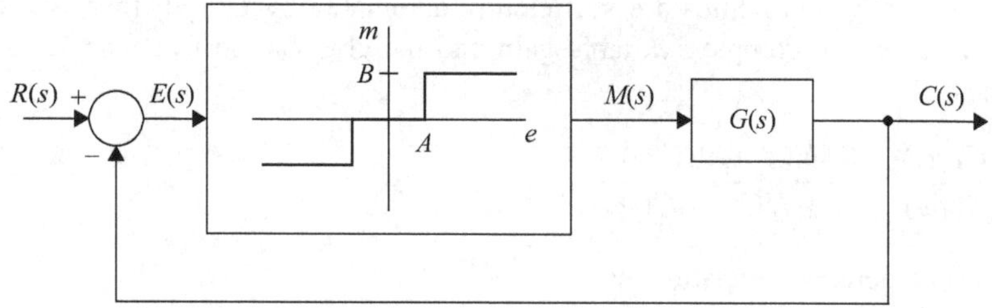

Figure 10.2 Block diagram of Problem 10.1

Resolution

$$G(j\omega) = \frac{10}{(\omega^2+1)^{\frac{3}{2}}} e^{j(-3\arctan\omega)} \tag{10.7}$$

$$N(X) = \frac{4B}{\pi X}\sqrt{1-\left(\frac{A}{X}\right)^2} \tag{10.8}$$

$$1 + GN = 0 \tag{10.9}$$

and thus

$$\begin{cases} \frac{10}{(\omega^2+1)^{\frac{3}{2}}} \cdot \frac{4B}{\pi X}\sqrt{1-\left(\frac{A}{X}\right)^2} = 1 \\ -3\arctan\omega = -\pi \end{cases} \Rightarrow \begin{cases} \omega = \sqrt{3}\ \text{rad/s} \\ X = 1.06 \vee X = 3.00 \end{cases} \tag{10.10}$$

See Figure 10.3: the limit cycle is unstable if $X = 1.06$ and stable when $X = 3.00$. The rightmost point of the $N(X)$ curve is -0.79 (reached when $X = 1.41$). K_G only affects the magnitude of $G(j\omega)$, not its phase; when it changes, the curve will be scaled horizontally. There will be no limit cycle when it does not intersect $N(X)$. So the smallest value of K_G for which there will be an intersection is

$$\left.\frac{K_G}{(\omega^2+1)^{\frac{3}{2}}}\right|_{\omega=\sqrt{3}} = 0.79 \Rightarrow K_G = 6.32 \tag{10.11}$$

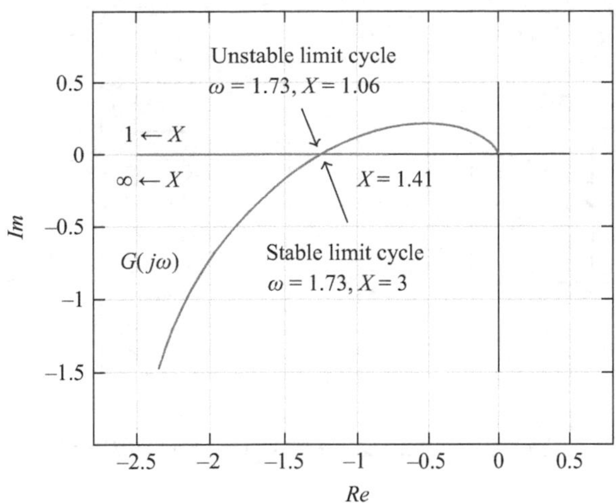

Figure 10.3 Nyquist plot and describing function for Problem 10.1

Problem 10.2 The saturation in Figure 10.4a, defined by parameters k and A, has the describing function $N = \frac{2k}{\pi}\left[\arcsin\left(\frac{A}{X}\right) + \frac{A}{X}\sqrt{1-\left(\frac{A}{X}\right)^2}\right]$. Knowing this, determine the describing function of the nonlinearity in Figure 10.4b.

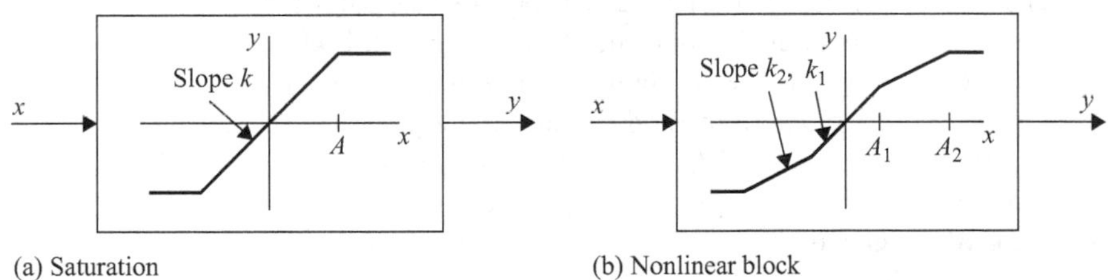

(a) Saturation (b) Nonlinear block

Figure 10.4 Blocks for Problem 10.2

Resolution See Figure 10.5.

$$N_1(X) = \frac{2(k_1 - k_2)}{\pi}\left(\arcsin\frac{A_1}{X} + \frac{A_1}{X}\sqrt{1-\left(\frac{A_1}{X}\right)^2}\right) \tag{10.12}$$

$$N_2(X) = \frac{2k_2}{\pi}\left(\arcsin\frac{A_2}{X} + \frac{A_2}{X}\sqrt{1-\left(\frac{A_2}{X}\right)^2}\right) \tag{10.13}$$

$$N(X) = N_1(X) + N_2(X). \tag{10.14}$$

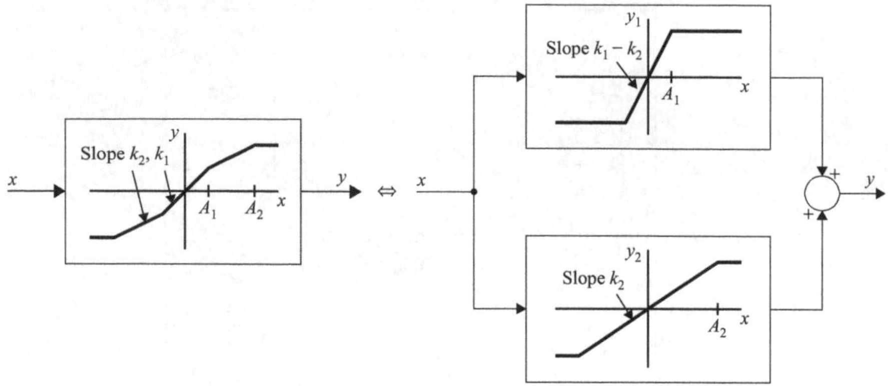

Figure 10.5 Block diagram for the solution of Problem 10.2

10.3 Proposed problems

Exercise 10.1 The control system in Figure 10.6 comprises nonlinearities N_1 and N_2, as well as linear, low-pass elements with transfer functions $G_1(s)$ and $G_2(s)$. Let N_{12} be the describing function of the two nonlinearities in series.

A) The describing function method can be adopted, and the existence of limit cycles can be studied from the Nyquist plot of characteristic equation, $N_{12}G_1(j\omega)G_2(j\omega) + 1 = 0$, where $N_{12} = N_1N_2$.

B) The describing function method can be adopted, and the existence of limit cycles can be studied from the Nyquist plot of characteristic equation, $N_{12}G_1(j\omega)G_2(j\omega) + 1 = 0$, where, in the general case, $N_{12} \neq N_1N_2$.

C) The describing function method cannot be adopted, and the existence of limit cycles cannot be studied from the Nyquist plot of the passing characteristic equation.

D) None of the above.

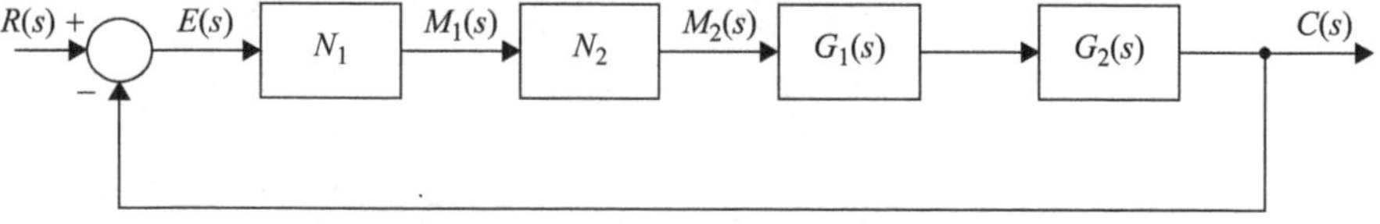

Figure 10.6 Block diagram of Exercise 10.1

Exercise 10.2 The describing function method allows determining approximately:

A) The stable nodes.
B) The unstable nodes.
C) The saddle points.
D) The limit cycles.

Exercise 10.3 It can be said of the describing function method that:

A) It allows reckoning a plant's response for any kind of input signal.
B) It allows reckoning limit cycles only if the plant is a low-pass filter.
C) It allows reckoning responses of linear systems only (and cannot be applied to nonlinear plants).
D) None of the above.

Exercise 10.4 Let N_1 and N_2 be the describing functions of two nonlinear elements, and N_T be the describing function of both elements in series, as seen in Figure 10.7. Then, in general, we have:

A) $N_T = N_1 \times N_2$
B) $N_T \neq N_1 \times N_2$.

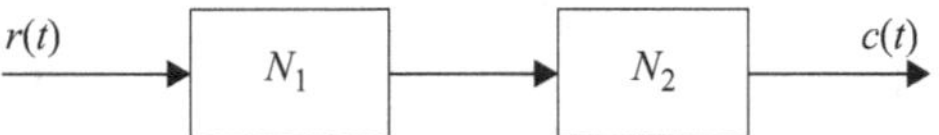

Figure 10.7 Block diagram of Exercise 10.4

Exercise 10.5 In the plant of Figure 10.8, the nonlinearity is a dead zone with the describing function $N = k - \frac{2k}{\pi}\left[\arcsin\left(\frac{A}{X}\right) + \left(\frac{A}{X}\right)\sqrt{1-\left(\frac{A}{X}\right)^2}\right]$. There is consequently an unstable limit cycle with amplitude X and frequency ω given approximately by:

A) $X = 2.235, \omega = 1.414$ rad/s
B) $X = 2.235, \omega = 0.707$ rad/s
C) $X = 2.235, \omega = 1.000$ rad/s
D) $X = 2.235, \omega = 2.000$ rad/s.

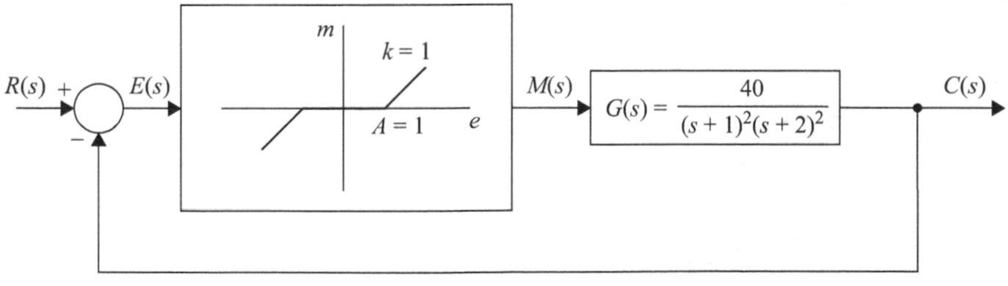

Figure 10.8 Block diagram of Exercise 10.5

Exercise 10.6 Consider the control system in Figure 10.9, where $N = \dfrac{4M}{\pi X}$ and $G(s) = \dfrac{K}{s(s+1)^2}$. It can be seen that there is a limit cycle with amplitude X and frequency ω given approximately by:

A) $X = \dfrac{4MK}{\pi}, \omega = \dfrac{1}{2}$ rad/s

B) $X = MK\pi, \omega = 2$ rad/s

C) $X = \dfrac{2MK}{\pi}, \omega = 1$ rad/s

D) None of the above.

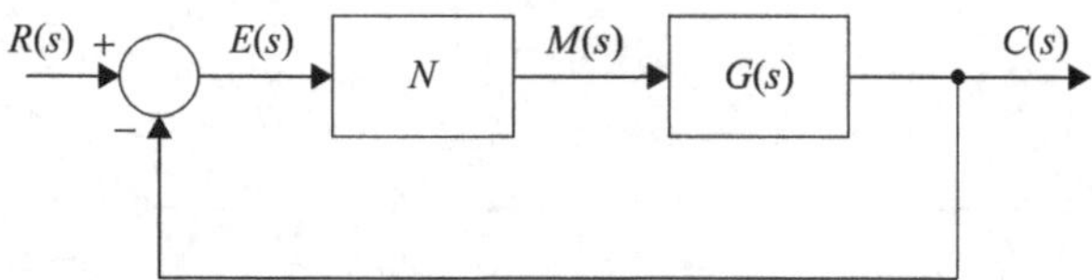

Figure 10.9 Block diagram of Exercise 10.6

Exercise 10.7 Consider the control system in Figure 10.10 where $N = \dfrac{4M}{\pi X}$ and $G(s) = \dfrac{K}{(s+1)^3}$. It can be seen that there is a limit cycle with amplitude X and frequency ω given approximately by:

A) $X = \dfrac{MK}{\pi}, \omega = \dfrac{1}{3}$ rad/s

B) $X = \dfrac{2MK}{\pi}, \omega = 3$ rad/s

C) $X = \dfrac{MK}{2\pi}, \omega = 3^{1/2}$ rad/s

D) None of the above.

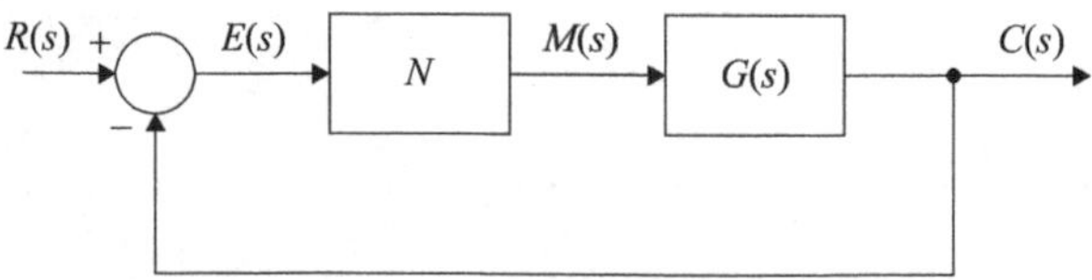

Figure 10.10 Block diagram of Exercise 10.7

Exercise 10.8 In the plant of Figure 10.11, N is the describing function of a relay with hysteresis nonlinearity, where $|N| = \dfrac{4B}{\pi X}$ and $\arg\{N\} = \arcsin\left(\dfrac{A}{X}\right)$. Let $G(s) = \dfrac{Ke^{-sT}}{s}$, $K = \pi, T = 1, A = 1$, and $B = \frac{1}{4}$. There is consequently a limit cycle with amplitude X and frequency ω given approximately by:

A) $X = 2.535, \omega = 0.394$ rad/s
B) $X = 0.394, \omega = 2.535$ rad/s
C) $X = 1.353, \omega = 0.739$ rad/s
D) $X = 0.739, \omega = 3.14$ rad/s.

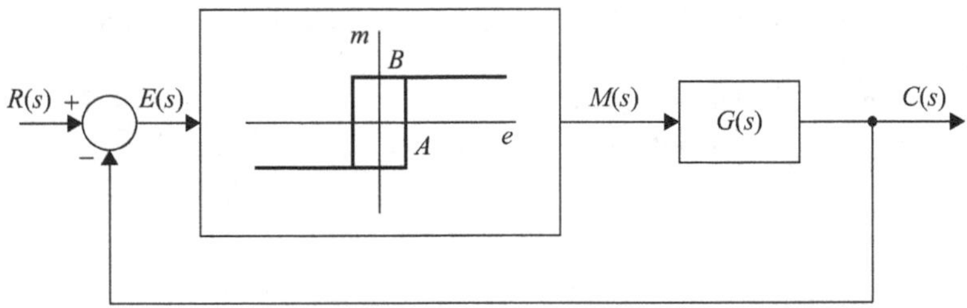

Figure 10.11 Block diagram of Exercise 10.8

Exercise 10.9 In the plant of Figure 10.12, $N = \dfrac{4M}{\pi X}$, $G(s) = \dfrac{K}{(s+1)^4}$, $M = \pi$ and $K = 2$. There is consequently a limit cycle with amplitude X and frequency ω given approximately by:

A) $X = 1/2, \ \omega = 1/2$ rad/s

B) $X = 2, \ \omega = 1$ rad/s

C) $X = 2\pi, \ \omega = 2^{\frac{1}{2}}$ rad/s

D) None of the above.

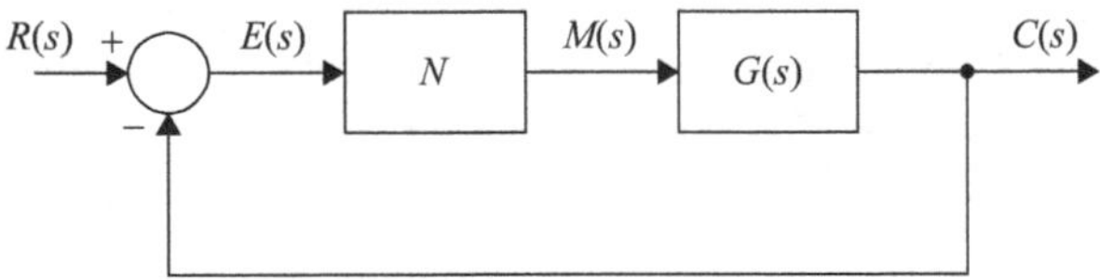

Figure 10.12 Block diagram of Exercise 10.9

Exercise 10.10 Consider the plant in Figure 10.13.

1. Let $R = 0, A = 4, k = 1$ and $K_1 = 20$. Find the frequency and the amplitude of the limit cycle.
2. Is this limit cycle stable or unstable?
3. Increasing gain K_1, how does the limit cycle change?

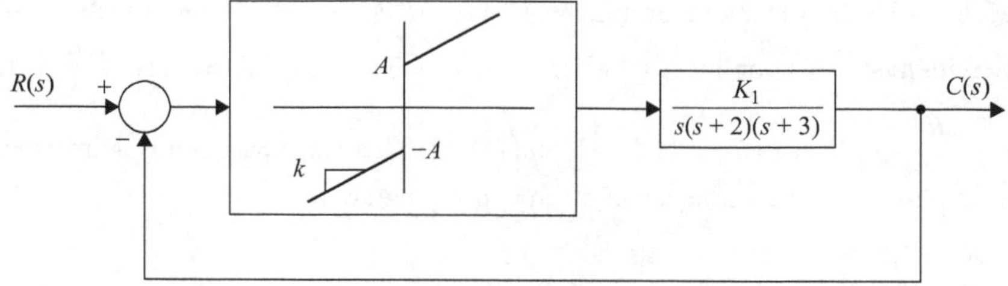

Figure 10.13 Block diagram of Exercise 10.10

Exercise 10.11 Consider the plant in Figure 10.14.

1. Let $R = 0, A = 4$ and $K_1 = 10$. Find the frequency and the amplitude of the limit cycle.
2. Is this limit cycle stable or unstable?
3. Can the limit cycle be eliminated changing gain K_1?

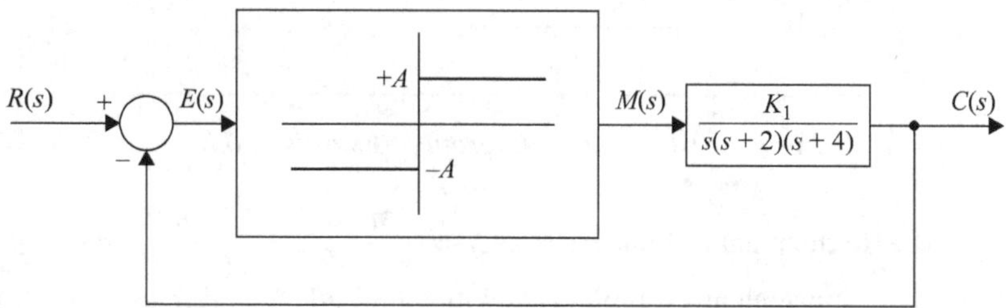

Figure 10.14 Block diagram of Exercise 10.11

Exercise 10.12 The control system in Figure 10.15a includes a nonlinear element N and a linear system with transfer function $G(s)$. Using the describing function method, the Nyquist plot in Figure 10.15b is obtained. Thus:

A) Only point A corresponds to a stable limit cycle.
B) Only point B corresponds to a stable limit cycle.
C) Neither point A nor point B corresponds to a stable limit cycle.
D) Both point A and point B correspond to a stable limit cycle.

Exercise 10.13 Consider the plant in Figure 10.16.

1. Let $R = 0$ and $M = 4$. Find the frequency and the amplitude of the limit cycle.
2. Is this limit cycle stable or unstable?

Exercise 10.14 Consider the plant in Figure 10.17.

1. Let $R = 0, A = 1$ and $K = 1$. Find the frequency and the amplitude of the limit cycle.
2. Is this limit cycle stable or unstable?

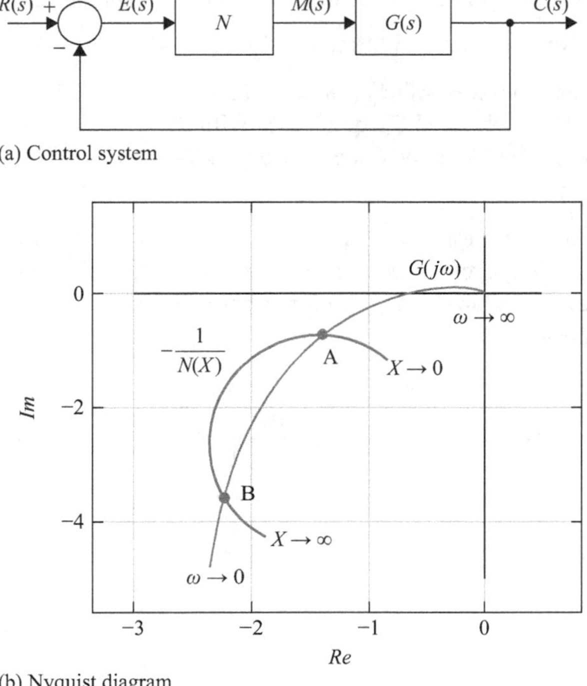

Figure 10.15 Block diagram and Nyquist plot of Exercise 10.12

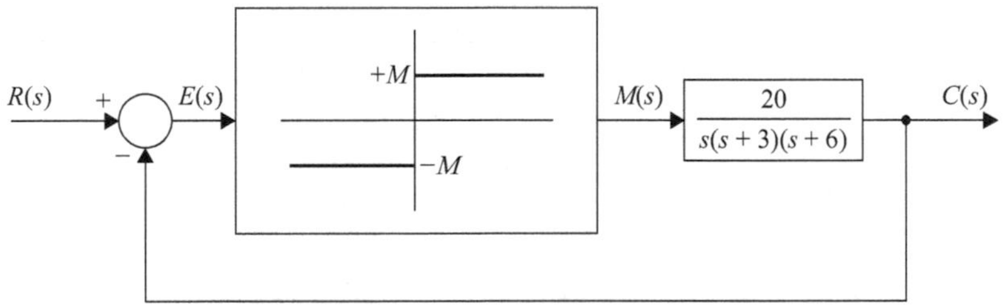

Figure 10.16 Block diagram of Exercise 10.13

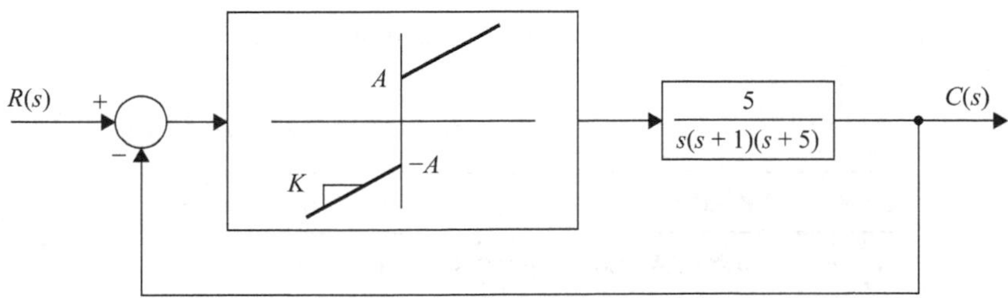

Figure 10.17 Block diagram of Exercise 10.14

Exercise 10.15 The control system in Figure 10.18a includes a nonlinear block $m = f(e)$ and a linear system with transfer function $G(s) = \frac{10}{s(s+1)(s+2)}$. Block $m = f(e)$ can assume two values: $m = e^3$ and $m = e^{1/3}$. The Nyquist plot in Figure 10.18b can be obtained for both cases with the describing function method. For the two cases, points A, B and C are shown as in Table 10.1. Given this, in point A, we have:

A) A stable limit cycle for $m = e^3$ and an unstable limit cycle for $m = e^{1/3}$.
B) An unstable limit cycle for $m = e^3$ and a stable limit cycle for $m = e^{1/3}$.
C) No limit cycle (stable or unstable) for either value of m.
D) None of the above.

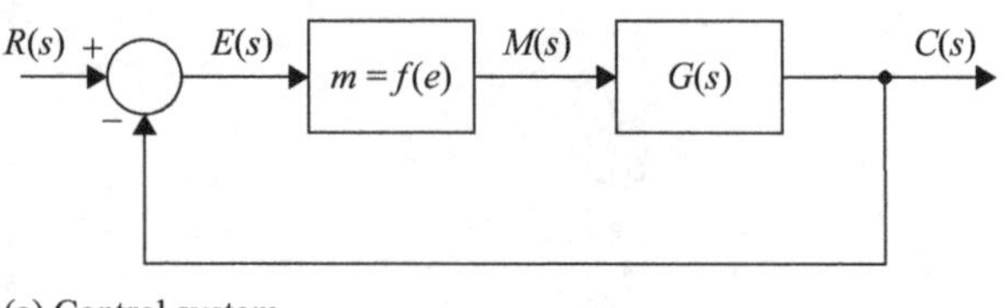

(a) Control system

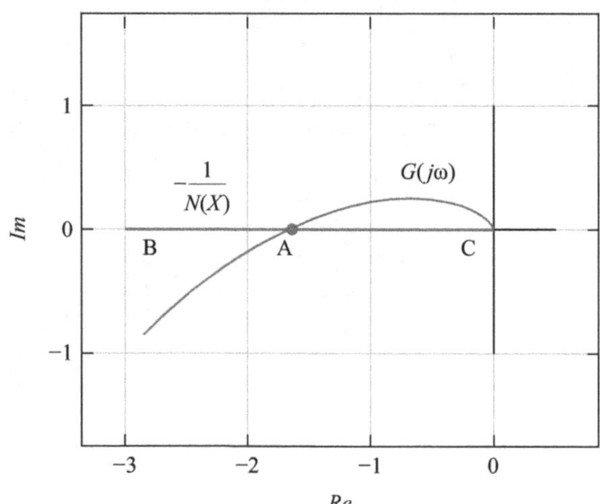

(b) Nyquist diagram

Figure 10.18 Block diagram and Nyquist plot of Exercise 10.15

Table 10.1 Behavior of points A, B and C in Exercise 10.15

Case	Point A	Point B	Point C
$m = e^3$	$\omega_A \approx 1.41$ rad/s, $X_A \approx 0.89$	$X_B \to 0$	$X_C \to \infty$
$m = e^{1/3}$	$\omega_A \approx 1.41$ rad/s, $X_A \approx 2.69$	$X_B \to \infty$	$X_C \to 0$

Exercise 10.16 The control system in Figure 10.19a comprises a nonlinear block (a saturation with $A = 1$ and a unit ramp) and a linear system with transfer function $G(s) = \dfrac{1}{s(s+1)}e^{-2s}$. The describing function method shows that in point P (seen in Figure 10.19b) we have $\omega_A \approx 0.54$ rad/s, $X_A \approx 2$. Thus:

A) Point P corresponds to a stable limit cycle.
B) Point P corresponds to an unstable limit cycle.
C) Point P does not correspond to a limit cycle (stable or unstable).
D) None of the above.

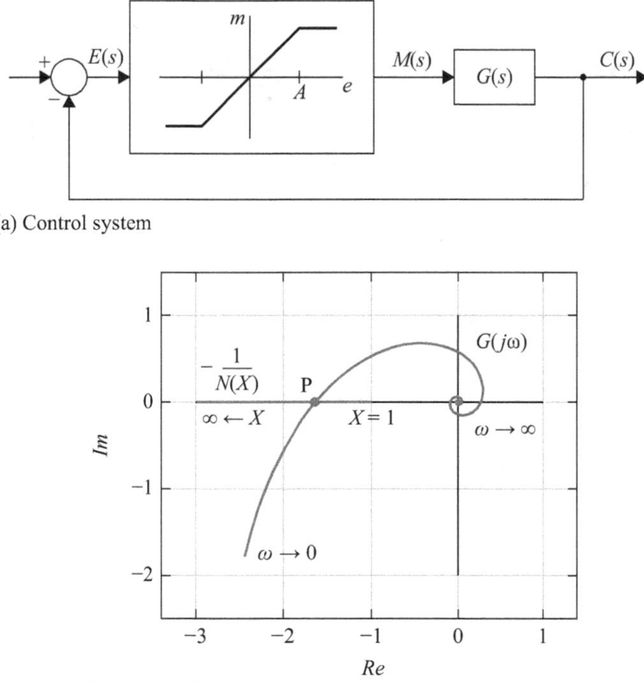

Figure 10.19 Block diagram and Nyquist diagram of Exercise 10.16

Exercise 10.17 The plant in Figure 10.20a includes a backlash nonlinearity with $A = 1$ and a unit slope ramp and a linear system with transfer function $G(s) = \dfrac{1.3}{s(s+1)^2}$. From the describing function method, it is known that in point P, $\omega_P \approx 0.68$ rad/s, $X_P \approx 3.1$, and that in point Q, $\omega_Q \approx 0.36$ rad/s, $X_Q \approx 1.34$ (see Figure 10.20b). Thus:

A) There is a stable limit cycle in P and an unstable limit cycle in Q.
B) There is an unstable limit cycle in P and a stable limit cycle in Q.
C) Points P and Q correspond to no limit cycle (stable or unstable).
D) None of the above.

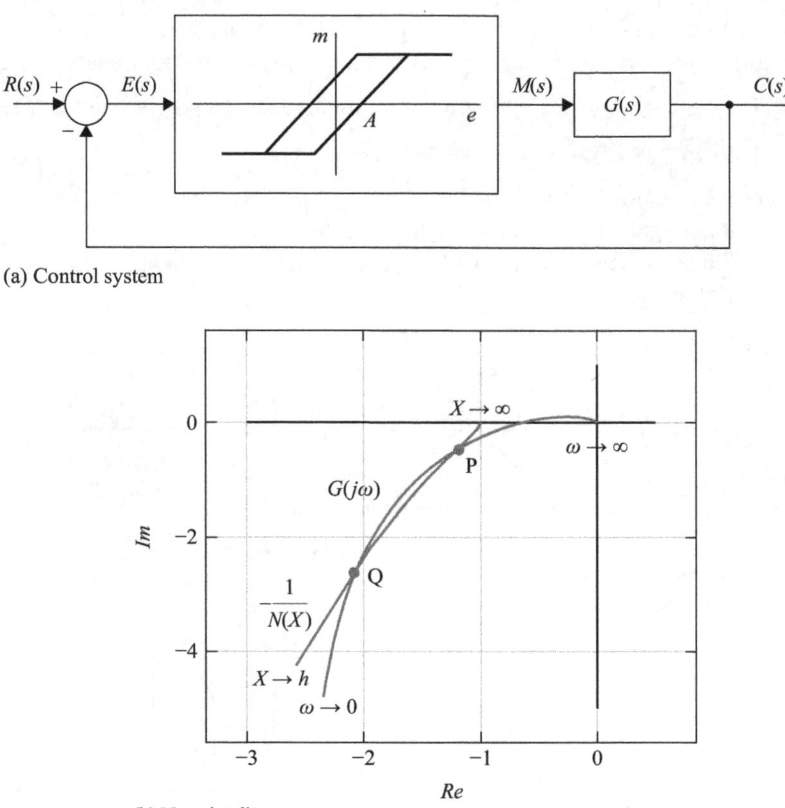

Figure 10.20 Block diagram and Nyquist diagram of Exercise 10.17

Exercise 10.18 Apply the describing function method to the plant in Figure 10.21. Then:

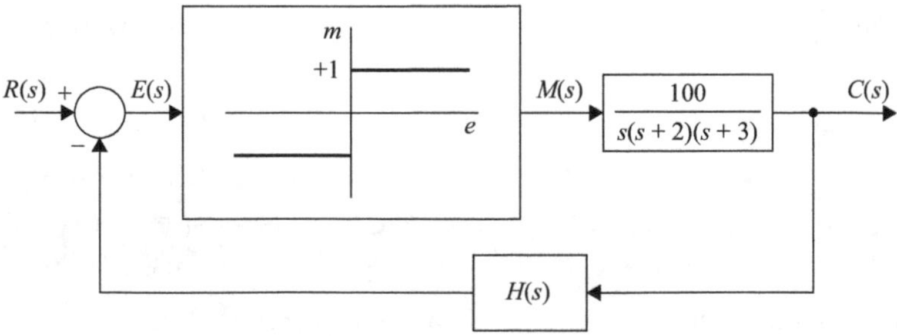

Figure 10.21 Block diagram of Exercise 10.18

1. For $H(s) = 1$:
   A) There is a limit cycle with frequency $\omega = \sqrt{6}$ rad/s.
   B) There is a limit cycle with frequency $\omega = \dfrac{40}{3\pi}$ rad/s.
   C) There is a limit cycle with frequency $\omega = \dfrac{40}{3\pi}\sqrt{6}$ rad/s.
   D) None of the above.
2. Again for $H(s) = 1$:
   E) There is a limit cycle with amplitude $X = \sqrt{6}$.
   F) There is a limit cycle with amplitude $X = \dfrac{40}{3\pi}$.
   G) There is a limit cycle with amplitude $X = \dfrac{40}{3\pi}\sqrt{6}$.
   H) None of the above.
3. For $H(s) = s + a$:
   I) There no longer is a limit cycle when $0 < a \leq 5$.
   J) There no longer is a limit cycle when $5 \leq a$.
   K) No limit cycle ever exists, irrespective of the value of a.
   L) There is always a limit cycle, irrespective of the value of a.

Exercise 10.19 Consider a control system comprising the nonlinearity in Figure 10.22, with the describing function $N(X)$ and a low-pass linear plant $G(s)$. The describing function is:

A) $N = \dfrac{B}{\pi X}$

B) $N = \dfrac{4B}{\pi X}\sqrt{1 - \left(\dfrac{A}{X}\right)^2}$

C) $N = \dfrac{4B}{\pi X} \arg\left(-\arcsin\left(\dfrac{A}{X}\right)\right)$

D) $N = A + \dfrac{4B}{\pi X}$.

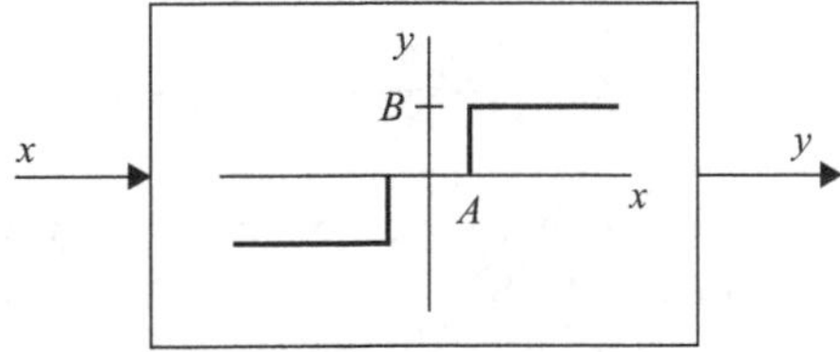

Figure 10.22 Nonlinearity of Exercise 10.19

Exercise 10.20 Consider a control system comprising the nonlinearity in Figure 10.23, with the describing function $N(X)$ and a low-pass linear plant $G(s)$. The describing function is:

A) $N = \dfrac{4B}{\pi X}$

B) $N = \dfrac{4B}{\pi X}\sqrt{1 - \left(\dfrac{A}{X}\right)^2}$

C) $N = \dfrac{4B}{\pi X} \arg\left(-\arcsin\left(\dfrac{A}{X}\right)\right)$

D) $N = A + \dfrac{4B}{\pi X}$.

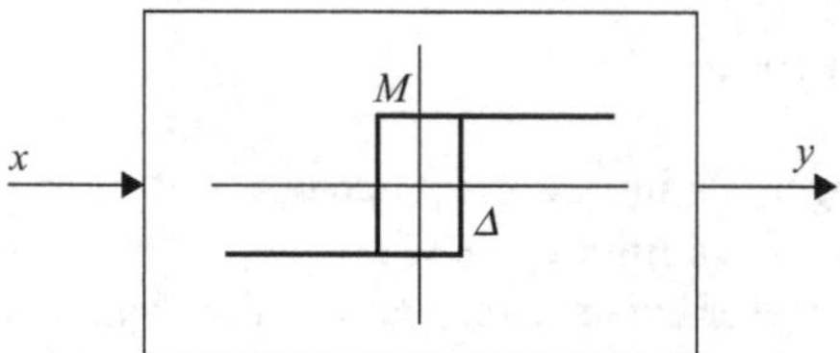

Figure 10.23 Nonlinearity of Exercise 10.20

Exercise 10.21 Figure 10.24 shows two nonlinear elements (a relay and a dead zone) in series. This is equivalent to a single system as follows:

A) a dead zone with $A = 0.5$
B) a relay with $M = 0.5$
C) a relay with $M = 1$
D) None of the above.

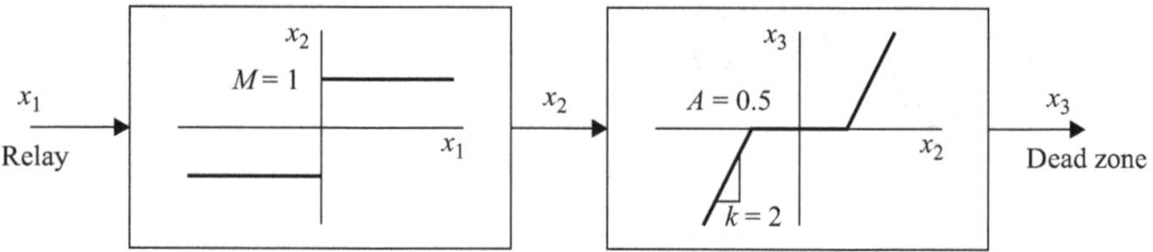

Figure 10.24 Block diagram of Exercise 10.21

Exercise 10.22 The two relay nonlinearities in series, seen in Figure 10.25, are equivalent to a single system, as seen in:

A) Figure 10.26
B) Figure 10.27
C) Figure 10.28
D) None of the above.

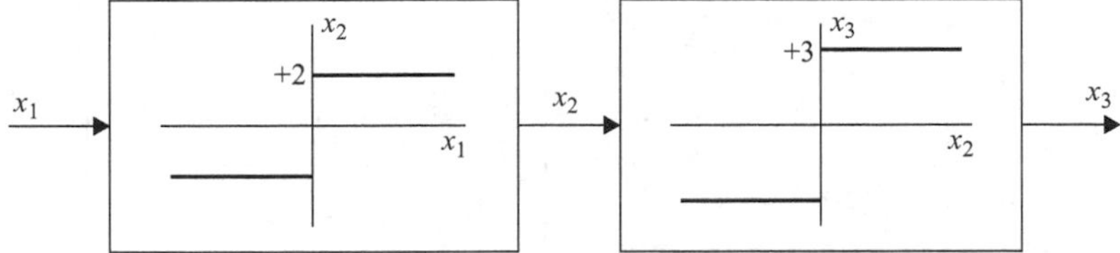

Figure 10.25 Block diagram of Exercise 10.22

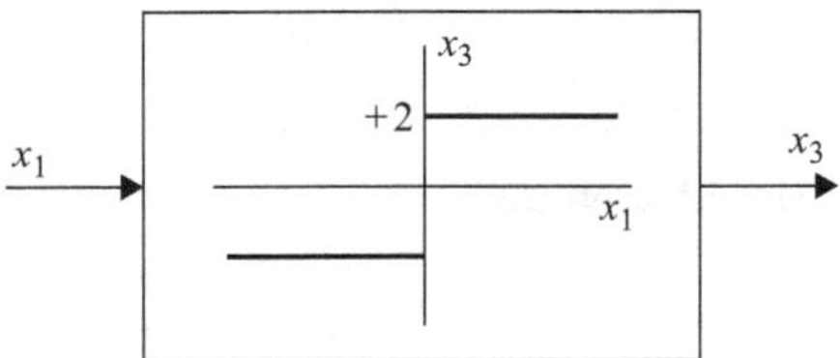

*Figure 10.26 Option **A)** of Exercise 10.22*

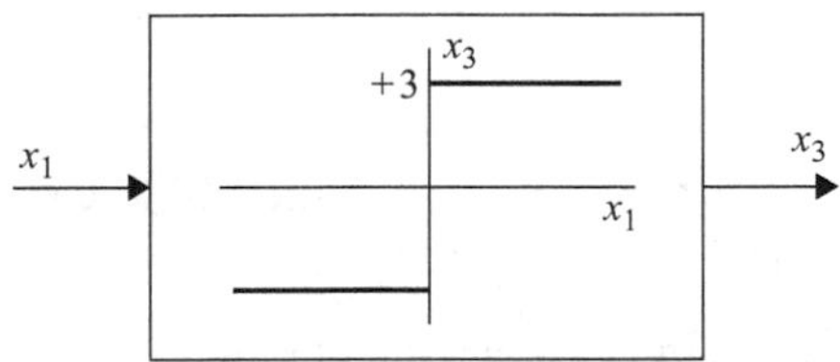

*Figure 10.27 Option **B)** of Exercise 10.22*

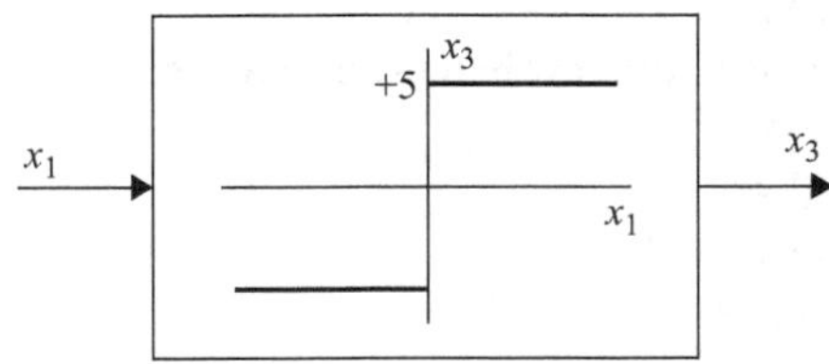

*Figure 10.28 Option **C)** of Exercise 10.22*

Exercise 10.23 Apply the describing function method to the plant in Figure 10.29. Then:

1. There is a limit cycle with frequency:
   A) $\omega = 5.66$ rad/s
   B) $\omega = 0.52$ rad/s
   C) $\omega = 2.50$ rad/s
   D) None of the above.

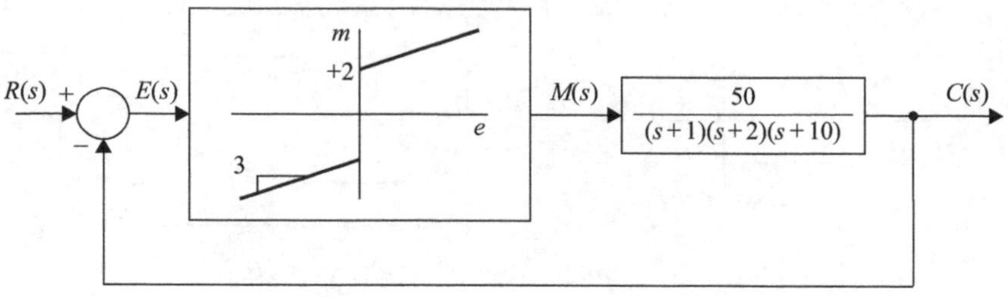

Figure 10.29 Block diagram of Exercise 10.23

2. There is a limit cycle with amplitude:
   E) $X = 5.66$
   F) $X = 0.52$
   G) $X = 2.50$
   H) None of the above.
3. The Nyquist plot shows that the limit cycle is:
   I) stable
   J) unstable.

Exercise 10.24 Apply the describing function method to the plant in Figure 10.30, with positive feedback and $0 \leq \zeta \leq 1$. Then:

1. For $\zeta = 0.105$ there is a limit cycle with amplitude:
   A) $X = 10.0$
   B) $X = 15.10$
   C) $X = 24.25$
   D) None of the above.
2. For $\zeta = 0.105$ there is a limit cycle with frequency:
   E) $\omega = 10.0$ rad/s
   F) $\omega = 15.10$ rad/s
   G) $\omega = 24.25$ rad/s
   H) None of the above.

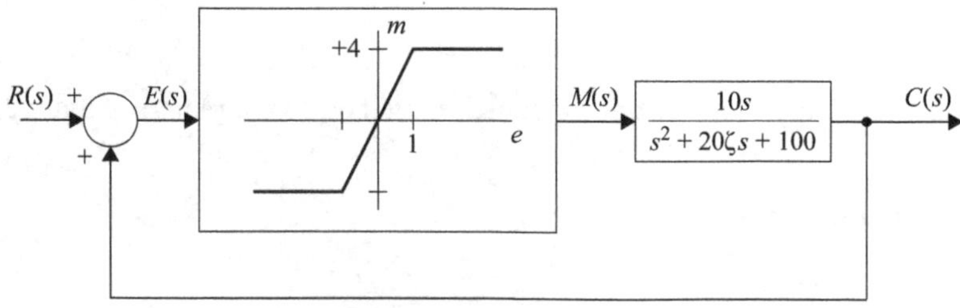

Figure 10.30 Block diagram of Exercise 10.24

3. Let $0 \leq \zeta \leq 1$. For each value of ζ there is a limit cycle with frequency ω and amplitude X as follows:
   I) Both ω and X do not depend on ζ.
   J) Both ω and X depend on ζ.
   K) ω does not depend on ζ, but X depends on ζ.
   L) X does not depend on ζ, but ω depends on ζ.

Exercise 10.25 Figure 10.31 shows two nonlinear elements (a relay and a dead zone) in parallel. This is equivalent to a single system as follows:

A) Figure 10.32
B) Figure 10.33
C) Figure 10.34
D) None of the above.

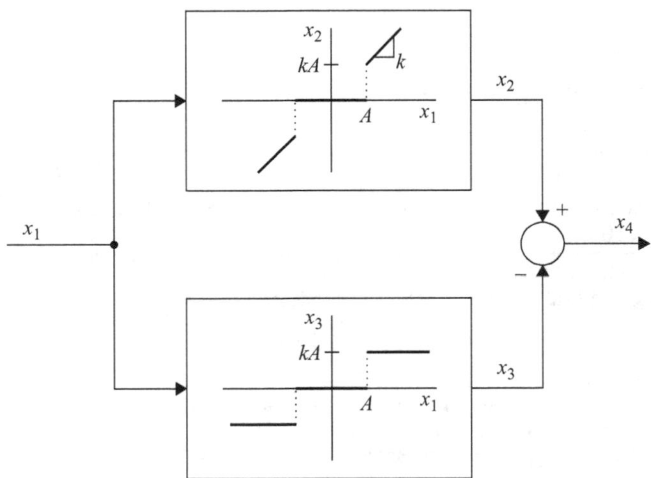

Figure 10.31 Block diagram of Exercise 10.25

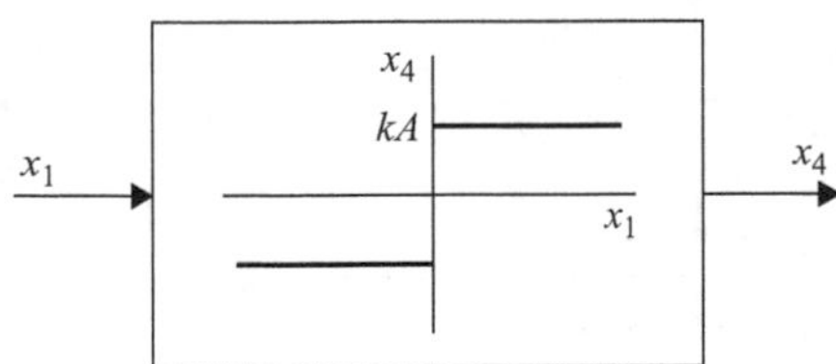

*Figure 10.32 Option **A)** of Exercise 10.25*

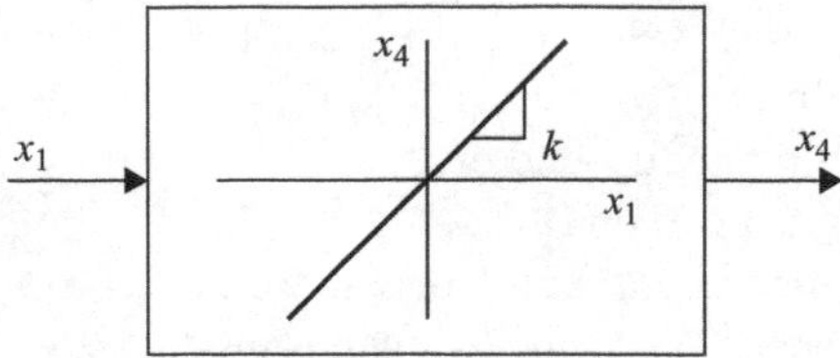

Figure 10.33 Option ***B)*** *of Exercise 10.25*

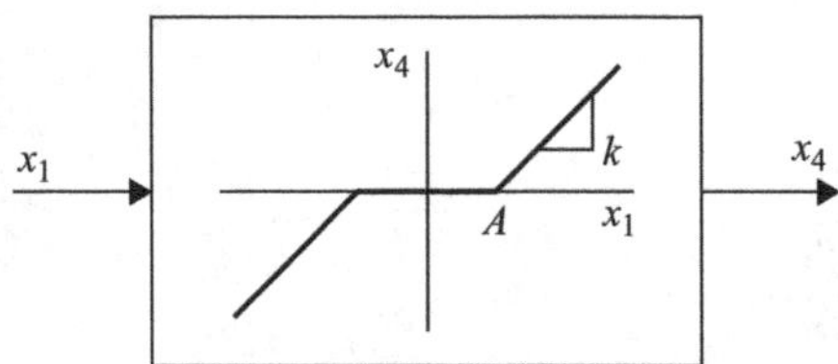

Figure 10.34 Option ***C)*** *of Exercise 10.25*

Exercise 10.26 In the control system of Figure 10.35a, the describing function of the nonlinear element (shown in Figure 10.35b) is $N(X) = \frac{2k}{\pi}\left[\arcsin\left(\frac{A}{X}\right) + \frac{A}{X}\sqrt{1-\frac{A^2}{X^2}}\right]$, for $X \geq A$, with slope $k = 2$ and $A = 1$. The transfer function of the plant is $G(s) = \frac{10}{s(s+1)(s+2)}$. Verify if there is a limit cycle. If there is, find its frequency and amplitude, and determine whether it is stable.

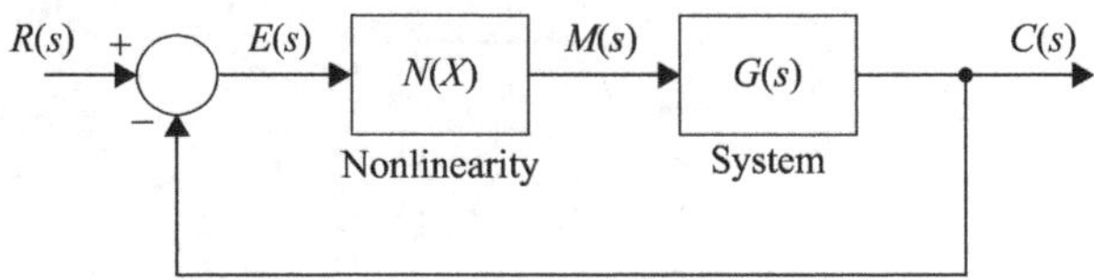

(a) Control system

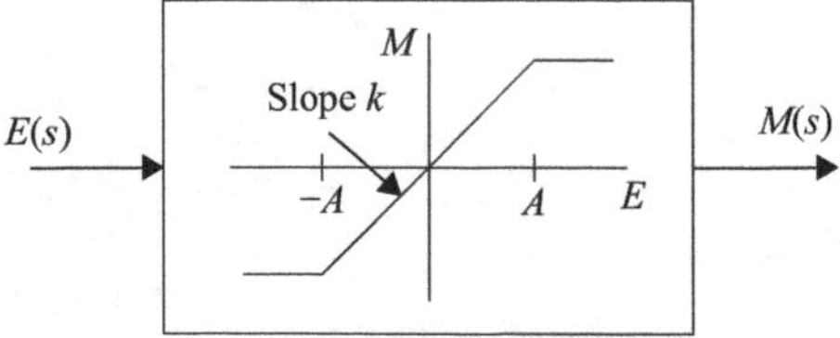

(b) Non linear element

Figure 10.35 Control system of Exercise 10.26

Exercise 10.27 The control system in Figure 10.36a comprises nonlinearity N and a linear plant with transfer function $G(s)$. The Nyquist plot in Figure 10.36b is obtained with the describing function method. Then:

A) Only point P corresponds to a stable limit cycle.
B) Only point Q corresponds to a stable limit cycle.
C) Neither point P nor point Q corresponds to a stable limit cycle.
D) Both point P and point Q correspond to a stable limit cycle.

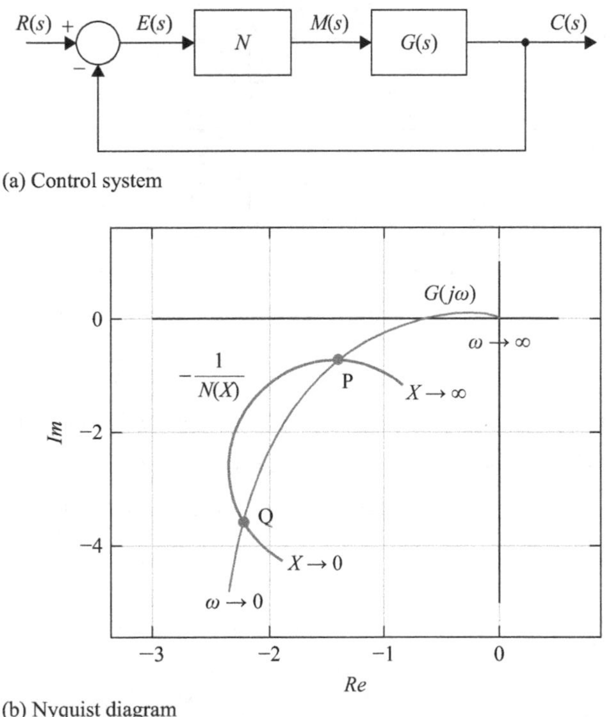

Figure 10.36 Block diagram and Nyquist plot of Exercise 10.27

Exercise 10.28 The plant in Figure 10.37 includes a backlash nonlinearity with $2A = 0.5$ and $B = 1$, $C = 1$. Let $G(s) = \dfrac{3}{s(s+1)^2}$. Use the describing function method to verify if this plant has a limit cycle. If it does, find its frequency and amplitude.

Exercise 10.29 Let $G(s) = \dfrac{e^{-2s}}{s(s+2)}$ in the plant of Figure 10.38. Use the describing function method to verify if this plant has a limit cycle. If it does, find its frequency and amplitude.

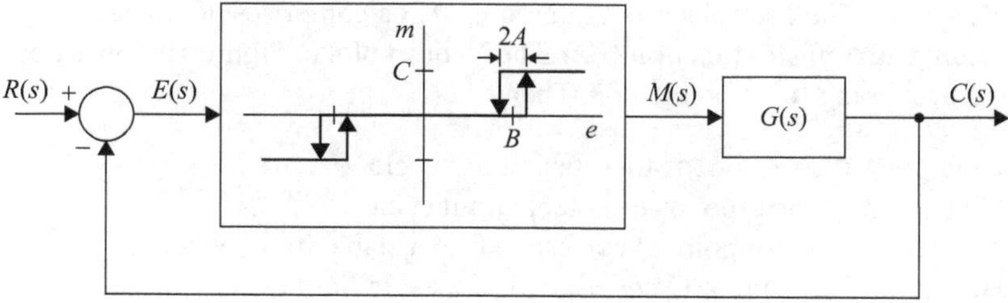

Figure 10.37 Block diagram of Exercise 10.28

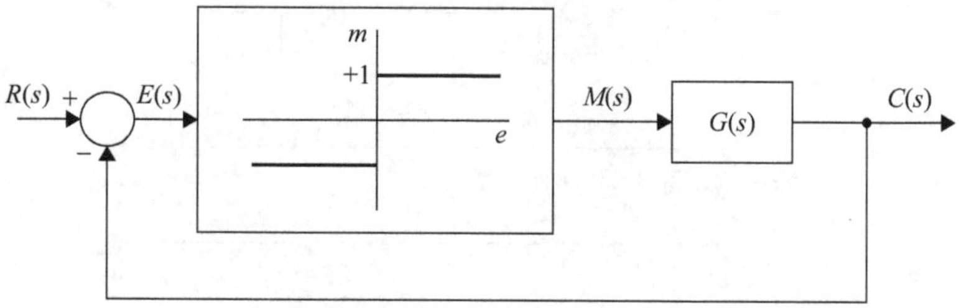

Figure 10.38 Block diagram of Exercise 10.29

Exercise 10.30 The plant in Figure 10.39 includes a backlash nonlinearity with $A = 1$ and $k = 1$. Let $G(s) = \dfrac{1.5}{s(s+1)^2}$. Use the describing function method to verify if this plant has a limit cycle. If it does, find its frequency and amplitude.

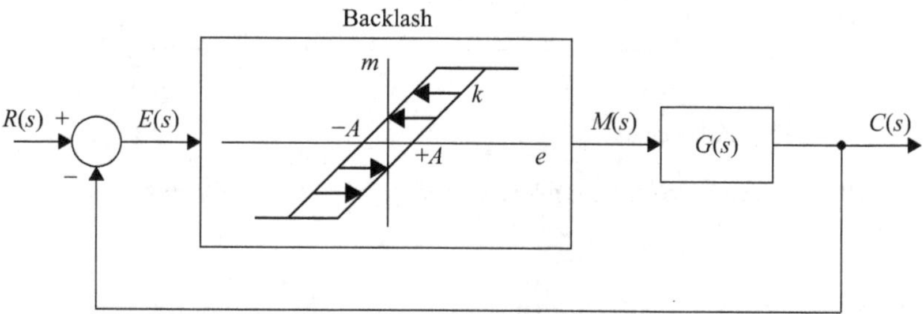

Figure 10.39 Block diagram of Exercise 10.30

Exercise 10.31 Let $G(s) = \dfrac{100}{s(s+2)(s+3)}$ in the plant of Figure 10.40. Use the describing function method to verify if this plant has a limit cycle. If it does, find its frequency and amplitude.

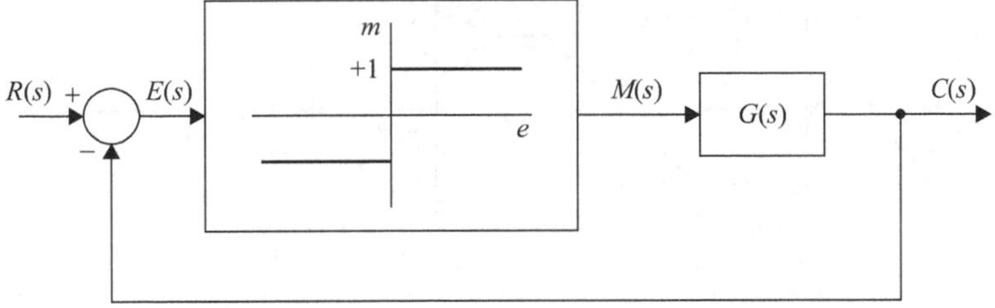

Figure 10.40 Nyquist plot and describing function of Exercise 10.31

Exercise 10.32 The control system in Figure 10.41 comprises a saturation nonlinearity with $A = 1$ and slope $k_N = 1$ and a linear plant with the transfer function $G(s) = \dfrac{K_G}{(s+1)^4}$. Use the describing function method to find:

1. approximate values for the amplitude X and the frequency ω of the limit cycle when $K_G = 10$;
2. the properties of the limit cycle (as seen in the Nyquist diagram);
3. the smallest K_G for which there no longer is a limit cycle.

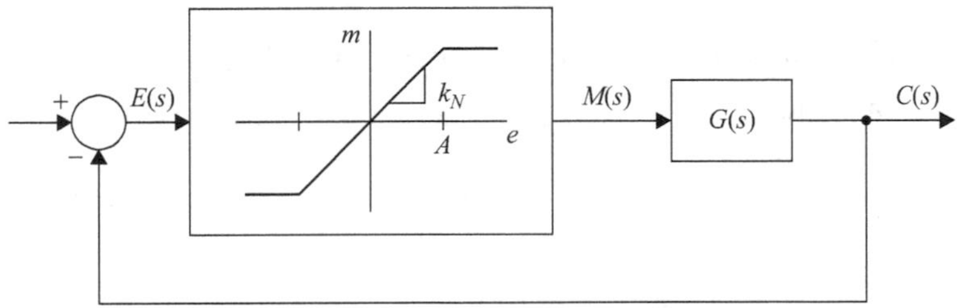

Figure 10.41 Block diagram of Exercise 10.32

Exercise 10.33 The control system in Figure 10.42 comprises a saturation nonlinearity with $A = 1$ and $k = 1$ and a linear plant with transfer function $G(s) = \dfrac{K_G(s+1)}{s^2(s+2)(s+3)}$. Use the describing function method to find:

1. approximate values for the amplitude X and the frequency ω of the limit cycle when $K_G = 10$;
2. the properties of the limit cycle (as seen in the Nyquist diagram);
3. the smallest K_G for which there no longer is a limit cycle.

Exercise 10.34 Consider the control system in Figure 10.43.

1. Use the describing function method to find approximate values for the amplitude X and the frequency ω of the limit cycle.
2. Use the Nyquist diagram to find whether the limit cycle is stable or unstable.

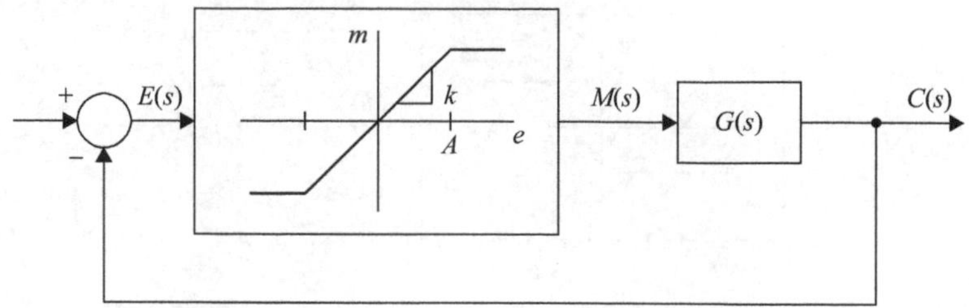

Figure 10.42 Block diagram of Exercise 10.33

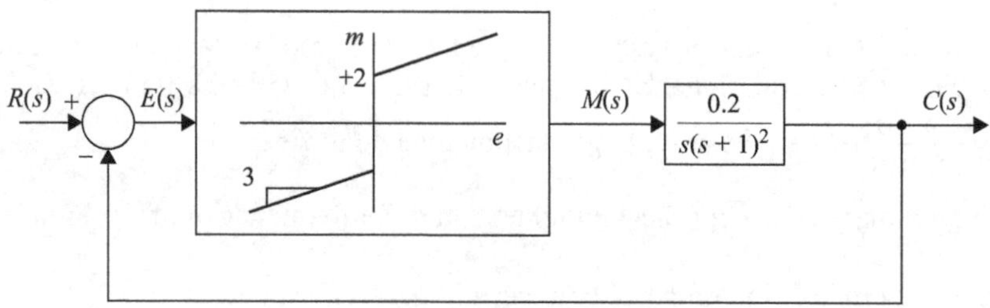

Figure 10.43 Block diagram of Exercise 10.34

Exercise 10.35 Consider the control system in Figure 10.44.

1. Use the describing function method to find approximate values for the amplitude X and the frequency ω of the limit cycle.
2. Use the Nyquist diagram to find whether the limit cycle is stable or unstable.

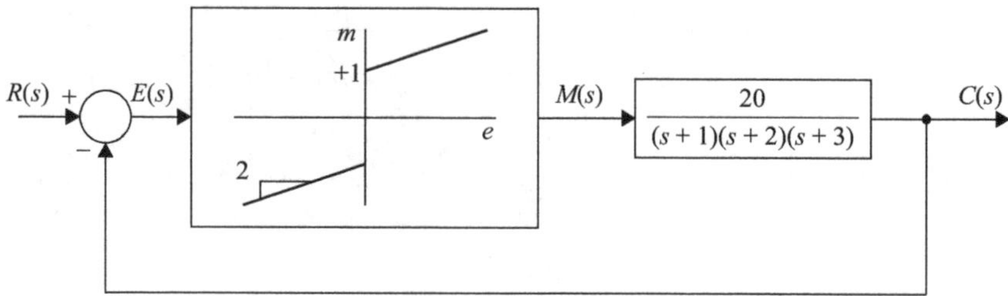

Figure 10.44 Block diagram of Exercise 10.35

10.4 Describing function method using computer packages

In this section, we consider Exercise 10.17, involving a nonlinear block with backlash and the linear system $G(s) = \frac{1.3}{s(s+1)^2}$. We analyze the nonlinear system in the perspective of the describing function method using the computer packages MATLAB®, SCILAB™ and OCTAVE©.

10.4.1 MATLAB

This subsection describes some basic commands that can be adopted with the package MATLAB.

The dynamical system is represented by means of the transfer function and the describing function.

The complete code is as follows.

```
% Describing function method

% G(jw), transfer function
kmax = 100; % number of points
d1   = -1; % wmin = 10^d1
d2 = 1; % wmax = 10^d2
w = 1i*logspace(d1,d2,kmax);

G = 1.3./(w.*(w+1).^2);

ReG = real(G);
ImG = imag(G);

% N(X), describing function
rmax = 100; % number of points

% Parameters of backlash
h = 1; % width
k = 1; % slope
Xmin = h*1.15; % minimum amplitude of X
Xmax = 100; % maximum amplitude of X

X = linspace(Xmin,Xmax,rmax);

aux1 = 2*h-X;
aux2 = 2*(h.*(X-h)).^0.5;
N_im = -1i*4*k*h.*(X-h)./(pi*X.^2);
N_re = k/2*(1-2/pi*(atan2(aux1,aux2)+aux1.*aux2./X.^2));
N = N_re+N_im;
N1 = -1./N;

ReN1 = real(N1);
ImN1 = imag(N1);
```

```
% Plots, real versus imaginary
figure
plot(ReG,ImG)
hold on
plot(ReN1,ImN1,'r')

% Plot formatting
plot(-1.2131,-0.47806,'o')
text(-2,0.5,'X = 1.2, w = 0.29, stable limit cycle')

plot(-2.1841,-3.292,'o')
text(-2.2,-4,'X = 5.5, w = 0.81, unstable limit cycle')

text(-2.7,-3.4,'-1/N(X)')
text(-2.4,-10,'G(jw)')

title('Describing function method')
xlabel('Re')
ylabel('Im')
```

MATLAB creates the figure window represented in Figure 10.45.

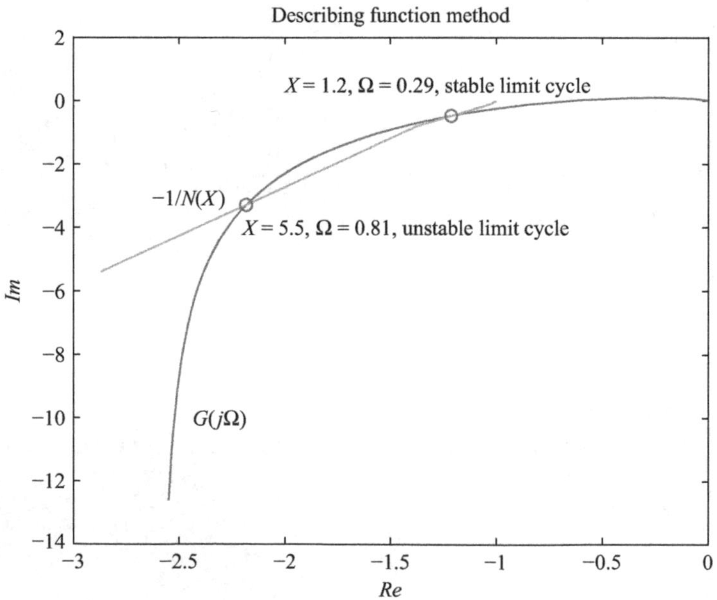

Figure 10.45 Plot of $G(j\omega)$ *and* $-\frac{1}{N(X)}$ *in the complex plane using MATLAB*

10.4.2 SCILAB

This subsection describes some basic commands that can be adopted with the package SCILAB.

The dynamical system is represented by means of the transfer function and the describing function.

The complete code is as follows.

```
//// Describing function method

// G(jw), transfer function
kmax = 100; // number of points
d1   = -1; // wmin = 10^d1
d2 = 1; // wmax = 10^d2
w = %i*logspace(d1,d2,kmax);

G = 1.3./(w.*(w+1).^2);

ReG = real(G);
ImG = imag(G);

// N(X), describing function
rmax = 100; // number of points

// Parameters of backlash
h = 1; // width
k = 1; // slope
Xmin = h*1.15; // minimum amplitude of X
Xmax = 100; // maximum amplitude of X

X = linspace(Xmin,Xmax,rmax);

aux1 = 2*h-X;
aux2 = 2*(h.*(X-h)).^0.5;
N_im = -%i*4*k*h.*(X-h)./(%pi*X.^2);
N_re = k/2*(1-2/%pi*(atan(aux1,aux2)+aux1.*aux2./X.^2));
N = N_re+N_im;
N1 = -1 ./N;

ReN1 = real(N1);
ImN1 = imag(N1);

// Plots, real versus imaginary
//figure
```

```
plot(ReG,ImG)
//hold on
set(gca(),"auto_clear","off")
plot(ReN1,ImN1,'r')

// Plot formatting
plot(-1.2131,-0.47806,"o")
xstring(-2,0.5,"X = 1.2, w = 0.29, stable limit cycle")

plot(-2.1841,-3.292,"o")
xstring(-2.1,-4,"X = 5.5, w = 0.81, unstable limit cycle")

xstring(-2.7,-3.4,"-1/N(X)")
xstring(-2.4,-10,"G(jw)")

title("Describing function method")
xlabel("Re")
ylabel("Im")
```

SCILAB creates the figure window represented in Figure 10.46.

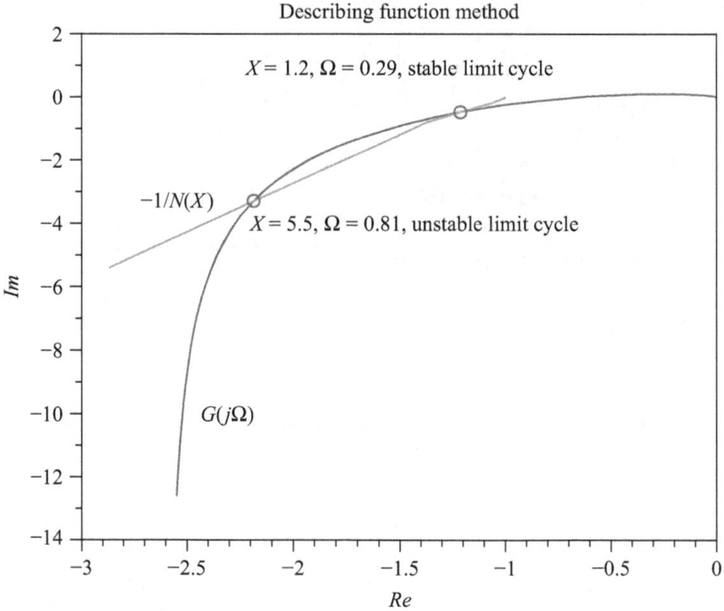

Figure 10.46 Plot of $G(j\omega)$ and $-\frac{1}{N(X)}$ in the complex plane using SCILAB

10.4.3 OCTAVE

This subsection describes some basic commands that can be adopted with the package OCTAVE. It is required to load package `control`.

The dynamical system is represented by means of the transfer function and the describing function.

The complete code is as follows.

```
% Describing function method

% G(jw), transfer function
kmax = 100; % number of points
d1  = -1; % wmin = 10^d1
d2 = 1; % wmax = 10^d2
w = 1i*logspace(d1,d2,kmax);

G = 1.3./(w.*(w+1).^2);

ReG = real(G);
ImG = imag(G);

% N(X), describing function
rmax = 100; % number of points

% Parameters of backlash
h = 1; % width
k = 1; % slope
Xmin = h*1.15; % minimum amplitude of X
Xmax = 100; % maximum amplitude of X

X = linspace(Xmin,Xmax,rmax);

aux1 = 2*h-X;
aux2 = 2*(h.*(X-h)).^0.5;
N_im = -1i*4*k*h.*(X-h)./(pi*X.^2);
N_re = k/2*(1-2/pi*(atan2(aux1,aux2)+aux1.*aux2./X.^2));
N = N_re+N_im;
N1 = -1./N;

ReN1 = real(N1);
ImN1 = imag(N1);
```

```
% Plots, real versus imaginary
figure
plot(ReG,ImG)
hold on
plot(ReN1,ImN1,'r')
% hold on

% Plot formatting
plot(-1.2131,-0.47806,'o')
text(-2,0.5,'X = 1.2, w = 0.29, stable limit cycle')
% hold on
plot(-2.1841,-3.292,'o')
text(-2.2,-4,'X = 5.5, w = 0.81, unstable limit cycle')

text(-2.7,-3.4,'-1/N(X)')
text(-2.4,-10,'G(jw)')

title('Describing function method')
xlabel('Re')
ylabel('Im')
```

OCTAVE creates the figure window represented in Figure 10.47.

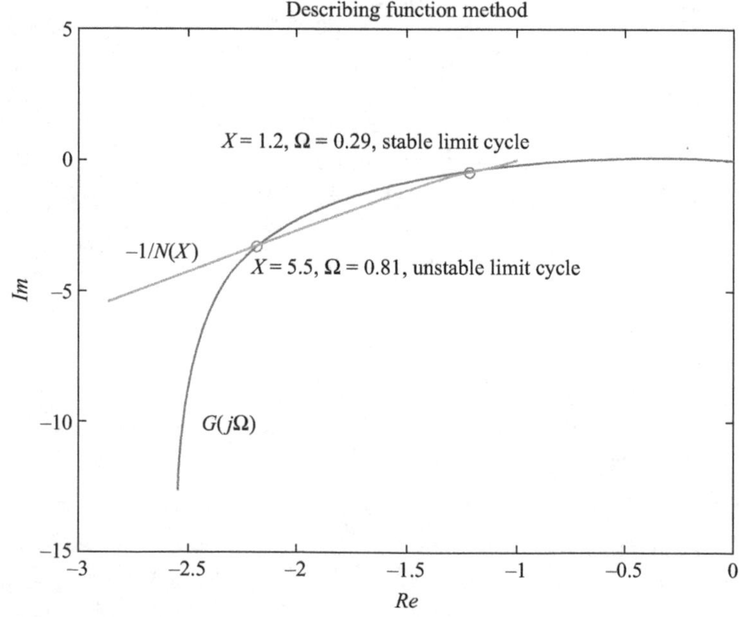

Figure 10.47 Plot of $G(j\omega)$ and $-\frac{1}{N(X)}$ in the complex plane using OCTAVE

Chapter 11
Analysis of nonlinear systems with the phase plane method

11.1 Fundamentals

Consider the second-order system:

$$\ddot{x} + f(x, \dot{x}) = 0 \tag{11.1}$$

The system response can be represented graphically by the locus of $\dot{x}(t)$ versus $x(t)$, that is, parametrized in t. The pair $\{x(t), \dot{x}(t)\}$ corresponds to the coordinates of a point in the so-called phase plane (PP). As time varies in the interval $t \in [0, \infty[$, this point describes a PP trajectory. A family of PP trajectories is called a phase portrait. By means of the PP technique, we can analyze the time response of linear and nonlinear second-order systems to general input functions [28,33–35].

11.1.1 List of symbols

f_1, f_2	analytic functions in the state variables
t	time
x_1, x_2, or $(x, \dot{x})$	state-space variables of a second-order system
λ_1, λ_2	roots of the characteristic equation

11.1.2 Phase plane method preliminaries

The PP method was introduced by Poincaré to determine graphically the solution of the second-order system:

$$\begin{cases} \dfrac{dx_1}{dt} = f_1(x_1, x_2) \\ \dfrac{dx_2}{dt} = f_2(x_1, x_2) \end{cases} \tag{11.2}$$

where $\{x_1, x_2\}$ represent the states of the system and $\{f_1, f_2\}$ are linear, or nonlinear, functions of the states.

A major class of second-order systems can be represented in the following form:

$$\begin{cases} \dfrac{dx_1}{dt} = x_2 \\ \dfrac{dx_2}{dt} = f(x_1, x_2) \end{cases} \tag{11.3}$$

that yields expression (11.1) for $x_1 = x$ and $x_2 = \dot{x}$.

Equation (11.2) can be written as the following ratio:

$$\frac{dx_2}{dx_1} = \frac{f_2(x_1, x_2)}{f_1(x_1, x_2)} \tag{11.4}$$

that has for solution $x_2 = \phi(x_1)$.

As that solution is unique, except at the singular points, $\dot{x} = 0$, or

$$\begin{cases} f_1(x_1, x_2) = 0 \\ f_2(x_1, x_2) = 0 \end{cases} \tag{11.5}$$

then the trajectories in the PP do not intersect each other, except at the singular points, where an infinite number of trajectories can converge, or diverge.

For constructing the PP trajectories, we can use:

- analytical methods
- numerical methods
- experimental methods

The analytical methods rely on the integration of the system differential equations to obtain $x_2 = \phi(x_1)$. Often we can also explore the symmetries exhibited by the trajectories in the PP to complete the graph.

The dynamical system $\ddot{x} + f(x, \dot{x}) = 0$ can be written as $\frac{d\dot{x}}{dx} = -\frac{f(x,\dot{x})}{\dot{x}}$, where the symmetries are:

- symmetry about the x-axis: $f(x, \dot{x}) = f(x, -\dot{x})$.
- symmetry about the $\dot{x}$-axis: $f(x, \dot{x}) = -f(-x, \dot{x})$.
- symmetry about the origin: $f(x, \dot{x}) = -f(-x, -\dot{x})$.

11.1.3 Singular points

Suppose that for the system (11.2) we have:

- $f_1(x_1, x_2)$ and $f_2(x_1, x_2)$ are analytic in the variables x_1 and x_2;
- point $(0, 0)$ is a singular, or equilibrium, point, so that $f_1(0, 0) = 0$ and $f_2(0, 0) = 0$.

Expanding $f_1(x_1, x_2)$ and $f_2(x_1, x_2)$ in a Taylor's series, we obtain:

$$\begin{cases} \dfrac{dx_1}{dt} = a_1 x_1 + b_1 x_2 + a_{11} x_1^2 + a_{12} x_1 x_2 + a_{22} x_2^2 + \cdots \\ \dfrac{dx_2}{dt} = a_2 x_1 + b_2 x_2 + b_{11} x_1^2 + b_{12} x_1 x_2 + b_{22} x_2^2 + \cdots \end{cases} \tag{11.6}$$

if we neglect the nonlinear terms, then we obtain:

$$\begin{cases} \dfrac{dx_1}{dt} = a_1x_1 + b_1x_2 \\ \dfrac{dx_2}{dt} = a_2x_1 + b_2x_2 \end{cases} \tag{11.7}$$

that is equivalent to:

$$\ddot{x} + a\dot{x} + bx = 0 \tag{11.8}$$

where $a = -a_1 - b_2$ and $b = a_1b_2 - a_2b_1$. Therefore, we have:

- if the roots of the linear characteristic equation have negative real part, then $x \to 0$ when $t \to \infty$;
- if at least one root is zero, then the linear characteristic equation is not sufficient to extract conclusions.

For a characteristic equation of the form $\lambda^2 + a\lambda + b = 0$ $(b \neq 0)$, the locus of the roots $\{\lambda_1, \lambda_2\}$ in the s-plane determines the type of singular point (see Appendix A).

11.1.4 Limit cycles

A limit cycle corresponds to an isolated closed curve in the PP. A closed trajectory means that the motion has a periodic nature. An isolated trajectory indicates the limiting behavior of the cycle, with nearby trajectories possibly converging to, or diverging from, it.

Limit cycles can be classified as Figure 11.1:

- stable—trajectories converge to the limit cycle
- unstable—trajectories diverge from the limit cycle
- semistable—some trajectories converge and others diverge

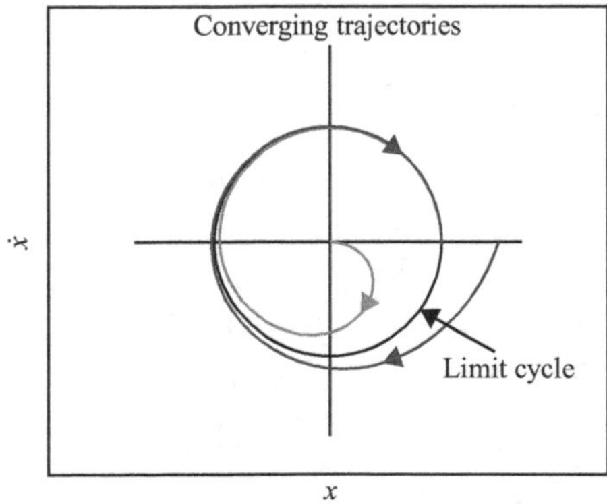

Figure 11.1 Limit cycles in the PP

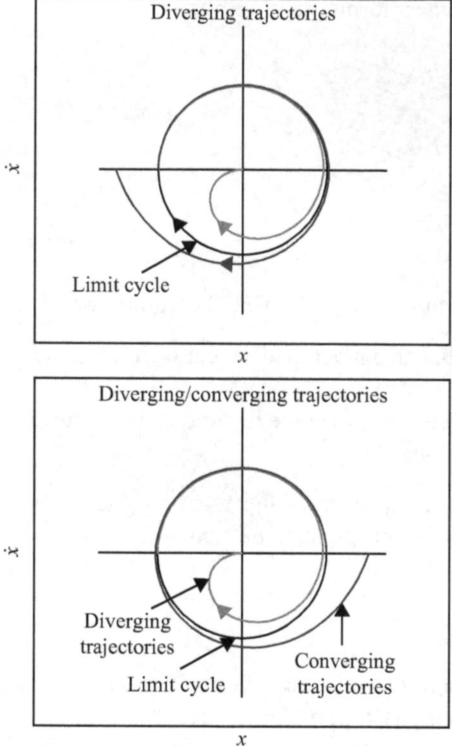

Figure 11.1 (Continued)

11.2 Solved problems

Problem 11.1 Consider the plant $\ddot{x} + \cos(x)\dot{x} + x = 0$. Figure 11.2 shows its representation on phase plane $(x, \dot{x})$.

1. Point A is:
   A) a stable focus.
   B) an unstable focus.
   C) a saddle point.
   D) a center.
2. Point B is:
   E) a stable limit cycle.
   F) an unstable limit cycle.
   G) a semistable limit cycle.
   H) none of the above.
3. Point C is:
   I) a stable limit cycle.
   J) an unstable limit cycle.

K) a semistable limit cycle.
L) none of the above.

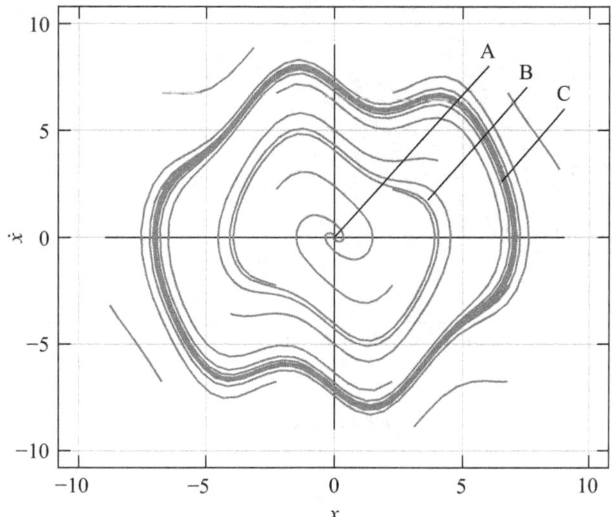

Figure 11.2 Phase plane of Problem 11.1

Resolution The correct answers are options **A)**, **F)** and **I)**.

Problem 11.2 Consider the nonlinear plant $\begin{bmatrix} \dot{x}_1 \\ \dot{x}_2 \end{bmatrix} = \begin{bmatrix} -1 & 2 \\ -2 & -1 \end{bmatrix} \begin{bmatrix} x_1 \\ x_2 \end{bmatrix} + \begin{bmatrix} x_1^2 \\ 0 \end{bmatrix}$.

1. Find its equilibrium points.
2. Linearize it around $(x_1, x_2) = (0, 0)$.
3. Analyze the type of this equilibrium point.
4. From its state plane (x_1, x_2) shown in Figure 11.3, find the plant's singular points, limit cycles and zones of stability and instability.

Resolution The singular points are $x_1 = 0$, $x_2 = 0$ and $x_1 = 5$, $x_2 = -10$.
Linearizing around the origin,

$$\begin{bmatrix} \dot{x}_1 \\ \dot{x}_2 \end{bmatrix} = \begin{bmatrix} -1 & 2 \\ -2 & -1 \end{bmatrix} \begin{bmatrix} x_1 \\ x_2 \end{bmatrix} \tag{11.9}$$

The characteristic equation is $(\lambda + 1)(\lambda + 1) + 4 = 0 \Leftrightarrow \lambda = -1 \pm 4j$; this is a stable focus.

In Figure 11.4, A is an unstable zone, B is a limit cycle, C is a stable zone and D is a singular point.

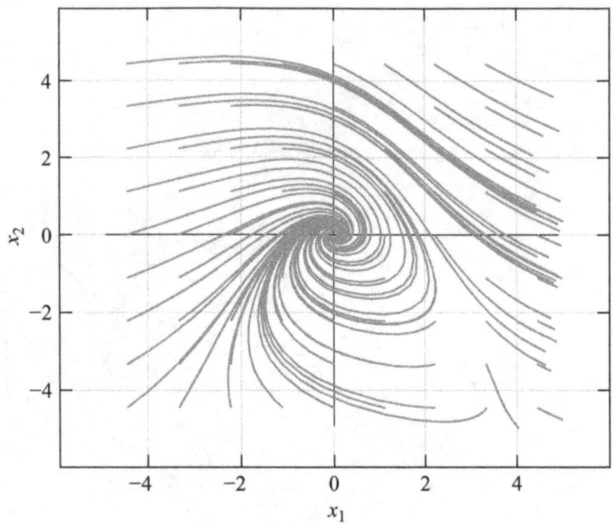

Figure 11.3 Phase plane of Problem 11.2

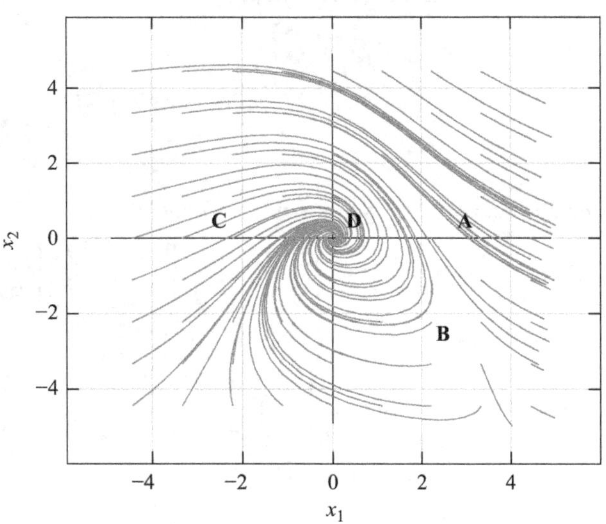

Figure 11.4 Solution of Problem 11.2

Problem 11.3 Consider a plant described by the nonlinear differential equation $\frac{d^2x}{dt^2} + \frac{dx}{dt} + x \cdot (1 + x) = 0$.

1. Are there symmetries in the phase plane?
2. Find the singular points on the phase plane.
3. Analyze how the plant behaves in a neighborhood of those singular points.

Resolution Let $x = x_1$, $\dot{x} = \dot{x}_1 = \frac{dx_1}{dt} = x_2$, and thus $\ddot{x} = -\dot{x} - x - x^2 \Leftrightarrow \frac{dx_2}{dt} = -\dot{x}_1 - x_1 - x_1^2$. Consequently,

$$-\frac{\dot{x}_1 + x_1 + x_1^2}{\dot{x}_1} = \frac{\frac{dx_2}{dt}}{\frac{dx_1}{dt}} = \frac{dx_2}{dx_1} = \frac{d\dot{x}_1}{dx_1} \tag{11.10}$$

Our plant is now in the form $\frac{d\dot{x}_1}{dx_1} = -\frac{f(x_1,\dot{x}_1)}{\dot{x}_1}$ with $f(x_1,\dot{x}_1) = \dot{x}_1 + x_1 + x_1^2$; and since $f(x_1,\dot{x}_1) \neq f(x_1,-\dot{x}_1)$ and $f(x_1,\dot{x}_1) \neq -f(-x_1,\dot{x}_1)$, there are no symmetries in the phase plane.

Singular points are found as

$$\begin{cases} \dot{x}_1 = 0 \\ \dot{x}_1 + x_1 + x_1^2 = 0 \end{cases} \Rightarrow \begin{cases} \dot{x}_1 = 0 \\ x_1 = 0 \end{cases} \vee \begin{cases} \dot{x}_1 = 0 \\ x_1 = -1 \end{cases} \tag{11.11}$$

Linearizing around $(x_1,\dot{x}_1) = (0,0)$, we get $\ddot{x} + \dot{x} + x = 0$ and the characteristic equation is $\lambda^2 + \lambda + 1 = 0 \Rightarrow \lambda = -\frac{1}{2} \pm j\frac{\sqrt{3}}{2}$, so this is a stable focus.

Around $(x_1,\dot{x}_1) = (-1,0)$, we use the variable transform $x = y - 1 \Rightarrow \dot{x} = \dot{y} \Rightarrow \ddot{x} = \ddot{y}$ and get $\ddot{y} + \dot{y} + (y-1) + (y-1)^2 = \ddot{y} + \dot{y} + y^2 - y = 0$, linearized as $\ddot{y} + \dot{y} - y = 0$. The characteristic equation is $\lambda^2 + \lambda - 1 = 0 \Rightarrow \lambda = \frac{-1\pm\sqrt{5}}{2}$, so this is a saddle point.

Problem 11.4 Consider a plant described by the nonlinear differential equation $\frac{d^2x}{dt^2} + \frac{dx}{dt} + \cos(x) = 0$.

1. Are there symmetries in the phase plane?
2. Find singular points on the phase plane.
3. Analyze how the plant behaves in a neighborhood of those singular points.

Resolution Let $x = x_1$, $\dot{x} = \dot{x}_1 = \frac{dx_1}{dt} = x_2$, and thus $\ddot{x} = -\dot{x} - \cos x \Leftrightarrow \frac{dx_2}{dt} = -\dot{x}_1 - \cos x_1$. Consequently,

$$-\frac{\dot{x}_1 + \cos x_1}{\dot{x}_1} = \frac{\frac{dx_2}{dt}}{\frac{dx_1}{dt}} = \frac{dx_2}{dx_1} = \frac{d\dot{x}_1}{dx_1} \tag{11.12}$$

Our plant is now in the form $\frac{d\dot{x}_1}{dx_1} = -\frac{f(x_1,\dot{x}_1)}{\dot{x}_1}$ with $f(x_1,\dot{x}_1) = \dot{x}_1 + \cos x_1$; and since $f(x_1,\dot{x}_1) \neq f(x_1,-\dot{x}_1)$ and $f(x_1,\dot{x}_1) \neq -f(-x_1,\dot{x}_1)$, there are no symmetries in the phase plane.

Singular points are found as

$$\begin{cases} \dot{x}_1 = 0 \\ \dot{x}_1 + \cos x_1 = 0 \end{cases} \Rightarrow \begin{cases} \dot{x}_1 = 0 \\ x_1 = \pm\frac{\pi}{2} + 2k\pi \end{cases} \tag{11.13}$$

Linearizing around $(x_1,\dot{x}_1) = (\frac{\pi}{2} + 2k\pi, 0)$, we get $\cos x \approx -x + \frac{\pi}{2} \Rightarrow \ddot{x} + \dot{x} - x + \frac{\pi}{2} = 0$. Using the variable transform $y = x - \frac{\pi}{2} \Rightarrow \dot{y} = \dot{x} \Rightarrow \ddot{y} = \ddot{x}$, this becomes $\ddot{y} + \dot{y} - y = 0$, and the characteristic equation is $\lambda^2 + \lambda - 1 = 0 \Rightarrow \lambda = \frac{-1\pm\sqrt{5}}{2}$, so this is a saddle point.

Around $(x_1, \dot{x}_1) = (-\frac{\pi}{2} + 2k\pi, 0)$ we get $\cos x \approx x + \frac{\pi}{2} \Rightarrow \ddot{x} + \dot{x} + x + \frac{\pi}{2} = 0$. Using the variable transform $y = x + \frac{\pi}{2} \Rightarrow \dot{y} = \dot{x} \Rightarrow \ddot{y} = \ddot{x}$, this becomes $\ddot{y} + \dot{y} + y = 0$, and the characteristic equation is $\lambda^2 + \lambda + 1 = 0 \Rightarrow \lambda = -\frac{1}{2} \pm j\frac{\sqrt{3}}{2}$, so this is a stable focus.

11.3 Proposed problems

Exercise 11.1 Consider the phase plane $(x, \dot{x})$ in Figure 11.5a, corresponding to the dynamic behavior of a simple pendulum with friction. Is the time response in Figure 11.5b a possible time response of the pendulum?

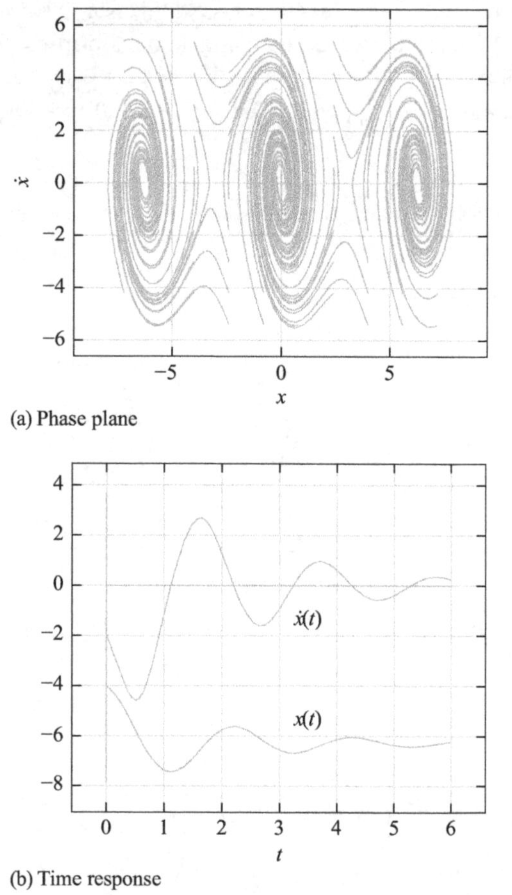

Figure 11.5 Phase plane and time response of Exercise 11.1

Exercise 11.2 Consider the plant $\ddot{x} + \omega^2 x = 0$. Figure 11.6 shows its representation on phase plane $(x, \dot{x})$. Then:

A) $B = A$
B) $B = A/\omega$
C) $B = A\omega$
D) $B = \omega$.

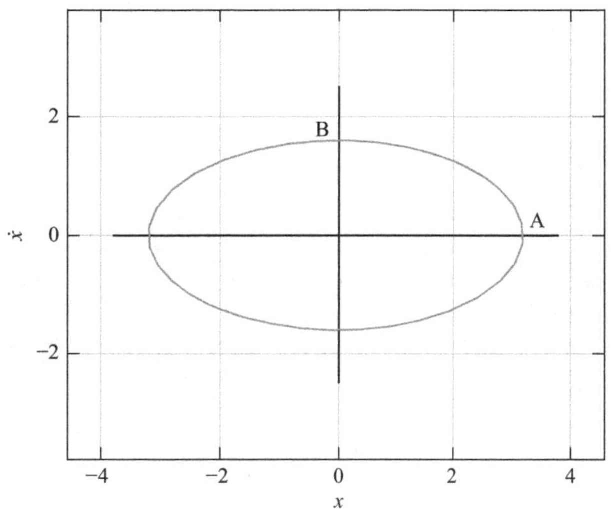

Figure 11.6 Phase plane of Exercise 11.2

Exercise 11.3 Consider the plant $\ddot{x} + f(x, \dot{x}) = 0$ and its representation on phase plane $(x, \dot{x})$. For two points of the same trajectory, $A = (x_A, \dot{x}_A)$ and $B = (x_B, \dot{x}_B)$, obtained at two close time instants t_A and t_B, we have:

A) $t_A - t_B \approx 2\dfrac{x_A - x_B}{\dot{x}_A - \dot{x}_B}$

B) $t_A - t_B \approx 2\dfrac{\dot{x}_A - \dot{x}_B}{x_A - x_B}$

C) $t_A - t_B \approx 2\dfrac{x_A - \dot{x}_A}{x_B - \dot{x}_B}$

D) $t_A - t_B \approx 2\dfrac{\dot{x}_A - x_A}{\dot{x}_B - x_B}$.

Exercise 11.4 Consider a mass–spring–damper mechanical system corresponding to the linear model $M\ddot{x} + B\dot{x} + Kx = 0$, $M, B, K \in \mathbb{R}$. Analyzing the trajectories in the phase plane, we can say that:

A) There can only be one limit cycle (for particular values of M, B, K).
B) There can be several limit cycles (for particular values of M, B, K).
C) There can be no limit cycles (whatever the values of M, B, K).
D) None of the above.

Exercise 11.5 Consider the plant $\ddot{x} + f(x, \dot{x}) = 0$ and its phase plane representation $(x, \dot{x})$. Trajectories on $(x, \dot{x})$ will be symmetric with respect to the x-axis if:

A) $f(x, \dot{x}) = f(-x, \dot{x})$
B) $f(x, \dot{x}) = -f(-x, \dot{x})$
C) $f(x, \dot{x}) = -f(x, -\dot{x})$
D) $f(x, \dot{x}) = f(x, -\dot{x})$.

Exercise 11.6 Consider a plant with model $\ddot{x} - \left(1 - x^2\right)\dot{x} + x = 0$. Figure 11.7a shows the output $x(t)$ and its derivative $\mathrm{d}x(t)/\mathrm{d}t$ for a particular initial condition. Figure 11.7b shows the corresponding phase plane $(x, \dot{x})$. We can thus conclude that $a, b \in \mathbb{R}$, which are, respectively, the scales of the x-axis and the $\mathrm{d}x/\mathrm{d}t$-axis are:

A) $a = b = 1.0$
B) $a = b = 2.0$
C) $a = b = 3.0$
D) $a = b = 4.0$.

Exercise 11.7 Consider a mass–spring–damper mechanical system corresponding to the linear model $\ddot{x} + a\dot{x} + 2x = 0$, $a \in \mathbb{R}$. Analyzing the trajectories in the phase plane, we can say that there is at $(x, \dot{x}) = (0, 0)$ a singular point of the type center, for:

A) $a = 1$
B) $a = -1$
C) $a = 0$
D) $a = 2$.

Exercise 11.8 Consider a mechanical system, comprising a mass, a spring and a non-linear friction, corresponding to model $\ddot{x} - x^2(1 - x)\dot{x} + x = 0$. Figure 11.8 shows some trajectories on the phase plane $(x, \dot{x})$. Analyzing the different regions of the system's dynamics, we can say that:

1. In what concerns limit cycles,
   A) There is one unstable limit cycle
   B) There is one semistable limit cycle
   C) There is one stable limit cycle
   D) There is no limit cycle.

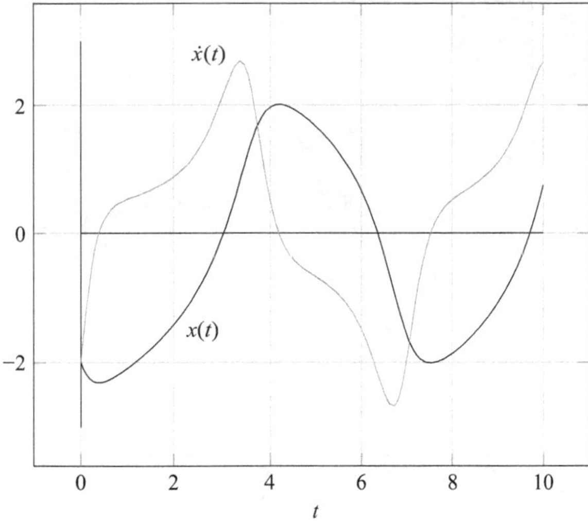

(a) Time response

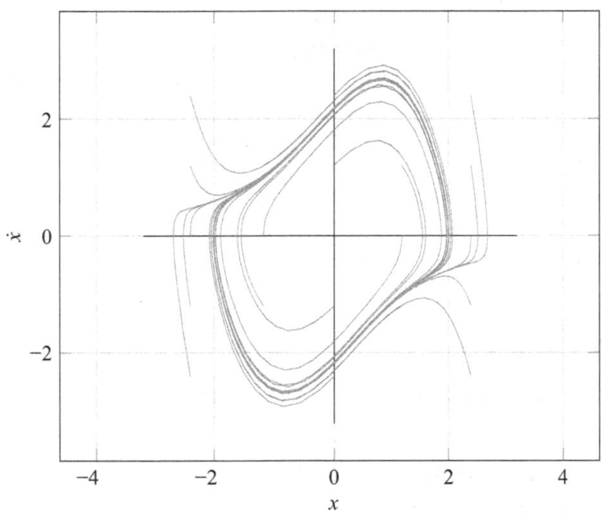

(b) Phase plane

Figure 11.7 Time response and phase plane of Exercise 11.6

2. In what concerns singular points,
   E) There is only one unstable singular point
   F) There is only one stable singular point
   G) There are no singular points
   H) None of the above.

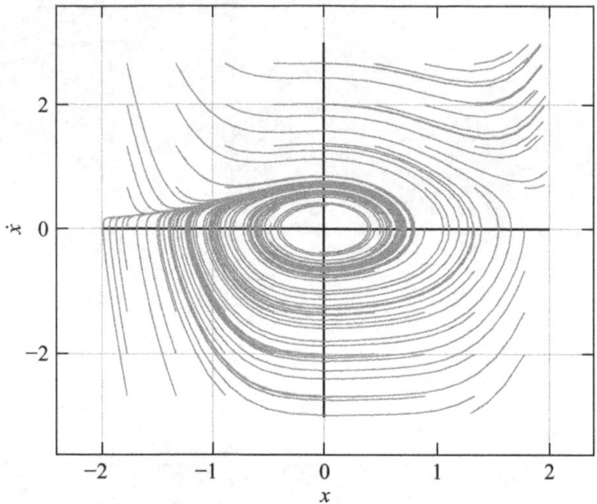

Figure 11.8 Phase plane of Exercise 11.8

Exercise 11.9 Consider a mass–spring–damper mechanical system corresponding to the linear model $\ddot{x} + 3\dot{x} + 2x = 0$. Analyzing the trajectories in the phase plane, we can say that there is at $(x, \dot{x}) = (0, 0)$ a singular point which is:

A) Stable node
B) Center
C) Saddle
D) There are no singular points.

Exercise 11.10 Consider a linear plant with transfer function $\dfrac{Y(s)}{R(s)} = \dfrac{1}{(s+1)(s-2)}$. Its representation on phase plane $(y, \dot{y})$ for several different initial conditions and $r(t) = 0$ leads to:

A) A center at $(y, \dot{y}) = (0, 0)$
B) A saddle point at $(y, \dot{y}) = (0, 0)$
C) A stable focus at $(y, \dot{y}) = (0, 0)$
D) An unstable focus at $(y, \dot{y}) = (0, 0)$.

Exercise 11.11 Consider the nonlinear plant of Figure 11.9a, consisting of inertia J, viscous friction B and a Coulomb friction. Also consider its representation on phase plane $(e, \dot{e})$ for several different initial conditions and $r(t) = 0$.

1. When $J = 1$, $B = 0$ and $F_c = 1$, the plot in Figure 11.9b is obtained. From zone A, corresponding to the steady-state error due to Coulomb friction, we can see that:
   A) $K = 0$
   B) $K = 2$

C) $K = 20$
D) $K = 1$.

2. When $J = 1, B = 2$ and $F_c = 1$, the plot in Figure 11.9c is obtained. From zone A, corresponding to the steady-state error due to Coulomb friction, we can see that:
   E) $K = 0$
   F) $K = 2$
   G) $K = 20$
   H) $K = 1$.
3. When $F_c = B = 0$ (that is to say, when there is no friction), the plant's phase plane representation (with $K \neq 0$) includes:
   I) A center at $(e, \dot{e}) = (0, 0)$
   J) A saddle point at $(e, \dot{e}) = (0, 0)$
   K) A stable focus at $(e, \dot{e}) = (0, 0)$
   L) An unstable focus at $(e, \dot{e}) = (0, 0)$.

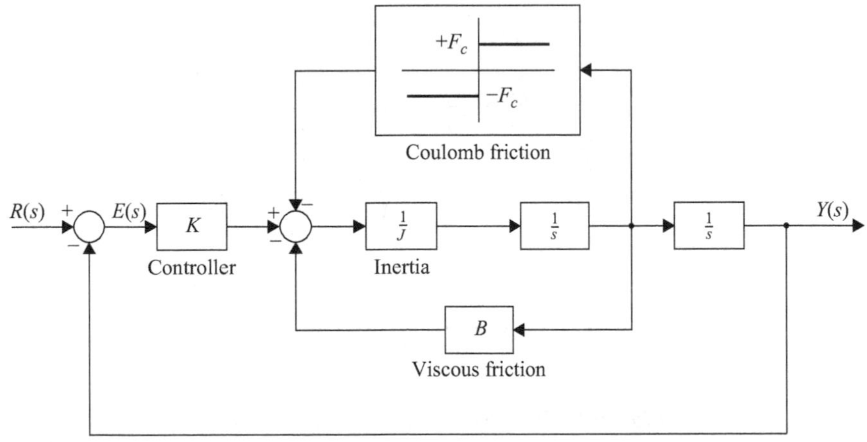

(a) Nonlinear plant

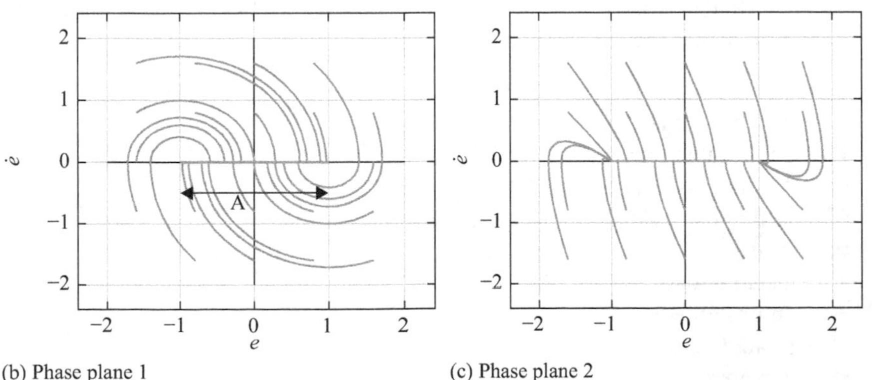

(b) Phase plane 1 (c) Phase plane 2

Figure 11.9 Plant and phase planes of Exercise 11.11

Exercise 11.12 Consider the plant $\ddot{x} + 0.5\dot{x} + x(x+2)(x-2) = 0$. Figure 11.10 shows its representation on phase plane $(x, \dot{x})$.

1. Points A are:
   A) stable focuses
   B) unstable focuses
   C) saddle points
   D) centers.
2. Point B is:
   E) a stable focuses
   F) an unstable focuses
   G) a saddle point
   H) a center.

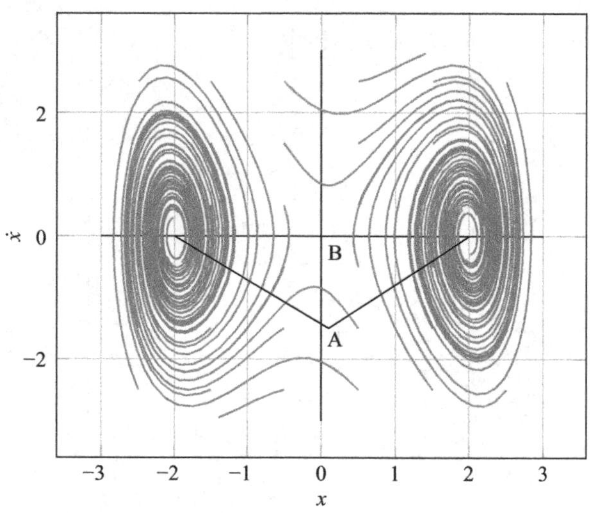

Figure 11.10 Phase plane of Exercise 11.12

Exercise 11.13 Consider the plant $\ddot{x} + \dot{x} + x - 2x^3 = 0$. Figure 11.11 shows its representation on phase plane $(x, \dot{x})$.

1. Point A is:
   A) a stable focus
   B) an unstable focus
   C) a saddle point
   D) a center.
2. Point B is:
   E) a stable focus
   F) an unstable focus
   G) a saddle point
   H) a center.

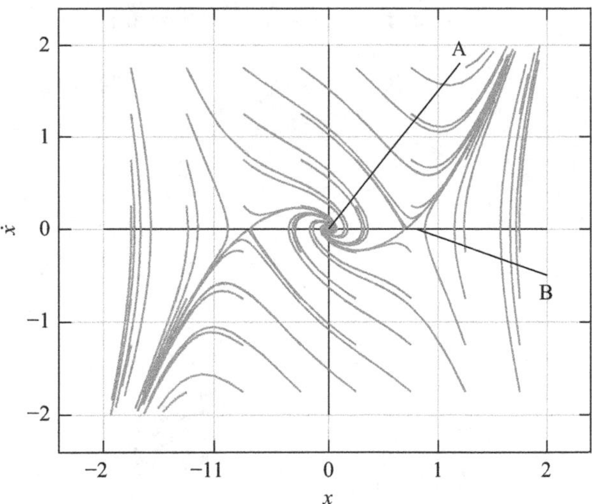

Figure 11.11 Phase plane of Exercise 11.13

Exercise 11.14 Consider the plant $f(\ddot{x}, \dot{x}, x) = 0$. Figure 11.12 shows its representation on phase plane $(x, \dot{x})$. Point (0,0) is:

A) a stable focus
B) an unstable focus
C) a saddle point
D) a center.

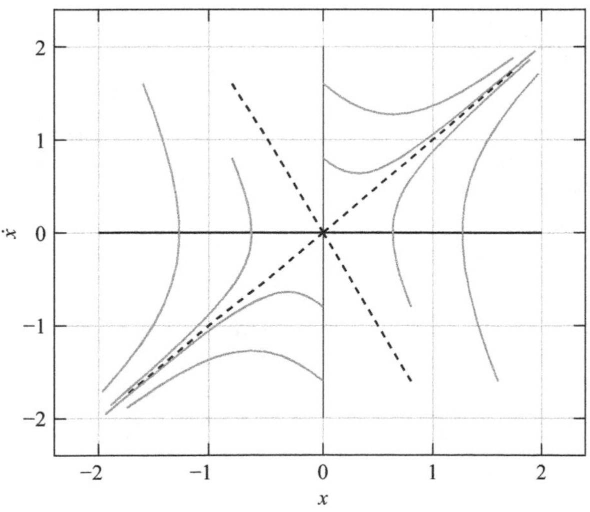

Figure 11.12 Phase plane of Exercise 11.14

Exercise 11.15 Consider the plant $\ddot{x} + \dot{x}^2 + x = 0$. Figure 11.13 shows its representation on phase plane $(x, \dot{x})$. Point A is:

A) a stable focus
B) an unstable focus
C) a saddle point
D) a center.

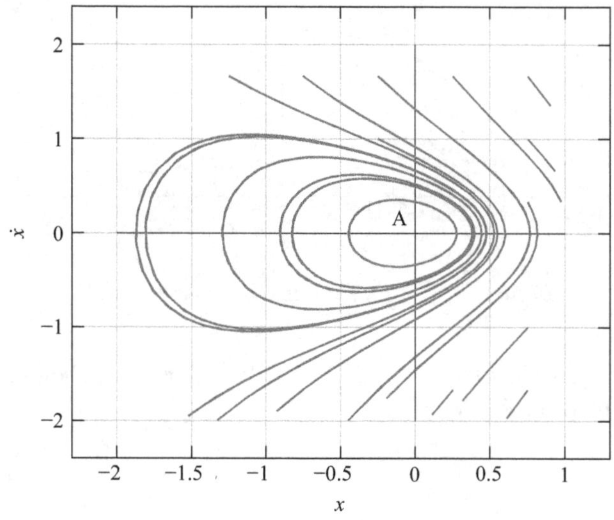

Figure 11.13 Phase plane of Exercise 11.15

Exercise 11.16 Consider the plant $\ddot{x} - x + x^3 = 0$. Figure 11.14 shows its representation on phase plane $(x, \dot{x})$.

1. Point A is:
   A) a stable focus
   B) an unstable focus
   C) a saddle point
   D) a center.
2. Point B is:
   E) a stable focus
   F) a limit cycle
   G) a saddle point
   H) a center.
3. Point C is:
   I) a stable focus
   J) a limit cycle
   K) a saddle point
   L) a center.

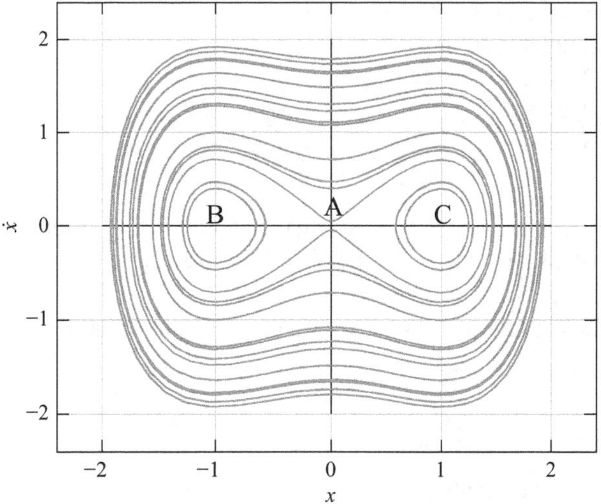

Figure 11.14 Phase plane of Exercise 11.16

Exercise 11.17 Consider the plant $\ddot{x} + \omega^2 x = 0$. Figure 11.15 shows its representation on phase plane $(x, \dot{x})$.

1. Point A is:
   A) a stable focus
   B) an unstable focus
   C) a saddle point
   D) a center.

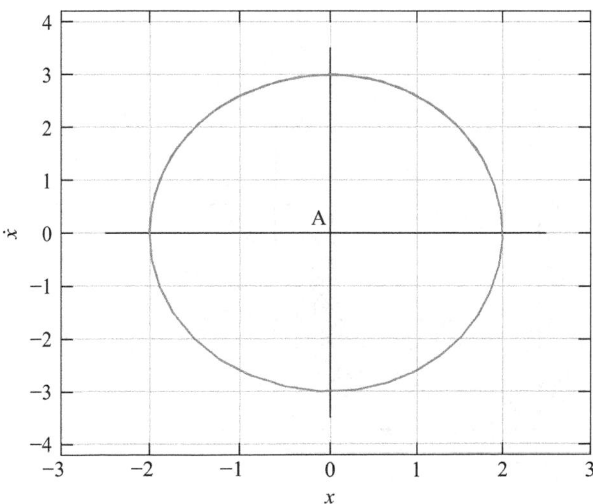

Figure 11.15 Phase plane of Exercise 11.17

2. It can be reckoned that
   E) $\omega = 3/2$ rad/s
   F) $\omega = 2$ rad/s
   G) $\omega = 2/3$ rad/s
   H) $\omega = 3$ rad/s.

Exercise 11.18 Consider the plant $\ddot{x} + \sin(x)\dot{x} + \cos(x) = 0$. Figure 11.16 shows its representation on phase plane $(x, \dot{x})$. Let $k = 0, \pm 1, \pm 2, \ldots$

1. The plant has saddle singular points on phase plane $(x, \dot{x})$ for:
   A) $(x, \dot{x}) = \left(\frac{\pi}{2} + 2k\pi,\ 0\right)$
   B) $(x, \dot{x}) = \left(-\frac{\pi}{2} + 2k\pi,\ 0\right)$
   C) There are no saddle points.
   D) None of the above.
2. The plant has unstable focus singular points on phase plane $(x, \dot{x})$ for:
   E) $(x, \dot{x}) = \left(\frac{\pi}{2} + 2k\pi,\ 0\right)$
   F) $(x, \dot{x}) = \left(-\frac{\pi}{2} + 2k\pi,\ 0\right)$
   G) There are no unstable focus points.
   H) None of the above.

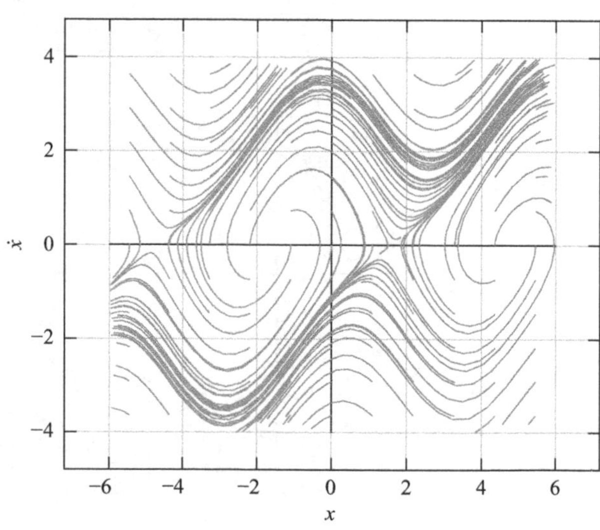

Figure 11.16 Phase plane of Exercise 11.18

Exercise 11.19 Consider a plant described by the nonlinear differential equation $\ddot{x} + \cos(x) = 0$.

1. Analyze symmetries of trajectories on the phase plane.
2. Find singular points on the phase plane.
3. Analyze how the plant behaves in a neighborhood of those singular points.

Exercise 11.20 Consider a plant described by the nonlinear differential equation $\ddot{x} + 0.2\dot{x} + 3x + x^2 = 0$.

1. Find singular points on the phase plane.
2. Analyze how the plant behaves in a neighborhood of those singular points.

Exercise 11.21 Consider a plant described by the nonlinear differential equation $\ddot{x} - x + x^3 = 0$.

1. Find singular points on the phase plane.
2. Analyze how the plant behaves in a neighborhood of those singular points.

Exercise 11.22 Consider a plant described by the nonlinear differential equation $\ddot{x} + \dot{x} + x(x+3)\,(x-3) = 0$.

1. Find singular points on the phase plane.
2. Analyze how the plant behaves in a neighborhood of those singular points.

Exercise 11.23 Consider second-order plant $\begin{cases} \dot{x}_1 = x_2 \\ \dot{x}_2 = x_1 - 0.4x_2\left(1 - x_1^2\right) \end{cases}$.

1. Find the singular points in state plane (x_1, x_2).
2. Analyze the plant's behavior around singular points.

Exercise 11.24 Consider the plant $\begin{cases} \dot{x}_1 = -x_2\left(1 + x_2^2\right) \\ \dot{x}_2 = x_1\left(1 + x_1^2\right) - 0.4x_2 \end{cases}$.

1. Find the singular points on state space (x_1, x_2).
2. Analyze the system's behavior in the vicinity of such points.

Exercise 11.25 Consider a plant described by the nonlinear differential equation $\frac{d^2x}{dt^2} + 2 \cdot \frac{dx}{dt} + x \cdot (x+1) \cdot (x-3) = 0$.

1. Find singular points on the phase plane.
2. Analyze how the plant behaves in a neighborhood of those singular points.

11.4 Phase plane analysis using computer packages

In this section, we consider Exercise 11.6 with the dynamical system $\ddot{x} - \left(1 - x^2\right)\dot{x} + x = 0$ and the phase plane plot using the computer packages MATLAB®, SCILAB™ and OCTAVE©.

11.4.1 MATLAB

This subsection describes some basic commands that can be adopted with the package MATLAB.

The model of the dynamical system is represented by means of the function `dynsys`. The simulation is performed using the numerical solver `ode45`. The trajectory in the phase plane plot is accomplished using the command `plot`.

The complete code is as follows.

```
%%%% Phase plane %%%%

% Time specification
t = 0:0.01:10;

% Initial conditions
x10 = 1;
x20 = 1;
x0 = [x10; x20];

% System simulation
[t,x] = ode45(@dynsys,t,x0);

% Plots trajectory
plot(x(:,1),x(:,2))

% Format plot
xlabel('x'); % Inserts the x-axis label
ylabel('dx/dt'); % Inserts the y-axis label
title('Phase plane'); % Inserts the title in the plot

grid
```

```
% Declare function name, inputs, and outputs

function y = dynsys(t,x)
        y = [x(2); -x(1)+(1-x(1)^2)*x(2)];% Dynamical
            % system
end
```

MATLAB creates the figure window represented in Figure 11.17.

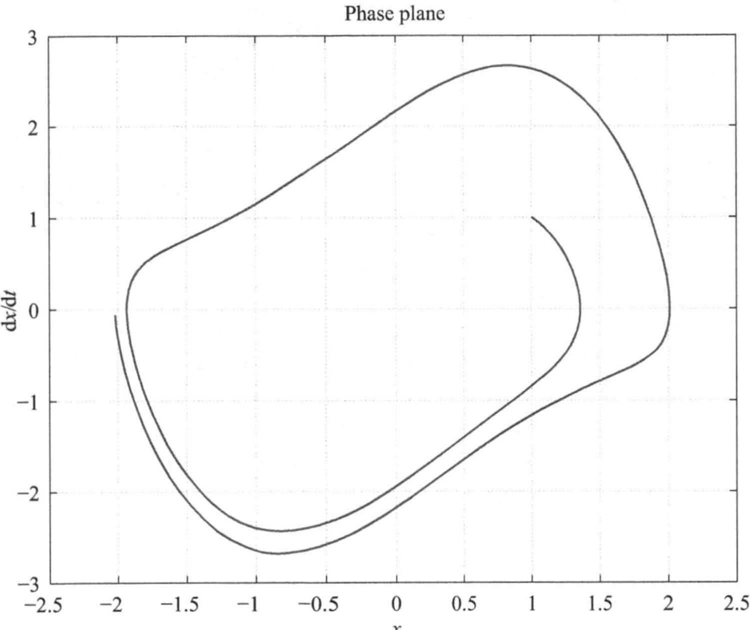

Figure 11.17 Trajectory in the phase plane for the dynamical system $\ddot{x} - \left(1 - x^2\right)\dot{x} + x = 0$ *using MATLAB*

11.4.2 SCILAB

This subsection describes some basic commands that can be adopted with the package SCILAB.

The model of the dynamical system is represented by means of the function `dynsys`. The simulation is performed using the numerical solver `ode`. The trajectory in the phase plane plot is accomplished using the command `plot`.

The complete code is as follows.

```
//// Phase plane ////
// Dynamical system
function y = dynsys(t,x)
    y = [x(2); -x(1)+(1-x(1)^2)*x(2)];
endfunction

// Time specification
t = 0:0.01:10;

// Initial conditions
```

```
x10 = 1;
x20 = 1;
x0 = [x10; x20];

// System simulation
x = ode(x0,0,t,dynsys);

// Plots trajectory
plot(x(1,:),x(2,:))

// Format plot
xlabel('x'); // Inserts the x-axis label
ylabel('dx/dt'); // Inserts the y-axis label
title('Phase plane'); // Inserts the title in the plot

xgrid
```

SCILAB creates the figure window represented in Figure 11.18.

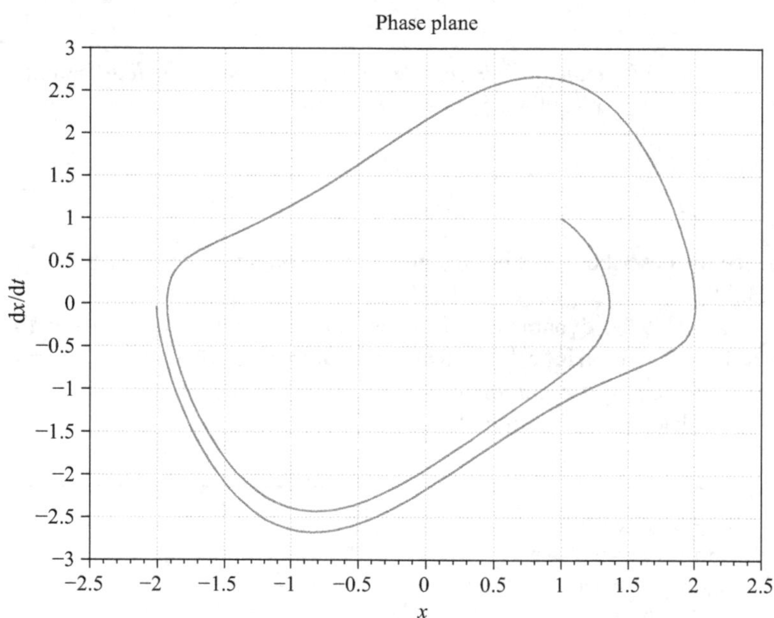

Figure 11.18 Trajectory in the phase plane for the dynamical system $\ddot{x} - (1 - x^2)\,\dot{x} + x = 0$ *using SCILAB*

11.4.3 OCTAVE

This subsection describes some basic commands that can be adopted with the package OCTAVE. It is required to load package odepkg.

The model of the dynamical system is represented by means of the function dynsys. The simulation is performed using the numerical solver ode45. The trajectory in the phase plane plot is accomplished using the command plot.

The complete code is as follows.

```
%%%% Phase plane %%%%

% Time specification
t = 0:0.01:10;

% Initial conditions
x10 = 1;
x20 = 1;

% System simulation
[t,x] = ode45(@dynsys,t,[x10 x20]);

% Plots trajectory
plot(x(:,1),x(:,2))

% Format plot
xlabel('x'); % Inserts the x-axis label
ylabel('dx/dt'); % Inserts the y-axis label
title('Phase plane'); % Inserts the title in the plot

grid
```

```
% Declare function name, inputs, and outputs

function y = dynsys(t,x)
        y = [x(2); -x(1)+(1-x(1)^2)*x(2)];% Dynamical
% system
end
```

OCTAVE creates the figure window represented in Figure 11.19.

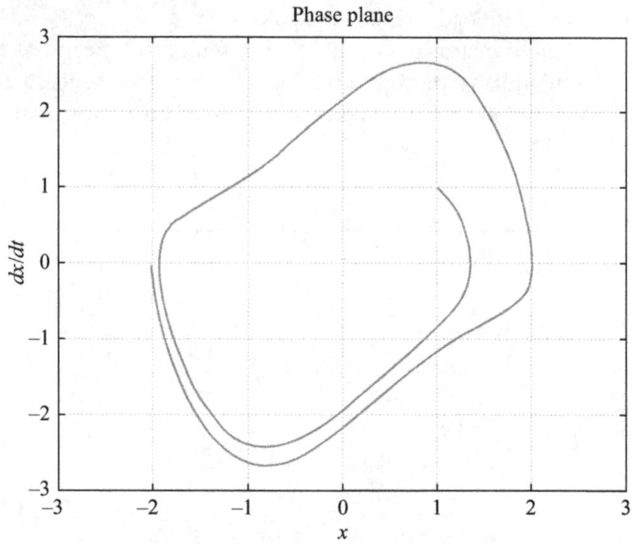

Figure 11.19 Trajectory in the phase plane for the dynamical system $\ddot{x} - \left(1 - x^2\right)\dot{x} + x = 0$ *using OCTAVE*

Chapter 12
Fractional order systems and controllers

12.1 Fundamentals

Derivatives and integrals can be extended to orders which are not integer. These can be used in differential equations to describe the dynamics of a system, or of a controller, in a more supple manner than with integer derivatives and integrals only.

12.1.1 List of symbols

c terminal of differentiation
${}_cD_t^\alpha$ derivative of order α, with terminals c and t
$E_{\alpha,\beta}(t)$ two-parameter Mittag-Leffler function (A.1)
$\mathscr{L}$ Laplace operator
s Laplace variable
t time
α differentiation order; commensurability order
Γ gamma function
ω_h upper limit of the frequency range where a CRONE approximation is valid
ω_l lower limit of the frequency range where a CRONE approximation is valid
$\omega_{p,m}$ mth pole of a CRONE approximation
$\omega_{z,m}$ mth zero of a CRONE approximation

12.1.2 Grünwald–Letnikov definition

From the definition of derivative $f'(t) = \lim_{h\to 0} \frac{f(t) - f(t-h)}{h}$, it can be shown by mathematical induction that

$$f^{(n)}(t) = \lim_{h\to 0} \frac{\sum_{k=0}^{n} (-1)^k \binom{n}{k} f(t - kh)}{h^n} \tag{12.1}$$

Combinations of a things b at a time are $\binom{a}{b} = \frac{a!}{b!(a-b)!}$. Function $\Gamma(x), x \in \mathbb{R}\backslash\mathbb{Z}_0^-$ generalizes the factorial since $\Gamma(k+1) = k!,\ k \in \mathbb{Z}_0^+$. Combinations become

$$\binom{a}{b} = \begin{cases} \frac{\Gamma(a+1)}{\Gamma(b+1)\Gamma(a-b+1)}, & \text{if } a, b, a-b \notin \mathbb{Z}^- \\ \frac{(-1)^b \Gamma(b-a)}{\Gamma(b+1)\Gamma(-a)}, & \text{if } a \in \mathbb{Z}^- \wedge b \in \mathbb{Z}_0^+ \\ 0, & \text{if } (b \in \mathbb{Z}^- \vee b - a \in \mathbb{N}) \wedge a \notin \mathbb{Z}^- \end{cases} \tag{12.2}$$

For cases not covered above, $\lim_{(\alpha,\beta)\to(a,b)} \binom{\alpha}{\beta} = \infty$ if $a \in \mathbb{Z}^- \wedge b \notin \mathbb{Z}_0^+$, or if $a \in \mathbb{Z}^- \wedge b - a \in \mathbb{N}$. Thanks to these generalized combinations we can define

$$f^{(\alpha)}(t)\Big|_c^t = \lim_{h\to 0^+} \frac{\sum_{k=0}^{\lfloor \frac{t-c}{h} \rfloor} (-1)^k \binom{\alpha}{k} f(t-kh)}{h^\alpha}, \quad \alpha \in \mathbb{R} \tag{12.3}$$

$$f^{(\alpha)}(t)\Big|_t^c = \lim_{h\to 0^+} \frac{\sum_{k=0}^{\lfloor \frac{c-t}{h} \rfloor} (-1)^k \binom{\alpha}{k} f(t+kh)}{h^\alpha}, \quad \alpha \in \mathbb{R} \tag{12.4}$$

Rather than $f^{(\alpha)}(t)\big|_c^t$, or $\frac{\mathrm{d}^\alpha f(t)}{\mathrm{d}t^\alpha}\Big|_c^t$, it is usual to write ${}_cD_t^\alpha f(t)$ (recurring to a functional operator D), and this will be the notation employed from here on. For $\alpha \in \mathbb{N}$, (12.3) becomes a right derivative and (12.4) a left derivative; in these cases, terminals c and t are useless. For $\alpha \in \mathbb{Z}^-$, (12.3) becomes an iterated Riemann integral with terminals (or integration limits) $c < t$; (12.4) has the terminals the other way round, $t < c$ (see (12.5) and (12.6)).

Natural order derivatives are local operators; all others are not local, since they depend on terminal c. In other words, if t is identified with time, the result of ${}_cD_t^\alpha f(t)$ has a memory of what happens to function f for some time before, since time c; and ${}_tD_c^\alpha f(t)$ is a noncausal operator, depending on future values of $f(t)$ up to time c.

12.1.3 Riemann–Liouville definition

Again using mathematical induction, it can be shown that

$$ {}_cD_x^{-n} f(t) = \overbrace{\int_c^x \cdots \int_c^x f(t)\,\mathrm{d}t \cdots \mathrm{d}t}^{n \text{ integrations}} = \int_c^x \frac{(x-t)^{n-1}}{(n-1)!} f(t)\,\mathrm{d}t, \quad n \in \mathbb{N} \tag{12.5}$$

$$ {}_xD_c^{-n} f(t) = \underbrace{\int_x^c \cdots \int_x^c f(t)\,\mathrm{d}t \cdots \mathrm{d}t}_{n \text{ integrations}} = \int_x^c \frac{(t-x)^{n-1}}{(n-1)!} f(t)\,\mathrm{d}t, \quad n \in \mathbb{N} \tag{12.6}$$

and that (if all the derivatives involved exist) the equality

$$ {}_cD_t^m\, {}_cD_t^n f(t) = {}_cD_t^{m+n} f(t) \tag{12.7}$$

holds in each of the three following cases:

$$m, n \in \mathbb{Z}_0^+ \tag{12.8}$$

$$m, n \in \mathbb{Z}_0^- \tag{12.9}$$

$$m \in \mathbb{Z}^+ \wedge n \in \mathbb{Z}^- \tag{12.10}$$

These theorems are, respectively, Cauchy's formula and the law of exponents, and can be used to define

$$
{}_cD_t^\alpha f(t) = \begin{cases} \displaystyle\int_c^t \frac{(t-\tau)^{-\alpha-1}}{\Gamma(-\alpha)} f(\tau)\,\mathrm{d}\tau, & \text{if } \alpha \in \mathbb{R}^- \\ f(t), & \text{if } \alpha = 0 \\ \dfrac{\mathrm{d}^{\lceil\alpha\rceil}}{\mathrm{d}t^{\lceil\alpha\rceil}}\, {}_cD_t^{\alpha-\lceil\alpha\rceil} f(t), & \text{if } \alpha \in \mathbb{R}^+ \end{cases} \tag{12.11}
$$

$$
{}_tD_c^\alpha f(t) = \begin{cases} \displaystyle\int_t^c \frac{(\tau-t)^{-\alpha-1}}{\Gamma(-\alpha)} f(\tau)\,\mathrm{d}\tau, & \text{if } \alpha \in \mathbb{R}^- \\ f(t), & \text{if } \alpha = 0 \\ (-1)^{\lceil\alpha\rceil}\dfrac{\mathrm{d}^{\lceil\alpha\rceil}}{\mathrm{d}t^{\lceil\alpha\rceil}}\, {}_tD_c^{\alpha-\lceil\alpha\rceil} f(t), & \text{if } \alpha \in \mathbb{R}^+ \end{cases} \tag{12.12}
$$

12.1.4 Equivalence of definitions and Laplace transforms

If $f(t)$ has $\max\{0, \lfloor\alpha\rfloor\}$ continuous derivatives, and $D^{\max\{0,\lceil\alpha\rceil\}}f(t)$ is integrable, then ${}_cD_t^\alpha f(t)$ exists according to both the Riemann–Liouville and Grünwald–Letnikov definitions, which provide the same result, and the corresponding Laplace transform is

$$
\mathscr{L}\left[{}_0D_t^\alpha f(t)\right] = \begin{cases} s^\alpha F(s), & \text{if } \alpha \in \mathbb{R}^- \\ F(s), & \text{if } \alpha = 0 \\ s^\alpha F(s) - \displaystyle\sum_{k=0}^{\lceil\alpha\rceil-1} s^k\, {}_0D_t^{\alpha-k-1} f(0), & \text{if } \alpha \in \mathbb{R}^+ \end{cases} \tag{12.13}
$$

12.1.5 Caputo definition

For $\alpha > 0$, (12.13) includes fractional derivatives of f in the initial conditions. These can be hard to determine. In order to have instead

$$
\mathscr{L}\left[{}_0D_t^\alpha f(t)\right] = \begin{cases} s^\alpha F(s), & \text{if } \alpha \in \mathbb{R}^- \\ F(s), & \text{if } \alpha = 0 \\ s^\alpha F(s) - \displaystyle\sum_{k=0}^{\lceil\alpha\rceil-1} s^{\alpha-k-1} D^k f(0), & \text{if } \alpha \in \mathbb{R}^+ \end{cases} \tag{12.14}
$$

we need to define

$$
{}_cD_t^\alpha f(t) = \begin{cases} \displaystyle\int_c^t \frac{(t-\tau)^{-\alpha-1}}{\Gamma(-\alpha)} f(\tau)\,\mathrm{d}\tau, & \text{if } \alpha \in \mathbb{R}^- \\ f(t), & \text{if } \alpha = 0 \\ {}_cD_t^{\alpha-\lceil\alpha\rceil} \dfrac{\mathrm{d}^{\lceil\alpha\rceil}}{\mathrm{d}t^{\lceil\alpha\rceil}} f(t), & \text{if } \alpha \in \mathbb{R}^+ \end{cases} \tag{12.15}
$$

$$
{}_tD_c^\alpha f(t) = \begin{cases} \displaystyle\int_t^c \frac{(\tau-t)^{-\alpha-1}}{\Gamma(-\alpha)} f(\tau)\,\mathrm{d}\tau, & \text{if } \alpha \in \mathbb{R}^- \\ f(t), & \text{if } \alpha = 0 \\ (-1)^{\lceil\alpha\rceil}\, {}_tD_c^{\alpha-\lceil\alpha\rceil} \dfrac{\mathrm{d}^{\lceil\alpha\rceil}}{\mathrm{d}t^{\lceil\alpha\rceil}} f(t), & \text{if } \alpha \in \mathbb{R}^+ \end{cases} \tag{12.16}
$$

which no longer corresponds to the law of exponents. If the Grünwald–Letnikov and Riemann–Liouville definitions yield, in the above-mentioned conditions, the same result ${}^{\mathrm{GL}}_{0}D_t^\alpha f(t) = {}^{\mathrm{RL}}_{0}D_t^\alpha f(t) = \mathscr{X}$, then the Caputo definition becomes

$$
{}^{\mathrm{C}}_{0}D_t^\alpha f(t) = \mathscr{X} - \sum_{k=0}^{\lceil\alpha\rceil-1} \frac{(t-c)^{-\alpha+k}\,\frac{\mathrm{d}^k f(c)}{\mathrm{d}t^k}}{\Gamma(k-\alpha+1)} \tag{12.17}
$$

12.1.6 Fractional transfer functions

Applying the Laplace transform to integro-differential equations of fractional order with zero initial conditions, fractional transfer functions are obtained, i.e., transfer functions with noninteger powers of s. Those of the form

$$
G(s) = \frac{\displaystyle\sum_{k=0}^{m} b_k s^{k\alpha}}{\displaystyle\sum_{k=0}^{n} a_k s^{k\alpha}}, \quad \alpha \in \mathbb{R}^+ \tag{12.18}
$$

are called commensurate or commensurable transfer functions; here α is the commensurate order.

Finding the frequency response of fractional transfer functions is done as for integer ones. Figure 12.1 shows the frequency responses of s^α and $\frac{1}{\left(\frac{s}{a}\right)^\alpha+1}$ as examples; $\frac{1}{\left(\frac{s}{a}\right)^{2\alpha}+2\zeta\left(\frac{s}{a}\right)^\alpha+1}$ can have zero, one or two resonance peaks and is better studied numerically.

The condition of stability of a fractional transfer function is the same of an integer one: roots of the denominator must all lie on the left side of the complex plane. For commensurable transfer functions, Matignon's theorem says that the roots σ_k, $k=1,\ldots,n$ of the polynomial $A(\sigma)=\sum_{k=0}^{n} a_k\sigma^k$ built with the denominator coefficients of transfer function (12.18) must verify $|\angle\sigma_k| > \alpha\frac{\pi}{2}$, $\forall\, k$ for $G(s)$ to be stable (see Figure 12.2).

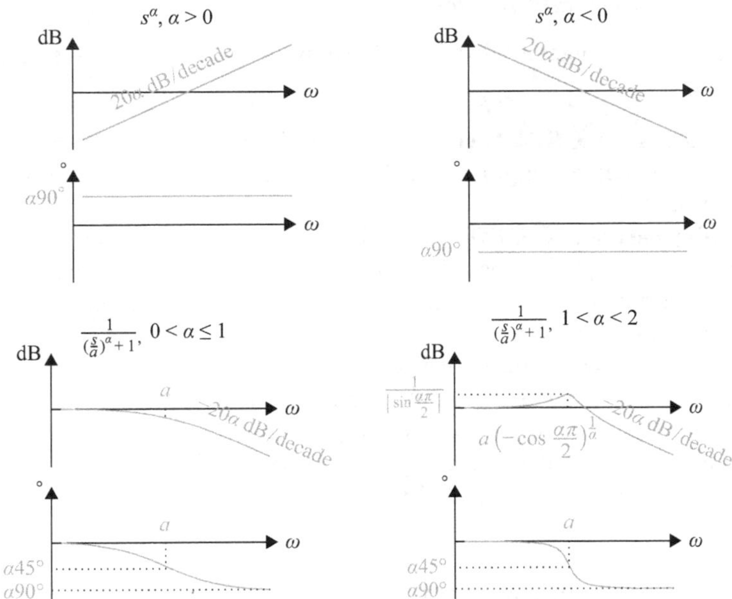

Figure 12.1 Bode diagrams. © 2013 IET. Reprinted, with permission, from Reference 36

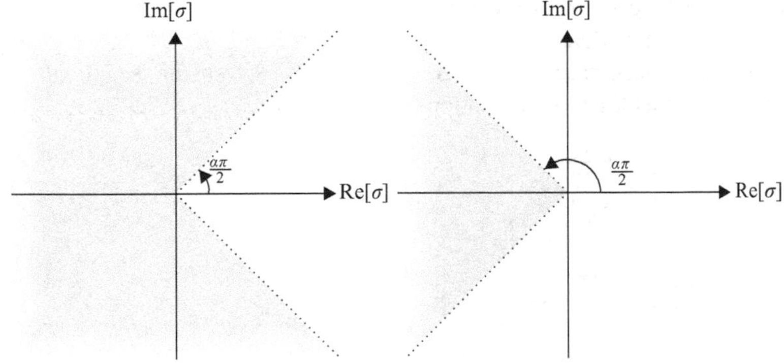

Figure 12.2 In gray: regions of the complex plane where the roots of $A(\sigma)$ may lie in the case of a stable commensurate transfer function; left: $0 < \alpha < 1$; right: $1 < \alpha < 2$. © 2013 IET. Reprinted, with permission, from Reference 36

The root-locus of fractional transfer functions controlled in closed loop with a proportional controller is found, in practice, numerically, sweeping the complex plane (see e.g., Reference 37 for a possible implementation). For commensurable transfer functions, the usual rules apply considering polynomials in s^α; the only difference for the integer case is that the zone of stability is that of the Matignon theorem.

12.1.7 Fractional controllers

Fractional transfer functions can model plants or controllers. Usual fractional controllers are:

- **First-generation CRONE controllers** are given by $C(s)=s^\alpha, \alpha>0$ so as to increase the phase margin. Actually, the phase increase is the same at all frequencies.
- **Second-generation CRONE controllers** are integer-order transfer functions with a phase adapted to that of the plant, so that the open loop behaves as s^α, with a constant phase. This means that the open-loop gain may change while the phase margin remains the same.
- **Fractional PID controllers** are given by $C(s)=P+\frac{I}{s^\lambda}+Ds^\mu,\ \lambda,\mu\in\mathbb{R}^+$. They are usually tuned to ensure a constant phase of the open loop, just as second-generation CRONE controllers, though the fixed transfer function does not always lend itself to this. There are tuning rules to try to achieve this for plants with particular types of responses [36,38].

12.1.8 Integer approximations

Fractional transfer functions are often implemented using integer-order approximations, of which the following two are usually the ones with better accuracy:

- **The CRONE approximation** works for s^α within the frequency range $[\omega_l, \omega_h]$ (though accuracy deteriorates near the limits), and has N stable real poles and N minimum-phase real zeros. More complex transfer functions must be approximated as linear combinations of powers of s.

$$s^\alpha \approx C\prod_{m=1}^{N}\frac{1+\frac{s}{\omega_{z,m}}}{1+\frac{s}{\omega_{p,m}}} \tag{12.19}$$

$$\omega_{z,m} = \omega_l\left(\frac{\omega_h}{\omega_l}\right)^{\frac{2m-1-\alpha}{2N}} \tag{12.20}$$

$$\omega_{p,m} = \omega_l\left(\frac{\omega_h}{\omega_l}\right)^{\frac{2m-1+\alpha}{2N}} \tag{12.21}$$

  The approximation works better when $\alpha\in\,]-1,1[$. Outside this range it is a good idea to make $s^\alpha = s^{\lceil\alpha\rceil}s^{\alpha-\lceil\alpha\rceil}$ or $s^\alpha = s^{\lfloor\alpha\rfloor}s^{\alpha-\lfloor\alpha\rfloor}$ and approximate the last term only. The latter form can be used also when $\alpha\in\,]-1,0[$ to ensure the effect of an integer integrator.
- **The Matsuda approximation** uses the gain of the frequency response $G(j\omega)$ of whatever transfer function G we want to implement. The response must be known at $N+1$ frequencies $\omega_0, \omega_1, \ldots, \omega_N$ (that do not need to be ordered), and

the approximation has $\lceil \frac{N}{2} \rceil$ zeros and $\lfloor \frac{N}{2} \rfloor$ poles (and will be causal only if N is even).

$$G(s) \approx d_0(\omega_0) + \cfrac{s-\omega_0}{d_1(\omega_1) + \cfrac{s-\omega_1}{d_2(\omega_2) + \cfrac{s-\omega_2}{d_3(\omega_3) + \ddots \cfrac{s-\omega_{N-1}}{d_N(\omega_N)}}}} \tag{12.22}$$

$$d_0(\omega) = |G(j\omega)| \tag{12.23}$$

$$d_k(\omega) = \frac{\omega - \omega_{k-1}}{d_{k-1}(\omega) - d_{k-1}(\omega_{k-1})}, \quad k = 1, 2, \ldots, N \tag{12.24}$$

Choosing frequencies wisely is important for success. When approximating s^α, frequencies can be logarithmically spaced in the frequency range of interest.

- **Discrete approximations** are better found discretizing one of the two above. But if an accurate impulse response is crucial, a finite impulse response filter with the coefficients set to the response desired will be a good option.

12.2 Solved problems

Problem 12.1 Use inverse Laplace transforms and the convolution theorem to find an analytical expression for the impulse response of a plant with transfer function $G(s) = \frac{3}{(s^{\frac{1}{2}}+1)(s^{\frac{1}{3}}+11)}$.

Resolution

$$\begin{aligned}
\mathscr{L}^{-1}\left[\frac{3}{(s^{\frac{1}{2}}+1)(s^{\frac{1}{3}}+11)}\mathscr{L}[\delta(t)]\right] &= 3\mathscr{L}^{-1}\left[\frac{1}{s^{\frac{1}{2}}+1}\right] * \mathscr{L}^{-1}\left[\frac{1}{s^{\frac{1}{3}}+11}\right] \\
&= 3\left[t^{-\frac{1}{2}}E_{\frac{1}{2},\frac{1}{2}}(-t^{\frac{1}{2}})\right] * \left[t^{-\frac{2}{3}}E_{\frac{1}{3},\frac{1}{3}}(-11t^{\frac{1}{3}})\right] \\
&= 3\int_0^t \frac{E_{\frac{1}{2},\frac{1}{2}}(-\sqrt{t-\tau})E_{\frac{1}{3},\frac{1}{3}}(-11\sqrt[3]{\tau})}{\sqrt{t-\tau}\sqrt[3]{\tau^2}}\,d\tau
\end{aligned}$$

Problem 12.2 Use Matignon's theorem to find whether the plant with transfer function $\frac{1}{s^{2/3} - 6s^{1/3} + 18}$ is stable.

Resolution The roots of the polynomial in $\alpha = \frac{1}{3}$ are $3 \pm 3j$, which fall on the stable part of the complex plane, defined by rays making $\pm 90° \times \frac{1}{3} = \pm 30°$ angles with the positive real axis.

Problem 12.3 Find the tautochrone curve, that is to say, the shape of the frictionless curve such that a mass m at rest thereupon, as seen in Figure 12.3, will always take the same time T to reach the bottom, irrespective of the initial height h from which it starts sliding down.

1. Let the curve surface be defined by curve $y(x)$, with its final point at $(x, y) = (0, 0)$. Let the acceleration of gravity be g, and the (ever increasing) velocity of the mass be V. Find an expression for the total mechanical energy of the mass at any height y.
2. Show that $V = \sqrt{2g(h-y)}$, at any height y.
3. Let ℓ be the distance traveled along the curve by the mass, from height y to the end of the curve, given by $\ell(y) = \int_0^y \sqrt{1 + \left(\frac{\mathrm{d}x}{\mathrm{d}y}\right)^2}\,\mathrm{d}y$. Show that $T(h)$ is given by $T(h) = \int_0^h \frac{1}{\sqrt{2g(h-y)}} \frac{\mathrm{d}\ell(y)}{\mathrm{d}y}\,\mathrm{d}y$.
4. If the curve is tautochrone, $T(h)$ does not depend on h. Show from the Riemann–Liouville definition of fractional derivatives that this means that ${}_0D_y^{-\frac{1}{2}} \frac{\mathrm{d}\ell(y)}{\mathrm{d}y} = \sqrt{\frac{2gT^2}{\pi}}$. (Take into account that $\Gamma\left(\frac{1}{2}\right) = \sqrt{\pi}$.)
5. Apply a $\frac{1}{2}$-order derivative to both sides of the last result and show that $\frac{\mathrm{d}\ell(y)}{\mathrm{d}y} = \underbrace{\frac{\sqrt{2g}T}{\pi}}_{a} y^{-\frac{1}{2}}$. (Take into account that, according to the law of exponents, ${}_0D_t^{\frac{1}{2}}\, {}_0D_t^{-\frac{1}{2}} f(t) = f(t)$.)
6. From this last result, show that $x(y) = \int_0^y \sqrt{\frac{a^2}{v} - 1}\,\mathrm{d}v$.

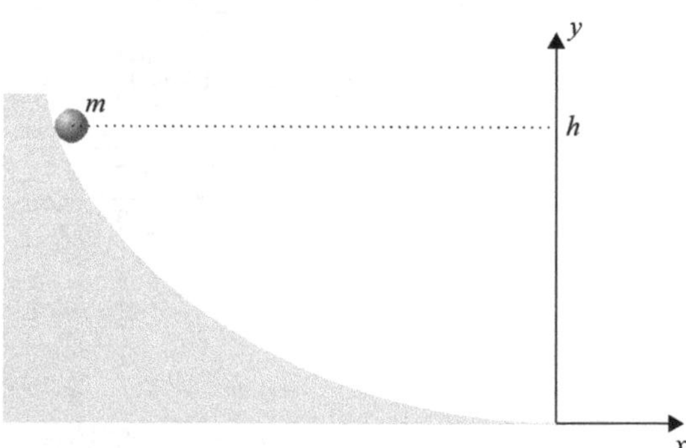

Figure 12.3 The tautochrone curve (the cycloid) of Problem 12.3. © 2013 IET. Reprinted, with permission, from Reference 36

7. Show that this last result is equivalent to the parametrization

$$x(\theta) = r\theta + r\sin\theta$$
$$y(\theta) = r - r\cos\theta$$

which is that of a cycloid, the curve described by a circle rolling on a horizontal straight line, with radius $r = \frac{a^2}{2} = \frac{gT^2}{\pi^2}$.

Resolution

1. The mass has a potential energy $E_p = mgy$ and a kinetic energy $E_k = \frac{1}{2}mV^2$; its mechanical energy is $E = E_p + E_k$.
2. At $y = h$, the mass is at rest and thus $E_k = 0$ while $E_p = mgh$. Since there is no friction E is conserved and so $mgh = mgy + \frac{1}{2}mV^2$, whence the result follows immediately.
3. ℓ increases with y, while V increases as y decreases. So $V = -\frac{d\ell(y)}{dt} = -\frac{d\ell(y)}{dy}\frac{dy}{dt}$. Equaling the two expressions for V, we get

$$\sqrt{2g(h-y)} = -\frac{d\ell(y)}{dy}\frac{dy}{dt} \Rightarrow \frac{dt}{dy} = -\frac{1}{\sqrt{2g(h-y)}}\frac{d\ell(y)}{dy}$$

Integrating this from the initial position of the mass h down to the bottom of the curve, $T(h) = \int_h^0 -\frac{1}{\sqrt{2g(h-y)}}\frac{d\ell(y)}{dy}\,dy$ is obtained, which is the same as the desired result.
4. Since T does not depend on h, the integral in the previous result can have any value of y as its upper limit of integration. Let us rewrite it as

$$\sqrt{2g}T = \underbrace{\int_0^y (y-\upsilon)^{\overbrace{-(-\frac{1}{2})-1}^{-1/2}} \frac{d\ell(\upsilon)}{d\upsilon}\,d\upsilon}_{\Gamma(\frac{1}{2})\,{}_0D_y^{-\frac{1}{2}}\frac{d\ell(y)}{dy}}$$

The desired result follows immediately.
5. This result is a straightforward application of formula 6 of Table A12 in the Appendix.
6. From the definition of ℓ, we know that $\frac{d\ell(y)}{dy} = \sqrt{1+\left(\frac{dx}{dy}\right)^2}$. Equaling this to the result above,

$$\sqrt{1+\left(\frac{dx}{dy}\right)^2} = ay^{-\frac{1}{2}} \Rightarrow \frac{dx}{dy} = \sqrt{\frac{a^2}{y} - 1} \tag{12.25}$$

from which the desired result follows immediately.

7. From the definition of the cycloid,

$$\theta = \arccos \frac{r-y}{r}$$

$$\Rightarrow x = r \arccos \frac{r-y}{r} + r \sin \arccos \frac{r-y}{r}$$

$$= r \arccos \frac{r-y}{r} + r \frac{\sqrt{r^2 - (r^2 - 2ry + y^2)}}{r}$$

$$= r \arccos \frac{r-y}{r} + \sqrt{2ry - y^2}$$

$$\Rightarrow \frac{dx}{dy} = r \frac{-1}{\sqrt{1 - \frac{(r-y)^2}{r^2}}} \left(-\frac{1}{r}\right) + \frac{1}{2} \left(2ry - y^2\right)^{-\frac{1}{2}} (2r - 2y)$$

$$= \frac{r}{\sqrt{r^2 - (r^2 - 2ry + y^2)}} + \frac{r-y}{\sqrt{2ry - y^2}}$$

$$= \frac{2r - y}{\sqrt{y(2r-y)}} = \sqrt{\frac{2r-y}{y}} = \sqrt{\frac{2r}{y} - 1}$$

This is the desired result, with radius r as mentioned.

Problem 12.4 Establish a correspondence between Nyquist plots A and B in Figure 12.4, Bode diagrams α and β in Figure 12.5, root-locus plots a and b in Figure 12.6, and the two transfer functions $G_1(s) = \frac{1}{s(s+1)^{1.5}(s+2)}$ and $G_2(s) = \frac{64.47+12.46s}{0.598+39.96s^{1.25}}$.

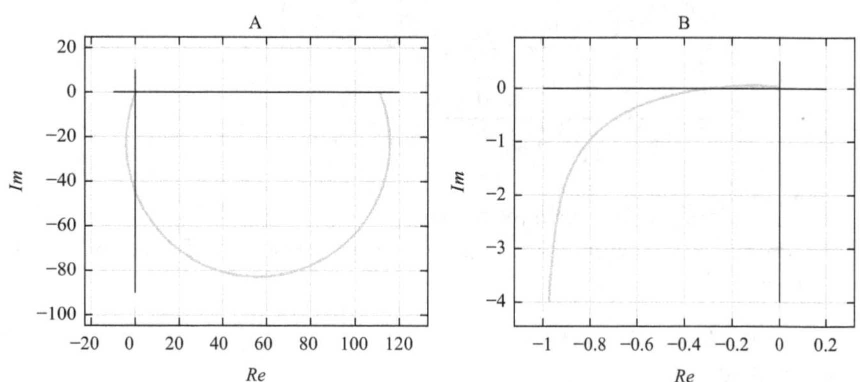

Figure 12.4 Nyquist diagrams of Problem 12.4

Resolution $G_1(s)$ is the only transfer function with a pole at the origin, so it must correspond to Nyquist plot B, Bode plot α, and root-locus plot a; $G_2(s)$ corresponds to the other ones.

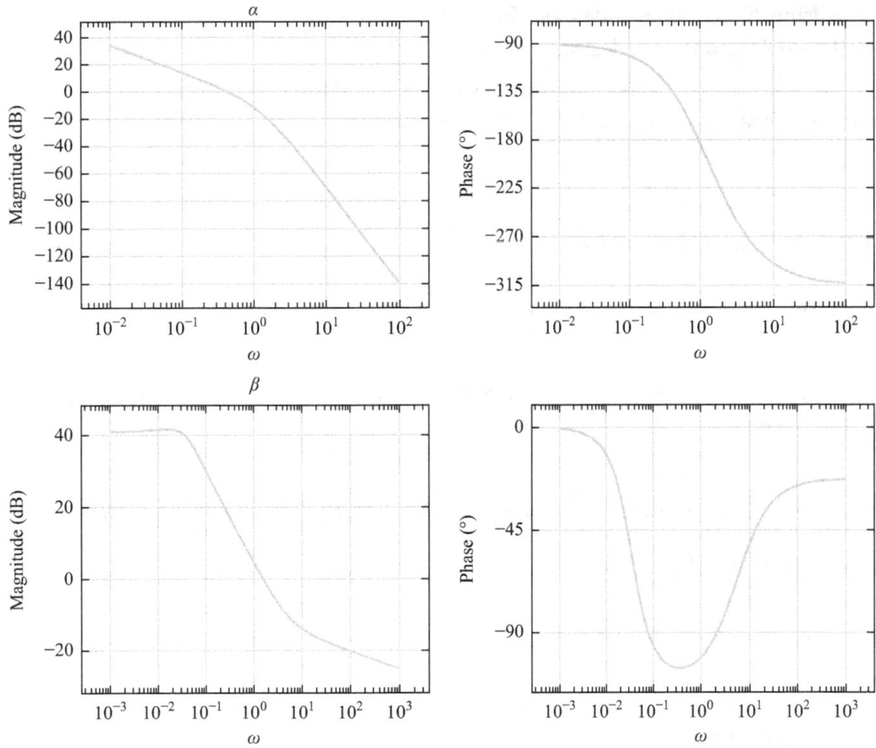

Figure 12.5 Bode diagrams of Problem 12.4

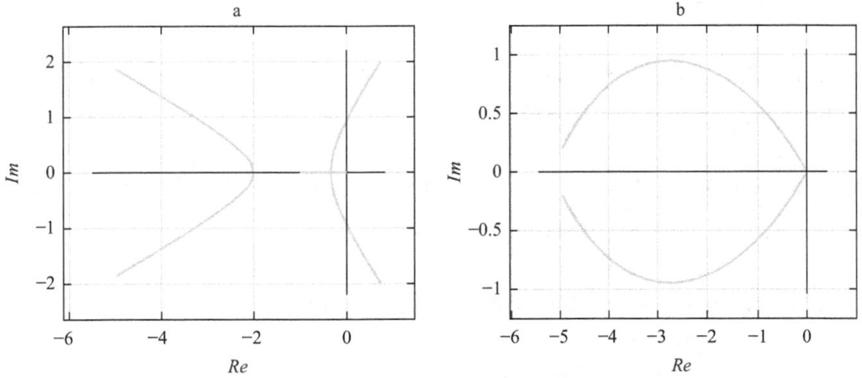

Figure 12.6 Root-locus plots of Problem 12.4

Problem 12.5 Find a fractional first-generation CRONE controller for a plant with transfer function $G(s) = \frac{s^2 + 12.2s + 6.31}{s^3 + 1.938s^2 + 0.1662s + 0.000631}$, ensuring a 90° phase margin in the [0.1 rad/s, 10 rad/s] range.

Resolution Since in the desired frequency range the phase of $G(s)$ is approximately $-135°$, and the desired phase is $-180° + 90° = -90°$, the controller will be $s^{\frac{-90° - (-135°)}{90°}} = s^{0.5}$. This can be multiplied by a gain different from 1, as long as the gain crossover frequency remains within the range.

12.3 Proposed problems

Exercise 12.1 Find the inverse Laplace transform of $\frac{6s^{\frac{1}{2}} - 3}{s + s^{\frac{1}{2}} - 2}$.

Exercise 12.2 Find the impulse and step responses of a plant with transfer function $G(s) = \frac{10}{s^{\frac{1}{2}} + 100}$.

Exercise 12.3 Consider a plant with transfer function $G(s) = \frac{1}{s^{\frac{1}{5}} + 20}$, and its time response $y(t) = t^{\frac{6}{5}} E_{\frac{1}{5}, \frac{11}{5}}(-20t^{\frac{1}{5}})$.

A) $y(t)$ is the response of $G(s)$ when the input is an impulse.
B) $y(t)$ is the response of $G(s)$ when the input is a unit step.
C) $y(t)$ is the response of $G(s)$ when the input is a ramp with slope 1.
D) None of the above.

Exercise 12.4 Use Matignon's theorem to find whether the plant with transfer function $\dfrac{1}{s - 6s^{1/2} + 18}$ is stable.

Exercise 12.5 Use Matignon's theorem to find whether the plant with transfer function $\dfrac{1}{s^{4/3} - 6s^{2/3} + 18}$ is stable.

Exercise 12.6 Use Matignon's theorem to find whether the plant with transfer function $\dfrac{s^{4/3} - 2}{s^{8/3} + 6s^{4/3} + 18}$ is stable.

Exercise 12.7 Consider the following six transfer functions:

$$G_1(s) = \frac{1}{\left(\frac{s}{10}\right)^{\frac{1}{2}} + 1}$$

$$G_2(s) = \frac{1}{\left(\frac{s}{10}\right)^{\frac{3}{2}} + 1}$$

$$G_3(s) = \frac{1}{\left(\frac{s}{10}\right)^3 + 4\left(\frac{s}{10}\right)^{\frac{3}{2}} + 1}$$

$$G_4(s) = \frac{1}{\left(\frac{s}{10}\right)^{\frac{4}{3}} + 4\left(\frac{s}{10}\right)^{\frac{2}{3}} + 1}$$

$$G_5(s) = \frac{1}{\left(\frac{s}{10}\right)^{3.9} + 4\left(\frac{s}{10}\right)^{1.95} + 1}$$

$$G_6(s) = \frac{1}{\left(\frac{s}{10}\right)^{\frac{4}{3}} - 0.2\left(\frac{s}{10}\right)^{\frac{2}{3}} + 1}$$

1. Which transfer functions have frequency responses with resonance peaks?
2. Plot the Nyquist and Bode diagrams of these transfer functions.
3. Which of these transfer functions correspond to the root-locus diagrams in Figure 12.7?

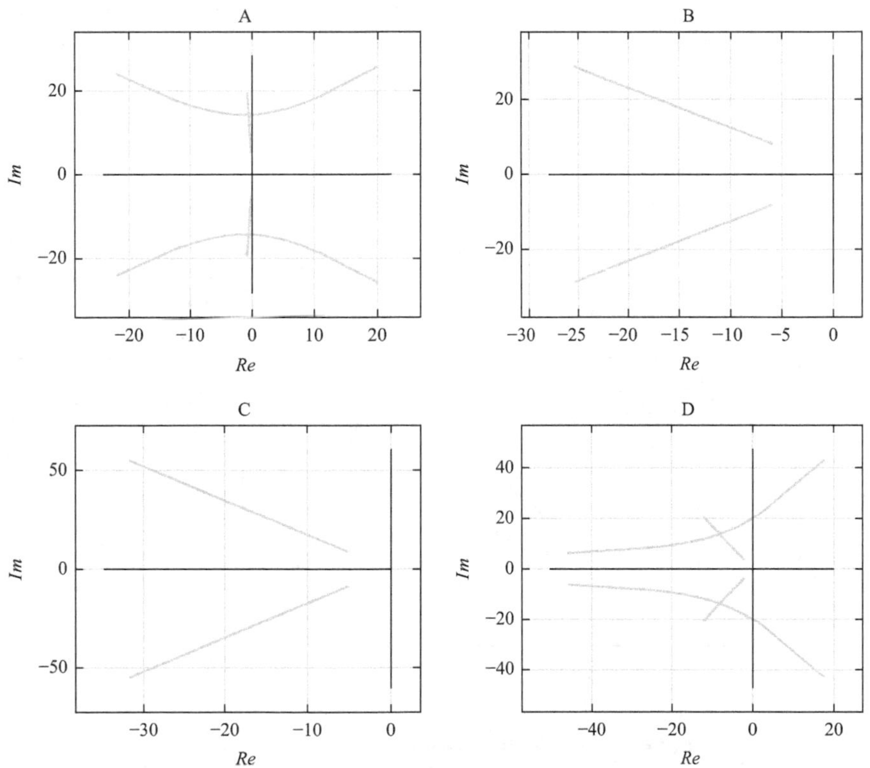

Figure 12.7 Root-locus plots of Exercise 12.7

Exercise 12.8 Establish a correspondence between Nyquist plots A–D in Figure 12.8, Bode plots α–δ in Figure 12.9, root-locus plots a–d in Figure 12.10, and transfer functions

$$G_1(s) = \frac{1}{s^{1.3} + 1}$$

$$G_2(s) = \frac{1}{s(s^{1.3} + 1)}$$

$$G_3(s) = \frac{1}{s^{0.6}(s^{1.3} + 1)}$$

$$G_4(s) = \frac{s^{1.1} + 2}{s^{1.9} + 1}$$

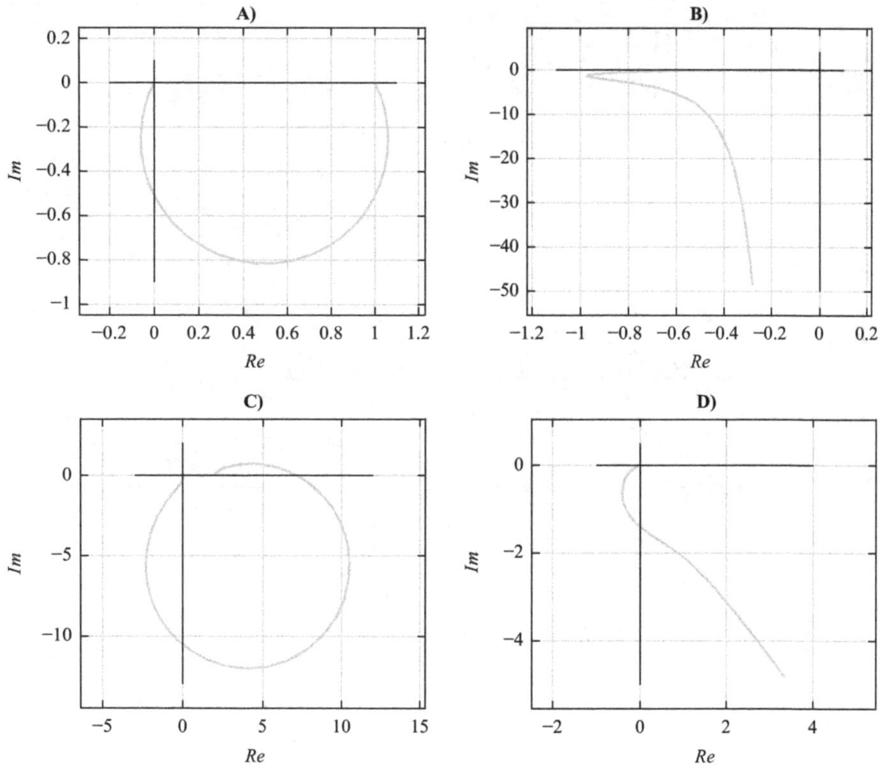

Figure 12.8 Nyquist diagrams of Exercise 12.8

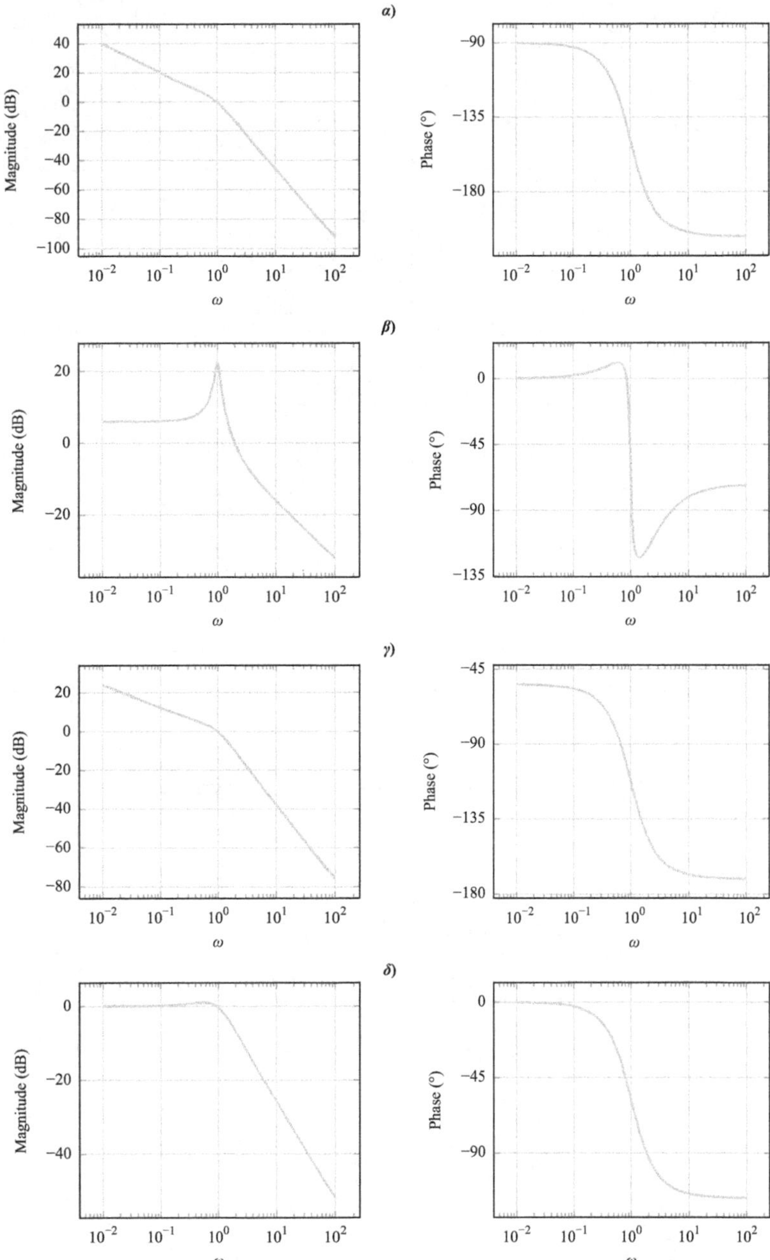

Figure 12.9 Bode diagrams of Exercise 12.8

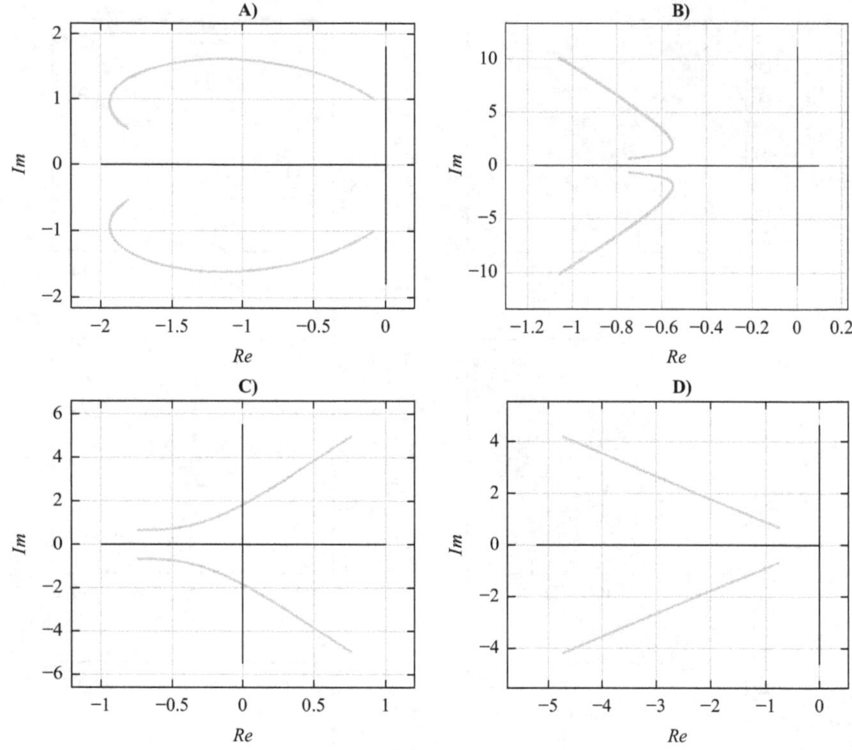

Figure 12.10 Root-locus plots of Exercise 12.8

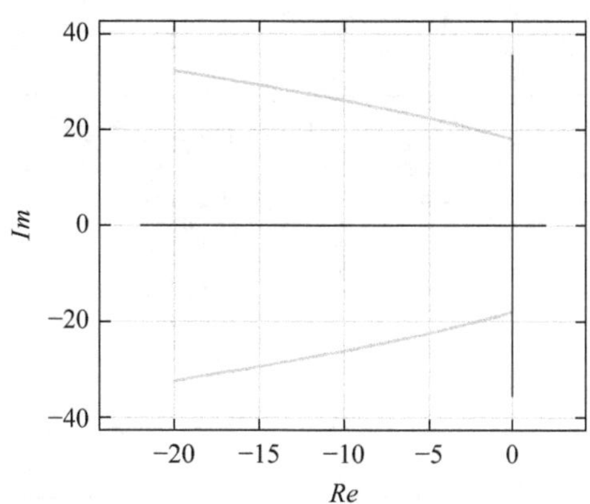

Figure 12.11 Root-locus plot of Exercise 12.9

Exercise 12.9 The root-locus plot in Figure 12.11 corresponds to transfer function:

A) $\dfrac{s^{4/3} - 2}{s^{8/3} + 6s^{4/3} + 18}$

B) $\dfrac{1}{s^{4/3} - 6s^{2/3} + 18}$

C) $\dfrac{1}{s - 6s^{1/2} + 18}$

D) None of the above.

Exercise 12.10 Find CRONE approximations for the transfer functions $G_1(s)$, $G_2(s)$ and $G_3(s)$ of Exercise 12.7. Let ω_l be three decades before the asymptotic gain-crossover frequency, and ω_h three decades after. Use five poles and five zeros to approximate s^α.

Exercise 12.11 The following controller achieves a nearly constant $-60°$ open-loop phase in the [10 rad/s, 200 rad/s] frequency range for plant $\frac{100s+50}{(s^2+0.1s+0.2)(\frac{s}{5000}+1)}$:

A) $C(s) = s^{-\frac{1}{3}}$

B) $C(s) = \frac{1}{s^{\frac{1}{3}}+1}$

C) $C(s) = 0.1s^{\frac{1}{3}}$

D) None of the above.

Exercise 12.12 The following controller achieves a nearly constant $-130°$ open-loop phase in the [10 rad/s, 1 000 rad/s] frequency range for plant $\frac{1}{s(s+0.01)}$:

A) $C(s) = s^{-0.56}$

B) $C(s) = \frac{s^{0.56}}{s^{0.56}+1\,000}$

C) $C(s) = \frac{s^{0.56}}{s^{0.56}+10}$

D) $C(s) = s^{0.44}$

Exercise 12.13 Suppose we want to control a plant with transfer function $G(s) = \frac{1}{4.32s^2+19.18s+1}$. Which of the following fractional PID controllers keeps the phase of the open loop equal to $-115°\pm10°$ in the [0.1 rad/s, 100 rad/s] frequency range?

A) $0.088 + \frac{6.52}{s^{0.68}} + 2.59s^{0.7}$

B) $3.22 + \frac{9.07}{s^{0.54}} + 3.12s^{0.66}$

C) $7 + \frac{12.4}{s^{0.6}} + 4.1s^{0.78}$

D) None of the above.

Exercise 12.14 Consider a fractional PID controller given by $C(s) = P + \frac{I}{s^\lambda} + Ds^\mu$.

1. Its phase at high frequencies is given by
   A) $-90° \times \lambda$
   B) $+90° \times \lambda$

C) $-90° \times \mu$
D) $+90° \times \mu$

2. Its phase at low frequencies is given by
   E) $-90° \times \lambda$
   F) $+90° \times \lambda$
   G) $-90° \times \mu$
   H) $+90° \times \mu$

12.4 Fractional control using computer packages

Unofficial packages for MATLAB® handling fractional order systems and controllers can be found in the Internet, namely the *CRONE®* toolbox, the *Ninteger* toolbox and the *Matrix approach to distributed-order ODEs and PDEs* toolbox (developed especially, as the name says, to deal with ordinary differential equations and partial differential equations of fractional order). Still all relevant calculations can be carried out using only simple commands.

12.4.1 MATLAB

12.4.1.1 Calculating a fractional derivative

Using the built-in gamma function, we can implement combinations (12.2) as

```
function out = combinations(a,b)

if isempty(a)
    out = a;
elseif isempty(b)
    out = b;
else
    if size(a)==1
        a = ones(size(b)) * a; % this ensures
        % better handling of cases like
        % combinations (-1,[0 1 2 3 4 5])
    end
    out = gamma(a+1) ./ (gamma(b+1) .* gamma(a-b+1));
    % most usual expression
    temp = (~(a-round(a))) .* (a<0); % this is where
    % an alternative expression will be used
    if sum(temp)
        outalt = (-1).^b .* gamma(b-a) ./ ...
            (gamma(b+1) .* gamma(-a));
        % alternative expression
        out(temp) = outalt(temp);
    end
end
```

This function is used by the code below to compute numerically the $\frac{3}{4}$-order derivative of $f(t) = 10\sin(t/2)$. This is done till $t = 30$ with a 0.05 step, using the Grünwald–Letnikov definition (12.3). The result is shown in Figure 12.12. It is easy to turn this code into a function that receives $f(t)$, the final time, the time step and α, returning the desired derivative. More accurate results for large times and small steps can be obtained as in Reference 36.

```
Ts = 0.05; % sampling time
tend = 30; % final time
t = 0 : Ts : tend; % time vector
f = 10 * sin(0.5*t); % function to differentiate
alpha = 3/4; % differentiation order
D = zeros(size(f)); % initialize the result with zeros
for index = 1 : length(D)
    k = 0 : index-1; % because t(index)/Ts == index-1
    temp = (-1).^k .* combinations(alpha,k) .* ...
        f(index-k) / Ts^alpha;
    temp(isnan(temp)) = 0; % in case something goes
    % wrong
    D(index) = sum(temp);
end
figure, plot(t, f, t, D)
```

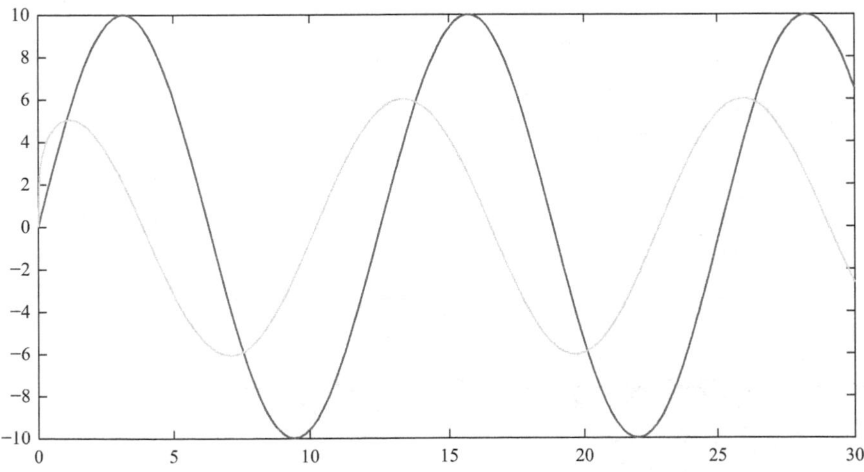

Figure 12.12 $10\sin(t/2)$ *and* ${}_cD_t^{\frac{3}{4}}10\sin(t/2)$

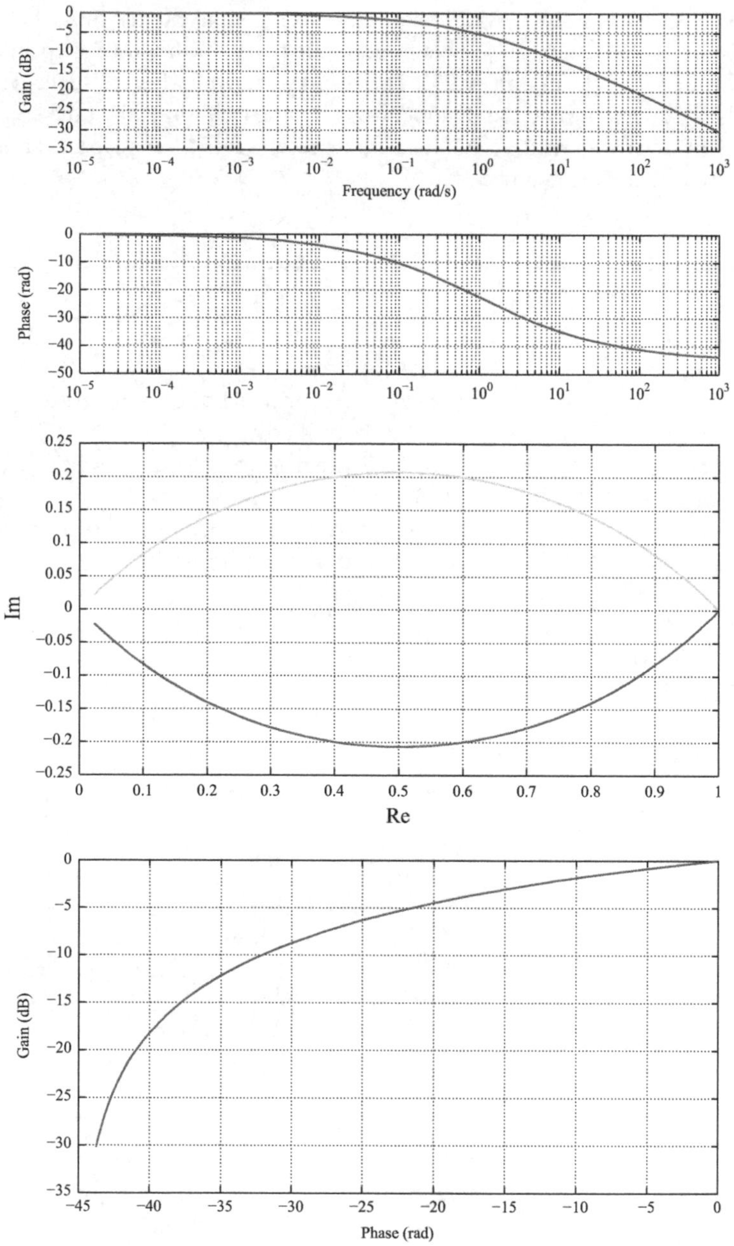

Figure 12.13 Frequency response of $G_1(s) = \frac{1}{(\frac{s}{10})^{\frac{1}{2}}+1}$ *from Exercise 12.7*

12.4.1.2 Calculating a frequency response

The frequency response of transfer function $G_1(s)$ from Exercise 12.7 can be obtained as follows. Results are shown in Figure 12.13.

```
w = logspace(-5, 3, 81); % frequency range,
    % 10 frequencies per decade
G1 = 1 ./ ( (w*1i).^0.5 + 1 );
gain = 20*log10(abs(G1)); % in dB
phase = angle(G1) * 180 / pi; % in degrees
% Bode diagram
figure, subplot(2,1,1), semilogx(w, gain)
grid on, xlabel('frequency / rad \cdots^-1'), ...
   ylabel('gain / dB')
subplot(2,1,2), semilogx(w, phase)
grid on, ylabel('phase / rad')
% Nyquist diagram
figure, plot(real(G1),imag(G1), real(G1),-imag(G1))
grid on, xlabel('Re'), ylabel('Im')
% Nichols diagram
figure, plot(phase,gain)
grid on, xlabel('phase / rad'), ylabel('gain / dB')
```

12.4.2 SCILAB

12.4.2.1 Calculating a fractional derivative

The code for combinations (12.2) is practically identical:

```
function out = combinations(a,b)

if isempty(a)
    out = a;
elseif isempty(b)
    out = b;
else
    if length(a)==1
        a = ones(length(b)) * a; // this ensures
        // better handling of cases like
        // combinations (-1,[0 1 2 3 4 5])
    end
    out = gamma(a+1) ./ (gamma(b+1) .* gamma(a-b+1));
    // most usual expression
    temp = (~(a-round(a))) .* (a<0); // this is where
    // an alternative expression will be used
    if sum(temp)
```

```
        outalt = (-1).^b .* gamma(b-a) ./ ...
            (gamma(b+1) .* gamma(-a));
        // alternative expression
        out(temp) = outalt(temp);
    end
end
```

The $\frac{3}{4}$-order derivative of $f(t) = 10\sin(t/2)$ is found as follows.

```
Ts = 0.05; // sampling time
tend = 30; // final time
t = 0 : Ts : tend; // time vector
f = 10 * sin(0.5*t); // function to differentiate
alpha = 3/4; // differentiation order
D = zeros(1,length(f)); // initialize the result with
zeros for index = 1 : length(D)
    k = 0 : index-1; // because t(index)/Ts == index-1
    temp = (-1).^k .* combinations(alpha,k) .* ...
        f(index-k) / Ts^alpha;
    temp(isnan(temp)) = 0; // in case something
    // goes wrong
    D(index) = sum(temp);
end
figure, plot(t, f, t, D)
```

12.4.2.2 Calculating a frequency response

The frequency response of transfer function $G_1(s)$ from Exercise 12.7 is now obtained as follows.

```
w = logspace(-5, 3, 81); // frequency range,
    // 10 frequencies per decade
G1 = 1 ./ ( (w*sqrt(-1)).^0.5 + 1 );
gain = 20*log10(abs(G1)); // in dB
phase = atan(imag(G1),real(G1)) * 180 / (2*acos(0));
        // in degrees
// Bode diagram
figure, subplot(2,1,1), plot2d("ln",w, gain)
set(gca(),"grid",[1 1]);
xlabel('frequency / rad \cdots^-1'), ylabel ...
   ('gain / dB')
```

```
subplot(2,1,2), plot2d("ln",w, phase)
set(gca(),"grid",[1 1]);
ylabel('phase / rad')
// Nyquist diagram
figure, plot(real(G1),imag(G1), real(G1),-imag(G1))
set(gca(),"grid",[1 1]);
xlabel('Re'), ylabel('Im')
// Nichols diagram
figure, plot(phase,gain)
set(gca(),"grid",[1 1]);
xlabel('phase / rad'), ylabel('gain / dB')
```

12.4.3 OCTAVE

The code provided above for MATLAB works just the same.

Appendix A

Table A.1 Table of Laplace transforms

	x(t)	**X(s)**
1	$\delta(t)$	1
2	$1(t)$	$\frac{1}{s}$
3	t	$\frac{1}{s^2}$
4	t^2	$\frac{2}{s^3}$
5	e^{-at}	$\frac{1}{s+a}$
6	$1-e^{-at}$	$\frac{a}{s(s+a)}$
7	te^{-at}	$\frac{1}{(s+a)^2}$
8	$\sin(\omega t)$	$\frac{\omega}{s^2+\omega^2}$
9	$\cos(\omega t)$	$\frac{s}{s^2+\omega^2}$
10	$e^{-at}\sin(\omega t)$	$\frac{\omega}{(s+a)^2+\omega^2}$
11	$e^{-at}\cos(\omega t)$	$\frac{s+a}{(s+a)^2+\omega^2}$
12	$\frac{1}{b-a}\left(e^{-at}-e^{-bt}\right)$	$\frac{1}{(s+a)(s+b)}$
13	$\frac{1}{ab}\left(1+\frac{1}{a-b}\left(be^{-at}-ae^{-bt}\right)\right)$	$\frac{1}{s(s+a)(s+b)}$
14	$\frac{\omega}{\Xi}e^{-\zeta\omega t}\sin(\omega\Xi t)$	$\frac{\omega^2}{s^2+2\zeta\omega s+\omega^2}$
15	$-\frac{1}{\Xi}e^{-\zeta\omega t}\sin(\omega\Xi t-\phi)$	$\frac{s}{s^2+2\zeta\omega s+\omega^2}$
16	$1-\frac{1}{\Xi}e^{-\zeta\omega t}\sin(\omega\Xi t+\phi)$	$\frac{\omega^2}{s\left(s^2+2\zeta\omega s+\omega^2\right)}$

Note 1: $\Xi=\sqrt{1-\zeta^2}$; $\phi=\arctan\frac{\Xi}{\zeta}$.

Table A.2 Laplace transform properties

	x(t)	**X(s)**
1	$Ax_1(t) + Bx_2(t)$	$AX_1(s) + BX_2(s)$
2	$ax(at)$	$X\left(\frac{s}{a}\right)$
3	$e^{at}x(t)$	$X(s-a)$
4	$\begin{cases} x(t-a), & t>a \\ 0, & t<a \end{cases}$	$e^{-as}X(s)$
5	$\frac{dx(t)}{dt}$	$sX(s) - x(0)$
6	$\frac{d^2x(t)}{dt^2}$	$s^2X(s) - sx(0) - x'(0)$
7	$\frac{d^nx(t)}{dt^n}$	$s^nX(s) - s^{n-1}x(0) - \cdots - x^{(n-1)}(0)$
8	$-tx(t)$	$\frac{dX(s)}{ds}$
9	$t^2x(t)$	$\frac{d^2X(s)}{ds^2}$
10	$(-1)^nt^nx(t)$	$\frac{d^nX(s)}{ds^n}$
11	$\int_0^t x(u)\,du$	$\frac{1}{s}X(s)$
12	$\int_0^t \cdots \int_0^t x(u)\,du = \int_0^t \frac{(t-u)^{(n-1)}}{(n-1)!}x(u)\,du$	$\frac{1}{s^n}X(s)$
13	$x_1(t) * x_2(t) = \int_0^t x_1(u)\,x_2(t-u)\,du$	$X_1(s)\,X_2(s)$
14	$\frac{1}{t}x(t)$	$\int_s^\infty X(u)\,du$
15	$x(t) = x(t+T)$	$\frac{1}{1-e^{-sT}}\int_0^T e^{-su}X(u)\,du$
16	$x(0)$	$\lim_{s\to\infty} sX(s)$
17	$x(\infty) = \lim_{t\to\infty} x(t)$	$\lim_{s\to 0} sX(s)$

Table A.3 Block diagrams and the corresponding simplified blocks

	Transformation	Equation	Block diagram	Equivalent block diagram
1	Cascaded blocks	$Y = (P_1P_2)X$	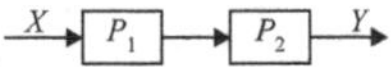	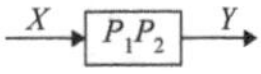
2	Combining blocks in parallel	$Y = P_1X \pm P_2X$	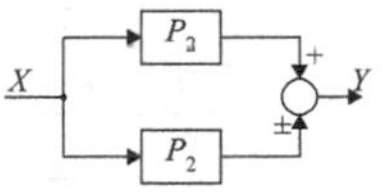	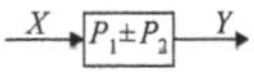
3	Removing a block from a forward loop	$Y = P_1X \pm P_2X$		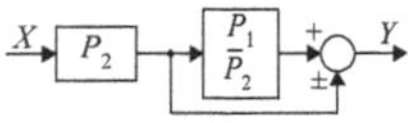
4	Eliminating a feedback loop	$Y = P_1(X \mp P_2Y)$	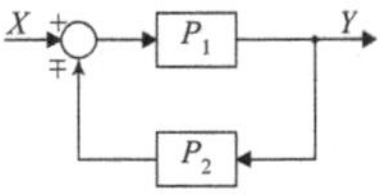	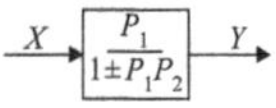
5	Removing a block from a feedback loop	$Y = P_1(X \mp P_2Y)$		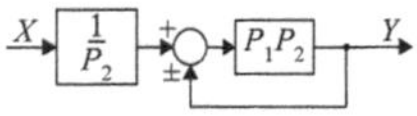
6	Rearranging summing junctions	$Z = W \pm X \pm Y$	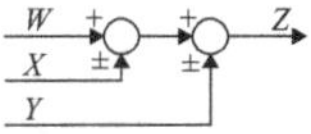	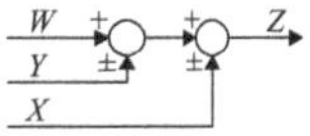 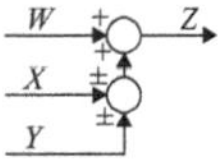
7	Moving a summing junction in front of a block	$Z = PX \pm Y$	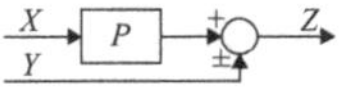	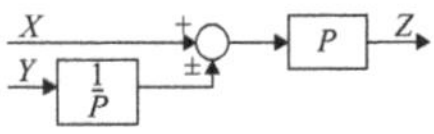
8	Moving a summing junction beyond a block	$Z = P(X \pm Y)$	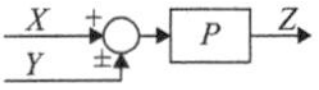	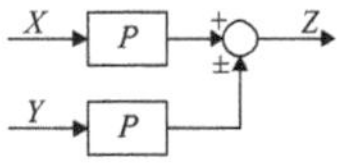
9	Moving a take-off point in front of a block	$Y = PX$	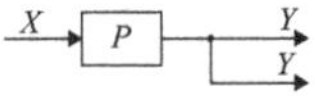	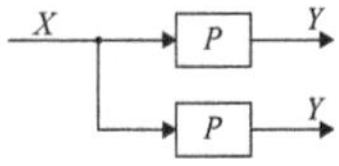
10	Moving a take-off point beyond a block	$Y = PX$	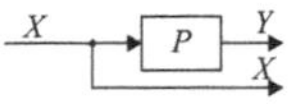	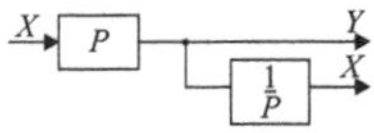
11	Moving a take-off point in front of a summing junction	$Z = W \pm X$	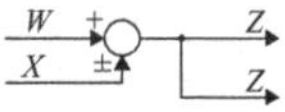	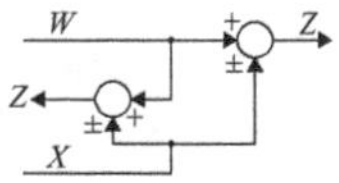
12	Moving a take-off point beyond a summing junction	$Z = X \pm Y$	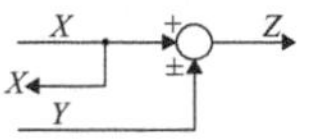	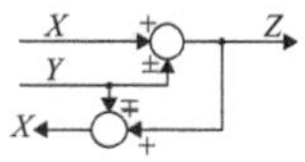

Table A.4 Exact and asymptotic Bode diagrams of elemental factors

Magnitude	**Phase**	**Factor**
		$(j\omega)$
		$(j\omega)^{-1}$
		$(1+j\omega T)$
		$(1+j\omega T)^{-1}$
		$\left[1+2\zeta\left(j\frac{\omega}{\omega_n}\right)+\left(j\frac{\omega}{\omega_n}\right)^2\right]$

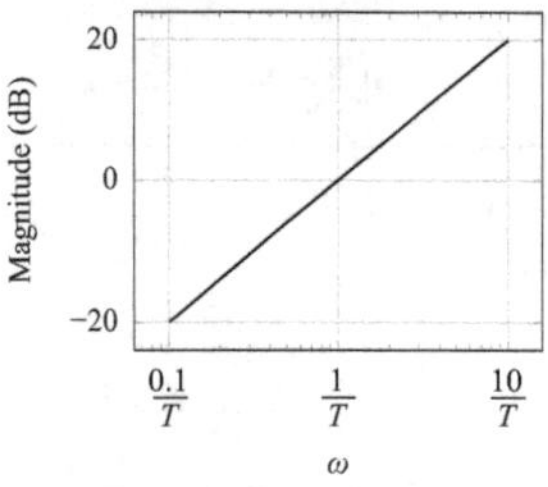

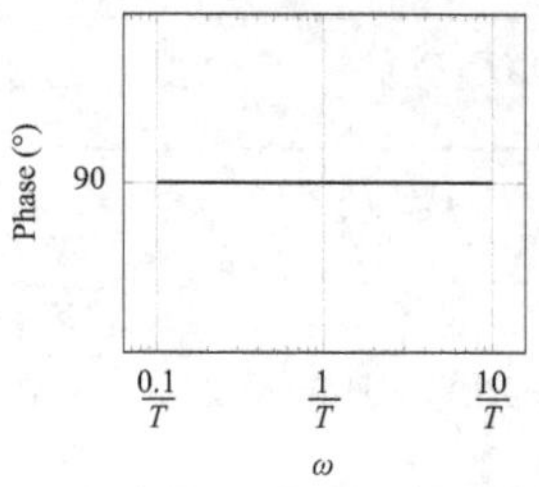

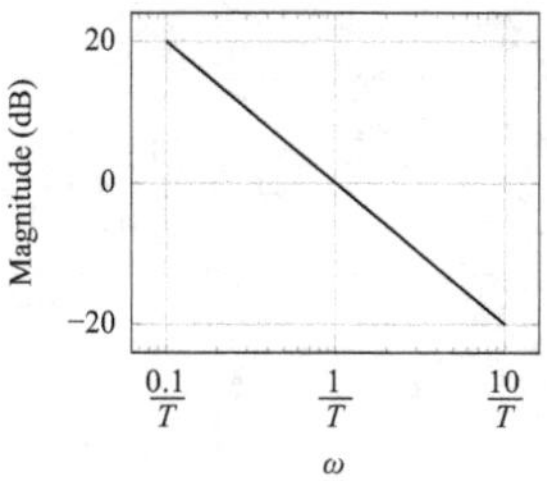

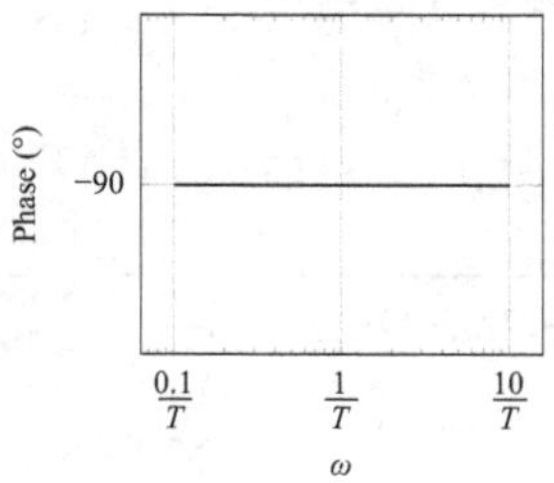

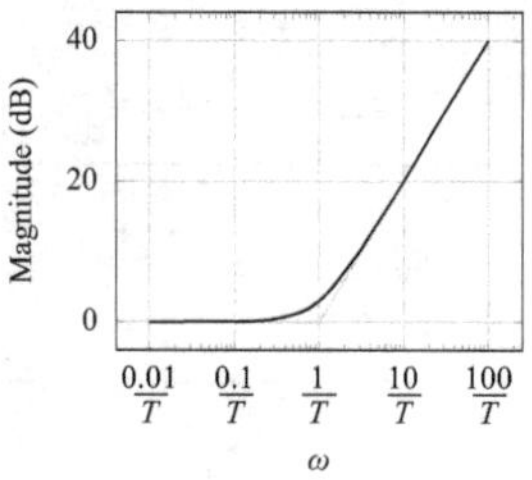

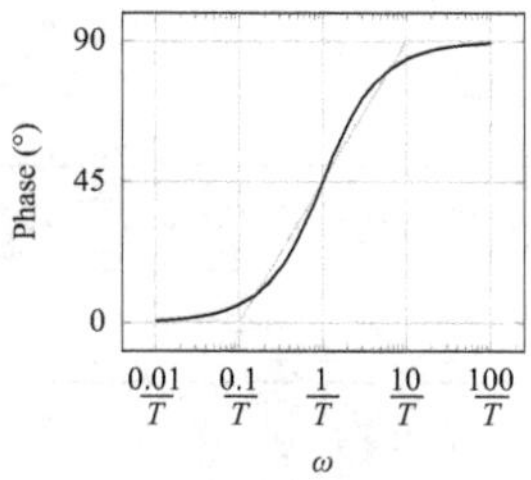

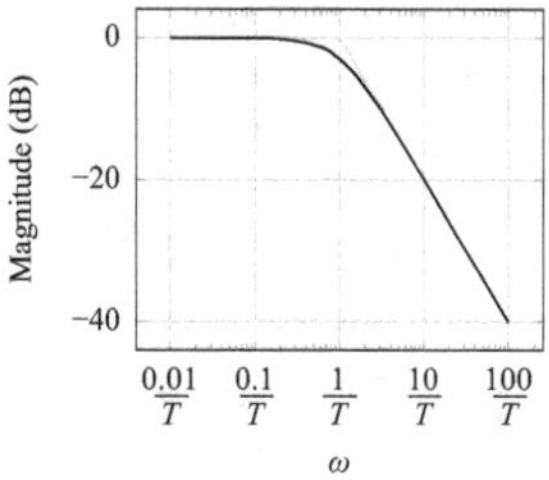

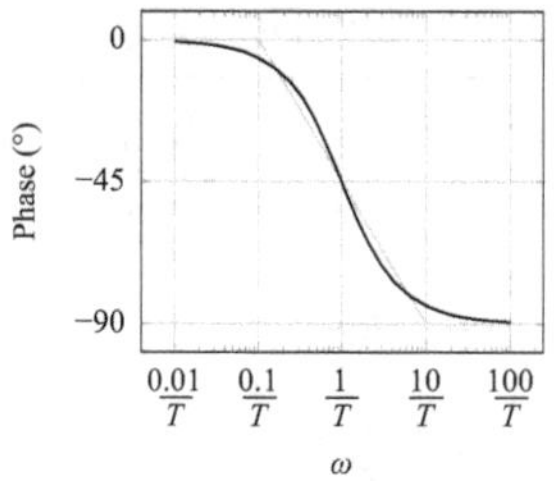

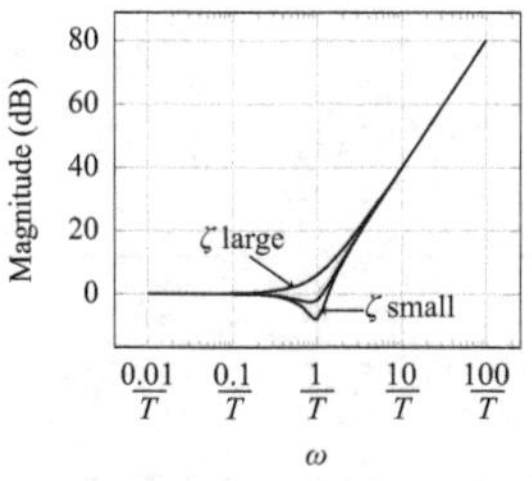

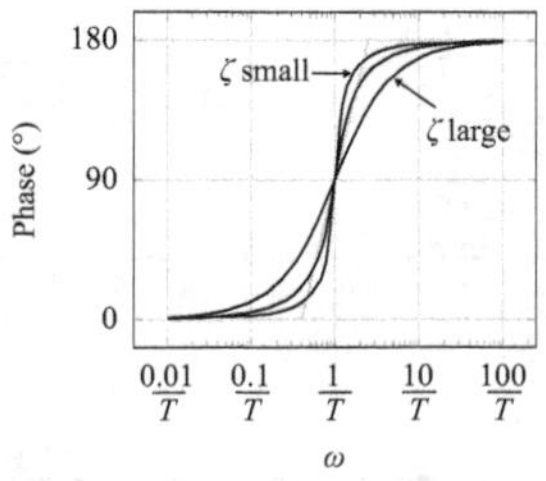

(Continues)

Table A.4 (Continued)

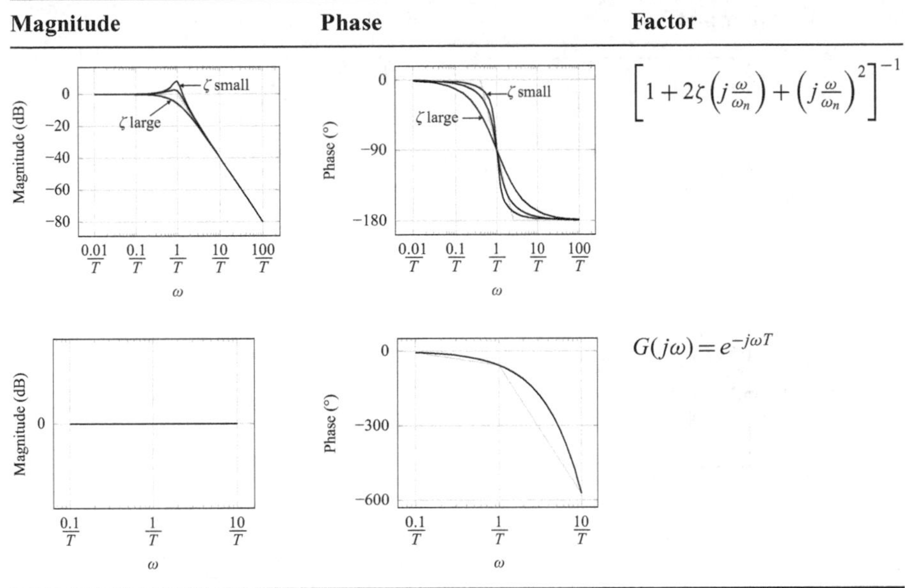

Table A.5 Polar plots of common transfer functions

Polar plot	Factor
Im; $\omega \rightarrow \infty$; $\omega = 0$; 0; Re	$(j\omega)$
Im; 0; $\omega \rightarrow \infty$; $\omega = 0$; Re	$(j\omega)^{-1}$

(*Continues*)

Table A.5 (Continued)

Polar plot	**Factor**
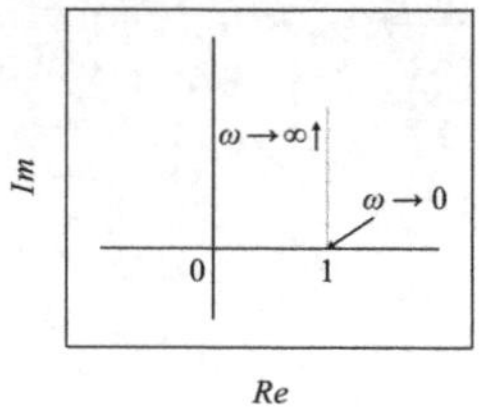	$(1 + j\omega T)$
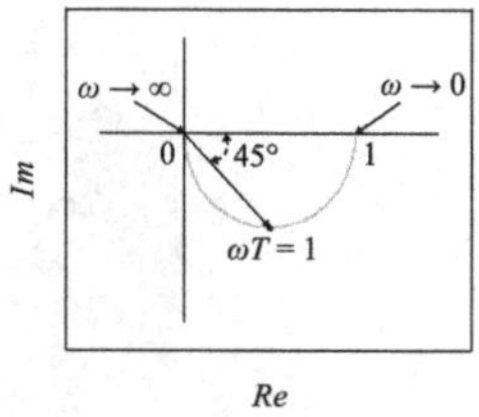	$(1 + j\omega T)^{-1}$
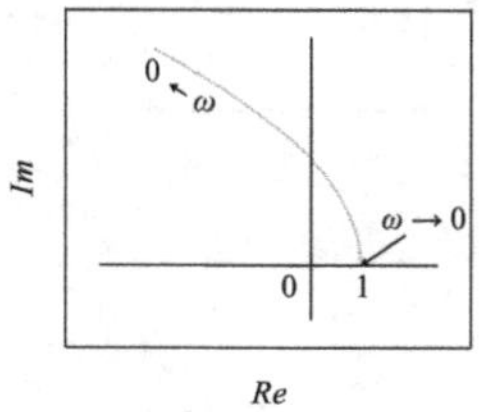	$\left[1 + 2\zeta\left(j\frac{\omega}{\omega_n}\right) + \left(j\frac{\omega}{\omega_n}\right)^2\right]$
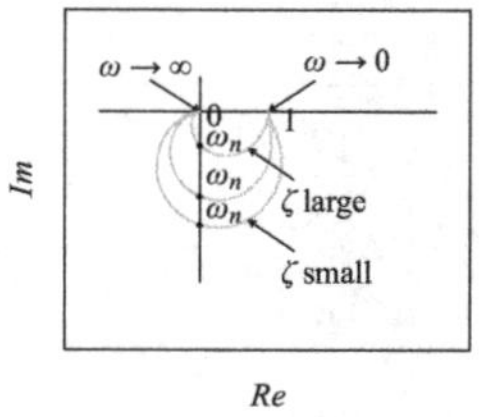	$\left[1 + 2\zeta\left(j\frac{\omega}{\omega_n}\right) + \left(j\frac{\omega}{\omega_n}\right)^2\right]^{-1}$
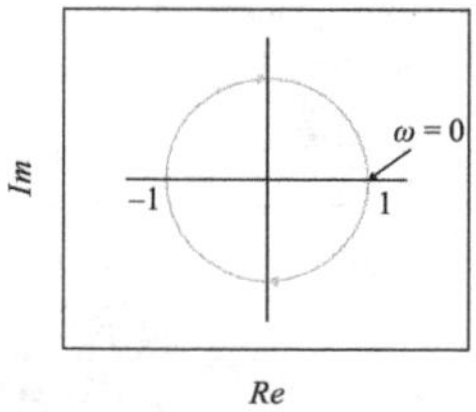	$e^{-j\omega T}$

Table A.6 Nichols plots of common transfer functions

Polar plot	**Factor**
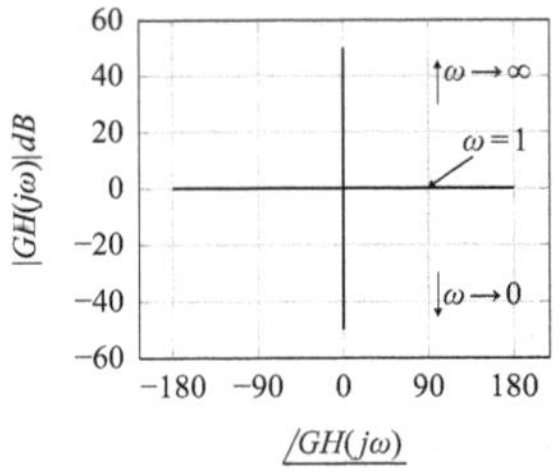	$(j\omega)$
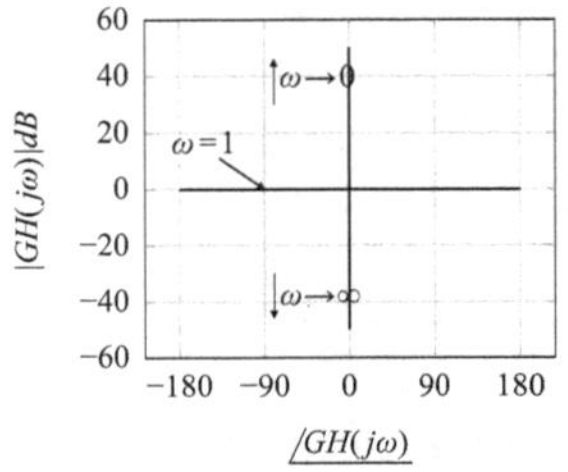	$(j\omega)^{-1}$
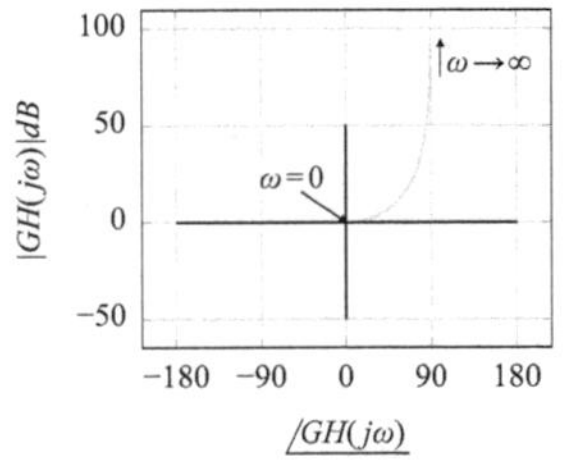	$(1+j\omega T)$
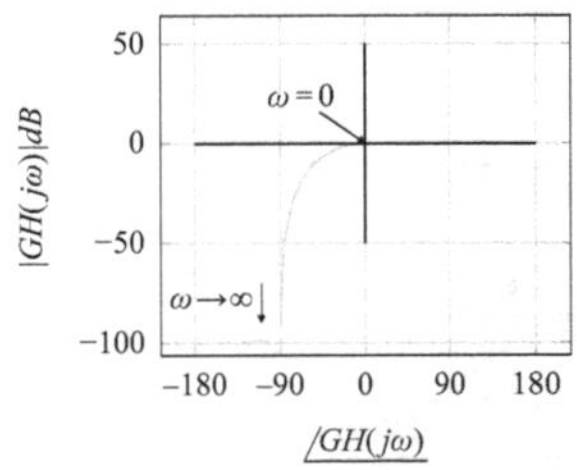	$(1+j\omega T)^{-1}$

(*Continues*)

Table A.6 (Continued)

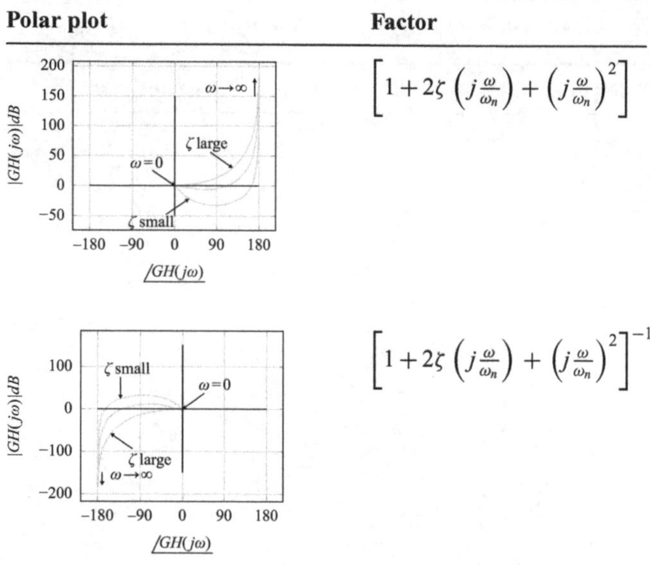

Polar plot	Factor
	$\left[1+2\zeta\left(j\frac{\omega}{\omega_n}\right)+\left(j\frac{\omega}{\omega_n}\right)^2\right]$
	$\left[1+2\zeta\left(j\frac{\omega}{\omega_n}\right)+\left(j\frac{\omega}{\omega_n}\right)^2\right]^{-1}$

Table A.7 Table of the z-transforms of common functions

	$x(t)$	$X(z)$
1	$\delta(t)$	1
2	a^t	$\frac{z}{z-a^h}$
3	t	$\frac{hz}{(z-1)^2}$
4	t^2	$\frac{h^2 z(z+1)}{(z-1)^3}$
5	e^{-at}	$\frac{z}{z-e^{-ah}}$
6	$1-e^{-at}$	$\frac{\left(1-e^{-ah}\right)z}{(z-1)\left(z-e^{-ah}\right)}$
7	te^{-at}	$\frac{he^{-ah}z}{\left(z-e^{-ah}\right)^2}$
8	$\sin(\omega t)$	$\frac{\sin(\omega h)z}{z^2-2\cos(\omega h)z+1}$
9	$\cos(\omega t)$	$\frac{z^2-\cos(\omega h)z}{z^2-2\cos(\omega h)z+1}$
10	$e^{-at}\sin(\omega t)$	$\frac{e^{-ah}\sin(\omega h)z}{z^2-2e^{-ah}\cos(\omega h)z+e^{-2ah}}$
11	$e^{-at}\cos(\omega t)$	$\frac{z^2-e^{-ah}\cos(\omega h)z}{z^2-2e^{-ah}\cos(\omega h)z+e^{-2ah}}$

Note: h is the sampling period.

Table A.8 Table of z-transform properties

	x(t)	**X(z)**
1	$ax_1(t) + bx_2(t)$	$aX_1(z) + bX_2(z)$
2	$a^{-t}x(t)$	$X(a^h z)$
3	$x(t - kh)$	$z^{-k}X(z),\ \forall t,\ x(t - kh) = 0\ \text{if}\ t < kh$
4	$x(t + kh)$	$z^k X(z) - z^k x(0) - z^{k-1}x(h) - \cdots - zx[(k-1)h],\ t \geq 0,$ $x(t + kh) = 0\ \text{if}\ t < 0$
5	$-tx(t)$	$hz\frac{dX(z)}{dz}$
6	$x(0)$	$\lim_{z\to\infty} X(z)$
7	$\lim_{t\to\infty} x(t)$	$\lim_{z\to 1} [(z-1)X(z)]$

Table A.9 Table of describing functions for some common nonlinearities

Nonlinearity	**N(X) with** $f(\gamma) = \frac{2}{\pi}\left[\arcsin(\gamma) + \gamma\sqrt{1-\gamma^2}\right]$
Output, M, Input, −M	$N = \frac{4M}{\pi X}$
Output, k_2, k_1, Input, A	$N = \begin{cases} k_2 + (k_1 - k_2)f\left(\frac{A}{X}\right), & X > A \\ k_1, & X \leq A \end{cases}$
Output, A, k, Input, −A	$N = k + \frac{4A}{\pi X}$
Output, B, k, Input, −A, A, −B	$N = \begin{cases} k\left[1 - f\left(\frac{A}{X}\right)\right] + \frac{4B}{\pi X}\sqrt{1 - \left(\frac{A}{X}\right)^2}, & X > A \\ 0, & X \leq A \end{cases}$
Output, k, Input, A	$N = \begin{cases} kf\left(\frac{A}{X}\right), & X > A \\ k, & X \leq A \end{cases}$

(Continues)

Table A.9 (Continued)

Nonlinearity	**$N(X)$ with $f(\gamma) = \frac{2}{\pi}\left[\arcsin(\gamma) + \gamma\sqrt{1-\gamma^2}\right]$**
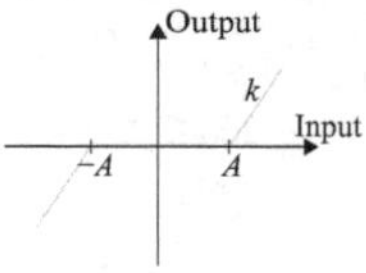 Output; k; Input; $-A$; A	$N = \begin{cases} k\left[1 - f\left(\frac{A}{X}\right)\right], & X > A \\ 0, & X \le A \end{cases}$
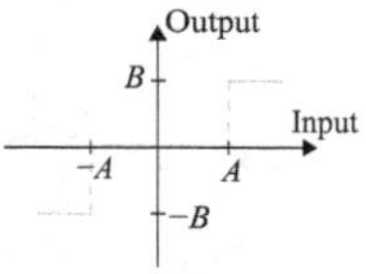 Output; B; Input; $-A$; A; $-B$	$N = \begin{cases} \frac{4B}{\pi X}\sqrt{1 - \left(\frac{A}{X}\right)^2}, & X > A \\ 0, & X \le A \end{cases}$
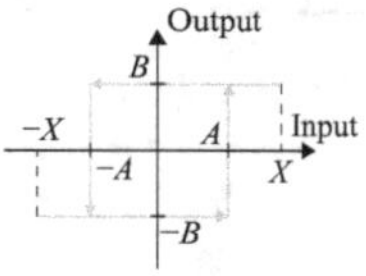 Output; B; $-X$; A; Input; $-A$; X; $-B$	$N = \begin{cases} \frac{4B}{\pi X}\sqrt{1 - \left(\frac{A}{X}\right)^2} - j\frac{4AB}{\pi X^2}, & X > A \\ 0, & X \le A \end{cases}$
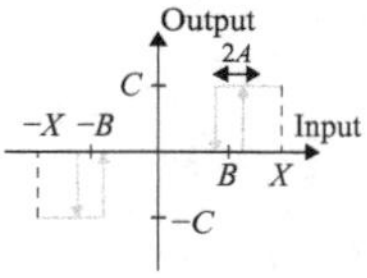 Output; $2A$; C; $-X$ $-B$; Input; B X; $-C$	$N = \begin{cases} \frac{2C}{\pi X}\left[\sqrt{1 - \left(\frac{B-A}{X}\right)^2} + \sqrt{1 - \left(\frac{B+A}{X}\right)^2}\right] - j\frac{4AC}{\pi X^2}, & X > A + B \\ 0, & X \le A \end{cases}$
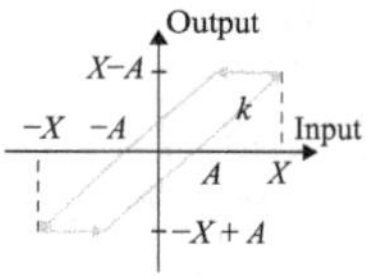 Output; $X-A$; $-X$ $-A$; k; Input; A X; $-X+A$	$N = \frac{k}{2}\left[1 - f\left(\frac{2 - \frac{X}{A}}{\frac{X}{A}}\right)\right] - j\frac{4kA(X-A)}{\pi X^2}, \ X > A$

Table A.10 Table of singular points in the phase plane

Roots	Phase plane	Name
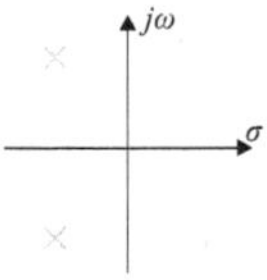	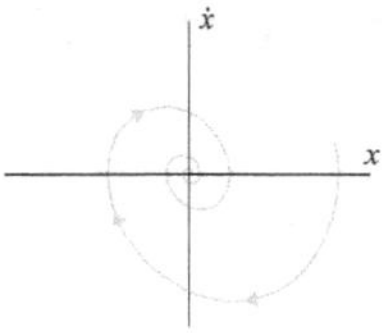	Stable focus
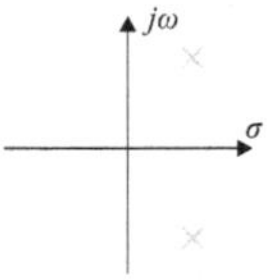	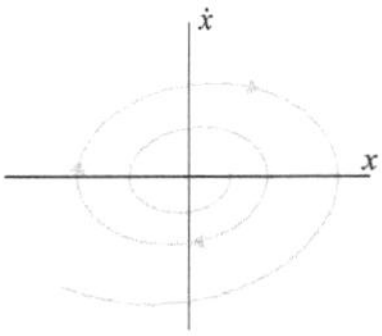	Unstable focus
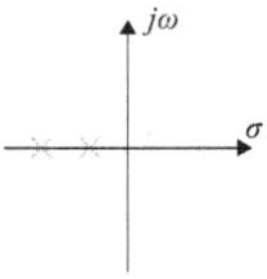	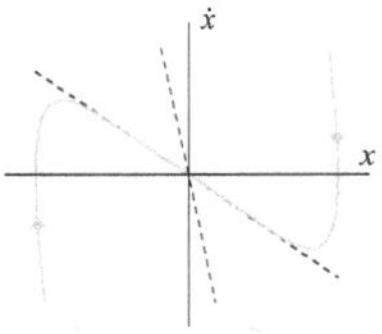	Stable node
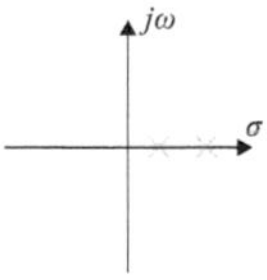	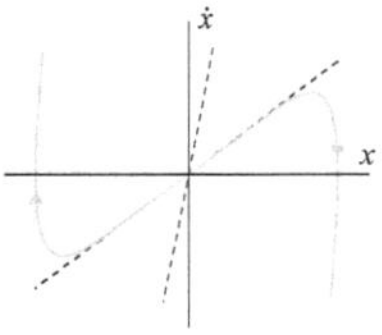	Unstable node
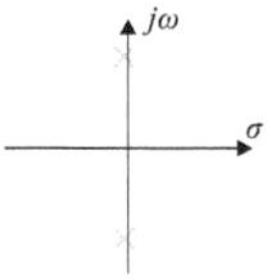	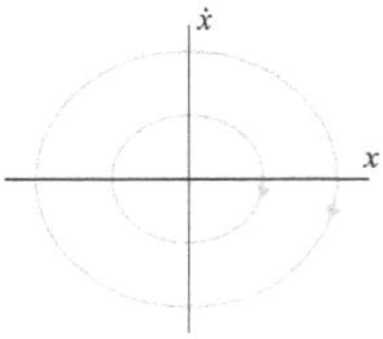	Center
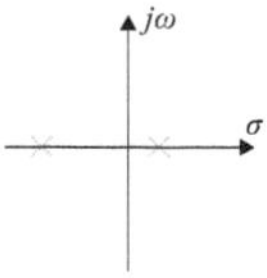	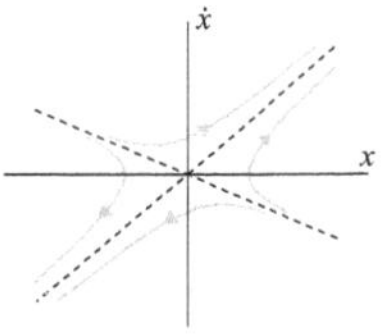	Saddle point

Table A.11 Table of Laplace transforms of fractional systems

	x(t)	**Asymptotic for $t \to +\infty$ x(t)**	**X(s)**
1	$\dfrac{t^{\alpha-1}}{\Gamma(\alpha)}$		$\dfrac{1}{s^\alpha}$
2	$t^{\alpha-1}E_{\alpha,\alpha}(-at^\alpha)$	$\dfrac{(-a)^{\frac{1-\alpha}{\alpha}}}{\alpha}e^{t(-a)^{1/\alpha}}$	$\dfrac{1}{s^\alpha + a}$
3	$t^{\alpha-\beta-1}E_{\alpha,\alpha-\beta}(-at^\alpha)$	$\dfrac{(-a)^{\frac{1-\alpha+\beta}{\alpha}}}{\alpha}e^{t(-a)^{1/\alpha}}$	$\dfrac{s^\beta}{s^\alpha + a}$
4	$t^{-\beta}E_{1,1-\beta}(-at)$	$(-a)^\beta e^{-ta}$	$\dfrac{s^\beta}{s+a}$
5	$\dfrac{t^{\alpha(k+1)-1}}{\Gamma(k+1)}\dfrac{\mathrm{d}^k}{\mathrm{d}(-at^\alpha)^k}E_{\alpha,\alpha}(-at^\alpha)$		$\dfrac{1}{(s^\alpha + a)^{k+1}}$
6	$\dfrac{t^{\alpha(k+1)-\beta-1}}{\Gamma(k+1)}\dfrac{\mathrm{d}^k}{\mathrm{d}(-at^\alpha)^k}E_{\alpha,\alpha-\beta}(-at^\alpha)$		$\dfrac{s^\beta}{(s^\alpha + a)^{k+1}}$
7	$\dfrac{t^{k-\beta}}{\Gamma(k+1)}\dfrac{\mathrm{d}^k}{\mathrm{d}(-at)^k}E_{1,1-\beta}(-at)$		$\dfrac{s^\beta}{(s+a)^{k+1}}$

Note 1: The two-parameter Mittag-Leffler function is defined as

$$E_{\alpha,\beta}(t) = \sum_{k=0}^{+\infty} \frac{t^k}{\Gamma(\alpha k + \beta)}, \quad \alpha, \beta > 0 \tag{A.1}$$

Note 2: The approximations in the second column above were found thanks to the asymptotic approximation

$$E_{\alpha,\beta}(t) \approx \frac{1}{\alpha} t^{\frac{1-\beta}{\alpha}} e^{t^{1/\alpha}} \tag{A.2}$$

valid when $t \to +\infty$.

Table A.12 Table of fractional derivatives

Grünwald–Letnikoff or Riemann–Liouville definitions				
1	${}_cD_t^{\alpha}$	$H(t-a)$	=	$\begin{cases} \frac{(t-\max\{c,a\})^{-\alpha}}{\Gamma(1-\alpha)}, & \text{if } t>a \\ 0, & \text{if } c\le t\le a\wedge c\neq a \end{cases}, \ c\in\mathbb{R}\vee c=-\infty$
2	${}_cD_t^{\alpha}$	$\delta(t-a)$	=	$\begin{cases} \frac{(t-a)^{-\alpha-1}}{\Gamma(-\alpha)}, & \text{if } t\ge a\ge c \\ 0, & \text{if } (c\le t<a\wedge c\neq a)\vee a<c \end{cases}, \ c\in\mathbb{R}\vee c=-\infty$
3	${}_0D_t^{\alpha}$	t^{λ}	=	$\frac{\Gamma(\lambda+1)}{\Gamma(\lambda-\alpha+1)}t^{\lambda-\alpha}, \ \lambda>-1$
4	${}_cD_t^{\alpha}$	$(t-c)^{\lambda}$	=	$\frac{\Gamma(\lambda+1)}{\Gamma(\lambda-\alpha+1)}(t-c)^{\lambda-\alpha}, \ \lambda>-1$
5	${}_0D_t^{\alpha}$	k	=	$\frac{k}{\Gamma(1-\alpha)}t^{-\alpha}$
6	${}_0D_t^{\frac{1}{2}}$	k	=	$\frac{k}{\sqrt{t\pi}}$
7	${}_{-\infty}D_t^{\alpha}$	$e^{\lambda t}$	=	$\lambda^{\alpha}e^{\lambda t}, \ \lambda>0$
8	${}_cD_t^{\alpha}$	$e^{\lambda t}$	=	$e^{\lambda c}(t-c)^{-\alpha}E_{1,1-\alpha}(\lambda(t-c)), \ \lambda\neq 0$
9	${}_0D_t^{\alpha}$	$e^{\lambda t}$	=	$t^{-\alpha}E_{1,1-\alpha}(\lambda t), \ \lambda\neq 0$
10	${}_{-\infty}D_t^{\alpha}$	$\sin(\lambda t)$	=	$\lambda^{\alpha}\sin\left(\lambda t+\frac{\alpha\pi}{2}\right), \ \lambda>0, \alpha>-1$
11	${}_{-\infty}D_t^{\alpha}$	$\cos(\lambda t)$	=	$\lambda^{\alpha}\cos\left(\lambda t+\frac{\alpha\pi}{2}\right), \ \lambda>0, \alpha>-1$

Caputo definition				
12	${}_cD_t^{\alpha}$	$H(t-a)$	=	$\begin{cases} \frac{(t-\max\{c,a\})^{-\alpha}}{\Gamma(1-\alpha)}, & \text{if } t>a \\ 0, & \text{if } c\le t\le a\wedge c\neq a \end{cases}, \ c\in\mathbb{R}\vee c=-\infty$
13	${}_cD_t^{\alpha}$	$\delta(t-a)$	=	$\begin{cases} \frac{(t-a)^{-\alpha-1}}{\Gamma(-\alpha)}, & \text{if } t\ge a\ge c \\ 0, & \text{if } (c\le t<a\wedge c\neq a)\vee a<c \end{cases}, \ c\in\mathbb{R}\vee c=-\infty$
14	${}_0D_t^{\alpha}$	t^{λ}	=	$\frac{\Gamma(\lambda+1)}{\Gamma(\lambda-\alpha+1)}t^{\lambda-\alpha}, \ \lambda>-1$
15	${}_cD_t^{\alpha}$	$(t-c)^{\lambda}$	=	$\frac{\Gamma(\lambda+1)}{\Gamma(\lambda-\alpha+1)}(t-c)^{\lambda-\alpha}, \ \lambda>-1$
16	${}_0D_t^{\alpha}$	k	=	0
17	${}_{-\infty}D_t^{\alpha}$	$e^{\lambda t}$	=	$\lambda^{\alpha}e^{\lambda t}, \ \lambda>0$
18	${}_cD_t^{\alpha}$	$e^{\lambda t}$	=	$\lambda^{\lceil\alpha\rceil}e^{\lambda c}t^{\lceil\alpha\rceil-\alpha}E_{1,1+\lceil\alpha\rceil-\alpha}(\lambda(t-c)), \ \lambda\neq 0$
19	${}_0D_t^{\alpha}$	$e^{\lambda t}$	=	$\lambda^{\lceil\alpha\rceil}t^{\lceil\alpha\rceil-\alpha}E_{1,1+\lceil\alpha\rceil-\alpha}(\lambda t), \ \lambda\neq 0$
20	${}_{-\infty}D_t^{\alpha}$	$\sin(\lambda t)$	=	$\lambda^{\alpha}\sin\left(\lambda t+\frac{\alpha\pi}{2}\right), \ \lambda>0, \alpha>-1$
21	${}_{-\infty}D_t^{\alpha}$	$\cos(\lambda t)$	=	$\lambda^{\alpha}\cos\left(\lambda t+\frac{\alpha\pi}{2}\right), \ \lambda>0, \alpha>-1$

Solutions

1.1 We can simplify and combine blocks for $Y(s)/R(s)$ and for $Y(s)/d(s)$, as in Figure 1. So the correct answers are options **A)** and **F)**.

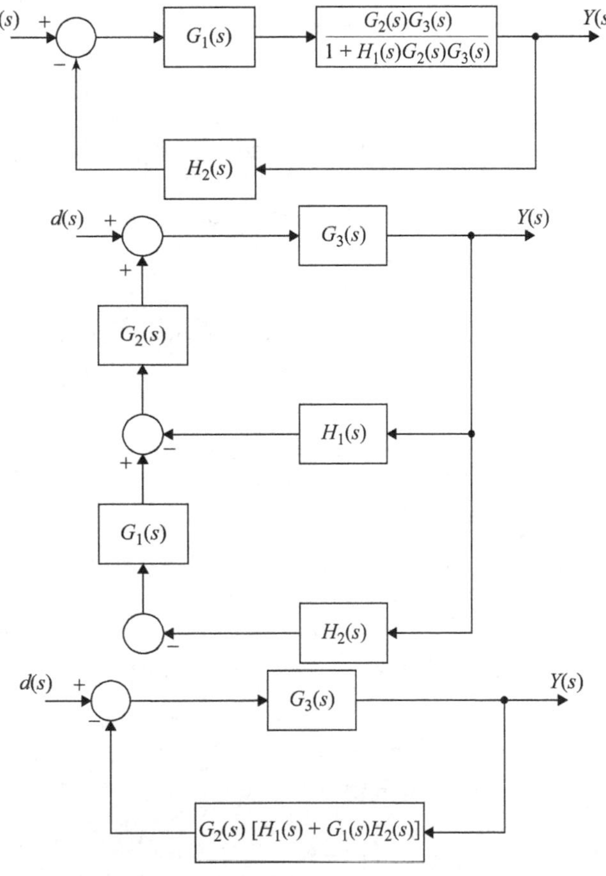

Figure 1 Resolution of Exercise 1.1

1.2 We can simplify and combine blocks, as in Figure 2, to get $Y(s)/R(s)$. Thus the correct answer is option **A)**.

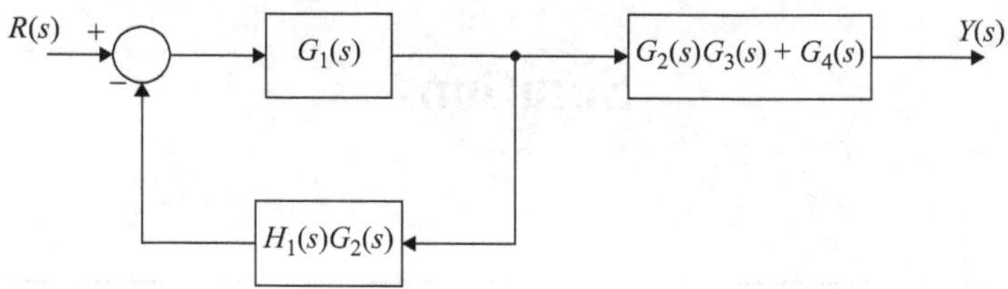

Figure 2 Resolution of Exercise 1.2

1.3 From Figure 3, it can be seen that the correct answer is option **B)**.

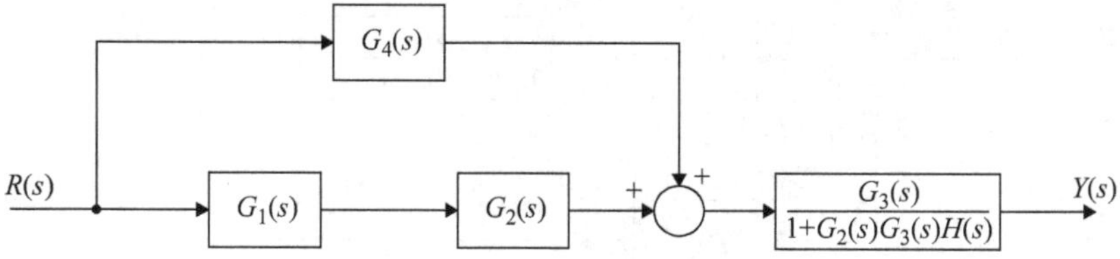

Figure 3 Resolution of Exercise 1.3

1.4 The correct answer is option **C)**.

1.5 We can simplify and combine blocks as in Figure 4. Thus the correct answer is option **C)**.

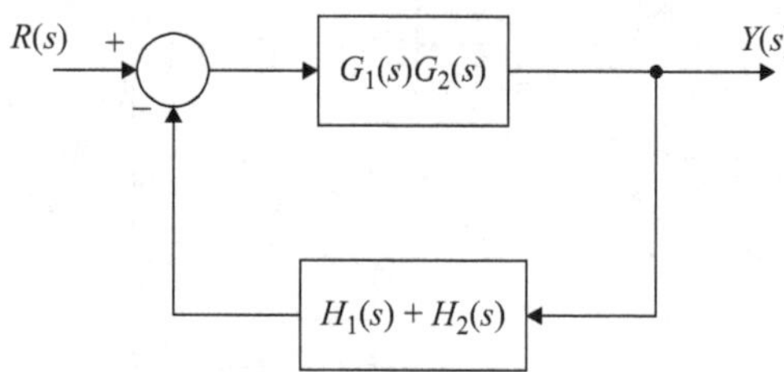

Figure 4 Resolution of Exercise 1.5

1.6 The correct answer is option **A)**.

1.7 The correct answer is option **D)**.

1.8 $$\frac{Y(s)}{R(s)} = \frac{G_1 + G_2}{1 + (G_1 + G_2)(G_3 - G_4)}$$

1.9 $$\frac{Y(s)}{R(s)} = \frac{\frac{G_1 G_2 G_3}{1+G_1 G_2}}{1 + \frac{G_1 G_2 G_3}{1+G_1 G_2} \cdot \frac{H_1 H_2 H_3}{1+H_1 H_2}}$$

1.10 As the transfer function of the system is $\frac{Y(s)}{R(s)} = \frac{-G(s)}{1+G(s)H(s)} + \frac{G(s)}{1+G(s)H(s)} = 0$, the correct answer is option **D)**.

1.11 The correct answer is option **B)**.

1.12 The correct answer is option **A)**.

1.13 The correct answer is option **B)**.

1.14 The correct answers are options **C)** and **H)**.

1.15 $\frac{Y(s)}{R(s)} = \frac{G_1(G_2 - G_3)G_4}{1 + G_1(G_2 - G_3)G_4H_1H_2}$

1.16 $\frac{Y(s)}{R(s)} = \frac{G_1G_2 + G_2G_3}{1 + G_2H_2 + G_1G_2H_1}$

1.17 $\frac{Y(s)}{R(s)} = \frac{G_1G_2 + G_2G_3}{1 + G_2H_2 + G_1G_2H_1H_2}$

1.18 $\frac{Y(s)}{R(s)} = \frac{G_1G_2G_3}{1 + G_3H_3 + G_2G_3H_2H_3 + G_1G_2G_3H_1H_2H_3}$

1.19 $\frac{Y(s)}{R(s)} = G_3 + \frac{G_1G_2}{1 + G_2H_2(1 + G_1H_1)}$

2.1 Let V be the potential between L_1 and C_1.

$$\begin{cases} I_1 = I_2 + I_3 \\ R_1 + L_1s = \frac{V_i - V}{I_1} \\ \frac{1}{C_1s} + R_2s = \frac{V}{I_2} \\ L_2s + R_3 = \frac{V}{I_3} \end{cases}$$

$$\frac{I_3}{V_i} = \frac{sR_2C_1 + 1}{s^2L_1C_1(1 + R_2) + s[L_1 + C_1(R_1 + R_2 + R_1R_2)] + (1 + R_1)}$$

2.2 The correct answer is option **A)**.

2.3 The correct answer is option **C)**.

2.4 The correct answer is option **A)**.

2.5 $$\frac{I_2(s)}{V(s)} = \frac{R_2C_2s}{R_1C_1R_2C_2Ls^3 + (R_1C_1L + R_2C_2L)s^2 + (L + R_1C_1R_2 + R_1R_2C_2)s + (R_1 + R_2)}$$

2.6 $\frac{V_0(s)}{V_i(s)} = \frac{R_2}{R_1R_2Cs + (R_1 + R_2)}$

2.7 The correct answer is option **A)**.

2.8 The correct answer is option **C)**.

2.9 The correct answer is option **B)**.

2.10 The motor has no inductance L_a, and hence

$$\frac{T}{V} = \frac{J_T s + B_T}{J_T s + B_T + K_e}$$

where variables are defined as in Section 2.1.3.3. At the pulley,

$$J\dot{\omega}_m = T + r\left(K_c\left(\frac{v_M}{s} - \frac{v_c}{s}\right) + R_C(v_M - v_c)\right)$$

At the mass,

$$M\dot{v}_M = Mg - K_c\left(\frac{v_M}{s} - \frac{v_c}{s}\right) - R_C(v_M - v_c)$$

2.11 $$\frac{X_2(s)}{F(s)} = \frac{K_0}{s^4 M_1 M_2 + s^3(M_1 B_2 + M_2 B_1) + s^2(M_1 K_0 + M_1 K_2 + B_1 B_2 + M_2 K_0) + s(B_1 K_0 + B_1 K_2 + B_2 K_0) + K_0 K_2}$$

2.12 $$G(s) = \frac{K}{(s^2 M_1 + k)(s^2 M_2 + K) - K^2}$$

2.13 The correct answer is option **B)**.

2.14 The correct answer is option **D)**.

2.15 $$\begin{cases} M_1\ddot{x}_1 = f - K_1 x_1 + K_2(x_2 - x_1) \\ M_2\ddot{x}_2 = -K_2(x_2 - x_1) - B_2\dot{x}_2 - K_3 x_2 - B_3\dot{x}_2 \end{cases}$$

$$\frac{X_2}{F} = \frac{K_2}{s^4 M_1 M_2 + s^3(M_1 B_2 + M_1 B_3) + s^2(M_1 K_2 + M_1 K_3 + M_2 K_1 + M_2 K_2) + s(B_2 K_1 + B_2 K_2 + B_3 K_1 + B_3 K_2) + (K_1 K_2 + K_1 K_3 + K_2 K_3)}$$

2.16 $$\begin{cases} M_1\ddot{x}_1 = f - K_1 x_1 - B_1\dot{x}_1 + K_2(x_2 - x_1) \\ M_2\ddot{x}_2 = -K_2(x_2 - x_1) - B_2\dot{x}_2 - K_3 x_2 \end{cases}$$

$$\frac{X_2}{F} = \frac{K_2}{s^4 M_1 M_2 + s^3(M_1 B_2 + M_2 B_1) + s^2(M_1 K_2 + M_1 K_3 + B_1 B_2 + M_2 K_1 + M_2 K_2) + s(B_1 K_2 + B_1 K_3 + B_2 K_1 + B_2 K_2) + (K_1 K_2 + K_1 K_3 + K_2 K_3)}$$

2.17 $$G(s) = \frac{1}{s^2 M + s(B_1 + B_2) + (K_1 + K_2)}$$

2.18 $$\begin{cases} M_1\ddot{x}_1 = K_1(x_2 - x_1) + B_1(\dot{x}_2 - \dot{x}_1) \\ M_2\ddot{x}_2 = K_2(x_3 - x_2) + B_2(\dot{x}_3 - \dot{x}_2) - K_1(x_2 - x_1) - B_1(\dot{x}_2 - \dot{x}_1) \\ M_3\ddot{x}_3 = F - K_3 x_3 - B_3\dot{x}_3 - K_2(x_3 - x_2) - B_2(\dot{x}_3 - \dot{x}_2) \end{cases}$$

2.19 $\begin{cases} FR = T \\ F = M\ddot{x} \end{cases} \Rightarrow \dfrac{X(s)}{T(s)} = \dfrac{1}{s^2 MR}$

where F is the force exerted on the pulley by the belt, and assuming that static and dynamic friction between the belt and mass M prevents slippage at all times.

$$\frac{Y(s)}{T(s)} = \frac{1}{(M+m)R}\left(\frac{1}{s^2} - \frac{1}{s^2 + \frac{K}{m}}\right)$$

2.20 The correct answer is option **A)**.

2.21 The correct answer is option **A)**.

2.22 The correct answer is option **B)**.

2.23 The correct answer is option **A)**.

2.24 $\begin{cases} M\ddot{x} = F - Kx - B\dot{x} \\ J\ddot{\theta} = T - Fr \\ \theta r = x \end{cases} \Rightarrow M\ddot{x} = \dfrac{T - J\ddot{\theta}}{r} - Kx - B\dot{x}$

$$\Rightarrow \frac{x}{T} = \frac{\frac{1}{r}}{\left(M + \frac{J}{r^2}\right)s^2 + Bs + K}$$

where F is the force exerted on the pulley.

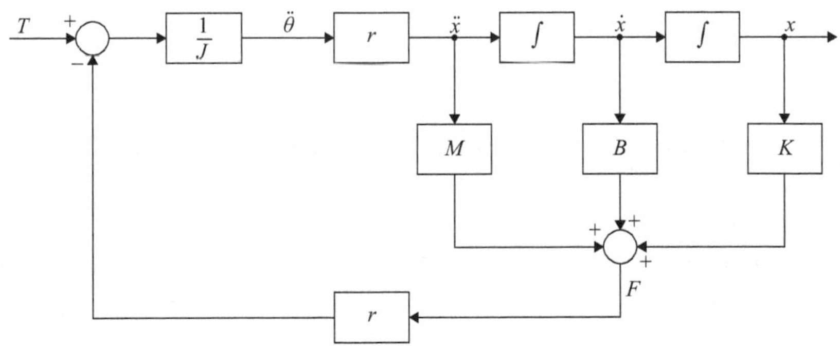

Figure 5 Block diagram of the resolution of Exercise 2.24

2.25 The correct answer is option **A)**.

2.26 The correct answer is option **B)**.

2.27 The correct answers are options **C)** and **E)**.

2.28 The correct answer is option **B)**.

2.29 The correct answer is option **C)**.

2.30

$$A_1\dot{h}_1 = q_i - q_1$$

$$q_1 = \frac{h_1}{R_1}$$

$$A_2\dot{h}_2 = q_1 - q_3 - q_2$$

$$q_3 = \frac{h_2 - h_3}{R_3}$$

$$A_3\dot{h}_3 = q_3$$

$$q_2 = \frac{h_2 - h_4}{R_2}$$

$$A_4\dot{h}_4 = q_2 - q_o$$

$$q_o = \frac{h_4}{R_4}$$

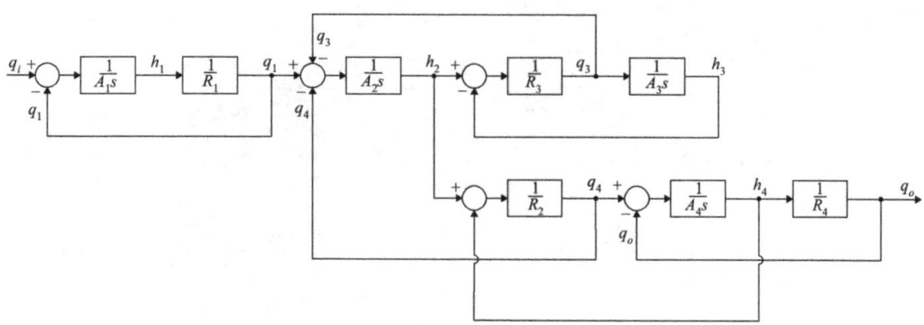

Figure 6 Block diagram of the resolution of Exercise 2.30

$$\frac{q_o}{q_i} = \frac{\dfrac{1}{R_2A_2s} + \dfrac{G_1G_3}{A_2s}}{(R_1A_1s+1)\left[\left(G_2 - \dfrac{G_1G_3}{R_2A_2A_4s^2}\right)(R_4A_4s+1) - \dfrac{1}{R_2A_4s} + \dfrac{1}{R_2^2A_2A_4s^2} + \dfrac{G_1G_3}{R_2A_2A_4s^2}\right]}$$

$$G_1 = \frac{1}{1 + \dfrac{1}{R_3A_2s} + \dfrac{1}{R_2A_2s} - \dfrac{1}{R_3A_2s}\dfrac{1}{R_3A_3s+1}}$$

$$G_2 = 1 + \frac{1}{R_2A_4s} - \frac{1}{R_2^2A_2A_4s^2}$$

$$G_3 = -\frac{1}{R_3R_2A_2s} - \frac{1}{R_2^2A_2s} + \frac{1}{R_2R_3A_2s}\frac{1}{R_3A_3s+1}$$

2.31 The correct answer is option **A)**.

2.32 The correct answer is option **B)**.

2.33 $T \cdot V = L \Leftrightarrow T = \dfrac{L}{V}$

$$Q_i = A_1 s H_1 + Q_1, \quad Q_1 = \frac{H_1 - H_2}{R_1}, \quad Q_1 = A_2 s H_2$$

$$Q_2 = kH_2, \quad Q_3 = e^{-sT} Q_2, \quad Q_3 = A_3 s H_3 + Q_o, \quad Q_o = \frac{H_3}{R_3}$$

2.34 See the circuits in Figure 7.

$$\frac{T_m}{T_0} = \frac{1}{1 + RCs}$$

$$\frac{T_m}{T_0} = \frac{1}{1 + (R_{fg} C_g + R_{gm} C_m + R_{fg} C_m)s + R_{fg} R_{gm} C_m C_g s^2}$$

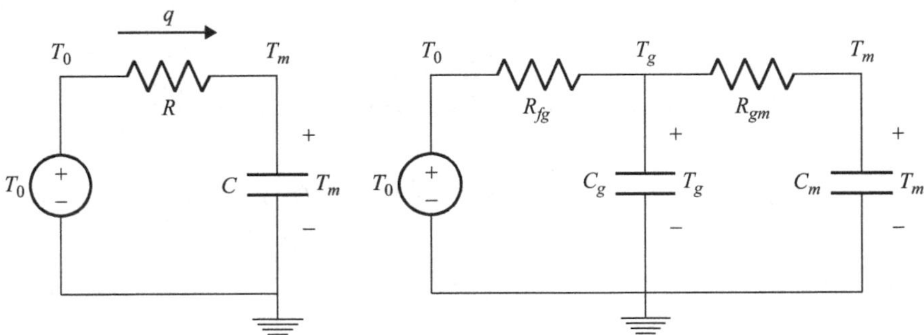

Figure 7 Electrical circuits of Exercise 2.34

2.35 The transfer function for the thermal phenomenon is

$$P - q = C_T \dot{T}_i \Rightarrow$$

$$P - \frac{T_i - T_o}{R_T} = C_T \dot{T}_i \Rightarrow$$

$$P - \frac{T}{R_T} = C_T \dot{T} \Rightarrow$$

$$PR_T = T + C_T Ts \Rightarrow$$

$$\frac{T}{P} = \frac{R_T}{1 + C_T s}$$

Since the transfer function of the circuit is $\frac{I}{V} = \frac{Cs}{1+RCs}$, and $P = RI^2$, no transfer function $\frac{T}{V}$ exists unless a linearization is employed, because of the nonlinear relation between P and V. The nonlinearity does not disappear when there is no capacitor, when $I = V/R$ but still $P = RI^2$.

2.36
$$\frac{T_i - T_b}{R_b} = C_b\frac{dT_b}{dt} + \frac{T_b - T_o}{R_b}$$
$$\frac{T_i - T_g}{R_g} = C_g\frac{dT_g}{dt} + \frac{T_g - T_o}{R_g}$$

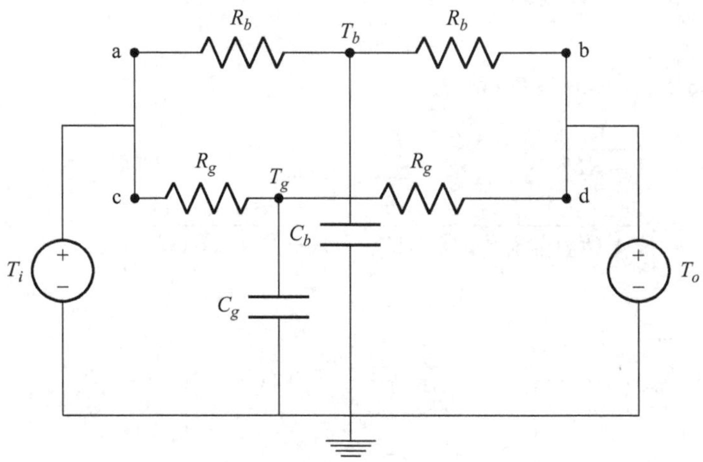

Figure 8 Wall and window equivalent circuit of Exercise 2.36

2.37
$$\frac{T_i - T_1}{R_1} = C_1\frac{dT_1}{dt} + \frac{T_1 - T_2}{R_1 + R_2}$$
$$\frac{T_1 - T_2}{R_1 + R_2} = C_2\frac{dT_2}{dt} + \frac{T_2 - T_3}{R_2 + R_3}$$
$$\frac{T_2 - T_3}{R_2 + R_3} = C_3\frac{dT_3}{dt} + \frac{T_3 - T_o}{R_3}$$

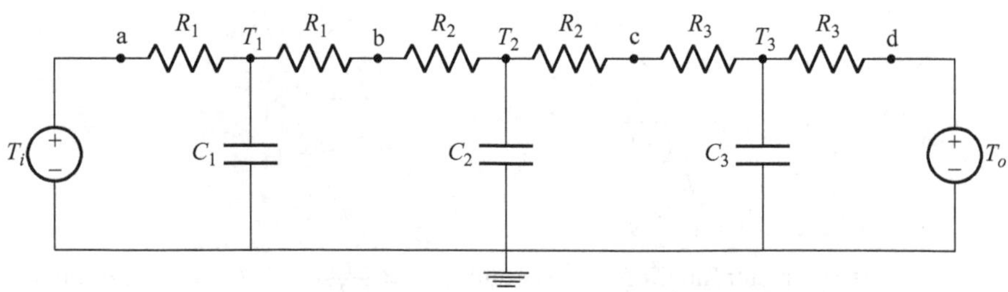

Figure 9 Wall equivalent circuit of Exercise 2.37

2.38 It is clear that $T_{sea} < T_{air} < T_{PTO}$. Let q_1 be the heat flux from the PTO to the air, q_2 the heat flux from the air to the sea, and (to simplify notation) $R = P\frac{1-\eta}{\eta}$. Then an energy balance on the PTO yields

$$
\begin{aligned}
R - q_1 &= C_{PTO}\dot{T}_{PTO} \Rightarrow \\
R - A_{PTO}h(T_{PTO} - T_{air}) &= C_{PTO}\dot{T}_{PTO} \Rightarrow \\
T_{PTO} - T_{air} &= \frac{R}{A_{PTO}h} - \frac{C_{PTO}}{A_{PTO}h}\dot{T}_{PTO}
\end{aligned}
$$

and an energy balance on the air yields

$$
\begin{aligned}
q_1 - q_2 &= C_{air}\dot{T}_{air} \Rightarrow \\
A_{PTO}h(T_{PTO} - T_{air}) - A_{air}h(T_{PTO} - T_{air}) &= C_{air}\dot{T}_{air}
\end{aligned}
$$

We have $\dot{T}_1 = \dot{T}_{PTO} - \dot{T}_{air}$ and $\dot{T}_2 = \dot{T}_{air} - \dot{T}_{sea} = \dot{T}_{air}$. Consequently,

$$
\begin{cases} T_1 = \frac{R}{A_{PTO}h} - \frac{C_{PTO}}{A_{PTO}h}(\dot{T}_1 + \dot{T}_2) \\ A_{PTO}hT_1 - A_{air}hT_2 = C_{air}\dot{T}_2 \end{cases} \Rightarrow \begin{cases} T_1 = \frac{1}{C_{PTO}s + A_{PTO}h}R - \frac{C_{PTO}}{C_{PTO}s + A_{PTO}h}T_2 \\ T_2 = \frac{A_{PTO}h}{C_{air}s + A_{air}h}T_1 \end{cases}
$$

From here we can arrive at

$$
\frac{T_1}{P} = \frac{1-\eta}{\eta}\,\frac{C_{air}s + A_{air}h}{C_{PTO}C_{air}s^2 + (C_{PTO}A_{PTO} + C_{air}A_{PTO} + C_{PTO}A_{air})hs + A_{PTO}A_{air}h^2}
$$

3.1 The correct answer is option **C)**.

3.2 The correct answer is option **D)**, since we must have $K > 2$.

3.3 The correct answer is option **A)**.

3.4 The correct answer is option **D)**, as $K > \frac{1}{2}$.

3.5 The correct answer is option **B)**.

3.6 The correct answer is option **C)**.

3.7 The correct answer is option **D)**, as there are no poles in the right half-plane.

3.8 There is no value of gain K leading to the time response given. $M_p = \frac{1.258}{0.95} \Rightarrow \omega_n = 3.94$; $t_p = 0.847 \Rightarrow \zeta = 0.5$; $K = \omega_n^2 - 2 = 13.52$. However, from the steady state, we have $\lim_{s\to 0} \frac{K}{s^2+3s+2+K} = 0.95 \Rightarrow K = 38$.

3.9 The correct answer is option **C)**: two, as can be seen from the Routh–Hurwitz table,

$$\begin{array}{c|ccc} s^5 & 2 & 4 & 2 \\ s^4 & 4 & 8 & 2 \\ \hline s^3 & \varepsilon & 4 & \\ s^2 & \frac{8\varepsilon-16}{\varepsilon} & 2 & \\ s & \frac{-2\varepsilon^2+32\varepsilon-64}{8\varepsilon-16} & & \\ 1 & 2 & & \end{array}$$

which, irrespective of the sign of ε, always has two changes of sign in the first column.

3.10 The correct answer is option **A)**.

3.11 The correct answer is option **C)**.

3.12 The correct answer is option **C)**.

3.13 $K > 1$.

3.14 The correct answer is option **D)**: none of the above, since we must have $K > \frac{1}{2}$.

3.15 The correct answer is option **C)**.

3.16 The correct answer is option **C)**.

3.17 The correct answer is option **A)**.

3.18 The correct answer is option **B)**.

3.19 The correct answer is option **B)**.

3.20 The correct answer is option **C)**.

3.21 $K > \frac{2}{3}$ when $H(s) = 1$, and $K < -\frac{1}{3}$ when $H(s) = s(s+1)$.

3.22 The correct answers are options **D)** (since $t_r = 0.258$ s), **E)**, and **J)**.

3.23 Since $2\zeta\omega_n = 2$, $\omega_n^2 = 9$, we get $\zeta = 0.333$, $\omega_n = 3$, from which can be seen that the correct answer is option **C)**.

3.24 $t_p = 0.83$ s, $y(t_p) = 0.18$, $y_{ss} = \frac{2}{16} = 0.125$.

3.25 The gain is $\frac{10}{4}$, and so $\omega_n^2 = 4$, $2\zeta\omega_n = 1$; hence the correct answers are options **C)** and **G)**.

3.26 The correct answer is option **B)**.

3.27 The correct answer is option **B)**.

3.28 The correct answer is option **C)**.

3.29 The correct answer is option **A)**.

3.30 The equation of motion is $f - kx - B\dot{x} = M\ddot{x}$, corresponding to transfer function $\frac{X(s)}{F(s)} = \frac{\frac{1}{K}\cdot\frac{K}{M}}{s^2+\frac{B}{M}s+\frac{K}{M}}$.

From steady-state response 0.03, we can obtain the gain: $\frac{0.03}{8.9} = \frac{1}{K} \Rightarrow K = 296.7$ N/m.

From the maximum overshoot $\frac{0.0029}{0.03} = 9.7\%$ we can obtain the damping coefficient: $e^{-\frac{\pi\zeta}{\sqrt{1-\zeta^2}}} = 0.097 \Rightarrow \zeta = 0.6$.

From here and from the peak time of 1 s, we can obtain the natural frequency: $1 = \frac{\pi}{\omega_n\sqrt{1-0.6^2}} \Rightarrow \omega_n = 3.93$ rad/s.

We can now find $3.93^2 = \frac{K}{M} \Rightarrow M = 19.2$ kg and $2 \times 0.6 \times 3.93 = \frac{B}{M} \Rightarrow B = 90.62$ Ns/m.
If the overshoot changes, we will now have $\zeta = 0.9$ and $\omega_n = 2.62$ rad/s; K is the same because the steady-state is the same, and $M = 43.25$ kg, $B = 203.97$ Ns/m.

3.31 The correct answer is option **D)**, since $\zeta = \frac{a}{\sqrt{a^2+b^2}}$.

3.32 The correct answer is option **A)**.

3.33 The correct answer is option **D)**, since $\zeta = 0.3$ and $\omega_n = 19.156$ rad/s.

3.34 $\frac{Y(s)}{U(s)} = \frac{100}{s^2+14s+100}$

3.35 The correct answer is option **D)**, since $t_r = 0.432$.

3.36 From $0.444 = e^{-\frac{\pi\zeta}{\sqrt{1-\zeta^2}}}$, we get $\zeta = 0.25$. Then, from $1.622 = \frac{\pi}{\omega_n\sqrt{1-\zeta^2}}$ we get $\omega_n = 2$ rad/s.

3.37 $\zeta = 0.2$, $\omega_n = 10$ rad/s.

3.38 $\zeta = 0.358$, $\omega_n = 2.804$ rad/s.

3.39 $\zeta = 0.447$, $\omega_n = 4.518$ rad/s.

3.40 The closed loop is $\frac{8}{s^2+6s+16}$, so $\omega_n = 4$ and $\zeta = \frac{3}{4}$. Hence $t_p = \frac{\pi}{4\sqrt{1-\frac{9}{16}}} = 1.19$ s, the steady-state output is $y_{ss} = \frac{8}{16} = 0.5$, the overshoot is $M_p = e^{-\frac{\frac{3}{4}\pi}{\sqrt{1-\frac{9}{16}}}} = 0.0284$, and thus $y(t_p) = 0.5 \times 1.0284 = 0.5142$, the rise time is $t_r = \frac{\pi - \arctan\frac{\sqrt{1-\frac{9}{16}}}{\frac{3}{4}}}{4\sqrt{1-\frac{9}{16}}} = 0.9142$ s, the settling time is $t_s = \frac{4}{3}$ s using the 2% criterion or $t_s = 1$ s using the 5% criterion.
As for the steady-state error when the input is a ramp, $e_{ss} = \frac{1}{\lim_{s\to 0} sG(s)} = \infty$.

3.41 There is clearly a delay in the response, and the response begins tangent to the horizontal axis, which means that there are no zeros; so the transfer function is of the form $\frac{Y(s)}{U(s)} = \frac{K\omega_n^2}{s^2 + 2\zeta\omega_n s + \omega_n^2} e^{-sT}$, with $\zeta < 1$ (i.e., underdamped).

Clearly $T = 0.5$. The overshoot is 25%, and so $0.25 = e^{-\frac{\zeta}{\sqrt{1-\zeta^2}}\pi} \Rightarrow \zeta = 0.4037$. The peak time is 1 s, and so $1 = \frac{\pi}{\omega_n\sqrt{1-\zeta^2}} \Rightarrow \omega_n = 3.4339$. Thus $G(s) = \frac{47.17\,e^{-0.5s}}{s^2+2.773s+11.79}$; it has no zeros, and its poles are $-1.3863 \pm 3.1416j$.

3.42 The closed loop is $\frac{4}{s^2+2s+4}$, so $\omega_n = 2$ and $\zeta = \frac{1}{2}$. Hence $t_p = \frac{\pi}{2\sqrt{1-\frac{1}{4}}} = 1.81$ s, the steady-state output is $y_{ss} = 1$, the overshoot is $M_p = e^{-\frac{\frac{1}{2}\pi}{\sqrt{1-\frac{1}{4}}}} = 0.163$, and thus $y(t_p) = 1.163$, the rise time is $t_r = \frac{\pi - \arctan\frac{\sqrt{1-\frac{1}{4}}}{\frac{1}{2}}}{2\sqrt{1-\frac{1}{4}}} = 1.21$ s, the settling time is $t_s = 4$ s using the 2% criterion or $t_s = 3$ s using the 5% criterion. As for the steady-state error when the input is a ramp, $e_{ss} = \frac{1}{\lim_{s\to 0} sG(s)} = \frac{4}{2} = 2$.

3.43 The correct answer is option **A)**.

3.44 The correct answer is option **D)**.

3.45 The correct answer is option **B)**.

3.46 The correct answers are options **A)** and **H)**.

3.47 The correct answer is option **C)**, since $|G(2j)| = 0.83$ and $\arg[G(2j)] = -85.24°$.

3.48 $\zeta = 0.5$, $\omega_n = 2$ rad/s, $t_p = 1.81$ s, $M_p = 0.163$, $y_{ss} = 0.5$.

3.49 $\zeta = 0.5$, $\omega_n = 36.28$ rad/s. We want $\zeta = 0.7$ and $\omega_n = 57.14$ rad/s, which cannot be obtained just by varying K.

3.50 The system is stable for $K > 0$. The desired steady-state error corresponds to $K = 20$.

3.51 The correct answer is option **A)**.

4.1 The correct answers are options **A)**, **F)** and **L)** (the root-locus does not cross the imaginary axis).

4.2 The correct answers are options **D)** and **E)**.

4.3 The correct answers are options **D)** and **G)**.

4.4 The correct answer is option **B)**.

4.5 The correct answer is option **D)**.

4.6 The correct answer is option **B)**.

4.7 The correct answer is option **C)**.

4.8 The correct answers are options **D)** ($\sigma = -4.5$), **F)** and **L)** (the root-locus does not cross the imaginary axis).

4.9 The correct answer is option **B)**.

4.10 The correct answer is option **C)**.

4.11 The correct answer is option **A)**.

4.12 The correct answer is option **C)**.

4.13 There are three asymptotes, with $\theta = \pm 60°$ and $\theta = 180°$, intersecting at $\sigma = \frac{-1.5-1.5+0}{3} = -3$. The departure angle of the upmost pole is $\phi = 180° - 90° - (90° + \arctan \frac{1.5}{2.78}) = -28.3°$. See Figure 10. Using the Routh–Hurwitz criterion we find that all poles are stable for $0 < K < 30$.

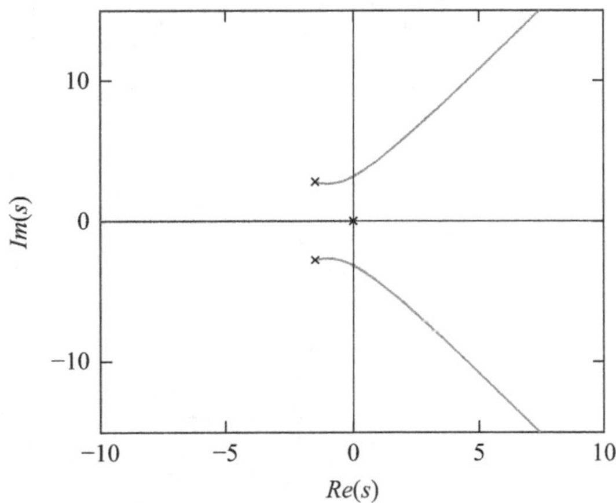

Figure 10 Root-locus of Exercise 4.13

4.14 The root-locus diverges at $\frac{\mathrm{d}}{\mathrm{d}s} \frac{s^2+2s}{s^2+10s+25} = 0 \Leftrightarrow s = -\frac{5}{4}$. All poles are stable for $K > 0$. See Figure 11.

4.15 The vertical asymptotes are located at $\sigma = \frac{-2-2+1}{2} = -1.5$, and the plant is stable for $K > 0$. See Figure 12.

4.16 There are two vertical asymptotes intersecting at $\sigma = \frac{-1-1}{2} = -1$. It is clear from the root-locus plot that the plant is stable for every $K \geq 0$. See Figure 13.

4.17 The root-locus converges at $\frac{\mathrm{d}}{\mathrm{d}s} \frac{s^3+3s^2}{s^2+3s+2} = 0 \Leftrightarrow s = -1.4$ (only real solution other than 0). The plant is stable for $K > 0$. See Figure 14.

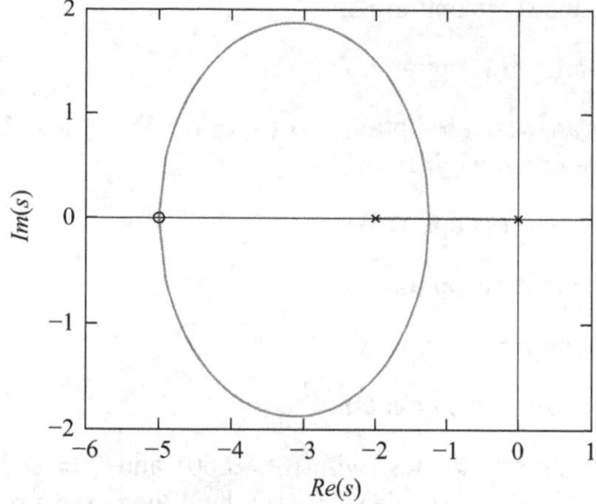

Figure 11 Root-locus of Exercise 4.14

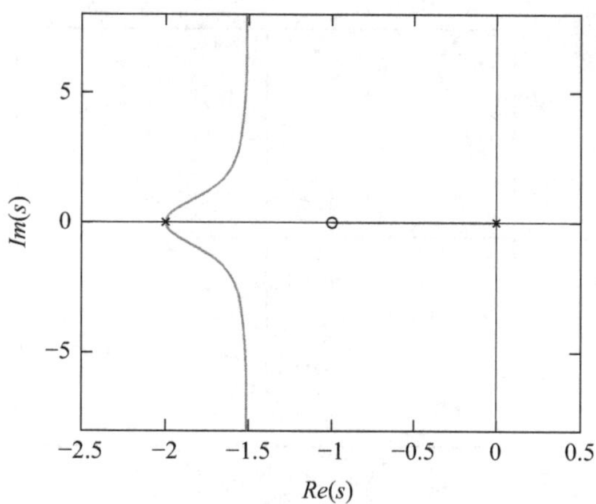

Figure 12 Root-locus of Exercise 4.15

5.1 See Figure 15.

5.2 See Figure 16. There are no gain crossover frequencies or phase crossover frequencies in any of these plants. This means that margins are ill-defined, and in such cases no conclusion can be obtained about stability in that way: the plant may or may not be stable. Indeed, stability is only ensured when *both* margins are known to be positive; so, even one ill-defined margin prevents conclusions about stability. In any case, plants $G_1(s)$, $G_2(s)$ and $G_3(s)$ *are* stable, as can be seen from their poles (or their root-locus).

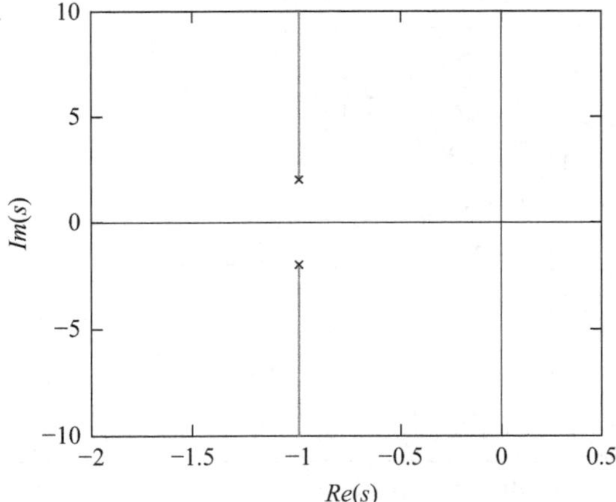

Figure 13 Root-locus of Exercise 4.16

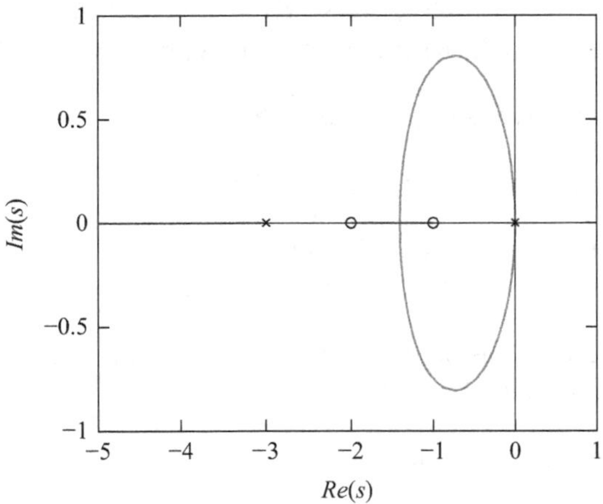

Figure 14 Root-locus of Exercise 4.17

5.3 From the phase $\arg G(j\omega) = -\frac{\pi}{2} - 2\arctan\omega$ we get

$$\pi + \arg G(j\omega) = \frac{\pi}{3} \Rightarrow \omega = 0.268 \text{ rad/s}$$

and from gain $|G(j\omega)| = \frac{K}{\omega(\omega^2+1)}$ we get

$$|G(j0.268)| = 1 \Rightarrow K = 0.287$$

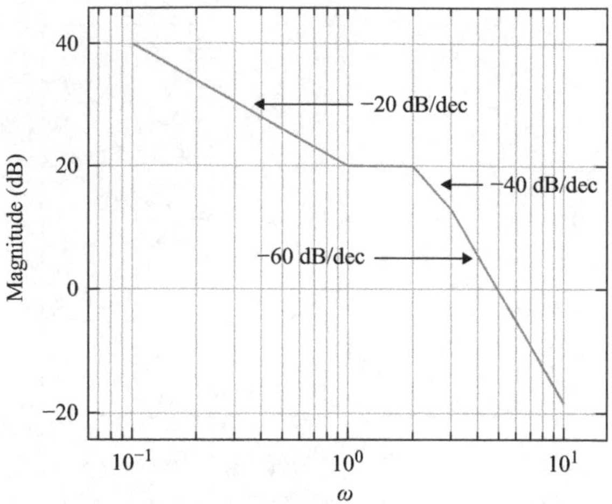

Figure 15 Resolution of Exercise 5.1

Neglecting the effect of the poles, $|G(j\omega)|_{\omega=1} = 20\log_{10} K$, from which we get $K = 0.567$. So the correct answers are options **A)** and **G)**.

5.4 The correct answers are options **B)** and **G)**.

5.5 The correct answer is option **B)**.

5.6 The correct answer is option **C)**, since it is the only one for which the phase at low frequencies is $-90°$ and the phase at high frequencies is $-180°$.

5.7 The correct answer is option **B)**, since it is the only one for which the phase at both low and high frequencies is $-180°$.

5.8 The correct answer is option **B)**.

5.9 The correct answer is option **B)**.

5.10 The correct answer is option **A)**.

5.11 The correct answer is option **D)**.

5.12 The correct answer is option **B)**.

5.13 The correct answer is option **D)**.

5.14 The correct answer is option **D)**.

5.15 The correct answer is option **D)**.

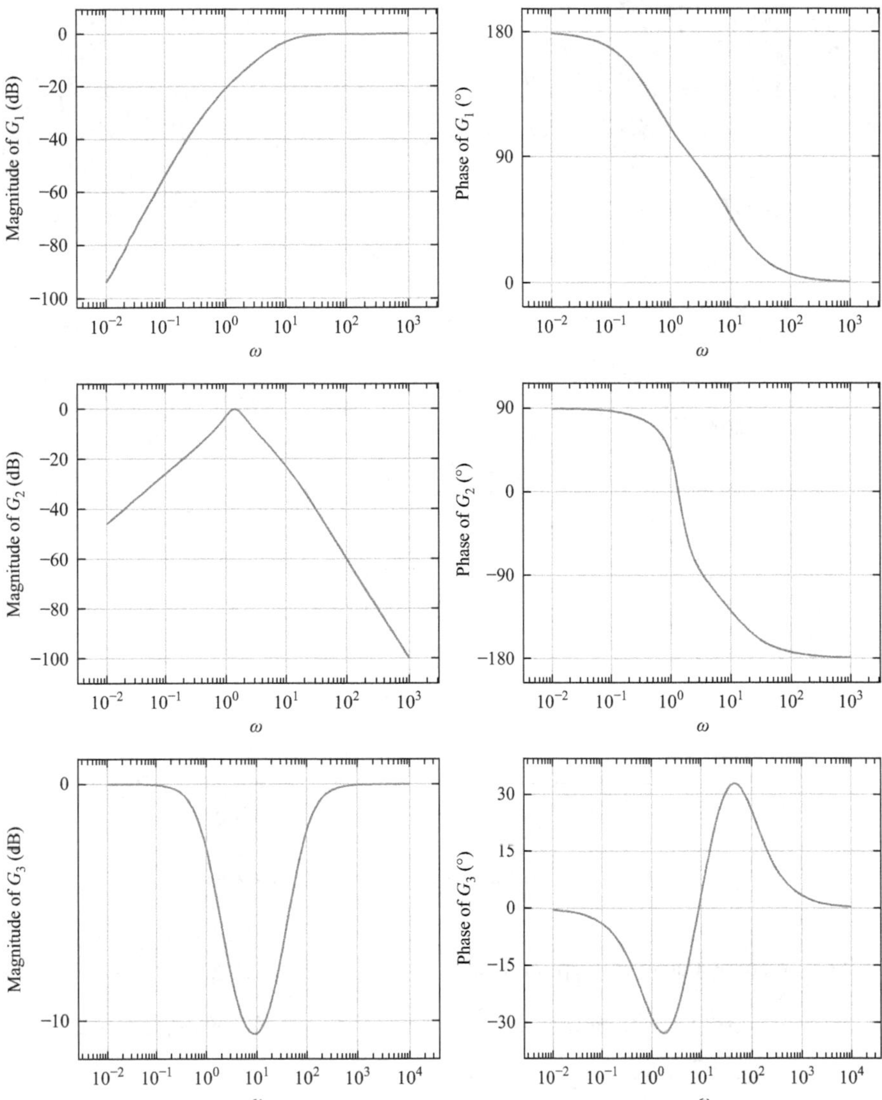

Figure 16 Resolution of Exercise 5.2

5.16 The correct answer is option **A)**.

5.17 The correct answer is option **D)**.

5.18 $GM = 1.75$ dB, $PM = 13.4°$.

5.19 $\begin{cases} \omega_n^2 = 16 \\ 2\zeta\omega_n = 4 \end{cases} \Rightarrow \begin{cases} \omega_n = 4 \text{ rad/s} \\ \zeta = 0.5 \end{cases}$

$$t_p = \frac{\pi}{4\sqrt{1-0.5^2}} = 0.907 \text{ s}$$

$$y(t_p) = 1 + e^{-\frac{0.5\pi}{\sqrt{1-0.5^2}}} = 1.163$$

$$t_r = \frac{e^{\frac{\arccos 0.5}{\tan \arccos 0.5}}}{4} = 0.458 \text{ s}$$

$$\omega_r = 4\sqrt{1 - 2 \times 0.5^2} = 2.828 \text{ rad/s}$$

$$M_r = \frac{1}{2\times 0.5\sqrt{1-0.5^2}} = 1.155$$

See Figure 17.

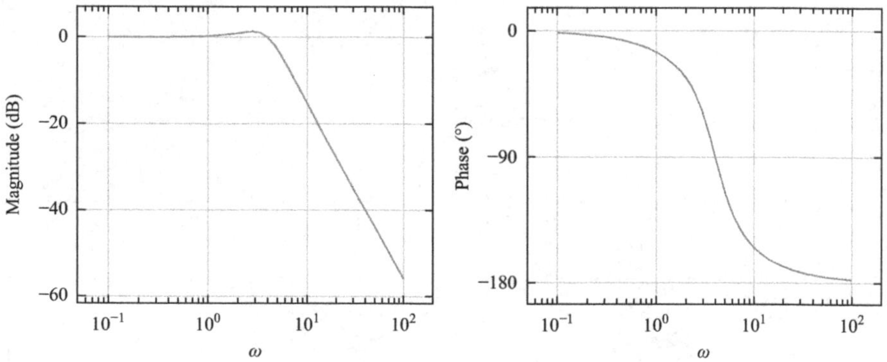

Figure 17 Resolution of Exercise 5.19

5.20 The correct answer is option **B)**.

5.21 From $10^{\frac{2.696}{20}} = 1.364 = \frac{1}{2\zeta\sqrt{1-\zeta^2}}$, we get $\zeta = 0.40$. Then, from $2.474 = \omega_n\sqrt{1 - 2\zeta^2}$ we get $\omega_n = 3$ rad/s.

5.22 See Figure 18. Since $Q_d(s) = \frac{1}{s}$ and

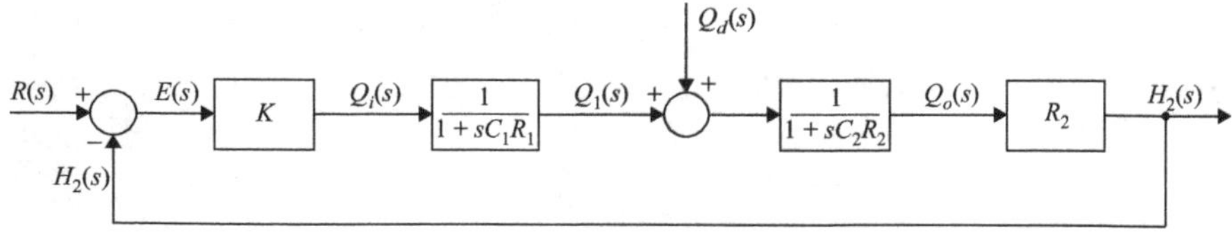

Figure 18 Resolution of Exercise 5.22

$$\frac{H_2}{Q_d} = \frac{(1 + sC_1R_1)R_2}{(1 + sC_1R_1)(1 + sC_2R_2) + KR_2}$$

we get $h_2(\infty) = \lim_{s\to 0} sH_2(s) = \frac{R_2}{1+KR_2}$.
The characteristic equation of $\frac{H_2}{Q_d}$ above is $s^2 + s\frac{C_1R_1+C_2R_2}{C_1C_2R_1R_2} + \frac{1+KR_2}{C_1C_2R_1R_2} = 0$, whence

$$\omega_n = \sqrt{\frac{1+KR_2}{C_1C_2R_1R_2}}$$

$$\zeta = \frac{\frac{C_1R_1+C_2R_2}{C_1C_2R_1R_2}}{2\sqrt{\frac{1+KR_2}{C_1C_2R_1R_2}}} = \frac{C_1R_1 + C_2R_2}{2\sqrt{1+KR_2}\sqrt{C_1C_2R_1R_2}}$$

5.23 The correct answer is option **D)**.

5.24 The correct answer is option **C)**.

5.25

$$G(j\omega) = \frac{30}{j\omega(j\omega+3)}e^{-0.1j\omega} = \frac{30}{\omega\sqrt{\omega^2+9}} \arg\left[-\frac{\pi}{2} - \arctan\frac{\omega}{3} - 0.1\omega\right]$$

$$\frac{30}{\omega_1\sqrt{\omega_1^2+9}} = 1 \Rightarrow \omega_1 = 5.08 \text{ rad/s}$$

$$PM = \pi + \left(-\frac{\pi}{2} - \arctan\frac{5.08}{3} - 0.1 \times 5.08\right) = 1.5^\circ$$

$$-\frac{\pi}{2} - \arctan\frac{\omega_\pi}{3} - 0.1\omega_\pi = -\pi \Rightarrow \omega_\pi = 5.2 \text{ rad/s}$$

$$GM = 20\log_{10}\frac{1}{\frac{30}{5.2\sqrt{5.2^2+9}}} = 0.51 \text{ dB}$$

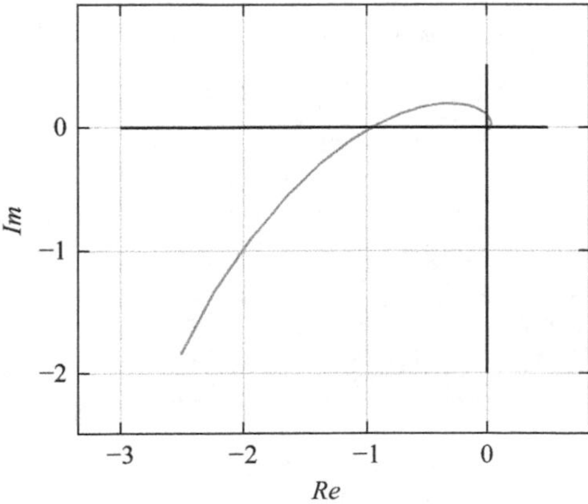

Figure 19 Nyquist plot of Exercise 5.25

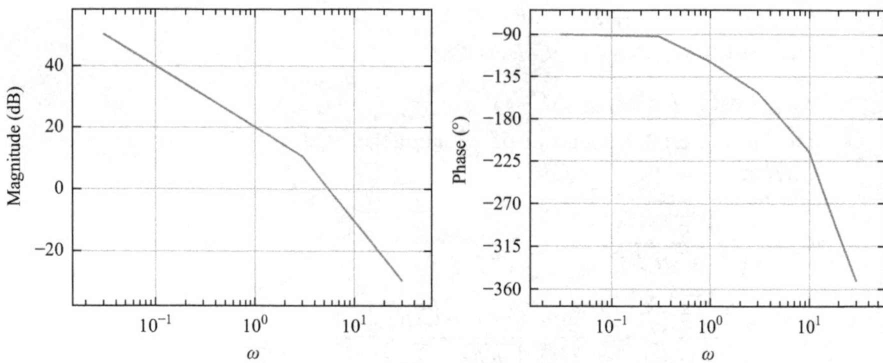

Figure 20 Bode diagram of Exercise 5.25

5.26

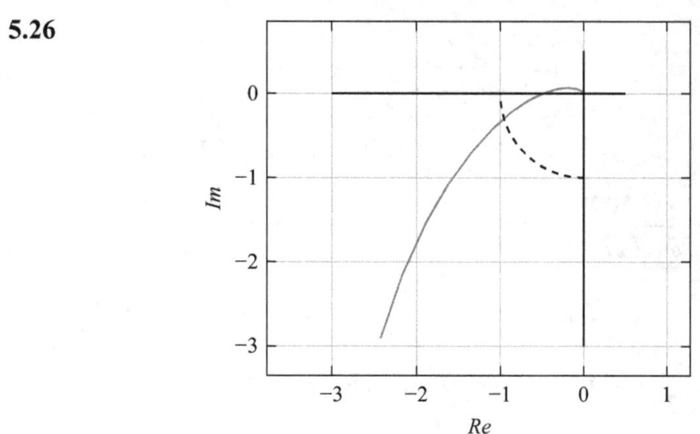

Figure 21 Nyquist plot of Exercise 5.26

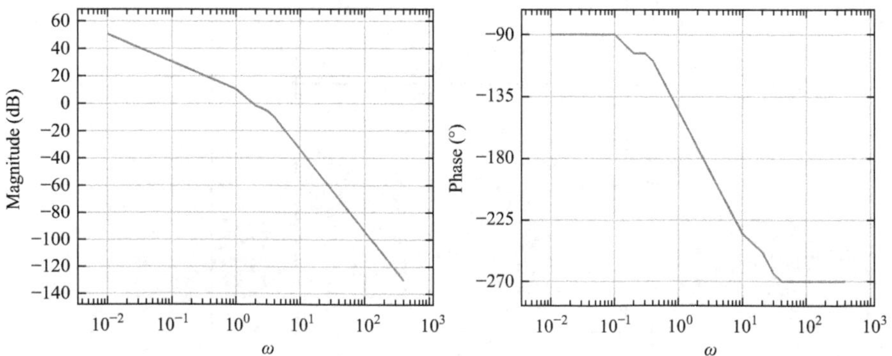

Figure 22 Bode diagram of Exercise 5.26

5.27

$$\begin{aligned} G(j\omega) &= \frac{2}{\omega}\arg\left[-270^\circ + 2\arctan\omega\right] \\ \frac{2}{\omega_1} &= 1 \Rightarrow \omega_1 = 2 \text{ rad/s} \\ PM &= \pi + (-270^\circ + 2\arctan 2) = 36.9^\circ \\ -270^\circ + 2\arctan\omega_\pi &= -180^\circ \Rightarrow \omega_\pi = 1 \text{ rad/s} \\ GM &= 20\log_{10}\frac{1}{2} = -6 \text{ dB} \end{aligned}$$

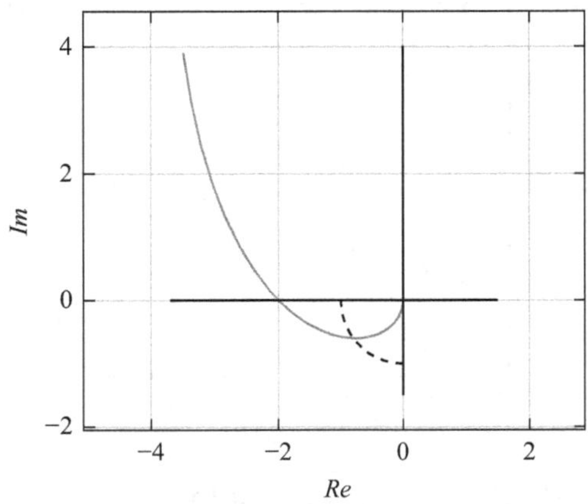

Figure 23 Nyquist plot of Exercise 5.27

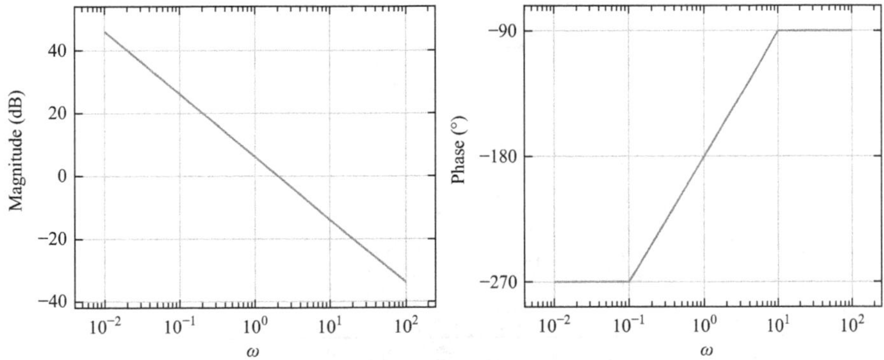

Figure 24 Bode diagram of Exercise 5.27

As the margins have opposite signs no conclusion about stability can be obtained in this way.

5.28 The correct answers are options **B)**, **G)**, **I)** and **N)**.

5.29 The correct answer is option **C)**.

5.30 The correct answers are options **B)** and **G)**.

5.31 A)—H)—K); B)—G)—I); C)—F)—J); D)—E)—L).

5.32 The root-locus diverges at −0.57 and converges at −6.51. Both the root-locus and the Nyquist plot show that the system is stable for all $K \geq 0$. See Figures 25–27.

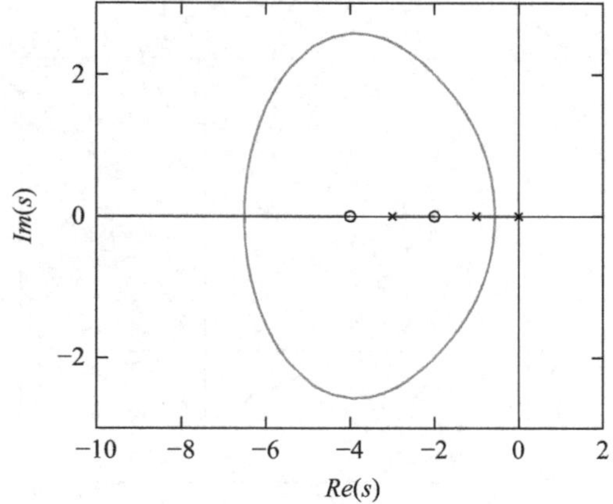

Figure 25 Root-locus of Exercise 5.32

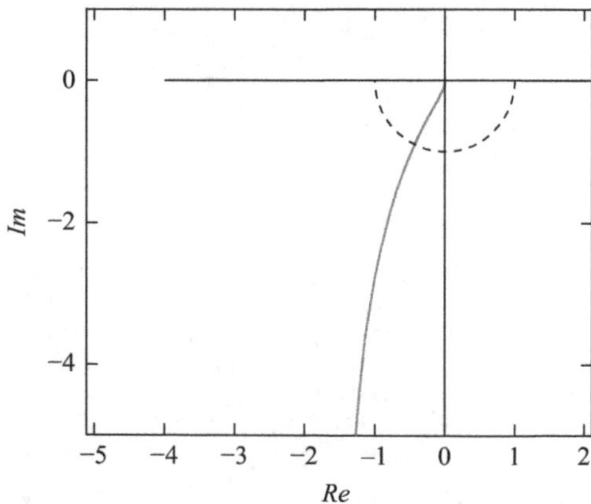

Figure 26 Nyquist diagram of Exercise 5.32

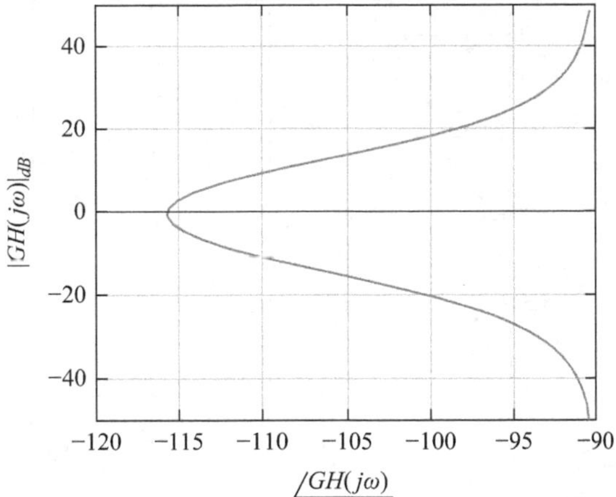

Figure 27 Nichols diagram of Exercise 5.32

5.33 There are two asymptotes with angle $\theta = \pm 90°$ at $\sigma = -2$; the root-locus diverges at -2.47. Both the root-locus and the Nyquist plot show that the system is stable for all $K \geq 0$. See Figures 28–30.

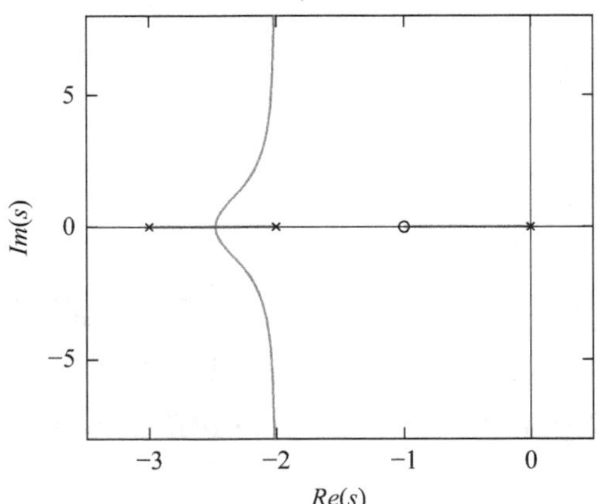

Figure 28 Root-locus of Exercise 5.33

5.34 The correct answers are options **A)**, **G)**, **J)**, **O)** and **Q)**.

5.35 The correct answers are options **C)**, **H)**, **J)**, **N)**, **Q)** and **X)**.

5.36 The correct answers are options **B)**, **E)**, **J)**, **M)** and **Q)**.

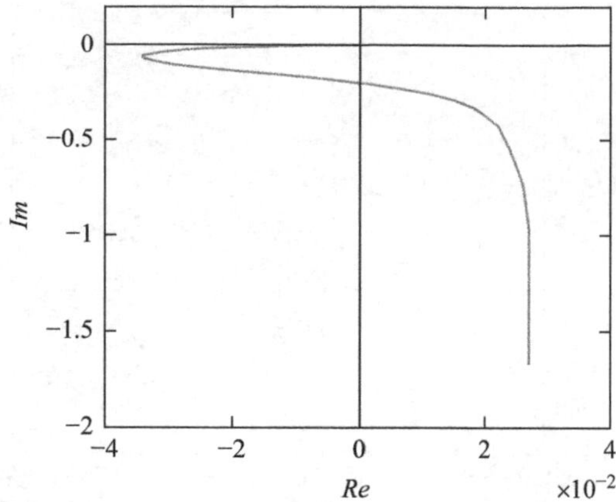

Figure 29 Nyquist diagram of Exercise 5.33

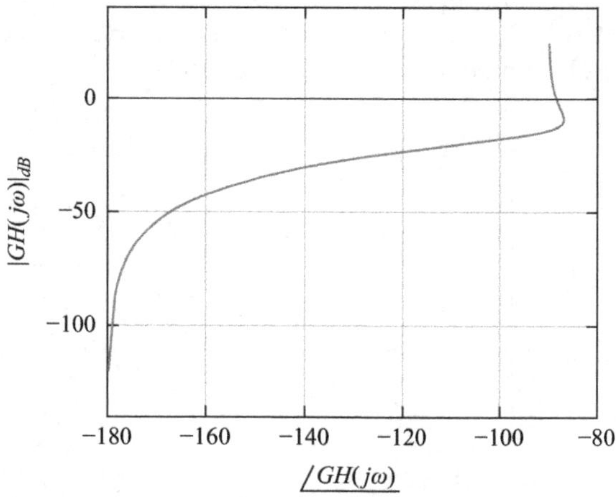

Figure 30 Nichols diagram of Exercise 5.33

5.37 The correct answers are options **B)**, **G)**, **K)**, **M)** and **Q)**.

5.38 The plant is stable for $K < 396$. The desired e_{ss} would require $K = 456$, for which the plant would not be stable. See Figures 31 and 32.

6.1 The correct answer is option **C)**.

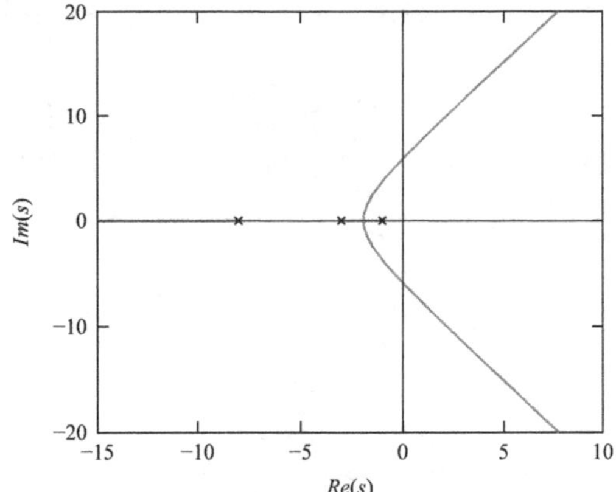

Figure 31 Root-locus of Exercise 5.38

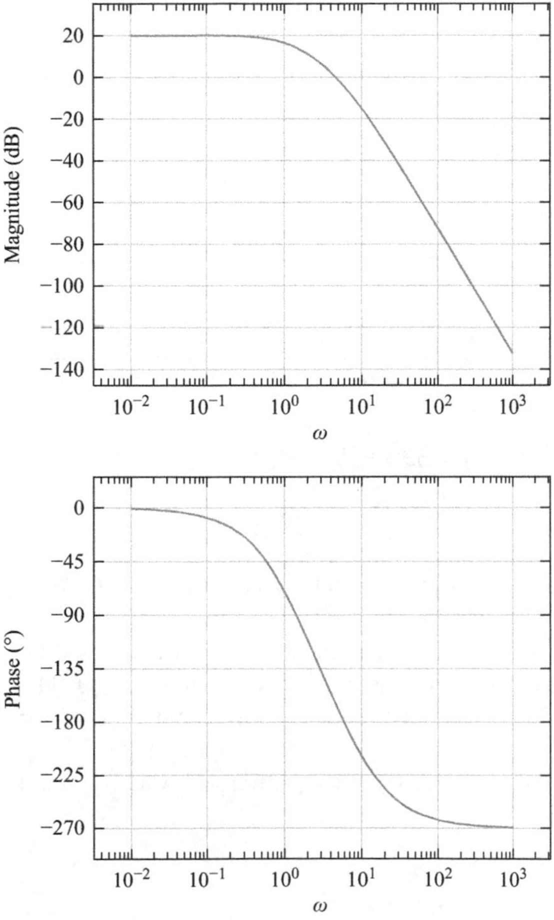

Figure 32 Bode diagram of Exercise 5.38

6.2

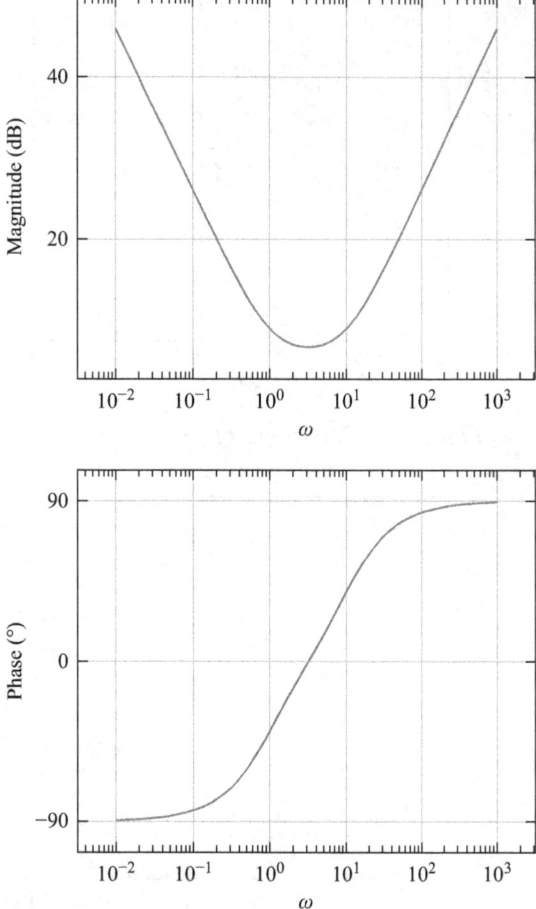

Figure 33 PID controller of Exercise 6.2

6.3 The correct answer is option **A)**: $K_P = 0.360$, $T_i = 6.176$, $T_d = 0.991$.

6.4 The dead time is $t_d = 3$, the response time is $T = 2$, and the gain is $K_G = 10$. From these values the controller parameters $K_P = 0.117$, $T_i = 5.132$ and $T_d = 0.854$ can be found. Thus the correct answer is option **C)**.

6.5 The closed-loop characteristic equation with variable gain K is given by

$$\frac{K}{(s+1)^4} + 1 = 0 \Leftrightarrow K + s^4 + 4s^3 + 6s^2 + 4s + 1 = 0$$

Using the Routh–Hurwitz criterion, we get

s^4	1	6	$K+1$
s^3	4	4	
s^2	5	$K+1$	
s^1	$\frac{16-4K}{5}$		
s^0	$K+1$		

From the two last lines, we see that the system is stable for $K \in]-1, 4[$. For $K = K_u = 4$, we find from the s^2 line that $5s^2 + 5 = 0 \Rightarrow s = \pm j$, and so the period of the critical oscillations is $P_u = \dfrac{2\pi}{|\pm j|} = 6.28$. From these values the controller parameters $K_P = 2.400$, $T_i = 3.142$ and $T_d = 0.785$ can be found. Thus the correct answer is option **C)**.

6.6 The closed-loop transfer function with variable gain K is $\dfrac{\frac{10K}{s(s+1)^2}}{1+\frac{10K}{s(s+1)^2}}$, and its characteristic equation is given by $s^3 + 2s^2 + s + 10K$. Using the Routh–Hurwitz criterion, we get

s^3	1	1
s^2	2	$10K$
s^1	$1-5K$	
s^0	$10K$	

From the two last lines, we see that the system is stable for $K \in]0, 0.2[$. For $K = K_u = 0.2$, we find from the s^2 line that $2s^2 + 2 = 0 \Rightarrow s = \pm j$, and so the period of the critical oscillations is $P_u = \dfrac{2\pi}{|\pm j|} = 6.28$. From these values the controller parameters $K_P = 0.120$, $T_i = 3.14$ and $T_d = 0.785$ can be found. Thus the correct answer is option **B)**.

6.7 Since $G(s) \approx \dfrac{3e^{-s}}{2s+1}$, the required controllers are $C_{PI}(s) = 0.6\left(1 + \dfrac{3}{10s}\right)$ and $C_{PID}(s) = 0.8\left(1 + \dfrac{1}{2s} + \dfrac{1}{2}s\right)$.

6.8 From

$$\frac{\Theta(s)}{\Theta_R(s)} = \frac{K_p(1+T_d s)\frac{L}{Js^2}}{1+K_p(1+T_d s)\frac{L}{Js^2}} = \frac{\frac{LK_P}{J}(1+T_d s)}{s^2 + s\frac{T_d K_P L}{J} + \frac{K_P L}{J}}$$

we obtain

$$\omega_n = \sqrt{\frac{K_pL}{J}}$$

$$2\zeta\omega_n = \frac{T_dK_pL}{J} \Rightarrow T_d = 1.4\sqrt{\frac{J}{K_pL}}$$

7.1 Applying the Laplace transform,

$$\begin{cases} sX(s) = \mathbf{A}X(s) + \mathbf{B}U(s) \\ Y(s) = \mathbf{C}X(s) + \mathbf{D}U(s) \end{cases} \Leftrightarrow \begin{cases} s\mathbf{I}X(s) - \mathbf{A}X(s) = \mathbf{B}U(s) \\ Y(s) = \mathbf{C}X(s) + \mathbf{D}U(s) \end{cases}$$

$$\Leftrightarrow \begin{cases} X(s)(s\mathbf{I} - \mathbf{A}) = \mathbf{B}U(s) \\ Y(s) = \mathbf{C}X(s) + \mathbf{D}U(s) \end{cases} \Leftrightarrow \begin{cases} X(s) = (s\mathbf{I} - \mathbf{A})^{-1}\mathbf{B}U(s) \\ Y(s) = \mathbf{C}(s\mathbf{I} - \mathbf{A})^{-1}\mathbf{B}U(s) + \mathbf{D}U(s) \end{cases}$$

$$\Leftrightarrow \begin{cases} X(s) = (s\mathbf{I} - \mathbf{A})^{-1}\mathbf{B}U(s) \\ \frac{Y(s)}{U(s)} = \mathbf{C}(s\mathbf{I} - \mathbf{A})^{-1}\mathbf{B} + \mathbf{D} \end{cases}$$

Thus, the correct answer is option **B)**.

7.2 $\mathbf{A} = \begin{bmatrix} -1 & 0 & 0 \\ 0 & -2 & 0 \\ 0 & 0 & -3 \end{bmatrix}$, $\mathbf{B} = \begin{bmatrix} 1 \\ 0 \\ 2 \end{bmatrix}$, $\mathbf{C} = \begin{bmatrix} 1 & 2 & 0 \end{bmatrix}$

The system is stable, non-controllable and non-observable. From $\frac{Y(s)}{U(s)} = \frac{1}{s+1}$, when $U(s) = \frac{1}{s}$ we get $Y(s) = \frac{1}{s}\frac{1}{s+1} = \frac{1}{s} - \frac{1}{s+1}$ and thus $y(t) = 1 - e^{-t}$.

7.3 $\mathbf{Q} = \begin{bmatrix} \mathbf{B} & \mathbf{AB} \end{bmatrix} = \begin{bmatrix} 1 & -2 \\ 3 & -6 \end{bmatrix}$, $\mathbf{R} = \begin{bmatrix} \mathbf{C} \\ \mathbf{CA} \end{bmatrix} = \begin{bmatrix} -2 & 1 \\ 4 & -2 \end{bmatrix}$

As the rank of **Q** is 1, the plant is not controllable. As the rank of **R** is 1, the plant is not observable.

7.4 $$\frac{Y(s)}{U(s)} = \mathbf{C}(s\mathbf{I}-\mathbf{A})^{-1}\mathbf{B} = \begin{bmatrix}1 & 0 & 0\end{bmatrix}\begin{bmatrix}\frac{1}{s-2} & 0 & 0\\ \frac{1}{(s-2)(s-3)} & \frac{1}{s-3} & 0\\ \frac{-(s-5)}{(s-2)(s-3)(s+5)} & \frac{2}{(s-3)(s-5)} & \frac{1}{s+5}\end{bmatrix}\begin{bmatrix}1\\0\\0\end{bmatrix}$$

$$= \frac{1}{s-2}$$

Thus, the correct answer is option **A)**.

$$\mathbf{Q} = \begin{bmatrix}\mathbf{B} & \mathbf{AB} & \mathbf{A}^2\mathbf{B}\end{bmatrix} = \begin{bmatrix}1 & 2 & 4\\ 0 & 1 & 5\\ 0 & -1 & 5\end{bmatrix}$$

$$\mathbf{R} = \begin{bmatrix}\mathbf{C}\\ \mathbf{CA}\\ \mathbf{CA}^2\end{bmatrix} = \begin{bmatrix}1 & 0 & 0\\ 2 & 0 & 0\\ 4 & 0 & 0\end{bmatrix}$$

As the rank of **Q** is 3, and the rank of **R** is 1, the correct answer is option **F)**.

7.5 $$s\mathbf{I}-\mathbf{A} = \begin{bmatrix}s+3 & 0\\ 1 & s+2\end{bmatrix}$$

$$(s\mathbf{I}-\mathbf{A})^{-1} = \frac{1}{(s+2)(s+3)}\begin{bmatrix}s+2 & 0\\ -1 & s+3\end{bmatrix} = \begin{bmatrix}\frac{1}{s+3} & 0\\ \frac{-1}{s+2}+\frac{1}{s+3} & \frac{1}{s+2}\end{bmatrix}$$

$$e^{\mathbf{A}t} = \begin{bmatrix}e^{-3t} & 0\\ -e^{-2t}+e^{-3t} & e^{-2t}\end{bmatrix}$$

Thus, the correct answer is option **D)**.

7.6 $$\frac{Y(s)}{U(s)} = \mathbf{C}(s\mathbf{I}-\mathbf{A})^{-1}\mathbf{B} = \begin{bmatrix}1 & 0 & 0\end{bmatrix}\begin{bmatrix}s & -1 & 0\\ 0 & s & -1\\ 15 & 11 & s+5\end{bmatrix}^{-1}\begin{bmatrix}0\\0\\1\end{bmatrix}$$

$$= \frac{1}{(s+3)(s^2+2s+5)}$$

Thus, the correct answer is option **C)**.

$$\mathbf{P} = \begin{bmatrix} 1 & 0 & 1 \\ -1 & 2 & -3 \\ -3 & -4 & 9 \end{bmatrix}$$

$$\mathbf{P}^{-1} = \frac{1}{8}\begin{bmatrix} 3 & -2 & -1 \\ 9 & 6 & 1 \\ 5 & 2 & 1 \end{bmatrix}$$

$$\mathbf{M} = \mathbf{P}^{-1}\mathbf{A}\mathbf{P} = \begin{bmatrix} -1 & 2 & 0 \\ -2 & -1 & 0 \\ 0 & 0 & -3 \end{bmatrix}$$

$$\mathbf{P}^{-1}\mathbf{B} = \frac{1}{8}\begin{bmatrix} -1 \\ 1 \\ 1 \end{bmatrix}$$

$$\mathbf{CP} = \begin{bmatrix} 1 & 0 & 1 \end{bmatrix}$$

Thus, the correct answer is option **F)**.

7.7 The correct answers are options **C)** and **G)**.

7.8

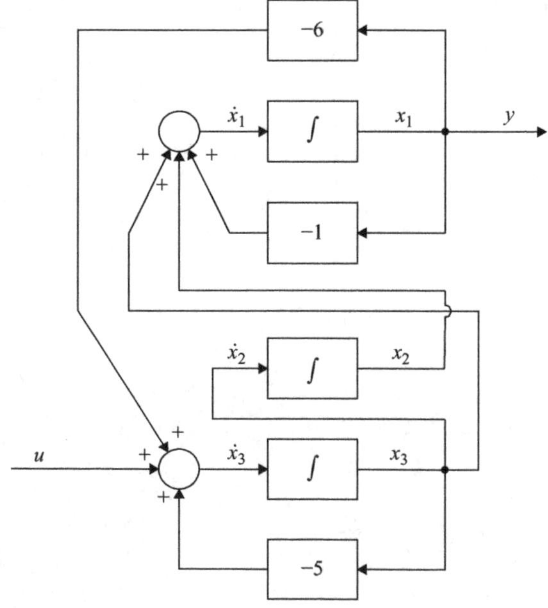

Figure 34 Block diagram of Exercise 7.8

$$\frac{Y(s)}{U(s)} = \mathbf{C}(s\mathbf{I} - \mathbf{A})^{-1}\mathbf{B} = \begin{bmatrix} 1 & 0 & 0 \end{bmatrix} \begin{bmatrix} s+1 & -1 & -1 \\ 0 & s & -1 \\ 6 & 0 & s+5 \end{bmatrix}^{-1} \begin{bmatrix} 0 \\ 0 \\ 1 \end{bmatrix}$$

$$= \begin{bmatrix} 1 & 0 & 0 \end{bmatrix} \frac{1}{s^3 + 6s^2 + 11s + 6} \begin{bmatrix} s^2 + 5s & s+5 & s+1 \\ 6 & s^2 + 6s + 11 & s+1 \\ -6s & -6 & s^2 + s \end{bmatrix} \begin{bmatrix} 0 \\ 0 \\ 1 \end{bmatrix}$$

$$= \frac{s+1}{s^3 + 6s^2 + 11s + 6}$$

$$\mathbf{Q} = \begin{bmatrix} \mathbf{B} & \mathbf{AB} & \mathbf{A}^2\mathbf{B} \end{bmatrix} = \begin{bmatrix} 0 & 1 & -5 \\ 0 & 1 & -5 \\ 1 & -5 & 19 \end{bmatrix}$$

As the rank of **Q** is 2, the plant is not controllable.

7.9 The correct answers are options **A)** and **H)**.

7.10 The correct answer is option **C)**.

7.11 The correct answer is option **B)**.

7.12 The correct answers are options **B)**, **D)**, **E)**, and **G)**.

7.13 The correct answers are options **A)** and **F)**.

7.14 The correct answers are options **C)**, **F)**, **K)**, **M)**, and **R)**.

7.15 The correct answers are options **C)** and **E)**.

7.16 The correct answers are options **B)**, **G)**, **I)**, and **P)**.

7.17 The correct answer is option **B)**.

7.18 The correct answers are options **C)** and **G)**.

7.19 The correct answer is option **A)**.

7.20 The eigenvalues of **A** are -1 and -2, and we get $\mathbf{A} = \begin{bmatrix} -e^{-t} + 2e^{-2t} & -e^{-t} + e^{-2t} \\ 2e^{-t} - 2e^{-2t} & 2e^{-t} - e^{-2t} \end{bmatrix}$. So the correct answer is option **D)**.

7.21 From the eigenvalues

$$\det[\mathbf{A} - \lambda\mathbf{I}] = \det\begin{bmatrix} 1-\lambda & 4 \\ 2 & 3-\lambda \end{bmatrix} = 0 \Leftrightarrow \lambda = 5 \vee \lambda = -1$$

it can be found that the correct answer is option **B)**.

7.22
$$\begin{bmatrix} \dot{x}_1 \\ \dot{x}_2 \end{bmatrix} = \begin{bmatrix} 3 & -2 \\ 2 & 5 \end{bmatrix}\begin{bmatrix} x_1 \\ x_2 \end{bmatrix} + \begin{bmatrix} -8 \\ 5 \end{bmatrix}u$$

$$y = \begin{bmatrix} -4 & 9 \end{bmatrix}\begin{bmatrix} x_1 \\ x_2 \end{bmatrix}$$

$$\mathbf{R} = \begin{bmatrix} \mathbf{C} \\ \mathbf{CA} \end{bmatrix} = \begin{bmatrix} -4 & 9 \\ 6 & 53 \end{bmatrix}$$

$$\mathbf{Q} = \begin{bmatrix} \mathbf{B} & \mathbf{AB} \end{bmatrix} = \begin{bmatrix} -8 & -34 \\ 5 & 9 \end{bmatrix}$$

As the rank of **R** is 2, the plant is observable. As the rank of **Q** is 2, the plant is controllable.

7.23 Controllable canonical form:

$$\begin{bmatrix} \dot{x}_1 \\ \dot{x}_2 \end{bmatrix} = \begin{bmatrix} 0 & 1 \\ -8 & -6 \end{bmatrix}\begin{bmatrix} x_1 \\ x_2 \end{bmatrix} + \begin{bmatrix} 0 \\ 1 \end{bmatrix}u$$

$$y = \begin{bmatrix} -5 & -2 \end{bmatrix}\begin{bmatrix} x_1 \\ x_2 \end{bmatrix} + u$$

Observable canonical form:

$$\begin{bmatrix} \dot{x}_1 \\ \dot{x}_2 \end{bmatrix} = \begin{bmatrix} -6 & 1 \\ -8 & 0 \end{bmatrix}\begin{bmatrix} x_1 \\ x_2 \end{bmatrix} + \begin{bmatrix} -2 \\ -5 \end{bmatrix}u$$

$$y = \begin{bmatrix} 1 & 0 \end{bmatrix}\begin{bmatrix} x_1 \\ x_2 \end{bmatrix} + u$$

Transfer function: $\dfrac{Y(s)}{U(s)} = \dfrac{(s+1)(s+3)}{(s+2)(s+4)}$

7.24
$$\begin{bmatrix}\dot{x}_1\\ \dot{x}_2\\ \dot{x}_3\end{bmatrix} = \begin{bmatrix}-\frac{1}{C_1R_1} & 0 & -\frac{1}{C_1}\\ 0 & 0 & \frac{1}{C_2}\\ \frac{1}{L} & -\frac{1}{L} & -\frac{R_2}{L}\end{bmatrix}\begin{bmatrix}x_1\\ x_2\\ x_3\end{bmatrix} + \begin{bmatrix}\frac{1}{R_1}\\ 0\\ 0\end{bmatrix}u$$

$$y = \begin{bmatrix}0 & 1 & 0\end{bmatrix}\begin{bmatrix}x_1\\ x_2\\ x_3\end{bmatrix}$$

$$(s\mathbf{I}-\mathbf{A})^{-1} = \frac{1}{s^3+5s^2+18s+12}\begin{bmatrix}s^2+3s+6 & 6 & -2s\\ 6 & s^2+5s+12 & 2(s+2)\\ 3s & -3(s+2) & s(s+2)\end{bmatrix}$$

$$\mathbf{C}(s\mathbf{I}-\mathbf{A})^{-1}\mathbf{B} = \frac{6}{s^3+5s^2+18s+12}$$

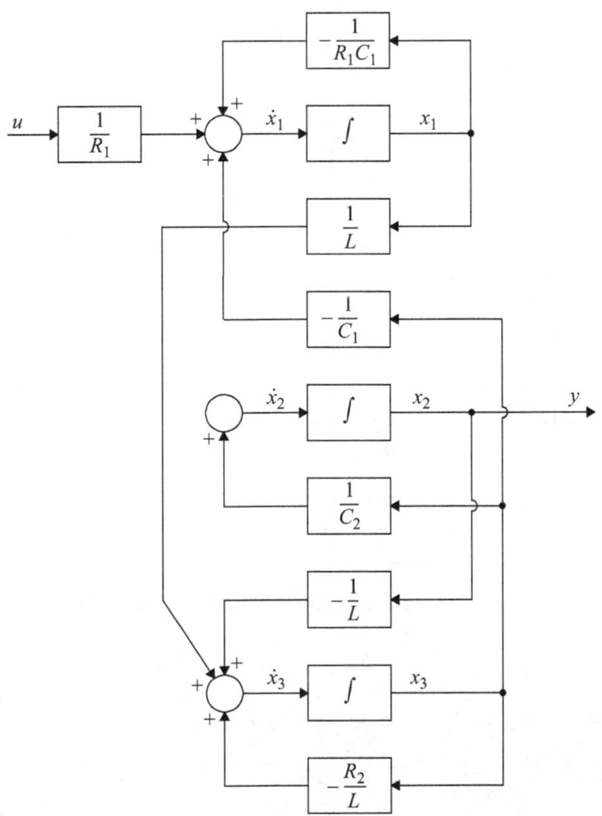

Figure 35 Block diagram of Exercise 7.24

7.25 $$\begin{bmatrix}\dot{x}_1\\ \dot{x}_2\end{bmatrix} = \begin{bmatrix}-\lambda_1 & 1\\ 0 & -\lambda_2\end{bmatrix}\begin{bmatrix}x_1\\ x_2\end{bmatrix} + \begin{bmatrix}0\\ 1\end{bmatrix}u$$

$$y = \begin{bmatrix}1 & 0\end{bmatrix}\begin{bmatrix}x_1\\ x_2\end{bmatrix}$$

$$\mathbf{Q} = \begin{bmatrix}\mathbf{B} & \mathbf{AB}\end{bmatrix} = \begin{bmatrix}0 & 1\\ 1 & -\lambda_2\end{bmatrix},$$

$$\mathbf{R} = \begin{bmatrix}\mathbf{C}\\ \mathbf{CA}\end{bmatrix} = \begin{bmatrix}1 & 0\\ -\lambda_1 & 1\end{bmatrix}$$

As the rank of **Q** is 2, the plant is controllable. As the rank of **R** is 2, the plant is observable.

$$\det(s\mathbf{I} - \mathbf{A}) = \det\begin{bmatrix}s+\lambda_1 & -1\\ 0 & s+\lambda_2\end{bmatrix} = (s+\lambda_1)(s+\lambda_2)$$

So the system is stable if $\lambda_1, \lambda_2 < 0$.

$$G(s) = \frac{1}{(s+\lambda_1)(s+\lambda_2)} = \frac{1}{\lambda_2 - \lambda_1}\left(\frac{1}{s+\lambda_1} - \frac{1}{s+\lambda_2}\right)$$

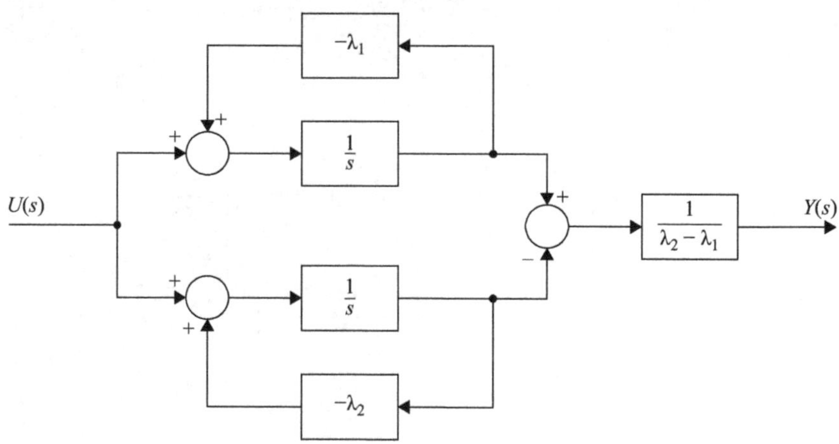

Figure 36 Block diagram of Exercise 7.25

7.26 $$\begin{bmatrix}\dot{i}_1\\ \dot{v}_2\\ \dot{i}_3\end{bmatrix} = \begin{bmatrix}-\frac{R_1}{L_1} & -\frac{1}{L_1} & -\frac{R_2}{L_2}\\ \frac{1}{C_2} & 0 & -\frac{1}{C_2}\\ \frac{R_2}{L_3} & \frac{1}{L_3} & -\frac{R_2+R_3}{L_3}\end{bmatrix}\begin{bmatrix}i_1\\ v_2\\ i_3\end{bmatrix} + \begin{bmatrix}\frac{1}{L_1}\\ 0\\ 0\end{bmatrix}u$$

$$y = \begin{bmatrix}R_2 & 1 & -(R_2+R_3)\end{bmatrix}\begin{bmatrix}i_1\\ v_2\\ i_3\end{bmatrix}$$

7.27 $\mathbf{A} = \begin{bmatrix} -1 & 0 & 0 \\ 0 & -2 & 0 \\ 0 & 0 & 1 \end{bmatrix}, \ \mathbf{B} = \begin{bmatrix} 1 \\ 0 \\ 2 \end{bmatrix}, \ \mathbf{C} = \begin{bmatrix} 2 & 3 & 4 \end{bmatrix}$

Since **A** has a positive eigenvalue, the system is unstable.

$$\mathbf{Q} = \begin{bmatrix} \mathbf{B} & \mathbf{AB} & \mathbf{A}^2\mathbf{B} \end{bmatrix} = \begin{bmatrix} 1 & -1 & 1 \\ 0 & 0 & 0 \\ 2 & 2 & 2 \end{bmatrix}$$

As the rank of **Q** is 2, the system is not controllable. This conclusion could also have been reached verifying that there is no way input u can affect state x_2. The transfer function can be found from the block diagram or alternatively from

$$\begin{aligned} \frac{Y(s)}{U(s)} &= \mathbf{C}(s\mathbf{I} - \mathbf{A})^{-1}\mathbf{B} = \begin{bmatrix} 2 & 3 & 4 \end{bmatrix} \begin{bmatrix} \frac{1}{s+1} & 0 & 0 \\ 0 & \frac{1}{s+2} & 0 \\ 0 & 0 & \frac{1}{s-1} \end{bmatrix} \begin{bmatrix} 1 \\ 0 \\ 2 \end{bmatrix} \\ &= \frac{10s + 6}{(s+1)(s-1)} \end{aligned}$$

7.28 The correct answer is option **A)**.

7.29 Since

$$\begin{cases} \dot{x} = -3x + 2u \\ y = 4x + 5u \end{cases}$$

the correct answer is option **D)**: none of the above.

7.30 The correct answers are options **B)** and **C)**.

7.31 $$\frac{d}{dt}\begin{bmatrix} i_1 \\ i_2 \\ V_1 \\ V_2 \end{bmatrix} = \begin{bmatrix} -\frac{R_1+R_2}{L_1} & \frac{R_2}{L_1} & -\frac{1}{L_1} & -\frac{1}{L_2} \\ \frac{R_2}{L_2} & -\frac{R_2}{L_2} & 0 & \frac{1}{L_2} \\ \frac{1}{C_1} & 0 & 0 & 0 \\ \frac{1}{C_2} & -\frac{1}{C_2} & 0 & 0 \end{bmatrix} \begin{bmatrix} i_1 \\ i_2 \\ V_1 \\ V_2 \end{bmatrix} + \begin{bmatrix} \frac{1}{L_1} \\ 0 \\ 0 \\ 0 \end{bmatrix} u$$

$$y = \begin{bmatrix} R_2 & -R_2 & 0 & 1 \end{bmatrix} \begin{bmatrix} i_1 \\ i_2 \\ V_1 \\ V_2 \end{bmatrix}$$

7.32 $$\begin{bmatrix}\dot{x}_1\\ \dot{x}_2\end{bmatrix}=\begin{bmatrix}-\frac{2R}{L_1} & -\frac{R}{L_1}\\ -\frac{R}{L_2} & -\frac{2R}{L_2}\end{bmatrix}\begin{bmatrix}x_1\\ x_2\end{bmatrix}+\begin{bmatrix}\frac{1}{L_1}\\ \frac{1}{L_2}\end{bmatrix}u$$

$$\begin{bmatrix}y_1\\ y_2\end{bmatrix}=\begin{bmatrix}0 & R\\ -R & -R\end{bmatrix}\begin{bmatrix}x_1\\ x_2\end{bmatrix}+\begin{bmatrix}0\\ 1\end{bmatrix}u$$

$$\begin{aligned}e^{\mathbf{A}t} &= \mathscr{L}^{-1}\{(s\mathbf{I}-\mathbf{A})^{-1}\}\\ &= \mathscr{L}^{-1}\left\{\frac{1}{s^2+2R\left(\frac{1}{L_1}+\frac{1}{L_2}\right)s+\frac{3R^2}{L_1L_2}}\begin{bmatrix}\left(s+\frac{2R}{L_2}\right) & -\frac{R}{L_1}\\ -\frac{R}{L_2} & \left(s+\frac{2R}{L_1}\right)\end{bmatrix}\right\}\end{aligned}$$

$$\frac{Y_1(s)}{U(s)}=\frac{1}{s^2+2R\left(\frac{1}{L_1}+\frac{1}{L_2}\right)s+\frac{3R^2}{L_1L_2}}\left(s\frac{R}{L_2}+\frac{R^2}{L_1L_2}\right)$$

The plant is both controllable and observable.

7.33 $$\begin{cases}A_1\dot{h}_1=q_i-q_1\\ q_1=\frac{h_1}{R_1}\\ A_2\dot{h}_2=q_1-q_o\\ q_0=\frac{h_2}{R_2}\end{cases}\Rightarrow\begin{cases}\dot{h}_1=\frac{1}{A_1}q_i-\frac{1}{A_1}q_1=-\frac{1}{A_1R_1}h_1+\frac{1}{A_1}q_i\\ \dot{h}_2=\frac{1}{A_2}q_1-\frac{1}{A_2}q_o=\frac{1}{A_2R_1}h_1-\frac{1}{A_2R_2}h_2\end{cases}$$

$$\begin{bmatrix}\dot{h}_1\\ \dot{h}_2\end{bmatrix}=\begin{bmatrix}-\frac{1}{A_1R_1} & 0\\ \frac{1}{A_2R_1} & -\frac{1}{A_2R_2}\end{bmatrix}\begin{bmatrix}h_1\\ h_2\end{bmatrix}+\begin{bmatrix}\frac{1}{A_1}\\ 0\end{bmatrix}u$$

$$y=\begin{bmatrix}0 & \frac{1}{R_2}\end{bmatrix}\begin{bmatrix}h_1\\ h_2\end{bmatrix}$$

$$\begin{aligned}G(s) &= \mathbf{C}(s\mathbf{I}-\mathbf{A})^{-1}\mathbf{B}=\begin{bmatrix}0 & 1\end{bmatrix}\begin{bmatrix}s+2 & 0\\ -1 & s+0.5\end{bmatrix}^{-1}\begin{bmatrix}1\\ 0\end{bmatrix}\\ &= \frac{1}{(s+2)(s+0.5)}\\ \mathbf{Q} &= \begin{bmatrix}\mathbf{B} & \mathbf{AB}\end{bmatrix}=\begin{bmatrix}1 & -2\\ 0 & 1\end{bmatrix}\\ \mathbf{R} &= \begin{bmatrix}\mathbf{C}\\ \mathbf{CA}\end{bmatrix}=\begin{bmatrix}0 & 1\\ 1 & -0.5\end{bmatrix}\end{aligned}$$

As the rank of **Q** is 2, and the rank of **R** is 2, the system is both observable and controllable.

7.34 $\begin{cases} \dot{x} = \mathbf{A}x + \mathbf{B}u \\ y = \mathbf{C}x \end{cases}$, $\mathbf{A} = \begin{bmatrix} -\frac{1}{RC} & 0 & -\frac{1}{C} \\ 0 & -\frac{1}{RC} & \frac{1}{C} \\ \frac{1}{L} & -\frac{1}{L} & 0 \end{bmatrix}$, $\mathbf{B} = \begin{bmatrix} \frac{1}{RC} \\ 0 \\ 0 \end{bmatrix}$, $\mathbf{C} = \begin{bmatrix} 0 & 1 & 0 \end{bmatrix}$

$$G(s) = \mathbf{C}(s\mathbf{I} - \mathbf{A})^{-1}\mathbf{B} = \begin{bmatrix} 0 & 1 & 0 \end{bmatrix} \begin{bmatrix} s+3 & 0 & 1 \\ 0 & s+3 & -1 \\ -2 & 2 & s \end{bmatrix}^{-1} \begin{bmatrix} 3 \\ 0 \\ 0 \end{bmatrix}$$

$$= \frac{6}{s^3 + 6s^2 + 13s + 12}$$

8.1 The correct answers are options **C)** and **F)**.

8.2 The correct answer is option **A)**.

8.3 Plotting the root-locus of plant $G_p(s) = \frac{20}{s(s+4)}$, we see that the poles will always be to the right of the desired location, as there is a vertical asymptote at $\sigma = -2$. To push them to the left, we need a lead controller, providing an additional pole clearly to the left, thus much faster than the desired dominant poles, and moving to the right as the gain increases. The controller will be given by (8.1) and with three unknowns K, a and b there are infinite solutions. If we fix $b = 20$, the closed loop becomes $\frac{G_c(s)G_p(s)}{1+G_c(s)G_p(s)} = \frac{20K(s+a)}{s^3+24s^2+(80+20K)s+20Ka}$, and since $-3 \pm 3j$ are poles, we get $(-3+3j)^3 + 24(-3+3j)^2 + (80+20K)(-3+3j) + 20Ka = 0 \Rightarrow a = 7 \wedge K = 2.3$. The step response of the closed loop consisting of $G_p(s)$ and $G_c(s) = 2.3\frac{s+7}{s+20}$ has a 5% overshoot and a 5% settling time under 0.6 s.

8.4 The root-locus converges/diverges at $\frac{d}{ds}\frac{s^2+4s+4}{s+4} = 0 \Leftrightarrow s = -6 \vee s = -2$. See Figure 37. Both poles of the closed-loop $\frac{K(s^3+8s^2+20s+16)}{s^2+(4+K)s+4(1+K)}$ are at -6 for $(-6)^2 + (4+K)(-6) + 4 + 4K = 0 \Leftrightarrow K = 8$; for lower values, poles are complex and the response is oscillatory. $(-2+2j)^2 + (4+K)(-2+2j) + 4 + 4K = 0$ has no real solutions: thus poles $-2 \pm 2j$ cannot be reached with a proportional controller.

8.5 Applying the law of Newton, $f - k_S x - B\dot{x} = M\ddot{x}$, and thus $\frac{X(s)}{F(s)} = \frac{1}{s^2M+sB+k_S}$. The closed-loop transfer function is given by $\frac{X(s)}{R(s)} = \frac{A\frac{X(s)}{F(s)}}{1+A\frac{X(s)}{F(s)}}$, where $R(s)$ is the reference; we want $\frac{A}{s^2M+sB+k_S} = -1 \Leftrightarrow \frac{k_S}{s^2M+sB+A} = -1$. Replacing values, this is $\frac{k_S}{(s+1)(s+5)} = -1$; for both closed-loop poles to meet at $s = -3$, we need $k_S = 4$, and so for $k_S \in]0, 4[$, there are no oscillations (of course k_S cannot be negative).

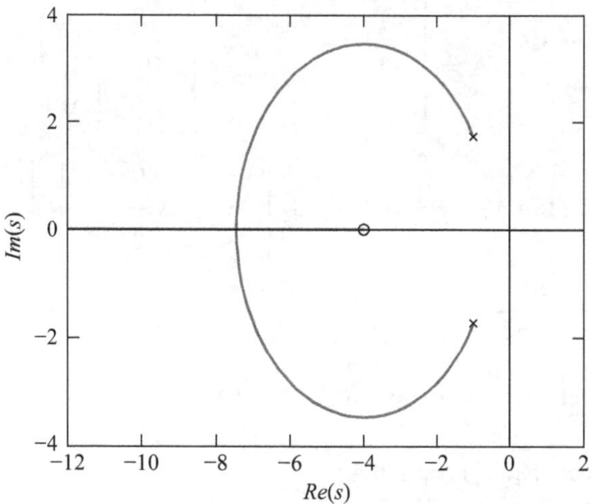

Figure 37 Root-locus of Exercise 8.4

8.6 The plant is always unstable if $H(s) = 1$, but if $H(s) = 1 + 0.5\,s$, it becomes stable for $K > 1.314$. See the root-locus plot in Figure 38.

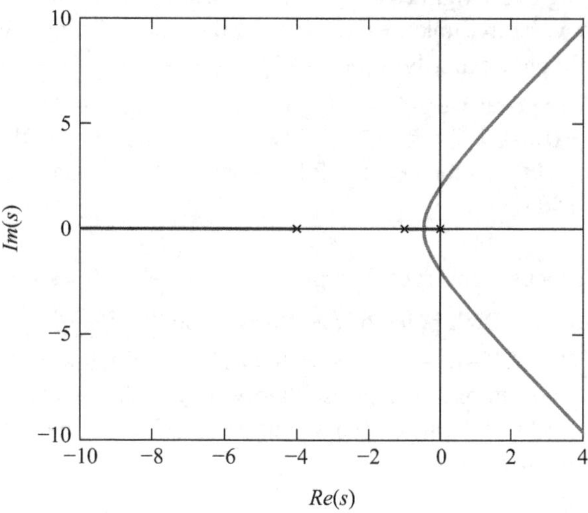

Figure 38 Root-locus of Exercise 8.6

8.7 The step response shows that there is a pole at the origin and a non-minimum phase zero: so, the root-locus of $\frac{H(s)}{\Delta(s)}$ when $K > 0$ is **D)**, and when $K < 0$ is **C)**. Thus $\frac{H(s)}{\Delta(s)} = K\frac{6-s}{s(s^2+4s+13)}$, and as there is one pole at the origin in the open loop when using proportional control, this means that a constant velocity, which corresponds to a ramp input, will be followed with a constant steady-state error, given by $\frac{13}{6K} = 0.9 \Rightarrow K = 2.4$. We need a lag controller with a steady-state gain of $\frac{0.9}{0.3} = 3$, and since the plant's gain crossover frequency is 1.17 rad/s and has a pole at the origin, we will place the controller's zero one decade and a half below at 0.04. This controller $C(s) = \frac{s+0.04}{s+\frac{0.04}{3}}$ fulfills the specifications, that could not be obtained with a proportional controller equal to 3 since the plant would become unstable.

8.8 $G(s)$ has a $-30°$ phase margin, so we want $\Delta PM = 50°$. For $\Delta PM = 55°$, we get $K = 10$, and, since the corresponding gain increase at frequency $\sqrt{ab}$ is 10 dB, we get $\omega^\star = 2.7$ rad/s. Hence, $a = 0.85$, $b = 8.5$, and $C(s) = 10\frac{s+0.85}{s+8.5}$, with which the phase margin is 18°. Notice that even though we increased ΔPM by 10% the phase margin is still below the nominal specification. Should the result have fallen below the acceptable tolerance, calculations would have to be redone with a larger ΔPM.
Poles at $-4 + 10j$ correspond to a 2%-settling time of 1 s. Fixing the compensator's zero at $a = 1$ and replacing the desired zeros at the closed loop's expression, we get at $C(s) = 174.4\frac{s+1}{s+15.7}$, with which the settling time is as expected.

8.9 We must have $\zeta\omega_n \geq \frac{1}{2}$ and $\zeta > 0.59$; see Figure 39. The controller is given by $C(s) = P + Ds$, and thus the closed loop is $\frac{\Theta(s)}{\Theta_{\text{ref}}(s)} = \frac{\frac{P}{12}+\frac{D}{12}s}{s^2+\frac{D+1}{12}s+\frac{P}{12}}$. The presence of the zero will cause a performance less satisfactory than that expected from the location of the poles. Thus, if we make $\zeta\omega_n = \frac{1}{2}$, in the limit of what is desired for the settling time, and choose a double real pole, by setting the discriminant to zero, we get

$$\frac{D+1}{12} = 1 \Rightarrow D = 11$$

$$1 - 4\frac{P}{12} = 0 \Rightarrow P = 3$$

but this controller $C(s) = 3 + 11s$ fails in what concerns the settling time (it is almost 8 s). A faster response must be sought: making e.g., $\zeta\omega_n = \frac{3}{4}$, we get

$$\frac{D+1}{12} = \frac{6}{4} \Rightarrow D = 17$$

$$\left(\frac{6}{4}\right)^2 - 4\frac{P}{12} = 0 \Rightarrow P = \frac{27}{4}$$

but while the settling time is now fine (it is about 5 s) the overshoot increases to 11%. A small increase of the open-loop gain will draw the two poles apart on the real axis, making the response less oscillatory, at the expense of a slightly larger (but acceptable) settling time: we can fulfill specifications with $C(s) = 1.05(6.75 + 17s)$.

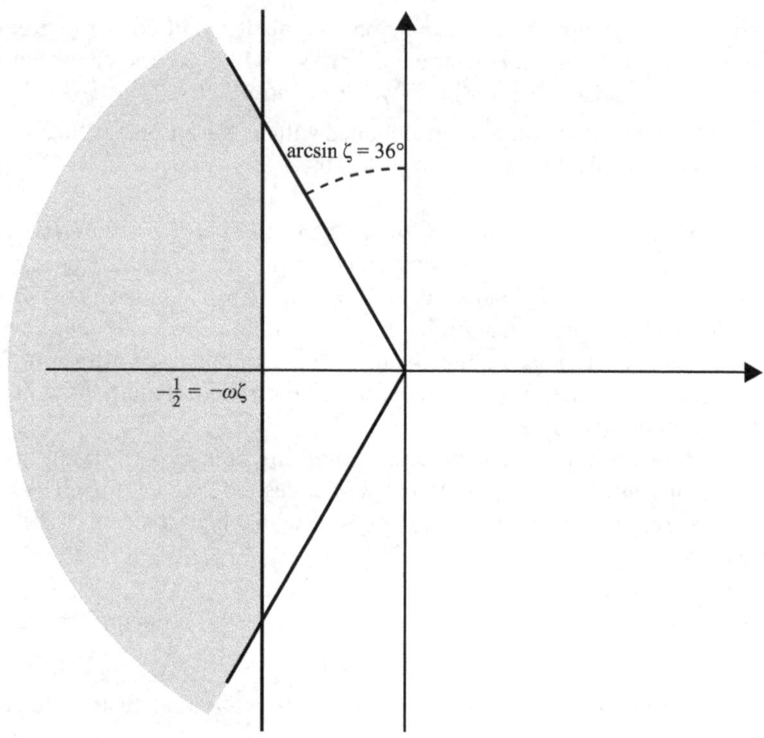

Figure 39 Specifications for Exercise 8.9

8.10 The WEC's transfer function is $\frac{Y(s)}{F(s)} = \frac{2\times10^{-6}}{0.6s^2+0.2s+1}$ and thus

$$\dot{\mathbf{x}} = \begin{bmatrix} 0 & 1 \\ -\frac{5}{3} & -\frac{1}{3} \end{bmatrix}\mathbf{x} + \begin{bmatrix} 0 \\ 1 \end{bmatrix} f$$

$$y = \begin{bmatrix} \frac{10}{3} \times 10^{-6} & 0 \end{bmatrix}\mathbf{x} + 0f$$

We have $a_1 = \frac{1}{3}$, $a_2 = \frac{5}{3}$, $b_0 = b_1 = 0$, and $b_2 = \frac{10}{3} \times 10^{-6}$. Specifications lead to $\zeta = 0.36$ and poles at $-2.14 \pm 5.57j$. The desired characteristic equation is $s^2 + 4.32s + 36 = 0$, so $\alpha_1 = 4.32$ and $\alpha_2 = 36$. Thus we make $\mathbf{K} = [34.33\ 3.99]$ and the new dynamics are $\frac{Y(s)}{v(s)} = \frac{\frac{10}{3}\times10^{-6}}{s^2+4.32s+36}$.

8.11 The correct answer is option **D)**, since $\mathbf{K} = [7 \quad 3]$.

8.12 $\mathbf{R} = \begin{bmatrix} -0.1667 & 0.3333 \\ 0.3889 & 0.2778 \end{bmatrix} \times 10^{-5}$

$$\mathbf{T} = \begin{bmatrix} 0 & 0 \\ -0.1667 & 0 \end{bmatrix} \times 10^{-5}$$

Let us suppose that the WEC's position is obtained with a sampling time T_s. When the input is a unit step,

$$\mathbf{U} = \begin{bmatrix} 1 & 1 & 1 & 1 & 1 & 1 & 1 & \dots \\ 0 & 0 & 0 & 0 & 0 & 0 & 0 & \dots \end{bmatrix}$$

When it is a unit slope ramp,

$$\mathbf{U} = \begin{bmatrix} 0 & T_s & 2T_s & 3T_s & 4T_s & 5T_s & 6T_s & \dots \\ 1 & 1 & 1 & 1 & 1 & 1 & 1 & \dots \end{bmatrix}$$

In both cases,

$$\mathbf{Y} = \begin{bmatrix} y(0) & y(T_s) & y(2T_s) & y(3T_s) & y(4T_s) & y(5T_s) & y(6T_s) & \dots \\ y'(0) & y'(T_s) & y'(2T_s) & y'(3T_s) & y'(4T_s) & y'(5T_s) & y'(6T_s) & \dots \end{bmatrix}$$

8.13 The correct answer is option **A)**.

8.14 The correct answer is option **D)**, because the matrices are, in fact, those of option **A)**, and clearly **R** cannot be inverted, or, in other words, the plant is not observable.

9.1 From $X(z) = 4\left(\frac{z}{z-1} - \frac{z}{z-\frac{1}{2}}\right)$, we get $x(k) = 4\left[1 - \left(\frac{1}{2}\right)^k\right]$, $k = 0, 1, 2, 3, \dots$.

9.2 The correct answer is option **A)**.

9.3 The correct answer is option **B)**.

9.4 The correct answer is option **D)**.

9.5 $X(s) = \frac{\frac{1}{2}}{(s+1)^2} + \frac{\frac{1}{4}}{s+1} + \frac{-\frac{1}{4}}{s+3}$

$$x(t) = \mathscr{L}^{-1}\{X(s)\} = \frac{1}{2}te^{-t} + \frac{1}{4}e^{-t} - \frac{1}{4}e^{-3t}$$

$$X(z) = \mathscr{Z}\{x(t)\} = \frac{1}{2}\frac{hze^{-h}}{(z-e^{-h})^2} + \frac{1}{4}\frac{z}{z-e^{-h}} - \frac{1}{4}\frac{z}{z-e^{-3h}}$$

9.6 The correct answer is option **A)**.

9.7 The correct answer is option **B)**.

9.8 The correct answer is option **A)**.

9.9 $X(z) = \frac{4}{35}\frac{z}{\left(z+\frac{1}{2}\right)^2} + \frac{96}{1\,225}\frac{z}{z+\frac{1}{2}} - \frac{4}{25}\frac{z}{z-2} + \frac{4}{49}\frac{z}{z-3} \Rightarrow$

$$x(k) = \frac{4}{35}(-2k)\left(-\frac{1}{2}\right)^k + \frac{96}{1\,225}\left(-\frac{1}{2}\right)^k - \frac{4}{25}2^k + \frac{4}{49}3^k$$

9.10 The correct answer is option **C)**.

9.11 $X(z) = -3 - 2\frac{z}{(z-1)^2} + 4\frac{z}{z-1} \Rightarrow x(kh) = -3\delta(k) - 2t + 4$

9.12
$$\sin\omega t = \frac{1}{2j}\left(e^{j\omega t} - e^{-j\omega t}\right) \Rightarrow$$

$$\begin{aligned}\mathscr{Z}(\sin\omega t) &= \mathscr{Z}\left[\frac{1}{2j}\left(e^{j\omega t} - e^{-j\omega t}\right)\right] \\ &= \frac{1}{2j}\left(\frac{z}{z-e^{j\omega h}} - \frac{z}{z-e^{-j\omega h}}\right) \\ &= \frac{z\sin(\omega h)}{z^2 - 2z\cos(\omega h) + 1}\end{aligned}$$

Thus the correct answer is option **C)**.

9.13 The correct answer is option **A)**.

9.14 Applying the $\mathscr{Z}$-transform, we get

$$z^2Y(z) - z^2y(0) - zy(1) + zY(z) - zy(0) + \frac{1}{8}Y(z) = \frac{z}{z-1}$$

Replacing the initial conditions, we get $Y(z) = \frac{8z}{(z-1)(8z^2+8z+1)} = 0.47\frac{z}{z-1} + 0.77\frac{z}{z+0.85} - 1.24\frac{z}{z+0.15}$ and hence $y(kh) = 0.47 + 0.77(-0.85)^k - 1.24(-0.15)^k$.

9.15 As $\frac{C(s)}{U(s)} = \frac{1-e^{-hs}}{s}$, the correct answer is option **D)**.

9.16 The correct answer is option **B)**.

9.17 The correct answer is option **C)**.

9.18 $X(z) = -\frac{z}{(z-1)^2} - \frac{z}{z-1} + \frac{z}{z-2} \Rightarrow x(kh) = -k - 1 + 2^k$

9.19 $X(z) = \frac{z}{z-\frac{1}{2}} + \frac{z}{z-\frac{1}{3}} \Rightarrow x(kh) = \left(\frac{1}{2}\right)^k + \left(\frac{1}{3}\right)^k$
Thus the correct answer is option **A)**. Dividing the numerator of $X(z)$ by its denominator, it can be seen that the correct answer is option **E)**.

9.20 The correct answers are options **A)** and **F)**.

9.21 The correct answer is option **D)**.

9.22 The correct answer is option **A)**.

9.23 The correct answers are options **C)** and **F)**.

10.1 The correct answer is option **B)**.

10.2 The correct answer is option **D)**.

10.3 The correct answer is option **B)**.

10.4 The correct answer is option **B)**.

10.5 The correct answer is option **A)**.

10.6 The correct answer is option **C)**.

10.7 The correct answer is option **C)**.

10.8 The correct answer is option **D)**.

10.9 The correct answer is option **B)**.

10.10 The stable limit cycle has amplitude 10.19 and frequency 2.45 rad/s and disappears when K_1 increases to the point where the two curves do not intersect.

10.11 The stable limit cycle has amplitude 1.061 and frequency 2.828 rad/s and does not disappear when K_1 changes.

10.12 The correct answer is option **A)**.

10.13 The stable limit cycle has amplitude 0.63 and frequency 4.24 rad/s.

10.14 The stable limit cycle has amplitude 0.255 and frequency 2.24 rad/s.

10.15 The correct answer is option **B)**.

10.16 The correct answer is option **A)**.

10.17 The correct answer is option **A)**.

10.18 The correct answers are options **A)**, **F)** and **J)**.

10.19 The correct answer is option **B)**.

10.20 The correct answer is option **B)**.

10.21 The correct answer is option **C)**.

10.22 The correct answer is option **B)**.

10.23 The correct answers are options **A)**, **F)** and **I)**.

10.24 The correct answers are options **C)**, **E)** and **K)**.

10.25 The correct answer is option **C)**.

10.26 There is a stable limit cycle with $X = 4.203$ and $\omega = 1.414$ rad/s.

10.27 The correct answer is option **B)**.

10.28
$$N(X) = \frac{2}{\pi X}\left[\sqrt{1-\left(\frac{0.75}{X}\right)^2}+\sqrt{1-\left(\frac{1.25}{X}\right)^2}\right]-j\frac{1}{\pi X^2}$$

$$1+GN = 0 \Rightarrow N = -\frac{1}{G}$$

$$G(j\omega) = \frac{3}{j\omega(j\omega+1)^2} \Rightarrow -\frac{1}{G} = \frac{2}{3}\omega^2 + j\frac{\omega(\omega^2-1)}{3}$$

and thus

$$\begin{cases} \frac{2}{\pi X}\left[\sqrt{1-\left(\frac{0.75}{X}\right)^2}+\sqrt{1-\left(\frac{1.25}{X}\right)^2}\right] = \frac{2}{3}\omega^2 \\ -\frac{1}{\pi X^2} = \frac{\omega(\omega^2-1)}{3} \end{cases} \Rightarrow \begin{cases} \omega = 0.882 \text{ rad/s} \\ X = 2.155 \end{cases}$$

See Figure 40: the limit cycle is stable.

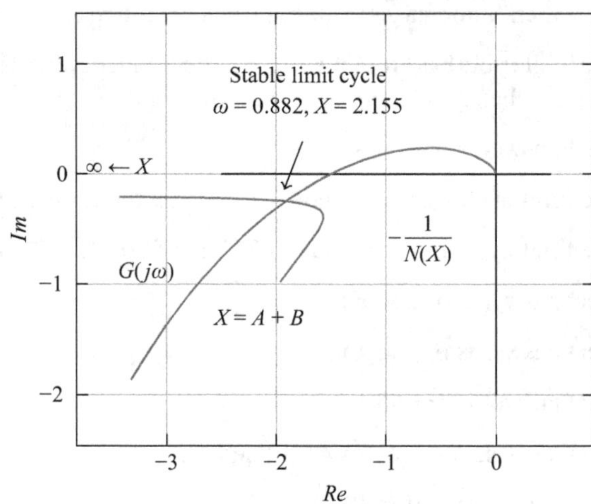

Figure 40 Nyquist plot and describing function of Exercise 10.28

10.29
$$N(X) = \frac{4}{\pi X}$$

$$1+GN = 0 \Rightarrow N = -\frac{1}{G}$$

$$G(j\omega) = \frac{e^{-j2\omega}}{j\omega(j\omega+2)} = \frac{1}{\omega\sqrt{\omega^2+4}}e^{j\left(-2\omega-\frac{\pi}{2}-\arctan\frac{\omega}{2}\right)}$$

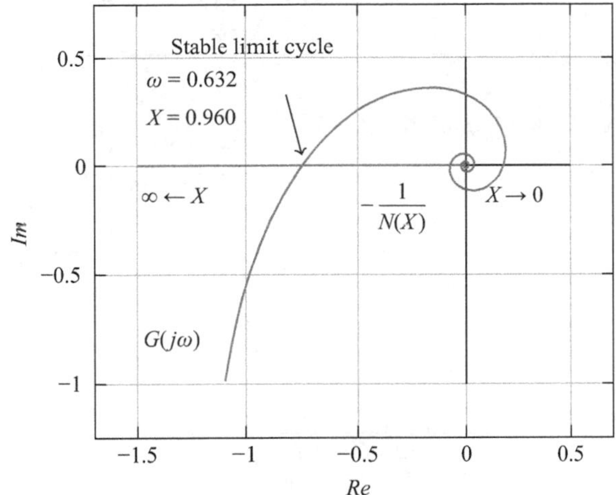

Figure 41 Nyquist plot and describing function of Exercise 10.29

and thus

$$\begin{cases} \frac{\pi X}{4} = \frac{1}{\omega\sqrt{\omega^2+4}} \\ -2\omega - \frac{\pi}{2} - \arctan\frac{\omega}{2} = -\pi \end{cases} \Rightarrow \begin{cases} \omega = 0.632 \text{ rad/s} \\ X = 0.960 \end{cases}$$

See Figure 41: the limit cycle is stable.

10.30 $$N(X) = \frac{1}{2}\left[1 - \frac{2}{\pi}\left(\arcsin\frac{2-X}{X} + \frac{2-X}{X}\cos\arcsin\frac{2-X}{X}\right)\right] - j\frac{4(X-1)}{\pi X^2}$$

$$1 + GN = 0 \Rightarrow N = -\frac{1}{G}$$

$$G(j\omega) = \frac{1.5}{j\omega(j\omega+1)^2}$$

There are two possible solutions: there is a stable limit cycle for $X = 1.18$, $\omega = 0.287$ rad/s, and an unstable limit cycle for $X = 5.45$, $\omega = 0.81$ rad/s. See Figure 42.

10.31 $$N = \frac{4}{\pi X}$$

$$1 + GN = 0 \Rightarrow N = -\frac{1}{G}$$

$$G(j\omega) = \frac{100}{j\omega(j\omega+2)(j\omega+3)} = \frac{100}{\omega(\omega^2+4)(\omega^2+9)} e^{j\left(-\frac{\pi}{2} - \arctan\frac{\omega}{2} - \arctan\frac{\omega}{3}\right)}$$

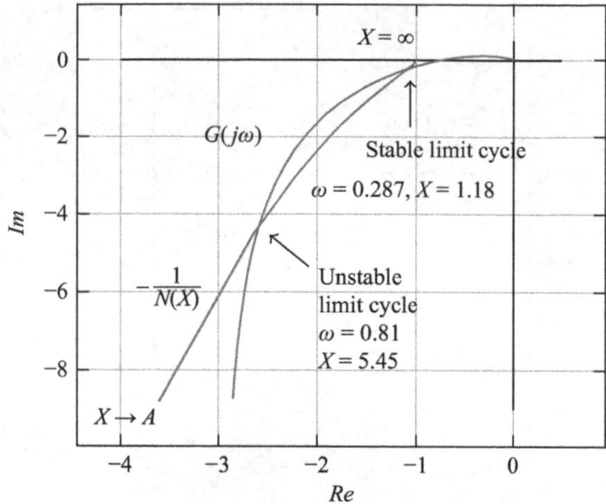

Figure 42 Nyquist plot and describing function of Exercise 10.30

and thus

$$\begin{cases} \frac{\pi X}{4} = \frac{100}{\omega(\omega^2+4)(\omega^2+9)} \\ -\frac{\pi}{2} - \arctan\frac{\omega}{2} - \arctan\frac{\omega}{3} = -\pi \end{cases} \Rightarrow \begin{cases} \omega = 2.449 \text{ rad/s} \\ X = 4.244 \end{cases}$$

See Figure 43: the limit cycle is stable.

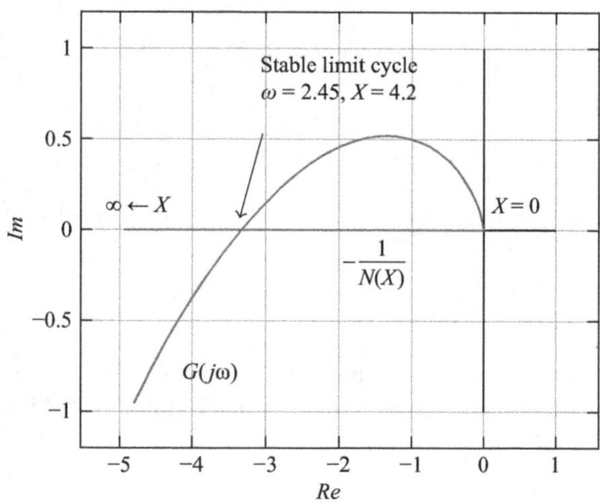

Figure 43 Nyquist plot and describing function of Exercise 10.31

10.32 There is a stable limit cycle with $X = 3.128$ and $\omega = 1$ rad/s; $K_G = 4$.

10.33 There is a stable limit cycle with $X = 2.475$ and $\omega = 1$ rad/s; $K_G = 5$.

10.34 There is a stable limit cycle with $X = 0.364$ and $\omega = 1$ rad/s.

10.35 There is a stable limit cycle with $X = 1.273$ and $\omega = 3.317$ rad/s.

11.1 The answer is yes.

11.2 The correct answer is option **C)**.

11.3 The correct answer is option **A)**.

11.4 The correct answer is option **C)**, because there are no limit cycles in linear systems.

11.5 The correct answer is option **D)**.

11.6 The correct answer is option **A)**.

11.7 A center corresponds to a characteristic equation with pure imaginary roots. This will be the case when $a = 0 \Rightarrow s^2 + 2 = 0$. So the correct answer is option **C)**.

11.8 The correct answers are options **A)** and **F)**.

11.9 The characteristic equation is $s^2 + 3s + 2 = 0$, with roots -1 and -2; since both are real and negative, the correct answer is option **A)**.

11.10 The correct answer is option **B)**.

11.11 The correct answers are options **D)**, **H)** and **I)**.

11.12 The correct answers are options **D)** and **G)**.

11.13 The correct answers are options **A)** and **G)**.

11.14 The correct answer is option **C)**.

11.15 The correct answer is option **D)**.

11.16 The correct answers are options **C)**, **H)** and **L)**.

11.17 The correct answers are options **D)** and **E)**.

11.18 The correct answers are options **A)** and **F)**.

11.19 The equation can be put in the form $\frac{\mathrm{d}\dot{x}}{\mathrm{d}x} = -\frac{\cos x}{\dot{x}} = -\frac{f(x,\dot{x})}{\dot{x}}$, where $f(x,\dot{x}) = \cos x$. Since $f(x,\dot{x}) = f(x,-\dot{x})$ there is symmetry about the x-axis. Since $f(x,\dot{x}) \neq -f(-x,\dot{x})$ there is no symmetry about the x-axis.

Singular points are $\cos x = 0 \wedge \dot{x} = 0 \Leftrightarrow \dot{x} = 0 \wedge x = \frac{\pi}{2}(2k+1)$, $k = 0, \pm 1, \pm 2, \ldots$.

For the case $x = 2k\pi + \frac{\pi}{2}$, we use the approximation $\cos x \approx \frac{\pi}{2} - x$ and, consequently, the variable change $y = \frac{\pi}{2} - x \Rightarrow \dot{y} = -\dot{x} \Rightarrow \ddot{y} = -\ddot{x}$. The equation becomes $-\ddot{y} + y = 0$, with characteristic equation $-s^2 + 1 = 0 \Leftrightarrow s = \pm 1$, so these are saddle points (see Figure 44).

For the case $x = 2k\pi + \frac{3\pi}{2}$, we use the approximation $\cos x \approx x - \frac{3\pi}{2}$ and, consequently, the variable change $y = x - \frac{3\pi}{2} \Rightarrow \dot{y} = \dot{x} \Rightarrow \ddot{y} = \ddot{x}$. The equation becomes $\ddot{y} + y = 0$, with characteristic equation $s^2 + 1 = 0 \Leftrightarrow s = \pm j$, so these are center points (see Figure 44).

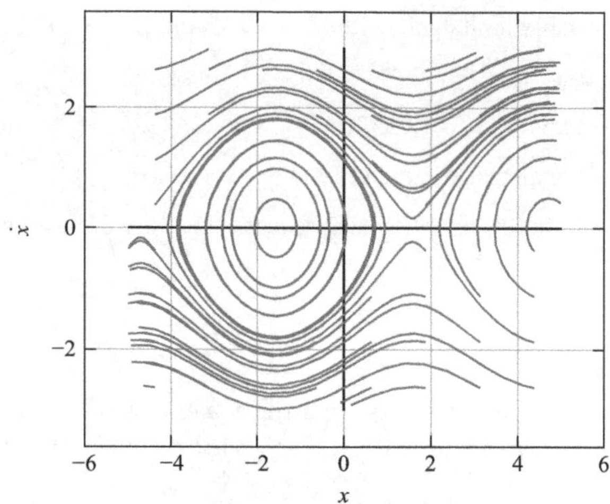

Figure 44 Phase plane of Exercise 11.19

11.20 The equation becomes $\frac{d\dot{x}}{dx} = -\frac{0.2\dot{x}+x(3+x)}{\dot{x}}$ and has two singular points. One is $(x, \dot{x}) = (0, 0)$; the characteristic equation is $s^2 + 0.2s + 3 = 0$; it is a stable focus. The other is $(x, \dot{x}) = (-3, 0)$; the characteristic equation is $s^2 + 0.2s - 3 = 0$; it is a saddle point.

11.21 The equation becomes $\frac{d\dot{x}}{dx} = \frac{x(1-x^2)}{\dot{x}}$ and has three singular points. One is $(x, \dot{x}) = (0, 0)$; the characteristic equation is $s^2 - 1 = 0$; it is a saddle point. The other two are $(x, \dot{x}) = (\pm 1, 0)$; in both the characteristic equation is $s^2 + 2 = 0$; these are center points.

11.22 The equation becomes $\frac{d\dot{x}}{dx} = -\frac{\dot{x}+x(x+3)(x-3)}{\dot{x}}$ and has three singular points. One is $(x, \dot{x}) = (0, 0)$; the linearized equation is $\ddot{x} + \dot{x} - 9x = 0$; the characteristic equation is $s^2 + s - 9 = 0$; it is a saddle point. The other two are $(x, \dot{x}) = (\pm 3, 0)$; in both the characteristic equation is $s^2 + s + 18 = 0$; these are stable focuses.

11.23 There is one singular point, $\frac{d\dot{x}_2}{d\dot{x}_1} = \frac{0}{0}$ at $x_1 = x_2 = 0$. The plant can be linearized around the origin: $\begin{cases} sX_1 = X_2 \\ sX_2 = X_1 - 0.4X_2 \end{cases}$; the characteristic equations come $s^2X_1 = X_1 - 0.4sX_1$, from which we get $s^2 + 0.4s - 1 = 0$ $\Leftrightarrow s = 0.82 \vee s = -1.22$. So the origin is a saddle point.

11.24 Since $\frac{dx_1}{dx_2} = -\frac{x_2\left(1+x_2^2\right)}{x_1\left(1+x_1^2\right)-0.4x_2}$, there is a singular point at $(x_1, x_2) = (0, 0)$. Linearizing around this point,

$$\begin{cases} \frac{dx_1}{dt} = -x_2 \\ \frac{dx_2}{dt} = x_1 - 0.4x_2 \end{cases} \Rightarrow \begin{cases} sX_1 = -X_2 \\ sX_2 = X_1 - 0.4X_2 \end{cases}$$

From here, we get the characteristic equation $s^2 + 0.4s + 1 = 0$, with roots $s = -0.2 \pm 0.98j$; it is a stable focus.

11.25 The equation becomes $\frac{d\dot{x}}{dx} = -\frac{2\dot{x}+x^3-2x^2-3x}{\dot{x}}$ and has three singular points. One is $(x,\dot{x}) = (0,0)$; the characteristic equation is $\lambda^2 + 2\lambda - 3 = 0$; it is a saddle point. Another one is $(x,\dot{x}) = (-1,0)$; the characteristic equation is $\lambda^2 + 2\lambda + 5 = 0$; it is a stable focus. The last one is $(x,\dot{x}) = (3,0)$; the characteristic equation is $\lambda^2 + 2\lambda + 12 = 0$; it is a stable focus as well.

12.1 $\mathscr{L}^{-1}\left[\frac{1}{s^{\frac{1}{2}}-1} + \frac{5}{s^{\frac{1}{2}}+2}\right] = t^{-\frac{1}{2}}E_{\frac{1}{2},\frac{1}{2}}(t^{\frac{1}{2}}) + 5t^{-\frac{1}{2}}E_{\frac{1}{2},\frac{1}{2}}(-2t^{\frac{1}{2}})$

12.2 The impulse response is $10t^{-\frac{1}{2}}E_{\frac{1}{2},\frac{1}{2}}(-100t^{\frac{1}{2}})$, and the step response is $10t^{\frac{1}{2}}E_{\frac{1}{2},\frac{3}{2}}(-100t^{\frac{1}{2}})$.

12.3 The correct answer is option **C)**.

12.4 The roots of the polynomial in $\alpha = \frac{1}{2}$ are $3 \pm 3j$, which fall on the limit of stability of the complex plane, defined by rays making $\pm 90° \times \frac{1}{2} = \pm 45°$ angles with the positive real axis.

12.5 The roots of the polynomial in $\alpha = \frac{2}{3}$ are $3 \pm 3j$, which fall on the unstable part of the complex plane, defined by rays making $\pm 90° \times \frac{2}{3} = \pm 60°$ angles with the positive real axis.

12.6 The roots of the polynomial in $\alpha = \frac{4}{3}$ are $-3 \pm 3j$, which fall on the stable part of the complex plane, defined by rays making $\pm 90° \times \frac{4}{3} = \pm 120°$ angles with the positive real axis.

12.7 Nyquist diagrams are given in Figure 45. Bode diagrams are given in Figures 46 and 47. The correspondence of root-locus plots is as follows: A—$G_5(s)$; B—$G_6(s)$; C—$G_2(s)$; D—$G_3(s)$; transfer functions $G_1(s)$ and $G_4(s)$ have no roots in the main Riemann sheet.

12.8 $G_1(s)$—**A**—δ—**d**
$G_2(s)$—**B**—α—**c**
$G_3(s)$—**D**—γ—**b**
$G_4(s)$—**C**—β—**a**

12.9 The correct answer is option **C)**.

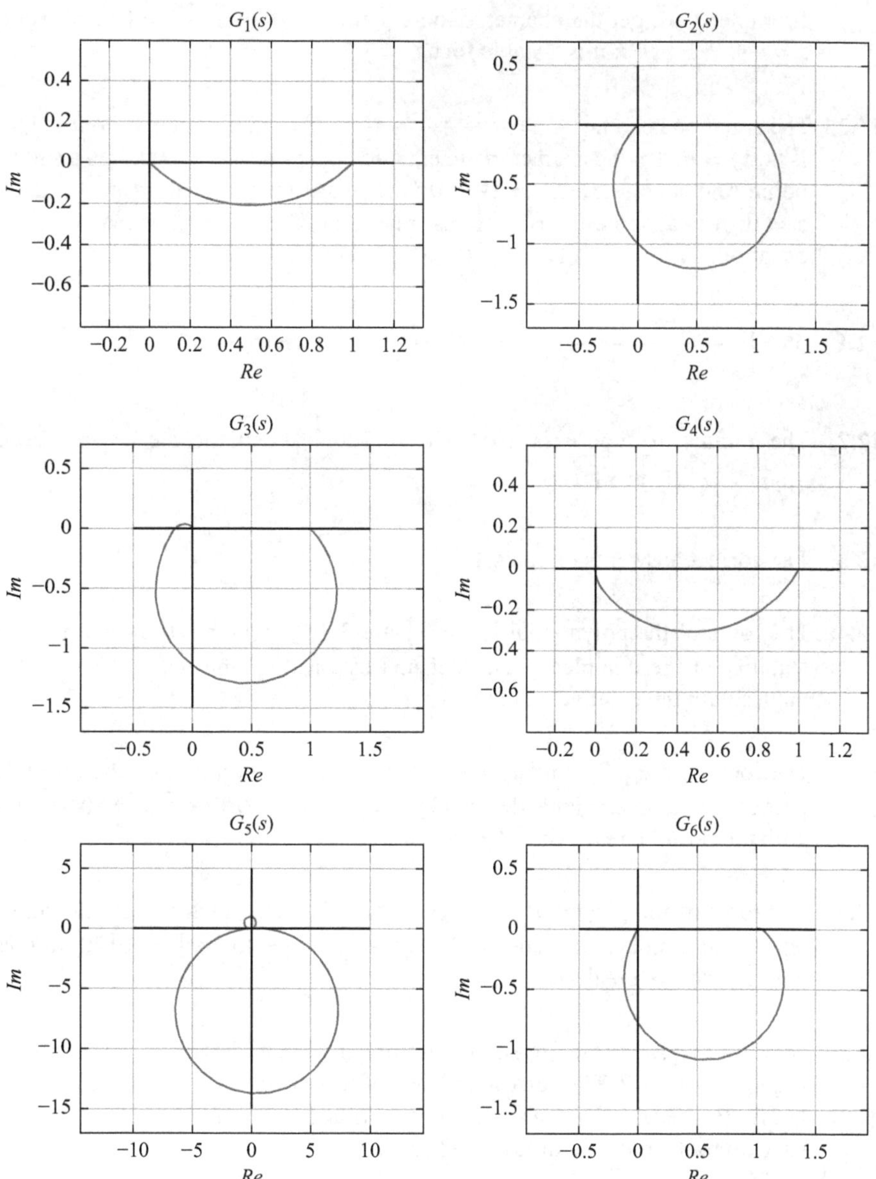

Figure 45 Nyquist diagrams of Exercise 12.7

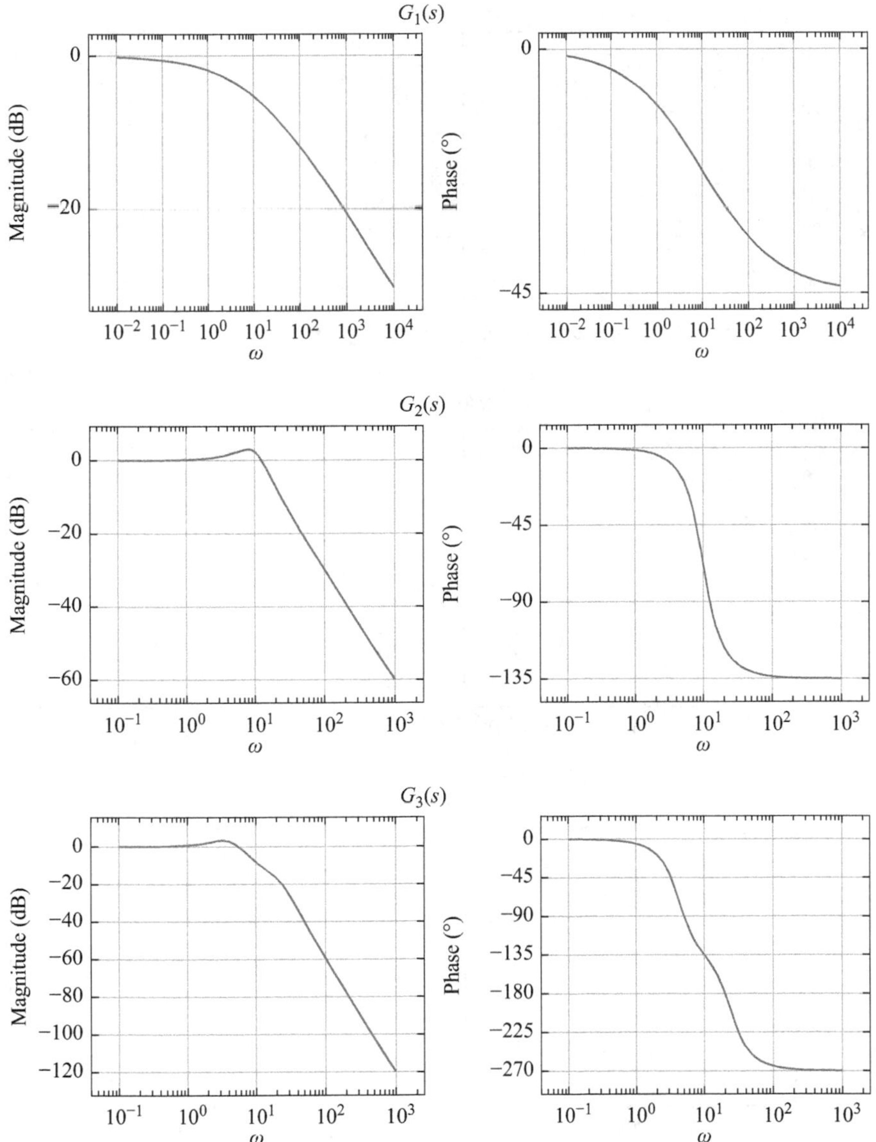

Figure 46 Bode diagrams of Exercise 12.7

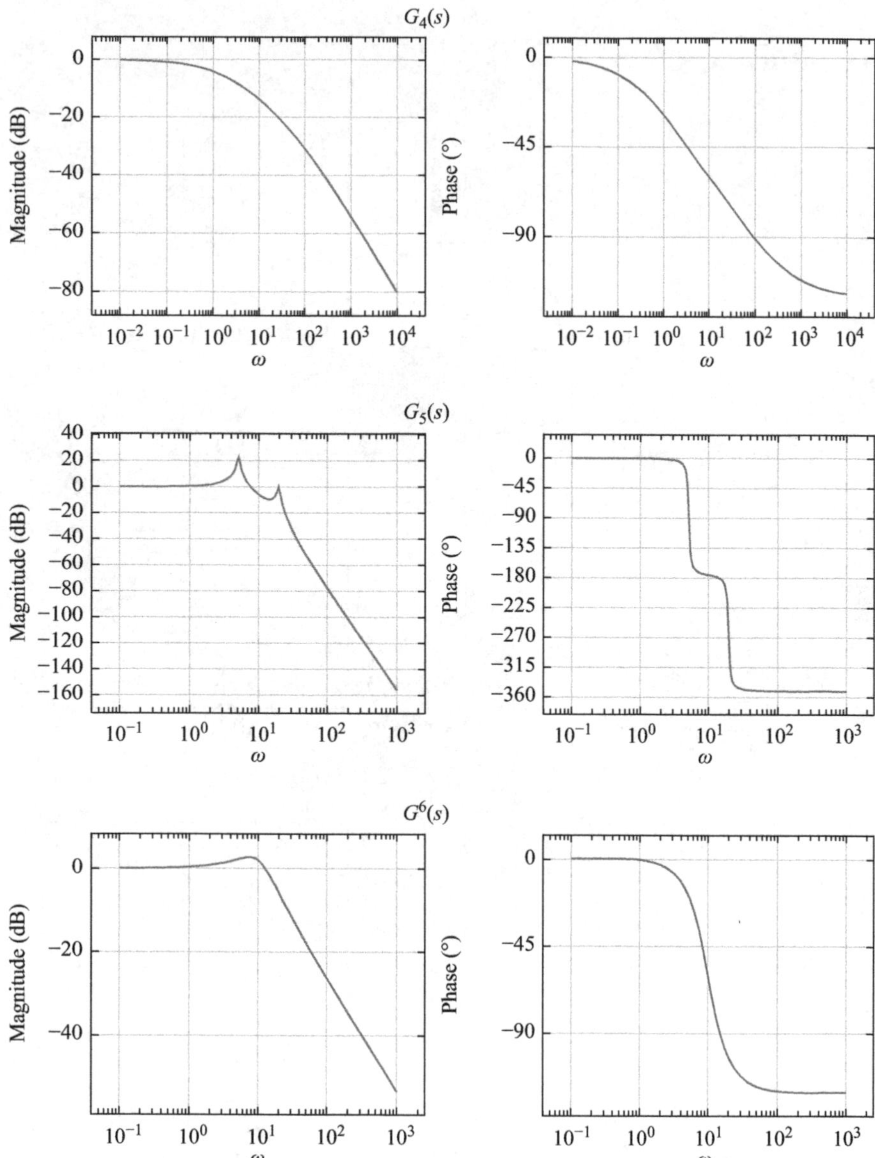

Figure 47 Bode diagrams of Exercise 12.7 (continued from Figure 46)

12.10 For all transfer functions, the asymptotic gain-crossover frequency is 10 rad/s, so $\omega_l = 0.01$ rad/s, $\omega_h = 10^4$ rad/s and $N = 6$. Thus

$$G_1(s) \approx \frac{1}{\frac{1}{\sqrt{10}}\frac{3.162s^5+4\,249s^4+3.389\times10^5s^3+1.698\times10^6s^2+5.349\times10^5s+10^4}{0.03162s^5+169.2s^4+5.371\times10^4s^3+1.072\times10^6s^2+1.344\times10^6s+10^5}+1}$$

$$= \frac{0.03065s^5+164s^4+5.206\times10^4s^3+1.039\times10^6s^2+1.303\times10^6s+9.693\times10^4}{s^5+1466s^4+1.559\times10^5s^3+1.559\times10^6s^2+1.466\times10^6s+10^5}$$

$$G_2(s) \approx \frac{1}{\frac{1}{\sqrt{10^3}}\frac{31.62s^5+1.067\times10^4s^4+2.138\times10^5s^3+2.692\times10^5s^2+2.13\times10^4s+100}{3.162\times10^{-5}s^5+0.6734s^4+851.2s^3+6.761\times10^4s^2+3.375\times10^5s+10^5}+1}$$

$$= \frac{3.162\times10^{-5}s^5+0.6734s^4+851.2s^3+6.761\times10^4s^2+3.375\times10^5s+10^5}{s^5+338.2s^4+7\,612s^3+7.612\times10^4s^2+3.382\times10^5s+10^5}$$

or, better still, approximating $s^{\frac{1}{2}}$ only,

$$G_2(s) \approx \frac{1}{\frac{s}{\sqrt{10^3}}\frac{3.162s^5+4\,249s^4+3.389\times10^5s^3+1.698\times10^6s^2+5.349\times10^5s+10^4}{0.03162s^5+169.2s^4+5.371\times10^4s^3+1.072\times10^6s^2+1.344\times10^6s+10^5}+1}$$

$$= \frac{0.3162s^5+1\,692s^4+5.371\times10^5s^3+1.072\times10^7s^2+1.344\times10^7s+10^6}{s^6+1\,344s^5+1.089\times10^5s^4+1.074\times10^6s^3+1.089\times10^7s^2+1.344\times10^7s+10^6}$$

As this approximation is better, the following result is obtained in the same manner as well:

$$G_3(s) \approx \frac{1}{\frac{s^3}{1\,000}+\frac{4s}{\sqrt{10^3}}\frac{3.162s^5+4\,249s^4+3.389\times10^5s^3+1.698\times10^6s^2+5.349\times10^5s+10^4}{0.03162s^5+169.2s^4+5.371\times10^4s^3+1.072\times10^6s^2+1.344\times10^6s+10^5}+1}$$

$$= \frac{1\,000s^5+5.349\times10^6s^4+1.698\times10^9s^3+3.389\times10^{10}s^2+4.249\times10^{10}s+3.162\times10^9}{s^8+5\,349s^7+1.711\times10^6s^6+5.088\times10^7s^5+1.403\times10^9s^4+8.495\times10^9s^3+3.603\times10^{10}s^2+4.253\times10^{10}s+3.162\times10^9}$$

12.11 The correct answer is option **C)**.

12.12 The correct answer is option **B)**.

12.13 The correct answer is option **B)**.

12.14 The correct answers are options **D)** and **E)**.

References

[1] D. V. Widder, *Laplace Transform (PMS-6)*, Princeton University Press, Princeton, NJ, 2015.

[2] K. Ogata, *Modern Control Engineering*, Prentice-Hall, Upper Saddle River, NJ, 2010.

[3] J. J. d'Azzo, C. D. Houpis, *Linear Control System Analysis and Design: Conventional and Modern*, McGraw-Hill, New York, NY, 1995.

[4] J. de Carvalho, *Dynamical Systems and Automatic Control*, Prentice-Hall, Englewood Cliffs, NJ, 1993.

[5] R. C. Dorf, R. H. Bishop, *Modern Control Systems*, Pearson, Upper Saddle River, NJ, 2011.

[6] B. Friedland, *Control System Design: An Introduction to State-Space Methods*, Dover Publications, Mineola, NY, 2012.

[7] G. Rizzoni, *Principles and Applications of Electrical Engineering*, McGraw-Hill, New York, NY, 2005.

[8] D. Rowell, D. N. Wormley, *System Dynamics: An Introduction*, Prentice Hall, Upper Saddle River, NJ, 1997.

[9] J. J. DiStefano, A. J. Stubberud, I. J. Williams, *Schaum's Outline of Feedback and Control Systems*, McGraw-Hill, New York, NY, 1997.

[10] W. R. Evans, "Graphical analysis of control systems," *Transactions of the American Institute of Electrical Engineers* 67 (1) (1948) 547–551.

[11] W. R. Evans, "Control system synthesis by root-locus method," *Transactions of the American Institute of Electrical Engineers* 69 (1) (1950) 66–69.

[12] A. M. Krall, "The root-locus method: A survey," *SIAM Review* 12 (1) (1970) 64–72.

[13] A. M. Eydgahi, M. Ghavamzedeh, "Complementary root locus revisited," *IEEE Transactions on Education* 44 (2) (2001) 137–143.

[14] E. Bahar, M. Fitzwater, "Numerical technique to trace the loci of the complex roots of characteristic equations," *SIAM Journal on Scientific and Statistical Computing* 2 (4) (1981) 389–403.

[15] C. I. Byrnes, D. S. Gilliam, J. He, "Root-locus and boundary feedback design for a class of distributed parameter systems," *SIAM Journal on Control and Optimization* 32 (5) (1994) 1364–1427.

[16] MATLAB®. Available from http://www.mathworks.com/ [Accessed May 2016].

[17] OCTAVE©. Available from http://www.gnu.org/software/octave/ [Accessed May 2016].

[18] SCILAB™. Available from http://www.scilab.org/ [Accessed May 2016].

[19] K. Ogata, *System Dynamics*, Vol. 3, Pearson, Upper Saddle River, NJ, 1998.

[20] G. F. Franklin, J. D. Powell, M. L. Workman, *Digital Control of Dynamic Systems*, Vol. 3, Addison-Wesley, Menlo Park, CA, 1998.

[21] K. H. Ang, G. Chong, Y. Li, "PID control system analysis, design, and technology," *IEEE Transactions on Control Systems Technology* 13 (4) (2005) 559–576.

[22] K. J. Åström, T. Hägglund, "The future of PID control," *Control Engineering Practice* 9 (11) (2001) 1163–1175.

[23] K. Åström, T. Hägglund, "Revisiting the Ziegler–Nichols step response method for PID control," *Journal of Process Control* 14 (6) (2004) 635–650.

[24] Y. Li, K. H. Ang, G. C. Y. Chong, "PID control system analysis and design," *IEEE Control Systems* 26 (1) (2006) 32–41.

[25] M. A. Johnson, M. H. Moradi, *PID Control*, Springer-Verlag, London, 2005.

[26] C. H. Houpis, G. B. Lamont, *Digital Control Systems*, McGraw-Hill, New York, NY, 1991.

[27] D. P. Atherton, *Nonlinear Control Engineering: Describing Function Analysis and Design*, Van Nostrand Reinhold, London, 1975.

[28] D. P. Atherton, *Stability of Nonlinear Systems*, Vol. 1, John Wiley & Sons, West Sussex, 1981.

[29] A. Bergen, R. Franks, "Justification of the describing function method," *SIAM Journal on Control* 9 (4) (1971) 568–589.

[30] R. S. Barbosa, J. T. Machado, "Describing function analysis of systems with impacts and backlash," *Nonlinear Dynamics* 29 (1–4) (2002) 235–250.

[31] A. Azenha, J. T. Machado, "On the describing function method and the prediction of limit cycles in nonlinear dynamical systems," *Systems Analysis Modelling Simulation* 33 (3) (1998) 307–320.

[32] A. M. Lopes, J. T. Machado, "Dynamics of the *N*-link pendulum: A fractional perspective," *International Journal of Control* (2016) 1–16. doi: 10.1080/00207179.2015.1126677.

[33] J. T. Machado, M. E. Mata, "Pseudo phase plane and fractional calculus modeling of western global economic downturn," *Communications in Nonlinear Science and Numerical Simulation* 22 (1) (2015) 396–406.

[34] A. M. Lopes, J. A. T. Machado, "State space analysis of forest fires," *Journal of Vibration and Control* (2015) doi: 10.1177/1077546314565687.

[35] A. M. Lopes, J. T. Machado, "Dynamic analysis of earthquake phenomena by means of pseudo phase plane," *Nonlinear Dynamics* 74 (4) (2013) 1191–1202.

[36] D. Valério, J. Sá da Costa, *An Introduction to Fractional Control*, Vol. 91, IET, London, 2013.

[37] J. A. T. Machado, "Root locus of fractional linear systems," *Communications in Nonlinear Science and Numerical Simulation* 16 (2011) 3855–3862.

[38] F. Padula, A. Visioli, "Tuning rules for optimal PID and fractional-order PID controllers," *Journal of Process Control* 21 (2011) 69–81.

Index

Page numbers in *italics* refer to the expositions at the beginning of the chapters. Page numbers in normal typeface refer to worked examples. Page numbers in **bold** refer to proposed exercises. For code in MATLAB®, SCILAB™ and OCTAVE©, see the Table of contents.

Ackermann formula, *250*, *253*
antenna, 254

block diagram, *1–3*, 3–9, **9–20**, 44–6, **60**, **65**, **91**, **106–7**, 143, **160**, *186*, *197–201*, 208–9, 212, 213–14, **216**, **218–20**, **222–3**, **225–6**, **228**, **230–5**, *249*, *252–3*, 254, **257–9**, *266*, *271*, 273–5, **276–9**, **281**, **283**, 290–2, **292–300**, **302–5**, **307–10**, *367*, **379–81**, **383–4**, **408**, **411–13**
Bode diagram, *136–7*, 141–7, **148–59**, **162**, **164**, **166**, **168–70**, **172**, *173–4*, *177–8*, *181–2*, **191**, *246–7*, *345*, 350–1, **353**, **355**, *361–2*, *368*, **394–6**, **398–9**, **403–4**, **427**, **429–30**

Caputo, *343–4*, *377*; *see also* fractional derivative
circuit: *see* electrical system
controllability, *207*, 208–10, 212–13, 215, **215–19**, **221**, **224**, **226–7**, **230–2**, **235–6**, *249*, *253*
controllability (discrete), *265*, *271–2*
CRONE approximation, *341*, *346*, **357**
CRONE control, *346*, 351–2

DC motor, 35, **52**, 207–8, 254
dead zone, *288*, 290, **293**, **302**, **305**
delay, *137*, 142, 163, 185, 187–8, *270*, **390**
describing function, *287–9*, **290–1**, 292–310, *311–16*, *373*, **421–5**
discrete time systems, *265–72*
discrete transfer function: *see* transfer function (discrete)

electrical system, *31–2*, 39, **47–52**

first order system, *74–6*, **92**, **94**, **97**, **103**
fractional calculus: *see* fractional derivative
fractional derivative, *341–4*, 348, *358*, *361*, *377*
frequency response, **102**, *135–6*, 142–7, **148–72**, *289*, *344–6*, 353–4, *360*, *362*

gain margin: *see* stability margins
Grünwald-Letnikoff: *see* fractional derivative

hysteresis, **295**, *374*

lag compensator, *246–8*, 254, **257**
Laplace transform, *1–3*, 43–4, 46, *74*, *200*, *206*, *267*, **276**, *343–4*, 347, **352**, *365–6*, *376*, *406*
lead compensator, *246–8*, 254–5, **257**, **260**
limit cycle, *265*, *287*, *289*, 290–1, **292–301**, **303–10**, *312*, *314*, *316*, *319*, 320–1, **326**, **332**, **421–5**

liquid level system, *36–8*, 42–4, **60–7**, **236**

Matignon's theorem, *344–5*, 347, **352**
Matsuda approximation, *346*
mechanical system, *29*, *32–5*, 40–1, **53–9**, **98**, **258**, **326**, **328**
missile, **259**
Mittag-Leffler function, *341*, *376*
model: *see* electrical system; liquid level system; mechanical system; thermal system

Nichols diagram, *139*, *141–2*, *145*, **164–5**, **167**, *173–4*, *176–9*, *181–2*, *184*, *361*, *363*, *371*
non-linear system: *see* describing function; phase plane
Nyquist diagram, *137–8*, *141–2*, 145–6, **160–70**, *173*, *175*, *177*, *179*, *181*, *183*, *289*, 290, 292, **296–300**, **304**, **307**, **309–10**, 350, **353**, **354**, *361*, *363*, **397–402**, **422–4**, **427–8**
Nyquist stability criterion, *139–40*, *287*, *289*

observability, *195*, *198*, *207*, 208, 210, 212–13, **215–19**, **222**, **224**, **226–7**, **230–1**, **235–6**, *245*, *250–2*
observability (discrete), *265*, *271*
operational amplifiers, *29*, *31–2*, **51**

phase margin: *see* stability margins
phase plane, *317–9*, 320–3, **324–35**, *336–40*, *375*, **426**
PID, *185–9*, 190, **191–3**, *246*, **404–5**
PID (fractional), *346*, **357**
plane, **259–60**
polar diagram: *see* Nyquist diagram
pole placement, *245–6*, *248–9*, *253*, 254, 256, **257**, **262**

response: *see* frequency response; time response

Riemann-Liouville: *see* fractional derivative
robot, **260**
root-locus, *115–18*, 119–24, **124–31**, *131–3*, 146, **164–72**, 254–6, **258–9**, **261**, *345*, 350–1, **353–4**, **356**, **390–3**, **400–1**, **403**, **415–16**, **427**
Routh's stability criterion, *84*

saturation, *288*, 291, **299**, **309**, *373*
second order system, *77–84*, 87–8, **94–6**, **99– 103**, **106**, **159**, *317–18*
stability margins, *140–1*, 142–4, **148–9**, **158–9**, **162–3**, **168**, **172**, *246–8*, 257, 260, *346*, *351*, *392*, **417**
state estimation, *253*
state-space, *195*, *197*, *200–2*, *206–7*, 209, 213, **215**, **218–20**, **222**, **224–8**, **230–2**, **234–7**, *238–43*, *248*, *268*, *270*, 274–5, **282**, *317*
steady-state error, *73*, *75–6*, *85*, 89, **102–7**, **172**, *185–6*, *245–6*, 254, **257**, **260**, **328–9**, **389–90**

tautochrone curve, 348
thermal systems, *38–9*, 45, **68–71**
time response, *73–84*, 87–9, **91**, **95–107**, *109–14*, *188–9*, *255*, *317*, **324**, **327**, **352**, **387**
transfer function, *1–3*, 4–9, **10–15**, **17–19**, *20–7*, *31*, 39–40, 42, 44–6, **47–51**, **53–6**, **60**, **65**, **68–9**, **71**, *74*, *77*, 87–9, **95–7**, **99–107**, *109*, *111*, *113–16*, 119, 124, **124–31**, *131–3*, *135–7*, *139*, 142–4, 146–7, **148–50**, 152–65, **169–71**, *173–84*, *186*, 190, 191–2, 195–6, *199–200*, 207, 209, 211–13, **215–19**, **221**, **224–5**, **229**, **231–2**, **234–6**, *237–43*, 245–6, *253*, 258–60, **262**, 269–70, **273**, **276**, **278–9**, **281**, **283**, *287*, 290, **292**, **296**, **298–9**, **306–7**, **309**, *311*, *313*, *315*, **328**, *344–6*, 350–1, **352–4**, **356–7**, *360*, *362*, *369*, *371*, **381**, **385**, **389–90**, **405**, **410**, **413**, **415**, **418**

transfer function (discrete), *265–7*, *269–70*, **273**, **276**, **278–9**, **281**, **283**
transfer function (fractional), *344–6*, 347, 350–1, **352–4**, **356–7**, *360*, *362*, **427**, **431**
transform: *see* Laplace transform, *Z* transform
transient response: *see* time response
tuning rules: *see* PID control

wave energy converter, 70–1, **262**

Z transform, *284–6*, *372–3*

Printed in the USA
CPSIA information can be obtained
at www.ICGtesting.com
JSHW011508221024
72173JS00005B/1240

9 781785 611742